GOD, ALIENS OR ARTIFICIAL INTELLIGENCE,

WHO WROTE THE TORAH CODE?

History revised after breaking the computer code in the first five books of the Bible

2nd Edition December 2025

Barry Steven Roffman

DEDICATION AND THANKS

This book is dedicated to my beloved, but deceased uncle, Eugene Roffman. Eugene was my father's brother. While I was a child, he lived across the street from my house. Just before he died at the age of 93, he let me know that he had back-engineered UFO's for the U.S. Army. More, he gave me a list of all scientists worldwide (including the Soviet Union in 1960) who worked with our government on this project. It was this information that led me with my son, Dr. David Alexander Roffman, to produce a 1,200-page report entitled Mars Correct, Critique of All NASA Mars Weather Data. Special thanks are also due to my older son, Major/Rabbi Robert Altair Roffman of the United States Air Force Reserve. Robert developed the Excel spreadsheet probability tool that allowed me to calculate the statistical significance of thousands of Torah Code matrices in this book and on my website. In short, Robert helped me to break open the Torah Code and show me what the Mind of God is all about,

Table of Contents

Preface
to God, Aliens, or Artificial Intelligence: Who Wrote the Torah Code?

I began my Torah Codes research in 1997 after reading Michael Drosnin's bestselling book *The Bible Code*. By the time I started writing this manuscript on December 30, 2024, I had amassed a vast collection of data, including thousands of Torah Code matrices, supporting the existence of a predictive Equidistant Letter Sequence (ELS) code in the Torah—the first five books of the Bible. Yet, after 28 years of research, the origin of the Code remained a mystery. When I began this new analysis, I considered three potential sources: God, extraterrestrials, or a military-grade artificial intelligence (AI) from the future. Such an AI would need the ability to transport data or people across time.

As a modern Orthodox Jew, I have a fairly strong understanding of traditional Jewish beliefs about the nature of God. While Christianity and Islam both trace their roots to Judaism, their concepts differ significantly. Christianity embraces a Trinitarian view of God, which Judaism, like Islam, rejects. In fact, the Quran considers this concept blasphemous. Specifically, Surah 4:171 states:

> *Believing in a Trinity (i.e., God being one of three) is a form of disbelief and blasphemy. It asserts that there is only one God, Allah, and that those who claim otherwise will face a painful punishment. The Quran also emphasizes that Jesus was a messenger of God, not God himself.*

However, while the Quran recognizes Jesus as the Messiah born of a virgin, Judaism rejects Jesus entirely, along with the New Testament, Muhammad, and the Quran. According to Jewish belief, any messiah must prove his legitimacy by ending all wars permanently. He must also prevent people from worshipping any god not accepted by Judaism. The Jewish understanding of God aligns with the concept of the Holy Spirit; therefore, Judaism sees no need to distinguish between God and the Holy Spirit.

In recent years, speculation has grown about whether aliens played a role in shaping human history. The Raëlian religion, for example, claims that extraterrestrials—called the Elohim—created humanity using advanced technology. This atheistic belief system views the Elohim as the original creators, later mistaken for gods. Part of this stems from the Hebrew word *Elohim*—the second word in the Bible, which appears to be plural due to its "-im" ending. Other groups, like Heaven's Gate, also believed in godlike aliens and the possibility of being transported to the heavens via spacecraft. These are often referred to as UFO religions, and they assert both ancient and modern contact with extraterrestrial intelligence.

In this book, I will present compelling evidence of ancient and modern ET contact. By compelling, I mean a detailed account of my direct interaction with a man claiming to be an extraterrestrial. This individual, who goes by the name Craig Ebrahimi, came into possession of a 27-pound meteorite that appears to contain 16 fossilized alien embryos. The object has undergone various chemical tests and is being studied by several PhDs since Craig discovered it in 2018. My investigation of Craig is covered in Chapter 5, *First Contact with an Extraterrestrial*. To explain why I take him seriously, Chapter 6, *Emails to and from an ET*, presents 38 pages of email correspondence between Craig, scientists studying his meteorite, and myself, spanning more than a year. These chapters also outline why I believe Craig was the individual who delivered the precursors of COVID-19 to the Chinese military through the Center for Infectious Disease Control in Tianjin, China. Chapter 7 explores NASA's alleged misinformation about Martian weather, while Chapter

8 details primitive life forms discovered on Mars.

When Drosnin published his first *Bible Code* book, he claimed that any story reported in *The New York Times* could be found encoded in the Torah. My findings support his claim. Decoding the Torah Code requires selecting the correct axis term—typically a transliteration of a person's full name, last name, or first initial and last name—usually seven to nine Hebrew letters in length. The axis term may also represent events like atomic holocaust, war, Islamic terrorism, or disasters. These terms are then linked to *a priori* keywords found in the open text or as ELS terms at equal skips of ±1 or the axis term's skip. Most matrices cover fewer than 1,000 letters (rows × columns) within the Torah's 304,805 characters. A matrix is usually followed by a spreadsheet showing the odds against such a configuration occurring by chance.

This book places special focus on what the Code reveals about influential politicians. Chapter 11 covers the Obama–Biden years, and Chapter 12 addresses President Trump. The individuals behind both Trump impeachment efforts are mostly named within the Code. Kamala Harris is the subject of Chapter 13. In total, I include over 125 matrices. Many of my regular readers come from military intelligence, particularly Fort Huachuca, Arizona. I have received threats for publishing this information. This book holds nothing back. While it highlights advanced military technology, it still considers the possible role of an imperfect Creator.

Initially, Chapter 1 explained the mathematical process of breaking the Code. But since most readers aren't fond of math, I moved this material to the back of the book as Annex A. Now, Chapter 1 opens with an exploration of a possible Jewish–ET connection and the early role of the Nephilim—likely reptilian beings. Chapter 2 presents evidence that reptilian aliens may control the Vatican. The book goes on to name possible reptilians within the U.S. Government, discuss NASA's lies about Mars, explore my primary alien contact, share over a year's worth of emails with him, and provide an in-depth Code-based analysis of those involved in attempts to bring down Trump. I also examine extraterrestrial influence on Israel's enemies and identify those who were assassinated as indicated in the Code. The Epilogue ties it all together and proposes who may have authored the Code.

To use the Torah Code, you must purchase CodeFinder software (CF) from http://www.biblecodesoftware.org/ for $47.95. I have no financial connection to it or its Australian developer, Kevin Acres. While other Bible Code software exists, none is as comprehensive or fast as CF. In 1997, it took me six months to find 41 code hits for "Ark of the Covenant." In a single computer pass over the 304,805 letters of the unwrapped Torah (Genesis 1:1 to Deuteronomy 34:12), there were six open-text and seven encoded hits. Wrapped searches produced an additional 28 hits. Today, I can locate them all in a second. With this tool, I offer the world its hidden history.

The Code reveals that most recent top Democrats were either aliens (reptilians), or people abducted and programed by them. This part of the invasion will be covered in detail in Chapter 3 where the ETs in the U.S. are named. But on Figure P-1, the axis term (first term sought) was **VLAD PUTIN.** To find him, Codefinder searched for all the letters in his name where the same number of letters was found between each letter. The software made more than one pass through the 304,805 letters of Torah. It then found the best spelling only once, at skip +34,416 letters. Then the program next looked for **REPTILIAN**, the invader alien species likely related to the giants (*Nephilim*) that the Book of Genesis (Genesis 6:4) says interbred with the daughters of man. The program found the 4-letter **REPTILIAN** encoded overlapping the 8-letter **VLAD PUTIN**, but **REPTILIAN** is at half the skip of **VLAD PUTIN.** Half of +34,416 letters is +17,708 letters. That is the skip of **REPTILIAN.** What is extraordinary here is that to show **VLAD PUTIN** and **REPTILIAN** requires only 16 letters (one column and 16 rows).

As the spreadsheet shows after Figure P-1 the odds against finding **VLAD PUTIN** and **REPTILIAN** on Figure P-1 are at least 20,082 to 1. The idea that a large number of top politicians in the United States are, according to the Torah Code, shape shifting reptilian aliens is, if true, horrifying. But if Putin also turns out to be one, that means that a very large part of our planet is likely on the way to extinction unless we expose them and remove them from power. This book, in large part, is meant to do that.

The Democrats, under Obama and Biden, did all in their power to weaken America's military. The Code does much to warn us about what they are. Likewise, Wikipedia states of Putin's war on Ukraine that, "According to BBC News Russian and the Mediazona news website, 125,681 Russian soldiers and contractors" were killed by 24 August 2025." Wikipedia goes on to say that, "The BBC stated that the actual death toll of Russian forces, counting only Russian servicemen and contractors (i.e. excluding DPR/LPR militia), was 189,051–273,073 by early August 2025." Why would Putin want to weaken his military that much? The Torah Code tells us that the militaries of both nations were deliberately weakened by the heads of their states. Was this to ease the conquest of Earth by aliens? There are a plethora of stories abounding in the news in October, 2025, that an interstellar object (3I/Atlas) is really a large alien spaceship. It is travelling about 152,000 miles per hour and accelerating. It brushed by Mars on October 3, 2025 at a distance from Mars of about 18 million miles. It will then pass Venus at a distance of about 60 million miles on November 3, 2025, and Earth on December 19 at a distance of about 167 million miles. Wikipedia says images by the Hubble Space Telescope suggest that the diameter of 3I/ATLAS's nucleus is between 0.2 and 3.5 miles, with the most likely diameter being less than 0.62 miles., Wikipedia is stressing the likelihood that it's a comet, but they are often censored by the government. There is a plethora of videos online warning that it could be spacecraft. Given Code warnings about ET invaders, we should be ready for potential course changes or visitors who arrive by other means.

ז
ו
ח
ל
ה
ד
ג
פ
ו
ו
ד
ט
ה
י
ע
ן

Term	Translation	Skip	Start	End
ולד פוטין	VLAD PUTIN	35416	Exodus 11:3.7	Genesis 27:7.37
זוחל	REPTILIAN	17708	Genesis 49:13.1	Exodus 33:20.12
ולד	VLAD	35416	Exodus 11:3.7	Leviticus 15:1.3
פוטין	PUTIN	35416	Numbers 7:31.12	Genesis 27:7.37

The ELS reference is 17,708 characters between rows.
The matrix starts at Genesis 49:13.1 and ends at Genesis 27:7.37.
It spans 265,621 characters of surface text, having 16 rows, 1 column and only 16 characters.
VLAD PUTIN R Factor .801 R Factor value = one chance in 6.424118514.

Military parade in Beijing to mark the 80th anniversary of the end of World War II
© Anadolu via Getty Images

Reptilian Feast? Odds against this Matrix are about 20,081 to 1.

Chilling hot mic exposes Putin and Xi's plans to harvest organs for immortality

Story by Jack Hobbs, Yelena Mandenberg ·

Russian President Vladimir Putin and Chinese President Xi Jinping were caught on a hot microphone talking about how they planned to use organ transplants to continue living for 150 years.

Both leaders, who were in the company of North Korea's Supreme Leader, Kim Jong-Un, allegedly spoke about how being 70 years old nowadays is technically "still being a child."

Figure P-1 is above and its probability spreadsheet is below. Odds against Figure P-1 are about 20,082 to 1.

	A	B	C	D	E	F	G	H
1	AXIS TERM VLAD PUTIN @ Skip +35,416	SKIP	NUMBER	DIVIDE BY	THE QUOTIENT	E QUOTIENT	Poisson Equation	EACH WORD
2	ELS RANK 1	USED	IN	304805	EQUALS	x 16 Letters	Probability for	OR PHRASE
3	R factor 0.801 = one chance in 6.324 to find axis term	ON	TORAH	LETTERS	FREQUENCY	Equals	the Word	HAS 1 CHANCE
4		FIGURE		IN THE	PER LETTER	WORD	Appearing	IN THE
5				TORAH		EXPECTANCY	At Least once	FOLLOWING
6								
7	REPTILIAN @ Skips 17,708 to 17,717	17708	6	304805	1.96847E-05	0.000314955	0.000314906	3175.55211
8	R FACTOR						0.158124804	6.324118514
9	COMBINED ODDS AGAINST THE MATRIX = H7*H8						4.97944E-05	20082.56789

Chapter 1
Jewish ET connection?

This chapter was originally inspired by a 2023 book (*US Army Insider Missions: Space Arks, Underground Cities & ET Contact*) by Dr. Michael Salla, in which he interviews a young, Hispanic member of the U.S. Army who goes by the pseudonym J.P. He claimed to be allowed to disclose great secrets about alien life so long as he did not reveal his real name. The link to buy the book is at https://www.amazon.com/US-Army-Insider-Missions-Underground-ebook/dp/B0C53DPZJC. The description of an alien flag may point to what I suspect about a link between Israel and ET. The ET flag is described on page 293. J.P.'s English (his third language) is not the best. It states:

> *It was like a blue and white flag; it was kind of weird, but it was similar to the Israeli flag, but it was different. It had two gold angels by the Star of David, but the flag was different; it was blue, white, and gold. The two angels were gold, and they were like (it) ... looks like the ark of the covenant type.*

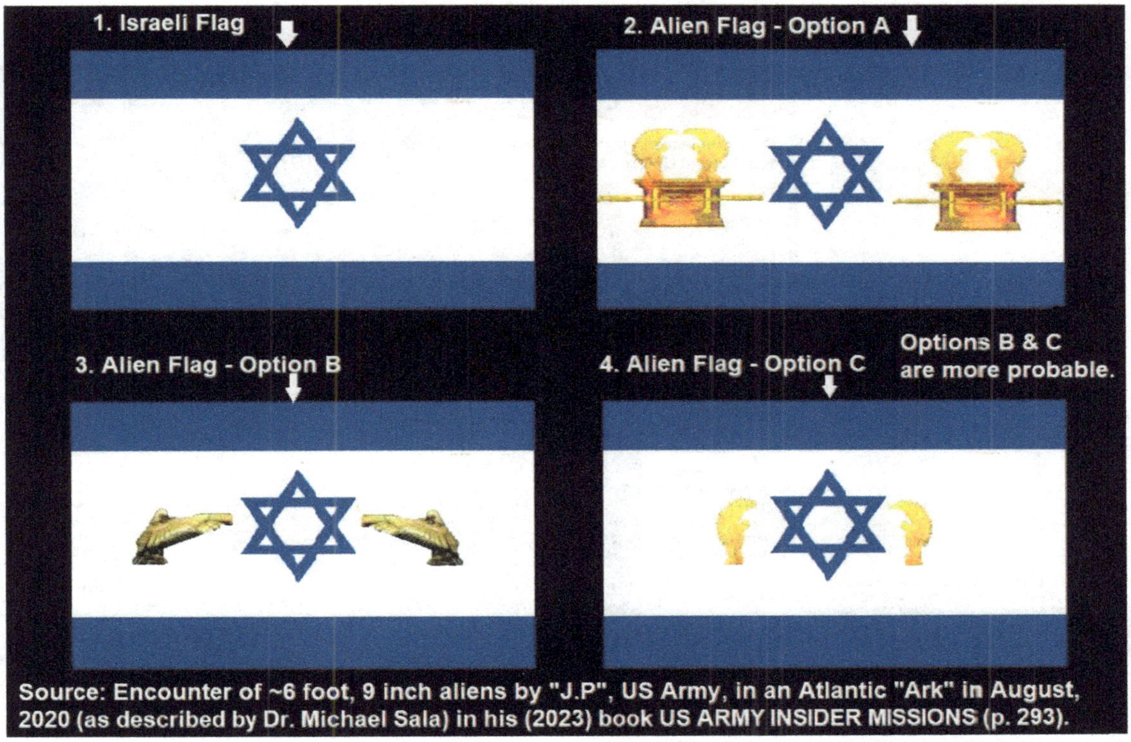

Fig 1-1 Israeli and ET flags

Figure 1-1 above. The Israeli version of their flag is shown with three versions of the alien flag seen. It is believed that the ark of the covenant shown in Option A is only shown due to less than perfect phrasing in J.P.'s description. Thus, Option B or C is more likely what was supposedly seen, but a photo or better description is needed to accurately describe the exact shape of the two angels allegedly depicted. However, in updating my manuscript in October, 2025 I must admit that Salla has allowed his reputation to be soiled by permitting pornographic advertising with his video advertising for which he charges up to $55 to view.

Further, J.P. is now out of the Army and revealing his identity. In publishing this warning, I have two caveats: (1) As is described later in this book and on my website, on August 1, 2016 an agent from the U.S. Army Intelligence Base at Fort Huachuca, Arizona threatened me about putting child pornography on my Torah Codes website unless I stopped publishing information about UFO and military secrets discussed in the Torah Code. I told him to go to hell and threatened to publish everything about this guy. Strangely he quickly grew unbelievably friendly and he even gave me information about the future which turned out to be correct. (2) Although Salla may not live up to what I thought about him originally, the questions I raise about ET-Jewish connects remain valid. The extraordinary birth of Isaac to a 90-year old mother and a 100-year old father is clearly not normal.

QUESTION: Were the ten lost tribes of Israel taken by aliens who then interbred with them? Even before the Jewish tribes came into existence, by the 6th chapter in the book of Genesis, there are references to beings who intermingle with — OK, who interbreed with — human beings, including "b'ney Elohim" and "Nephilim." While again, we are instructed not to belabor the meaning of these beings and incidents, lest our minds grow fuzzy and our teeth fall out, it isn't lost on some scholars that "Nephilim" means "fallen," as in "fallen from the sky." Either or both of these non-human species could well be our brothers from another planet (they are referred to as males).

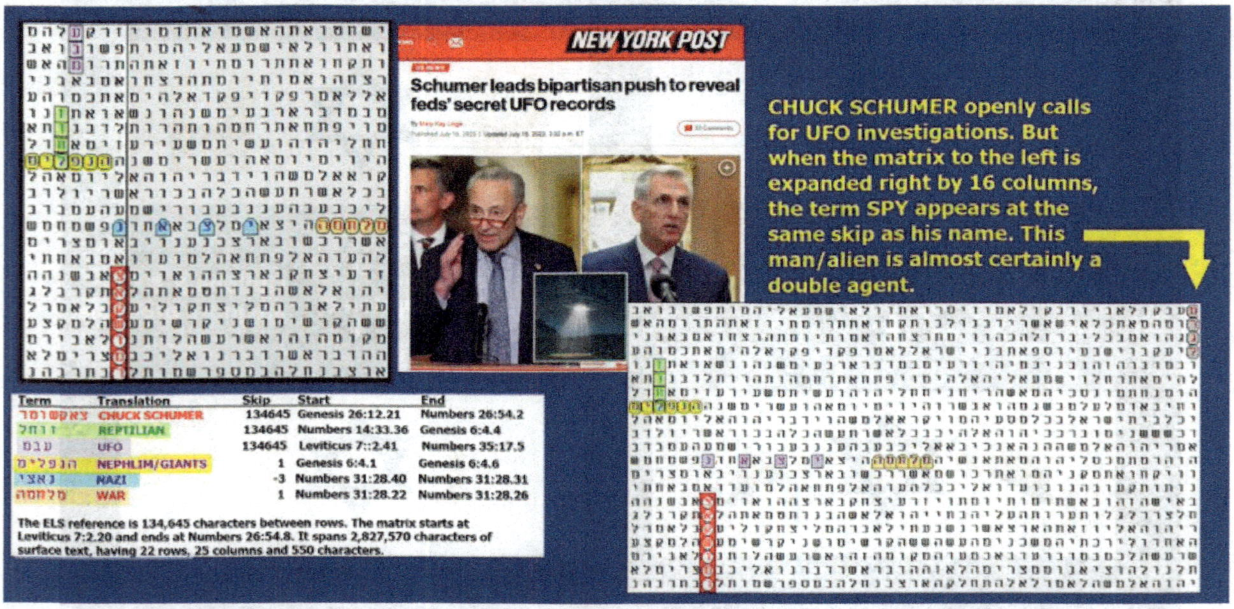

Fig 1-2 above: CHUCK SCHUMER, REPTILIAN And UFO are all at the same skip. REPTILIAN also shares a letter lamed with NEPHILIM in the open text. If this matrix was deliberately encoded, it was likely done so to warn us about Schumer and to inform us that the Nephilim found in the Bible were really Reptilians.

	A	B	C	D	E	F	G	H
1	AXIS TERM CHUCK SCHUMER @ SKIP +134645	SKIP	NUMBER	DIVIDE BY	THE QUOTIENT	E QUOTIENT	POISSON EQUATION	EACH WORD
2	ELS RANK 40	USED	IN	304805	EQUALS	X 164 LETTERS	PROBABILITY FOR	OR PHRASE
3		ON	TORAH	LETTERS	FREQUENCY	EQUALS	THE WORD	HAS 1 CHANCE
4		FIGURE		IN THE	PER LETTER	WORD	APPEARING	IN THE
5				TORAH		EXPECTANCY	AT LEAST ONCE	FOLLOWING OF
6								BEING HERE
7	REPTILIAN @ Skips +/- 1 and +/- 134645	134645	2	304805	6.56157E-06	0.001010482	0.001009972	990.1267076
8	NEPHILIM/GIANTS @ Skip +1 only (2 spellings)	1	4	304805	1.31231E-05	0.00580043	0.00578364	172.9015015
9	UFO @ Skips +/- 1 and +/- 134645	134645	173	304805	0.000567576	0.250868588	0.22187538	4.507034528
10	COMBINED SIGNIFICANCE = (H7*H8*H9)/40						5.18417E-05	19289.47616
11							#REF!	#REF!
12	AXIS TERM CHUCK SCHUMER @ SKIP +134645	SKIP	NUMBER	DIVIDE BY	THE QUOTIENT	E QUOTIENT	POISSON EQUATION	EACH WORD
13	ELS RANK 40	USED	IN	304805	EQUALS	X 550 LETTERS	PROBABILITY FOR	OR PHRASE
14		ON	TORAH	LETTERS	FREQUENCY	EQUALS	THE WORD	HAS 1 CHANCE
15		FIGURE		IN THE	PER LETTER	WORD	APPEARING	IN THE
16				TORAH		EXPECTANCY	AT LEAST ONCE	FOLLOWING OF
17								BEING HERE
18	REPTILIAN @ Skips +/- 1 and +/- 134645	134645	2	304805	6.56157E-06	0.003608865	0.003602361	277.5957553
19	NEPHILIM/GIANTS @ Skip +1 only (2 spellings)	1	4	304805	1.31231E-05	0.007217729	0.007191744	139.0483287
20	UFO @ Skips +/- 1 and +/- 134645	134645	173	304805	0.000567576	0.312166795	0.268140553	3.729387407
21	WAR @ Skip +1 only	1	36	304805	0.000118108	0.064959564	0.062894645	15.89960484
22	NAZI @ Skips -3 to -256	-3	452	304805	0.001482915	0.815603419	0.557627689	1.793311236
23	COMBINED SIGNIFICANCE = (h18*H19*H20*H21*H22)/40						9.74545E-06	102611.9884
24								
25	AXIS TERM CHUCK SCHUMER @ SKIP +134645	SKIP	NUMBER	DIVIDE BY	THE QUOTIENT	E QUOTIENT	POISSON EQUATION	EACH WORD
26	ELS RANK 40	USED	IN	304805	EQUALS	X 987 LETTERS	PROBABILITY FOR	OR PHRASE
27		ON	TORAH	LETTERS	FREQUENCY	EQUALS	THE WORD	HAS 1 CHANCE
28		FIGURE		IN THE	PER LETTER	WORD	APPEARING	IN THE
29				TORAH		EXPECTANCY	AT LEAST ONCE	FOLLOWING OF
30								BEING HERE
31	REPTILIAN @ Skips +/- 1 and +/- 134645	134645	2	304805	6.56157E-06	0.006476272	0.006455346	154.9103674
32	NEPHILIM/GIANTS @ Skip +1 only (2 spellings)	1	4	304805	1.31231E-05	0.012952543	0.01286902	77.70599326
33	SPY @ Skips +/- 1 and +/-134645	134645	11	304805	3.60886E-05	0.035619494	0.034992586	28.57748237
34	WAR @ Skip +1 only	1	36	304805	0.000118108	0.116572891	0.110034776	9.088035973
35	NAZI @ Skips -3 to -256	-3	452	304805	0.001482915	1.463637408	0.768606929	1.301055146
36	COMBINED SIGNIFICANCE = (h31*h32*h33*h34*h35)/40						9.83412E-06	101686.8318

The ten lost tribes were exiled from the Kingdom of Israel after its conquest by the Neo-Assyrian Empire c. 722 BCE. They were named as follows: Reuben, Simeon, Dan, Naphtali, Gad, Asher, Issachar, Zebulun, Manasseh, and Ephraim; (all but Judah and Benjamin, as well as some members of Levi, the priestly tribe, who did not have their own territory).

QUESTION: Does Judaism require belief in extraterrestrials? The Forward, a Jewish magazine, writes,

> *"Contemporary Rabbi Benjamin Blech once wrote,*
> *"What is fascinating ... is that a significant body of opinion in traditional sources not only validates the possibility for alien life on other planets but actually finds biblical and midrashic confirmation for this view, even as it suggests that belief in an all-powerful God forbids placing any limitations on the extent of his creative powers."*

In other words, a strong belief in an omnipotent God cannot logically exclude the possibility of other worlds populated by other creatures. Blech's colleague David Weintraub, professor of astronomy at Vanderbilt University and the author of Religions and Extraterrestrial Life: How Will We Deal With It?, He posits that Jewish theology may actually require a belief in extraterrestrials since there are no limits on the power of the Creator. He writes,

"[F]or Jews to say that no life beyond the Earth could possibly exist would be unacceptable, as such an idea would appear to place shackles on God's creative power. The universe belongs to God, and God can do what God wishes to do with the universe."

According to the Jewish sage, God draws the line at playing dice. The Forward once ran a short piece entitled "Putting Israel's UFO Frenzy Into Historical Perspective," after videos surfaced purporting to show UFOs hovering over the Jerusalem skyline. Unidentified flying objects are no stranger to the Holy Land. The Bible is full of accounts of sightings of heavenly beings. Indeed, a cornerstone of Jewish liturgy is the line from Psalms 19:2 — "The heavens declare the glory of God"— which some sages read to mean that we earthly inhabitants are not alone among the creatures in the universe who praise Hashem. In the Book of Judges (5:23), a passage reads,

"Curse Meroz, said the angel of the Lord, curse bitterly its inhabitants."

The 18th-century sage Rabbi Pinchas Eliyahu Horowitz claimed to have found Talmudic backing for his belief that Meroz was another planet, just one of many other worlds that God created.

Mid-20th-century kabbalist Aryeh Kaplan was also a professor of physics, and he often sought to find common ground between the two disciplines. For Rabbi Kaplan, there was no question that extraterrestrial life existed and that it was discussed in the Zohar. Kaplan wrote,

"The Midrash teaches us that there are seven earths. Although Ibn Ezra tries to argue that these refer to the seven continents, the Zohar clearly states that the seven are separated by a firmament and are inhabited. Although man does not inhabit them, they are the domain of intelligent creatures."

Israel's top scientist, Haim Eshed, reveals that people are on Mars. **HAIM ESHED** is encoded at the same skip as **MARS** (best spelling), **LIFE, ALIEN,** and **GRAY. PLEIADES** is one skip away, with **UFO** at the same absolute skip. See Figure 1-3. Odds against this matrix are about 22,162,538 to 1. The most significant *a priori* term is the best (5-letter) spelling of **MARS**, which was found at a special case skip against odds of about 411 to 1.

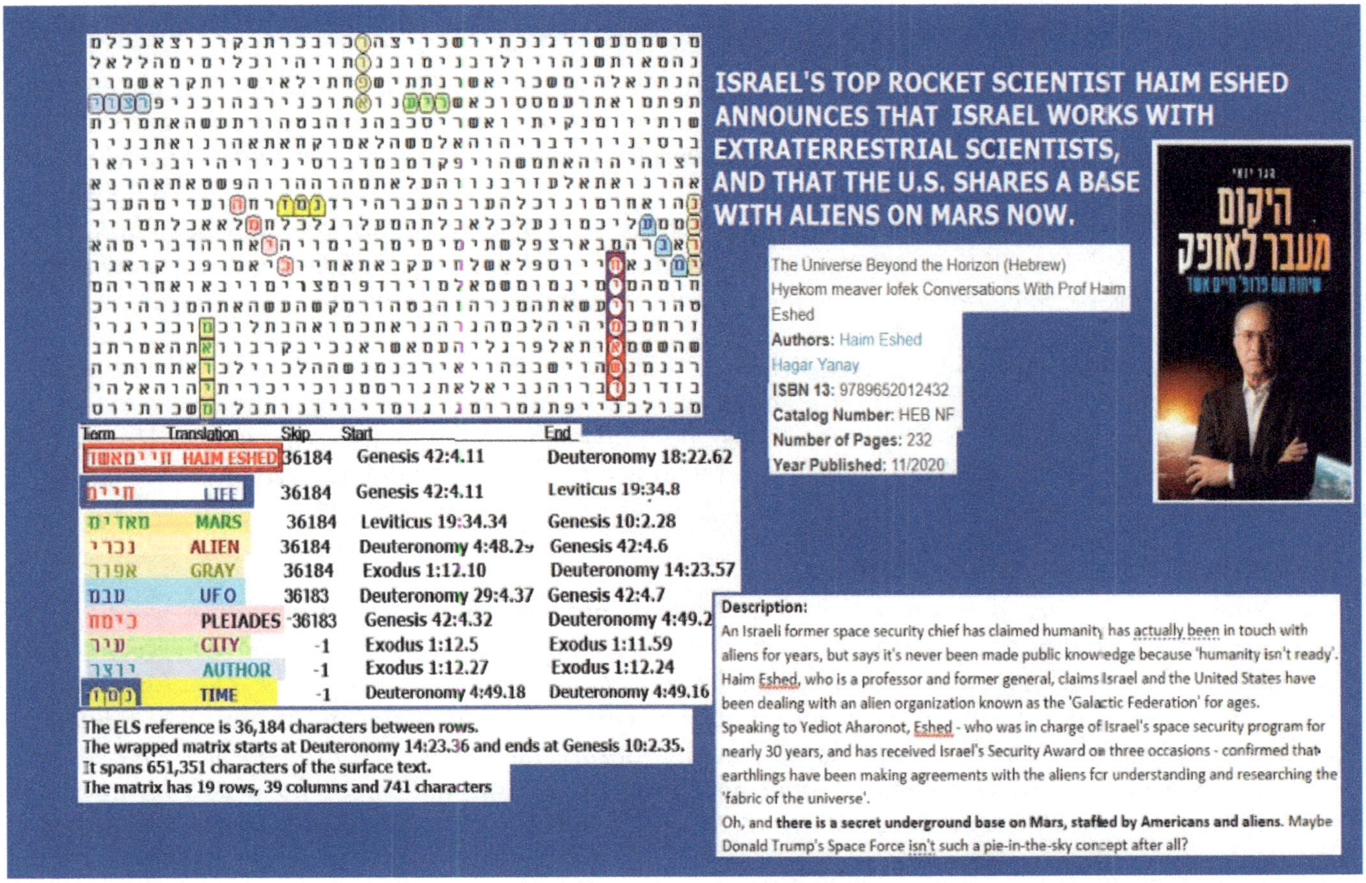

Figure 1-3 above: I have published information before about Haim Eshed. As head of the Israeli Space Agency for about 30 years, he really surprised the rest of the world when he announced that the Israelis are working with aliens. Still, when aliens start showing up elsewhere on Earth with a version of the Israeli flag to represent their nation, it's time to start asking some serious questions. What, exactly, is the relationship of Israel with ET? I have long suspected a link between our people and ET that began with the predicted conception and birth of Isaac, the first known human to be born into Judaism.

	A	B	C	D	E	F	G	H
1	Axis term HAIM ESHED @ Skip +36,184	SKIP	NUMBER	DIVIDE BY	THE QUOTIENT	E QUOTIENT	POISSON EQUATION	EACH WORD OR
2	ELS Rank 53	USED	IN	304805	EQUALS	X 741 LETTERS	PROBABILITY FOR	PHRASE HAS 1
3		ON	TORAH	LETTERS	FREQUENCY	EQUALS	THE WORD	CHANCE IN THE
4		FIGURE		IN THE	PER LETTER	WORD	APPEARING	IN THE FOLLOWING
5				TORAH		EXPECTANCY	AT LEAST ONCE	OF BEING HERE.
6								
7	MARS @ Skip +/- 1 and +/- 36184	36184	1	304805	3.28079E-06	0.002431062	0.00242811	411.8429826
8	GRAY @ Skips +/-1 and +/- 36184	36184	4	304805	1.31231E-05	0.00972425	0.009677122	103.3365054
9	ALIEN @ Skips +/-1 and +/- 36184	36184	19	304805	6.23349E-05	0.046190187	0.045139557	22.15346905
10	TIME @ Skips Skip +/- 1 and +/- 36184	-1	24	304805	7.87389E-05	0.0583455	0.056676027	17.64414435
11	LIFE @ Skips +/-1 and +/- 36184	36184	44	304805	0.000144355	0.106966749	0.10144445	9.857611742
12	AUTHOR @ Skips +/- 1 and +/- 36184 (2 synonyms)	-1	100	304805	0.000328079	0.243106248	0.215811309	4.633666727
13	CITY @ Skips +/- 1 and +/- 36184	-1	455	304805	0.001492758	1.10613343	0.669164309	1.494401279
14	UFO @ Skips +36183 to +36210	36183	1406	304805	0.004612785	3.41807385	0.967224495	1.03388614
15	PLEIADES @ Skips -36183 to -36506	36183	3101	304805	0.010173718	7.538724758	0.999467924	1.000532359
16	FINAL SIGNIFICANCE = (H7*H8*H9*H10*H11*H12*H13*H14*H15)/53						4.51212E-08	22162538.47

Who were the three men who visited Abraham and Sarah to predict the birth of Isaac? Were they angels or aliens? Let's examine the text.

18 1 The Lord appeared to Abraham near the great trees of Mamre while he was sitting at the entrance to his tent in the heat of the day. 2 Abraham looked up and saw three men (shelosha anashim = 3 men) standing nearby. When he saw them, he hurried from the entrance of his tent to meet them and bowed low to the ground.
3 He said, "If I have found favor in your eyes, my lord do not pass your servant by. 4 Let a little water be brought, and then you may all wash your feet and rest under this tree. 5 Let me get you something to eat so that you can be refreshed and then go on your way, now that you have come to your servant."

The word for *men* is used again in Genesis 18:16 and 18:22 when they started again to walk toward Sodom. It is also used in Genesis 19:5, 19:10 through Genesis 19:14. There it is in conjunction with the angels outside and inside of Lot's house. HaShem (The Name) is also intermittently used (in Genesis 18:1, 18:13, 18:17, 18:19, 18:20, 18:26, 18:33) to indicate God who is employing the landing party and who is also in direct contact with Abraham.

[1] Now the Lord was gracious to Sarah as he had said, and the Lord did for Sarah what he had promised. [2] Sarah became pregnant and bore a son to Abraham in his old age, at the very time God had promised him. [3] Abraham gave the name Isaac[a] to the son Sarah bore him. 4 When his son Isaac was eight days old, Abraham circumcised him, as God commanded him. 5 Abraham was a hundred years old when his son Isaac was born to him., [6] Sarah said, "God has brought me laughter, and everyone who hears about this will laugh with me." [7] And she added, "Who would have said to Abraham that Sarah would nurse children? Yet I have borne him a son in his old age."

With regards to this "landing party" of three, the initial use of the word *men* may indicate that if they were alien, they were likely a humanoid species like the Nordics or Pleiadians. The term for *angels* is used for two of them in Genesis 19:1 and Genesis 19:35. There are six other uses of the term for *angels* (*malachim*) in Torah; however, all of them (in Numbers 32:4, 32:7, Numbers 20:14, Numbers 21:21, Numbers 22:5, and Deuteronomy 2:26) are translated as *messengers.*

Could the correct interpretation for Genesis 19:1 and 19:15 really be *messengers* rather than *angels*? Yes. Sodom was destroyed because of its incredible immorality. Genesis 19:6-9 reads as follows:

Lot went out to them to the entrance and shut the door behind them, and he said, "I beg you, my brothers, do not act wickedly. See, now, I have two daughters who have never known a man. I shall bring them out to you and do to them as you please; but to these men do nothing inasmuch as they have come under the shelter of my roof." And they said, "Stand back!" Then they said, "This fellow came to sojourn and would act as a judge? Now we will treat you worse than them!" They pressed exceedingly upon the man (Lot), and they approached to break the door.

The message here to the world (especially to liberal, Democrat Americans) is clear. There will be consequences for sick behavior, be it bringing drag queens into kindergarten or enforcing transsexual actions on children even when their parents object. In Genesis, the punishment was sulfur and fire raining from heaven (Genesis 19:23), not unlike a weapon being dropped from orbit. On Earth, Putin (who hates homosexuals) threatens to use nuclear weapons against their supporters. Whatever wiped out Sodom and Gomorrah, the weapon was clearly dropped from above, and it was controlled via surveillance in Sodom.

Given that the translation "*angel*" is not certain, and that "messenger" works just as well, there is a high probability that the event is at least partially extraterrestrial in nature, even if the ETs here only think they are working for God.

It seems like a high percent of reported abductions by aliens involve some type of reproductive probe. The most humorous one that I ever heard involved my good friend, U.S. Navy Commander C. B. Scott Jones. Scott was the top aide of Senator Claiborne Pell (elected in 1960 for six terms, Chairman of the Senate Foreign Relations Committee for 17 years). In 2007, Scott invited me into his home in Kerrville, Texas, for two days and one night for an interview about what he knew about UFOs. He also wanted to know if the Torah Code was written by ETs. Scott said he represented the U.S. at UFO-related meetings with Chinese and Russian Government officials. During our talks, he mentioned a MUFON meeting in which two women told him that they saw him on a UFO with a huge number of abductees. Scott didn't remember the incident, but he asked about what happened. "Well," they said, "*You were naked, on a table, with a hybrid female alien over you, having sex with you!" Scott laughed and then told them, "I sure wish I could remember that, but the joke was on the ET. If she wanted to get pregnant, I was shooting with blanks because a few weeks before, I just had a vasectomy!"*

While Scott laughed at hearing the story from MUFON, Sarah laughed at hearing the report on her upcoming pregnancy. In fact, her son was named Yitzchak (Isaac), which means "One who laughs or rejoices." But there are serious implications of the event here.

1. If both Sarah and Abraham were abducted, then Abraham's sperm could have been put into Sarah's egg in a lab. In this case, God's promise to give Abraham a son was fully carried out (with the help of medical technology).

2. If Sarah's egg was fertilized, but by an alien sperm, then Isaac was a hybrid – Sarah's true son, but not the true son of Abraham. This would make Ishmael not only still Abraham's first-born son, but the only child with a legitimate claim to rights associated with being directly descended from Abraham. This route tends to back Muslim theology. But it also decreases the chance that the ELS Torah Code could survive changes to the Torah text made by Jews. For example, Isaac is only four letters in Hebrew, while Ishmael is five. Very large skips of the axis term, and just about all wrapped text matrices requiring more than one computer pass through the 304,805 letters of Torah, would be destroyed by making such switches. And yet I have found many thousands of matrices. This fact argues for Option 1, which fully fulfills the promises made.

3. An entirely alien embryo was implanted into Sarah. Again, the Torah Code would likely be destroyed, but it really doesn't do much for Islam because the Quran mentions Isaac 15 times and all in a positive light. If it turns out that Isaac had this kind of conception, then all Western religions are likely based on alien hoaxes designed to help them manipulate humans to serve alien agendas. I think that Option 1 here is correct, but largely based on my experience with the Torah Code since 1997. Unfortunately, there are few (possibly no) people on Earth who are experienced enough with the Code to move it into the realm of evidence that will be universally accepted by science. Short of a full disclosure, we will need either a time machine or software like that proposed in *PROJECT CREATOR DECODER* to arrive at a true answer. The software is described on my website **at** https://www.arkcode.com/photo4_113.html.

There are times when I'm not sure if the Code is trying to say that Netanyahu is an enemy of aliens or their friend. It can be both ways, depending on whether there is more than one species of aliens involved. Israeli Brigadier General Haim Eshed (who ran the Israeli space program for 30 years) said that aliens work alongside Israelis. If true, Netanyahu certainly knows about the presence of ETs here (and in Israel), and, like President Trump, he is keeping their presence secret. In a way, the situation may be like stories that Franklin Roosevelt knew about Japan's plans to attack Pearl Harbor but did nothing to prevent it because he wanted to get into the war. With Netanyahu, he might have known about Hamas' plans to attack on October 7, 2023, but did nothing about it because he wanted an excuse to obliterate Hamas (and Gaza).

Figure 1-4 has an axis term that reads ***CONQUEST OF GAZA***. The term ***ALIEN*** is on the top row with ***NETANYAHU*** three rows right under it, and ***ISRAEL*** is five rows directly under that. ***FROM A UFO*** is at the absolute skip of ***CONQUEST OF GAZA***. What kind of aliens are alluded to? I don't see reptilians, but they are allegedly shape-shifters, and ***SHIFTER*** is at skip -1. The matrix was found against odds of about 337,331 to 1. The anti-Semitic poster alleging a link between Reptilians and Jews was published by none other than Yair Netanyahu in 2017. He is Prime Minister Netanyahu's son!!!!!!!!!

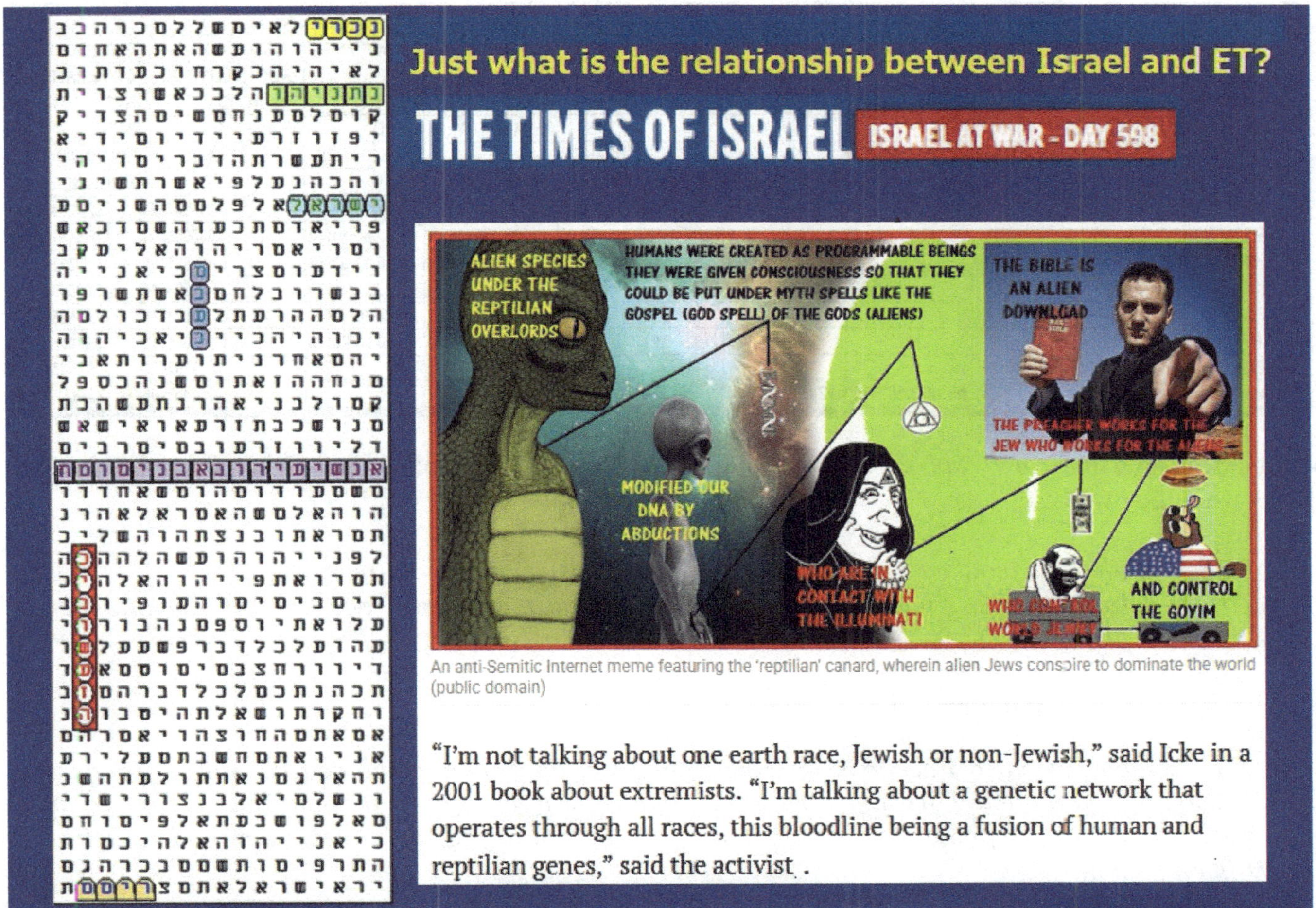

Term	Translation	Skip	Start	End
כיבושעזה	CONQUEST OF GAZA	54486	Numbers 5:30.61	Deuteronomy 13:15.21
נתניהו	NETANYAHU	1	Deuteronomy 12:21.59	Deuteronomy 12:21.64
נכרי	ALIEN	1	Exodus 21:8.33	Exodus 21:8.36
אנשיעירובאבנימומת	Men stone him so he dies	1	Deuteronomy 21:21.9	Deuteronomy 21:21.25
בעבם	IN A UFO	-54486	Deuteronomy 6:10.8	Exodus 14:4.53
ישראל	ISRAEL	1	Numbers 31:5.12	Numbers 31:5.16
ממיר	SHIFTER	-1	Exodus 14:30.49	Exodus 14:30.46

The ELS reference is 54,486 characters between rows.
The wrapped matrix starts at Exodus 21:8.33 and ends at Exodus 14:30.50.
It spans 2,124,971 characters of the surface text.
The matrix has 40 rows, 17 columns and contains 680 characters.

Figure 1-4 above includes a screenshot of a portion of the cartoon posted by Yair Netanyahu, September 8, 2017 (Facebook). Its spreadsheet is below. I suspect that the cartoon represents a leak in the Israeli Government, but as I make clear in Chapter 9 (Fighting radical Islam and ET), I think the Reptilians supplied Hamas with the technology to move so many troops (undetected and unopposed for many hours) into Israel. Probably, when General Eshed says that aliens work with the Israeli government, he meant Pleiadians, Nordics, or some other race that is basically human and at war with Reptilians. But Figures 1-5 and 1-6 should also be examined in response to this question.

	A	B	C	D	E	F	G	H
1	*AXIS* TERM CONQUEST GAZA @ Skip 54,486	SKIP	NUMBER	DIVIDE BY	THE QUOTIENT	E QUOTIENT	POISSON EQUATION	EACH WORD
2	ELS RANK 2	USED	IN	304805	EQUALS	X 680	PROBABILITY FOR	OR PHRASE
3		ON	TORAH	LETTERS	FREQUENCY	EQUALS	THE WORD	HAS 1 CHANCE
4	R factor 0.042= 1 CHANCE IN 1.10153931	FIGURE		IN THE	PER LETTER	WORD	APPEARING	IN THE
5	SHIFTER @			TORAH		EXPECTANCY	AT LEAST ONCE	FOLLOWING OF
6								BEING HERE
7	ALIEN @ Skip +1 only	1	10	304805	3.28079E-05	0.016075852	0.015947325	62.70644169
8	NETANYAHU @ Skip +1 only	1	16	304805	5.24926E-05	0.035694953	0.035065401	28.51813996
9	SHIFTER @ Skips +/- 1 and +/- 54486	-1	58	304805	0.000190286	0.09323994	0.089025106	11.23278646
10	AND DIE @Skip +1 only	1	138	304805	0.000452748	0.221846754	0.19896189	5.026088173
11	UFO @ Skips +/- 1 and +/- 54486	-54486	174	304805	0.000570857	0.279719821	0.244004474	4.098285509
12	ISRAEL@ Skip +1 only	1	591	304805	0.001938945	0.95008284	0.613291013	1.630547292
13	R factor 0.042= 1 CHANCE IN 1.10153931						0.90782053	1.10153931
14	combined odds against the matrix = (h7*h8*h9*h10*h11*h12)/2						2.96445E-06	337331.2484

What does the *Times of Israel* say about Yair Netanyahu's 2017 publication? Other than the cartoon, this is what they wrote:

> *'Alien reptile' and cloaked figure in Yair Netanyahu's meme have old-new origins,*
> *To demonize his father's political enemies, PM's son posted a meme 'rife with anti-Semitic themes' hidden in plain view*

By Matt Lebovi, 11 September 2017, 1:13 pm

> *The cartoon posted to Facebook by Yair Netanyahu on Friday contained no Stars of David, no swastikas, and no anti-Jewish text. However, hidden in plain sight within the since- removed meme were two of the most potent anti-Jewish narratives: the colonizing alien- lizard Jew and the scheming Illuminati Jew.*
>
> *The portrayal of Jews as other-than-human and engaged in a global conspiracy is not new, and can be traced to the notorious canard, "The Protocols of the Elders of Zion." More recently minted is the Netanyahu meme's green reptile-man, a stand-in for Jewish aspirations popularized by British conspiracy theorist David Icke.*
>
> *Since the early 1990s, Icke has regaled fellow travelers with tales of shape-shifting reptiles who dominate global affairs. The reptilian takeover is understood to be about Jews, who are capable of inhabiting and controlling people's minds. In recent years, Icke has packed stadiums across Europe with this mash-up of sci-fi horror and old-fashioned anti-Semitism. "People know how to decode the code about reptilians — whether Icke means it or not, it doesn't change that fact," said Jan Rathje of the Amadeu Antonio Stiftung, an anti-racism group in Germany. "It seems ridiculous, but the conclusion that Icke draws is that because he thinks the reptilians are pulling the strings behind it, therefore the Holocaust didn't happen – that's anti-Semitism," said Rathje.*
>
> *In the realm of reptilian domination, leaders ranging from the British royal family to Egypt's pharaohs have been used by Jews as "elite corporal hosts," also known as "reptilian overlords." In this "New World Order" conspiracy theory, gentile proxies — ranging from Jeb Bush to Kris Kristofferson — are foot soldiers in Jewry's quest for domination.*

"I'm not talking about one earth race, Jewish or non-Jewish," said Icke in a 2001 book about extremists. "I'm talking about a genetic network that operates through all races, this bloodline being a fusion of human and reptilian genes," said the activist." Icke has an 18- minute video posted at https://www.youtube.com/watch?v=vyCn0s8Y9ck. It claims that leading politicians are reptilian. He came to this conclusion without ever seeing one of my torah code matrices, just as I came to my conclusions about the real identity of politicians being reptilians without ever listening to Icke.

One of Icke's main targets is George Soros, the Hungarian-born Holocaust survivor and billionaire who, in the meme posted by Yair Netanyahu, dangles our planet as bait. Soros is positioned as the ultimate "feeder" of humanity to the extra-terrestrial Jew, that green alien holding an Occult symbol in front of the meme's third boogeyman. In his native Hungary, where Soros survived the Nazi occupation, the 95-year-old investor (in 2025) is seen as the proverbial "evil frog in the well," a demon bent on erasing borders and flooding the country with immigrants, drugs, and homosexuality.

FIGURE 1-5 below: The axis term is **REPTILIAN GOVERNMENT**. In the area with a white background, it only took 18 rows and eight columns (144 letters) to link **REPTILIAN GOVERNMENT** with **NETANYAHU** and **ISRAEL.** Odds against this match were about 4,658 to 1.

Just looking at the odds of **INVASION** to be at the same absolute skip as **REPTILIAN GOVERNMENT**, the matrix is 37 columns and 8 rows (296 letters) with odds against it at 1,030 to 1. Odds against the full Figure 1-6 that has **REPTILIAN GOVERNMENT** with **INVASION**, **NETANYAHU**, **PRESIDENT, HORRIBLE and THE SKY** are about 16,754,897 to 1.

Figures 1-6 and 1-7 back the idea that the October 7, 2023 Hamas invasion from Gaza was supported by reptilians. As for Netanyahu, I have long been his fan. But Figures 1-4 above, and 1- 5 and 1-6 below support the need for a serious investigation of Israeli Government and IDF. They failed to react in a timely manner. Figure 1-6 below has Gazin terrorist and murderer **SIRWAR** (as the axis term) at the same skip as **REPTILIAN**, with **REPTILIAN** touch **NETANYAHU** directly below it. Odds against Figure 1-6 (with **ENEMIES** also at skip +1) are about 7,690 to 1.

Popular Latest Newsletters

The Atlantic

U.S.

12 Million Americans Believe Lizard People Run Our Country

About 90 million Americans believe aliens exist. Some 66 million of us think aliens landed at Roswell in 1948. These are the things you learn when there's a lull in political news and pollsters get to ask whatever questions they want.

By Philip Bump

Term	Translation		Skip	Start	End
ממשלזוחל	REPTILIAN GOVERNMENT		-96412	Genesis 49:28.18	Genesis 9:16.37
פלישה	INVASION		96412	Numbers 1:44.5	Exodus 20:11.14
נוק	NUKE		96413	Exodus 20:11.48	Deuteronomy 31:28.15
השמים	THE SKY	Deuteronomy 9:10.85	1	Exodus 20:11.19	Exodus 20:11.23
נורא	HORRIBLE		1	Genesis 9:16.13	Genesis 9:16.16
ירח	MOON			Deuteronomy 31:27.70	Deuteronomy 31:27.68
נשיא	PRESIDENT		1	Numbers 1:44.25	Numbers 1:44.28
נתניהו	NETANYAHU		1	Deut 9:11.28	Deuteronomy 9:11.33
ישראל	ISRAEL		1	Numbers 27:20.32	Numbers 27:20.36

The ELS reference is 96.412 letters between rows.
The matrix starts at Deuteronomy 9:10.85 and ends at Genesis 49:28.19
The matrix spans 1,639.041 letters of the surface text.
The matrix has 18 rows, 37 columns and 666 letters.

REPTILIAN GOVERNMENT, UFO and INVASION are all at the same absolute skip. NUKE passes through and shares a letter *vav* with REPTILIAN GOVERNMENT. NETANYAHU is at skip +1 as is ISRAEL which appears four times on the full 666-letter matrix. MOON is at skip -1. There is some evidence for an alien base there.

REPTILIAN GOVERNMENT, NETANYAHU, ISRAEL and NUKE all appear in 144 letters. Given that Israeli Brigadier General Haim Eshed has said that aliens work with Israeli scientists, we need clarification of exactly what the relationship is for Netanyahu with reptilians. The is especially true because on October 7, 2023 the Israeli Defense Force took almost 6 hours before it began to reinforce the outnumbered Israelis in the communities bordering Gaza.

Figure 1-5 above with its spreadsheet below.

	A	B	C	D	E	F	G	H
1	AXIS TERM REPTILIAN GOVERNMENT @ Skip -96,412	SKIP	NUMBER	DIVIDE BY	THE QUOTIENT	E QUOTIENT	Poisson Equation	EACH WORD
2	ELS RANK 1	USED	IN	304805	EQUALS	x 666 Letters	Probability for	OR PHRASE
3	Moon and Nuke are not included below because they were found	ON	TORAH	LETTERS	FREQUENCY	Equals	the Word	HAS 1 CHANCE
4	*a posteriori*	FIGURE		IN THE	PER LETTER	WORD	Appearing	IN THE
5				TORAH		EXPECTANCY	At Least once	FOLLOWING
6								
7	INVASION @ Skips +/- 1 and +/- 96412	96412	1	304805	3.28079E-06	0.002185004	0.002182618	458.1653472
8	NETANYAHU @ Skip +1 only	1	16	304805	5.24926E-05	0.034960056	0.034356013	29.1069861
9	HORRIBLE @ Skip +1 only	1	23	304805	7.54581E-05	0.050255081	0.049013185	20.40267319
10	PRESIDENT @ Skip +1	1	68	304805	0.000223093	0.14858024	0.138069157	7.242747209
11	THE SKY @ Skip +1 only	1	81	304805	0.000265744	0.176985286	0.162207888	6.164928304
12	ISRAEL @ +1 only	1	591	304805	0.001938945	1.291337084	0.725097031	1.379125768
13	COMBINED ODDS AGAINST THE MATRIX = H7*H8*H9*H10*H11*H12						5.9684E-08	16754897.72
14								
15	AXIS TERM REPTILIAN GOVERNMENT @ Skip -96,412	SKIP	NUMBER	DIVIDE BY	THE QUOTIENT	E QUOTIENT	Poisson Equation	EACH WORD
16	ELS RANK 1	USED	IN	304805	EQUALS	x 144 Letters	Probability for	OR PHRASE
17	Moon is not included below because it was found	ON	TORAH	LETTERS	FREQUENCY	Equals	the Word	HAS 1 CHANCE
18	*a posteriori*	FIGURE		IN THE	PER LETTER	WORD	Appearing	IN THE
19				TORAH		EXPECTANCY	At Least once	FOLLOWING
20	NETANYAHU @ Skip +1 only	1	16	304805	5.24926E-05	0.007558931	0.007530434	132.7944667
21	ISRAEL @ +1 only	1	591	304805	0.001938945	0.279208018	0.243617455	35.08091345
22	COMBINED ODDS AGAINST THE MATRIX =H20*H21						0.000214659	4658.551194

Term	Translation	Skip	Start	End
סינוואר	SINWAR	-75073	Genesis 44:15.11	Numbers 22:30.31
נתניהו	NETANYAHU	1	Deuteronomy 30:7.2	Deuteronomy 30:7.7
זוחל	REPTILIAN	75073	Deuteronomy 32:36.51	Numbers 20:10.35
כלהאלות	ALL THESE CURSES	1	Deuteronomy 30:7.16	Deuteronomy 30:7.22
האלהעלאיביך	UPON YOUR ENEMIES	1	Deuteronomy 30:7.23	Deuteronomy 30:7.33

The ELS reference is 75,073 characters between rows.
The matrix starts at Numbers 22:30.21 and ends at Genesis 44:15.32.
The matrix spans 450,470 characters of the surface text.
The matrix has 7 rows, is 32 columns wide and contains a total of 224 characters.

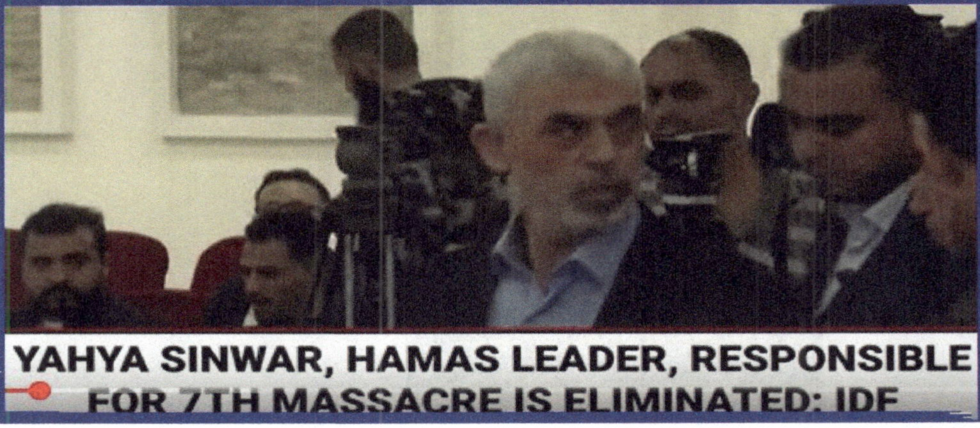

Figure 1-6 above: Axis term SINWAR paired with REPTILIAN and NETANYAHU. SINWAR and REPTILIAN are at the same absolute skip. REPTILIAN touches NETANYAHU at skip +1. SINWAR and REPTILIAN are named as YOUR ENEMIES. Odds against Figure 1- 6 existing are about 7,690 to 1.

	A	B	C	D	E	F	G	H
1	AXIS TERM SINWAR @ Skip -75,073	SKIP	NUMBER	DIVIDE BY	THE QUOTIENT	E QUOTIENT	POISSON EQUATION	EACH WORD
2	ELS RANK 55	USED	IN	304805	EQUALS	X 224	PROBABILITY FOR	OR PHRASE
3		ON	TORAH	LETTERS	FREQUENCY	EQUALS	THE WORD	HAS 1 CHANCE
4		FIGURE		IN THE	PER LETTER	WORD	APPEARING	IN THE
5				TORAH		EXPECTANCY	AT LEAST ONCE	FOLLOWING OF
6								BEING HERE
7	REPTILIAN @ SkipS +/- 1 and +/- only	75073	4	304805	1.31231E-05	0.002939584	0.002935268	340.6843968
8	NETANYAHU @ Skip +1 only	1	16	304805	5.24926E-05	0.011758337	0.011689478	85.54701781
9	ENEMIES @ Skips +1 only	1	32	304805	0.000104985	0.071389905	0.068901219	14.51353137
10	combined odds against the matrix = (h7*h8*h9)/55						0.000130027	7690.729288

Spreadsheet for Figure 1-6 is above. As a Jew, I am horrified to find any link between my people and reptilians. I would say that I never heard anything about it from any other Jew in my 78 years of life, but the truth is that my (now deceased) Uncle Eugene, shown on the top of all my website articles (along with my son and me), caused me to start all my Mars research. This was ten years after I started my Torah Codes research. My uncle let me know that he back-engineered UFOs for our government, and, shortly before he died, he asked me to find proof. He also gave me the list of all major scientists from around the world who (in 1960) worked with him on this project, including three from the (former) Soviet Union and one from Poland. See https://arkcode.com/photo5_29.html for the full list. From the U.S.S.R. were Dr. Alexander Ivanov (engineer, Radio Research Institute in Leningrad), Prof. Dr. Michael N. Micheev (Director, Physics Metal Institute in Moscow), and Pavel G. Miknevitch (from the All-Union Material Institute in Moscow). This revelation started on December 24, 1997, and continued in early June 2007 when I visited him after retiring from the Coast Guard (at the age of 60) and taking my family around the world. If my uncle did in any way work with ETs, I am sure that as a Jew, he was one of the few exceptions. I have never heard of any Jew who wants to control the world. Even in Israel, what people want is to live a simple life in an area about the size of New Jersey. In comparison, Muslims control most land from Morocco to Indonesia, and many of them are not satisfied with that.

Figure 1-7 below: Eugene Roffman (UFO back-engineer & mentor for this investigation).

Chapter 2
Et Infiltration in the Vatican

If you follow my work closely, you know that I strongly believe that Reptilians have taken over control of much of the U.S. Government. This suspicion is backed by many Torah Code matrices in Chapter 3. One of the better ones is on my matrices about Chris Wray of the FBI. But in glancing through my articles about Reptilians, I have seen suggestions that another Reptilian power center is in the Vatican. Figure 2-1, and the photos of the Paul VI Audience Hall at the Vatican, seem to back this idea.

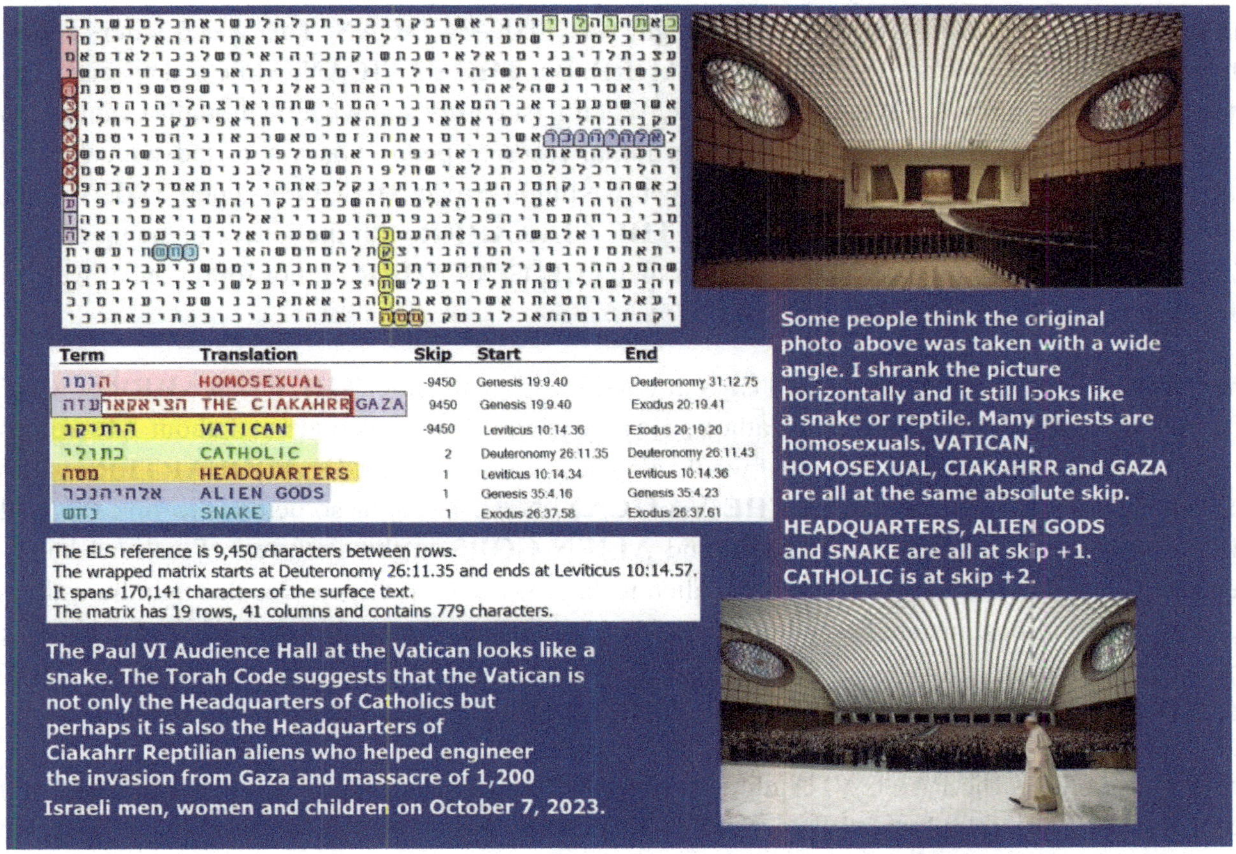

Figure 2-1 above: What leaps out at us with Figure 2-1 is that we have what at first looks like a 13-letter ELS. An ELS that long is almost never found. However, what we are really looking at are three terms that were all found in the same column. They are composed of 14 letters, but there is a one-letter overlap. This matrix was found against odds of 55,288 to 1.

	A	B	C	D	E	F	G	H
1	*AXIS TERM THE CIAKAHRR @ Skip +9450*	SKIP	NUMBER	DIVIDE BY	THE QUOTIENT	E QUOTIENT	POISSON EQUATION	EACH WORD
2	ELS RANK 6	USED	IN	304805	EQUALS	X 779	PROBABILITY FOR	OR PHRASE
3		ON	TORAH	LETTERS	FREQUENCY	EQUALS	THE WORD	HAS 1 CHANCE
4		FIGURE		IN THE	PER LETTER	WORD	APPEARING	IN THE
5				TORAH		EXPECTANCY	AT LEAST ONCE	FOLLOWING OF
6								BEING HERE
7	VATICAN @ Skips +/- 1 AND +/- 9450	9450	2	304805	6.56157E-06	0.005111465	0.005098423	196.1390652
8	GAZA @ Skips +/- 1 and +/- 9450	9450	21	304805	6.88965E-05	0.053670379	0.052255549	19.13672368
9	HOMO @ Skip +/-1 and +/- 9450	-9450	75	304805	0.000246059	0.191679927	0.17442893	5.732993935
10	SNAKE @ Skip +1 only	1	85	304805	0.000278867	0.21723725	0.195260978	5.121350982
11	HEADQUARTERS @ Skip +1 only	1	158	304805	0.000518364	0.403805712	0.332226151	3.009997851
12	INITIAL COMBINED SIGNIFICANCE = (H7*H8*H9*H10*H11)	-9450					3.01464E-06	331714.0719
13	ADJUSTED ODDS FOR AXIS TERM RANK 6 = H12/6						1.80879E-05	55285.67865

The first term sought was a variety of Reptilian known as **THE CIAKAHRR** (7 letters if we include an initial letter of *hey* for **THE**). The next term sought was **THE VATICAN**. This was 6 letters at the same absolute skip (opposite direction) as **THE CIAKAHRR.** But **THE VATICAN**, while at a special case skip, is not in the same column as **THE CIAKAHRR.** When I asked my son what he thought I should look for next, he suggested **HOMO** because of all the criminal cases of pedophilia associated with the Catholic priests. However, if his reasoning was fully correct, we might expect **HOMO** to touch **THE VATICAN**. In reality, it is found touching and overlapping **THE CIAKAHRR** with the letter *hey* being shared with **THE CIAKAHRR**. This might actually suggest that **THE CIAKAHRR** do not follow what we consider to be normal sexual practices on Earth.

The next term sought was **HEADQUARTERS**. It would work with either **THE CIAKAHRR** or with **THE VATICAN** if we are talking about the Catholic Headquarters. If we are concerned about ETs controlling us by shaping our main religions, then that would also be a major concern. **HEADQUARTERS** was found at skip +1, sharing the letter *hey* with **THE VATICAN**. Next, I sought something to link the Ciakahrr Reptilians with something ALIEN. I then found **ALIEN GODS** in the open text. Finally, I have often written that I thought Hamas was given some alien technology to help them break through the wall around Gaza, giving them almost instant access to over 1,200 Israelis that they could rape and or murder while the victims were home in their beds. So, I sought **GAZA.** It was right there in sequence with **THE CIAKAHRR**. This gave me **HOMOSEXUAL, THE CIAKAHRR,** and **GAZA** all in one column.

Figure 2-1 also shows the PAUL VI Auditorium at the Vatican. Typically, Catholic churches are in the shape of a cross. The rooms are full of statues of Jesus, or Mary, or saints. None of these are in this auditorium. Instead, the room is shaped like a snake head complete with fangs on stage. The word **SNAKE** is also on Figure 2-1 at skip +1. Very oddly, when you look up how the Biblical Being (the Serpent) associated with evil came to be the image that the Church wants to put out in association with the Pope, you often find excuses that demand you to forget what your eyes are plainly seeing. In fact, the image is serpent-like when looking from the back of the room to the stage at the front, and from the stage looking backward. It is often reinforced with the sculpture shown on Figure 2-2.

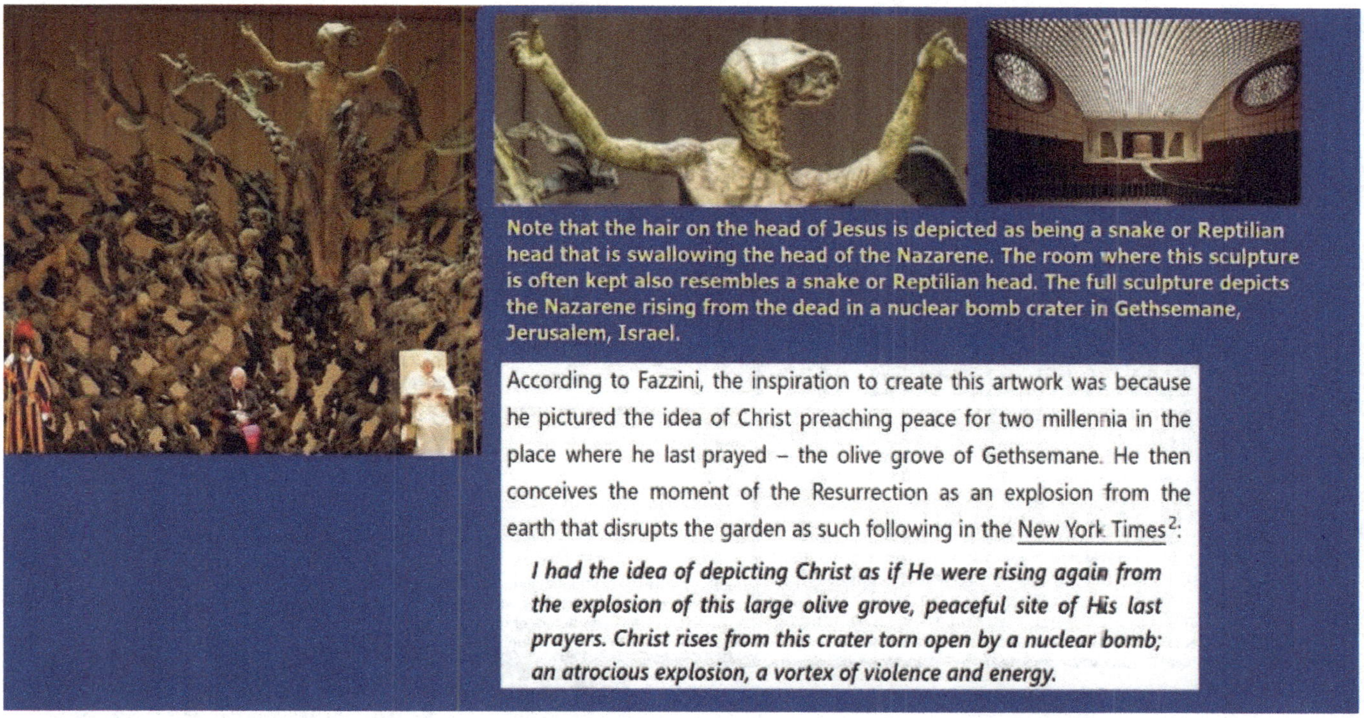

Figure 2-2 above: The idea that the Vatican is the Headquarters for Reptilian aliens is supported by the facts that the Vatican has built an enormous auditorium partly in Vatican City and partly in Rome – in the shape of a snake or Reptilian head, lacking all normal Catholic statues or crosses. Instead, it shows what happens after Jerusalem gets nuked. If this "artwork" shows a risen Nazarene, he is rising inside that head of a snake or reptilian creature that has just destroyed Jerusalem. As the Vatican allied with Nazis during World War 2, and as it helped Nazi war criminals escape via ratlines after World War 2, its choice of artwork here shows that its policy is still one of extermination for all Jews.

Reptilians have, quite possibly, served as the inspiration for gargoyles in Catholic churches in Europe. Wikipedia states, "*During the 12th Century, when gargoyles appeared in Europe, the Roman Catholic Church was growing stronger and converting many new people. Most of the population at this time was illiterate, so images were very important to convey ideas. Many early gargoyles depicted some version of a dragon, especially in France. In addition to serving as spouts for water, the gaping mouths of these gargoyles evoked the fearsome destructiveness of these legendary beasts, reminding the laity of the need for the church's protection.*" The reason why many gargoyles are depicted as these dragon-looking, unidentifiable monsters is because it is said the founding bishops of churches would rid their respective towns of these kinds of beasts. What does the Torah Code say about this? See Figure 2-3. Against odds of about 178 to 1, **GARGOYLE** is encoded in a 28-letter matrix with **REPTILIAN** at skip -1. The full matrix includes **FROM A STAR** and **FROM THE PLEIADES** at the same absolute skip as **GARGOYLE. UFO** is at skip -1. **PRESIDENTS** at skip +1 overlap **UFO** (two presidents are discussed in Chapter 3 as possibly being reptilian or abducted and programmed by reptilians – Obama and Biden). **GRAY** is one skip less than **GARGOYLE.** Grays supposedly help reptilians by abducting people for them. Against odds of about 23,038,267 to 1, this matrix tells us that this love of reptilians seen in the Pope John VI Auditorium is not new.

Based on Figure 2-3, it is likely that reptilians, always out to rule worlds, infiltrated the Catholic Church. Hitler was their natural (Satanic) ally during and after the Holocaust. Gargoyles are also found in Europe with the Church of England and the Episcopal Church.

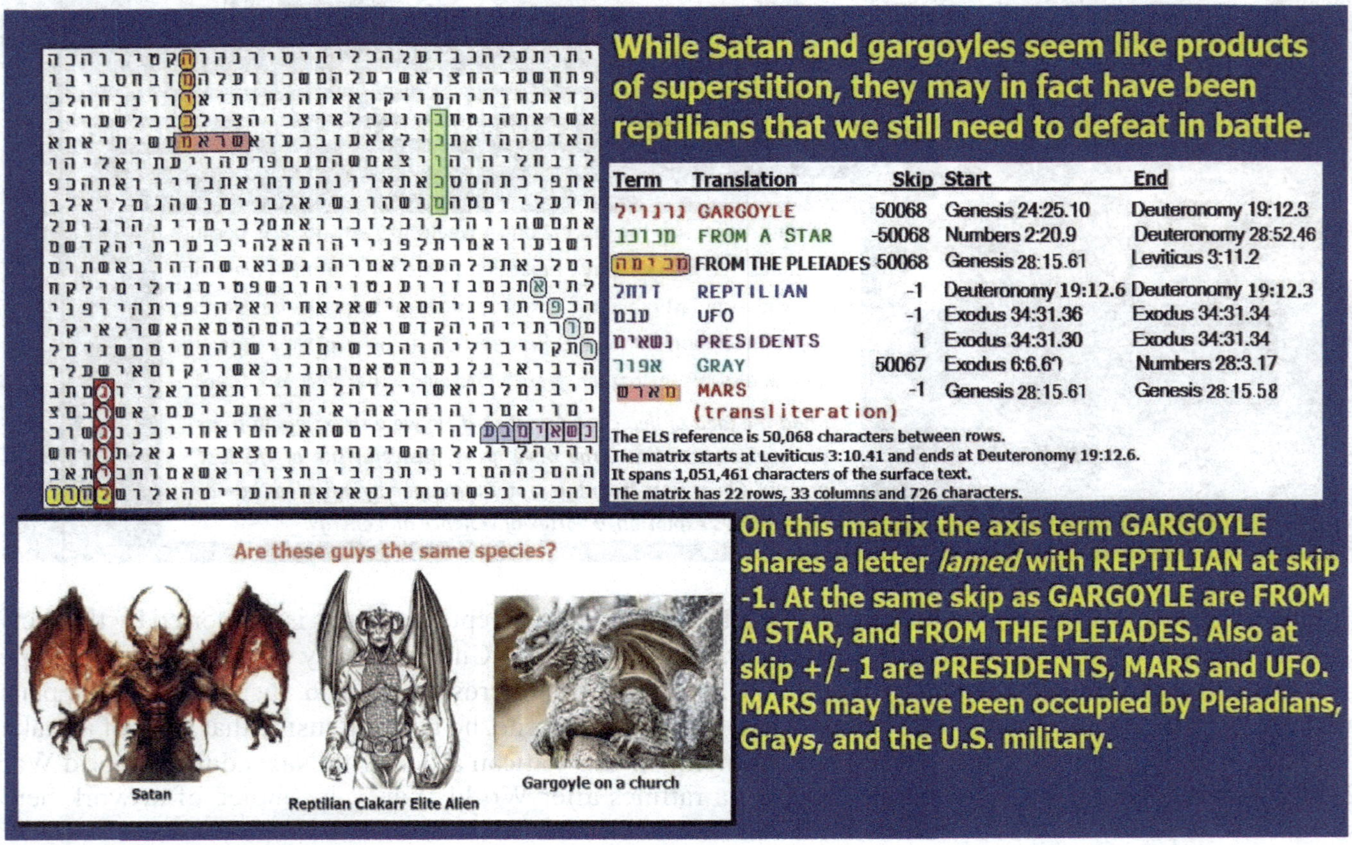

Figure 2-3. The Code gives us reason to believe that **REPTILIANS, GARGOYLES**, and **SATAN** all represent the same or related species. **REPTILIANS** and **GARGOYLES** met in 28 letters against odds of about 178 to 1.

	A	B	C	D	E	F	G	H
1	*AXIS* TERM GARGOYLE @ Skip+ 50,068	SKIP	NUMBER	DIVIDE B	THE QUOTIENT	E QUOTIENT	POISSON EQUATION	EACH WORD
2	ELS RANK 61	USED	IN	304805	EQUALS	x 726 LETTERS	PROBABILITY FOR	OR PHRASE
3		ON	TORAH	LETTERS	FREQUENCY	EQUALS	THE WORD	HAS 1 CHANCE
4		FIGURE		IN THE	PER LETTER	WORD	APPEARING	IN THE
5				TORAH		EXPECTANCY	AT LEAST ONCE	FOLLOWING OF
6								BEING HERE
7	REPTILIAN @ skips +/- 1 and +/- 50068	-1	2	304805	6.56157E-06	0.004763701	0.004752373	210.4211959
8	FROM A STAR @ Skips +/-1 and +/- 50068	-58068	2	304805	6.56157E-06	0.004763701	0.004752373	210.4211959
9	PRESIDENTS @ Skip +1 only	1	7	304805	2.29655E-05	0.016672955	0.01653473	60.47876052
0	FROM THE PLEIADES @ Skips +/-1 and +/- 50068	-58068	26	304805	8.53004E-05	0.061928118	0.06004955	16.65291411
1	MARS (2 transliterations with *samech* or *shin* end) @ Skips +/- 1 and +/-50068	-1	48	304805	0.000157478	0.114328833	0.1080354	9.256225282
2	UFO @ Skips +/- 1 and +/- 97832 only	-1	168	304805	0.000551172	0.400150916	0.329781108	3.032314388
3	GRAY @ Skips 50067 to 50380	50067	817	304805	0.002680402	1.945972015	0.857151695	1.166654638
4	Final odds against excluding the axis term = (h7*h8*h9*h10*h11*h12*h13)/61						4.17741E-08	23938267.32
5								
6	REPTILIAN @ skips +/- 1 AND +/- 50068	-1	1	304805	3.28079E-06	9.1862E-05	9.18578E-05	10886.39286
7	ELS 61 FACTORED IN = H16/61 (reptilian and gargoyle in 26 letters.						0.005603325	178.4654568
8								

Spreadsheet for Figure 2-3 above. Odds against this matrix were about 23,938,267 to 1. Although the match against **GARGOYLES REPTILIAN** at first seemed truly exceptional (10,886 to 1), this was without considering ELS rank 61 of the axis term (**GARGOYLE**). When the ELS rank of 61 is factored in, the match was actually against odds of about 178 to 1. That is still outstanding.

The general structure of the full matrix argues in favor of deliberate encoding, even though I generally look for axis terms that are 7 or 8 (really 8) letters long. Where I make exceptions, the axis term has Hebrew letters that are relatively rare. **GARGOYLE** has two letter *gimel*s. This is the 3rd rarest letter in Hebrew. Out of 304,805 letters in Torah, there are only 2,109 *gimel*s. The only letters that are fewer in number are *tet* and *samech*.

Chapter 3
ET Infiltration in the U.S. Government?

The axis term on Figure 3-1 is **ALIEN IN OUR GOVERNMENT. BIDEN** is at the same absolute skip. **OBAMA** is one skip away from **BIDEN.** These two men made up the core of alien infiltration of our government, but as this chapter makes clear, I can find matrices labeling most of their administrations as alien/Reptilian. The Torah Code matrix on Figure 3-1 is real, but the link to a humorous video about him is at (https://www.youtube.com/watch?v=Sa-rP-yCqPk). However, the link seemed to cause problems on my computer, so be careful with how you open it.

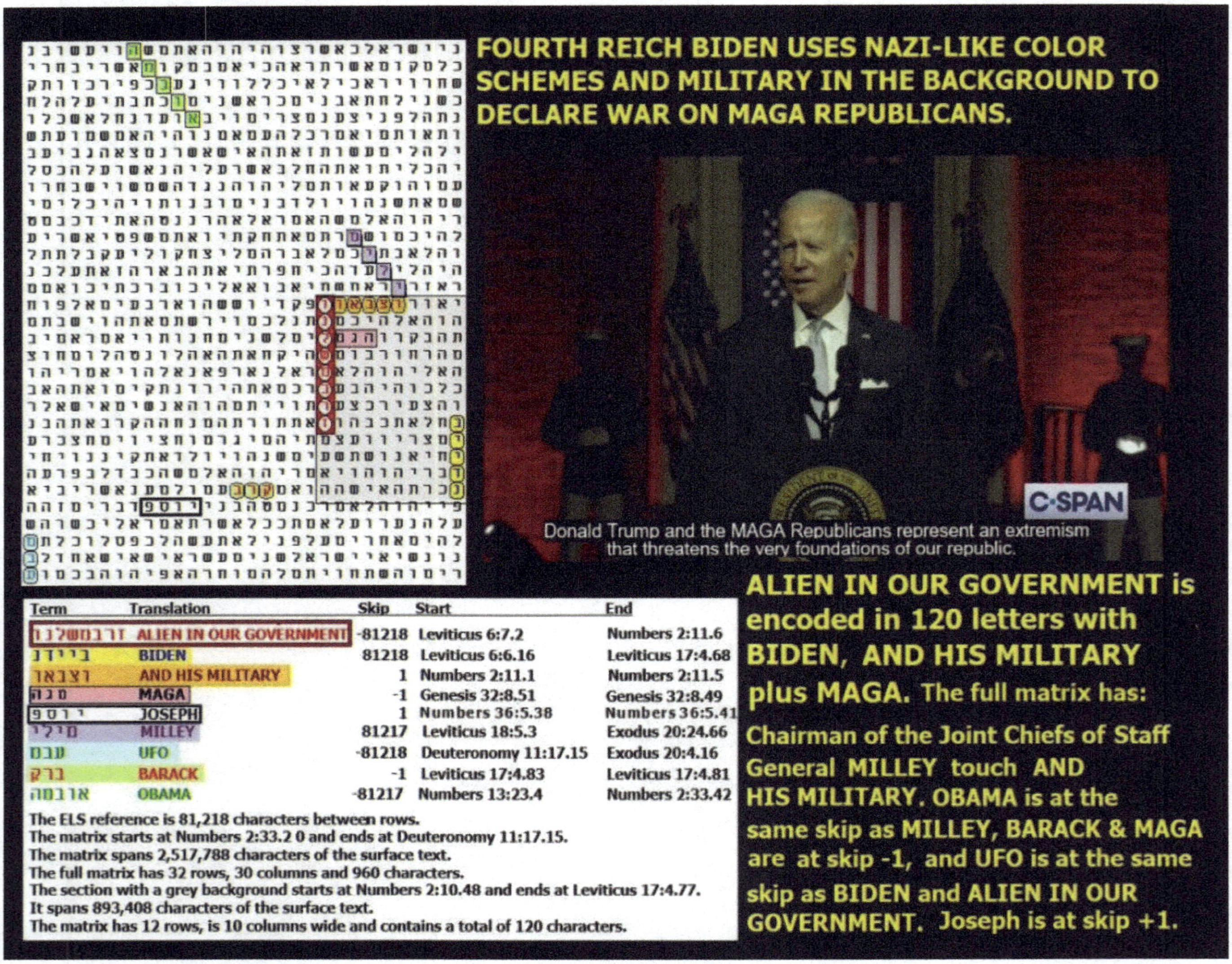

Figure 3-1 above and its statistical spreadsheets (below). While the full 960-letter matrix is of interest in light of the continuing **UFO/UAV** coverup, the incredible match between **AN ALIEN IN OUR GOVERNMENT, BIDEN, AND HIS MILITARY** and MAGA (Make America Great Again) in only 120 letters with **AN ALIEN IN OUR GOVERNMENT** and **BIDEN** at the same absolute skip was shocking.

Before looking for the axis term (**AN ALIEN IN OUR GOVERNMENT**), I expected that if it matched anyone, it would be **OBAMA. OBAMA** was on the full 960-letter matrix, but at skip 81,217, his name was one skip off the axis term (81,218). The odds against the match (1.78 to 1) meant that the match was easier to find than calling a coin toss correctly. But the odds against a match between **AN ALIEN IN OUR GOVERNMENT** and **BIDEN** (before considering ELS rank 2 of the axis term) were 2,540 to 1 or half of that (1,270 to 1) if we consider just the match but no other terms in the 120-letter box with it.

By far the most significant odds for Figure 3-1 were just for the 120-letter area shaded in grey. Odds against this section of the matrix were 166,704 to 1.

	A	B	C	D	E	F	G	H
1	Axis term ALIEN IN OUR GOVERNMENT @ Skip -81,218	SKIP	NUMBER	DIVIDE BY	THE QUOTIENT	E QUOTIENT	POISSON EQUATION	EACH WORD OR
2	ELS RANK 2. CALCULATION FOR 120-LETTER GREY AREA.	USED	IN	304805	EQUALS	X 120 LETTERS	PROBABILITY FOR	PHRASE HAS 1
3	R VALUE OF AXIS TERM = -0.197	ON	TORAH	LETTERS	FREQUENCY	EQUALS	THE WORD	CHANCE
4	(Not significant enough to count for odds)	FIGURE		IN THE	PER LETTER	WORD	APPEARING	IN THE FOLLOWING
5				TORAH		EXPECTANCY	AT LEAST ONCE	OF BEING HERE.
6	BIDEN @ Skips +/- 1 and +/- 81218	81218	1	304805	3.28079E-06	0.000393694	0.000393617	2540.541699
7	HIS MILITARY @ Skip +1	1	32	304805	0.000104985	0.012598219	0.012519193	79.87735193
8	MAGA (two 3-letter spellings) @ Skips +/- 1 and +/- 81218	-1	2383	304805	0.007818113	0.938173586	0.608658065	1.642958597
9	COMBINED ODDS AGAINST = (H6*H7*H8)/2						5.99865E-06	166704.2262
10								
12	Axis term ALIEN IN OUR GOVERNMENT @ Skip -81,218	SKIP	NUMBER	DIVIDE BY	THE QUOTIENT	E QUOTIENT	POISSON EQUATION	EACH WORD OR
13	ELS RANK 2	USED	IN	304805	EQUALS	X 960 LETTERS	PROBABILITY FOR	PHRASE HAS 1
14	Calculation for the full 960-letter matrix.	ON	TORAH	LETTERS	FREQUENCY	EQUALS	THE WORD	CHANCE
15		FIGURE		IN THE	PER LETTER	WORD	APPEARING	IN THE FOLLOWING
16				TORAH		EXPECTANCY	AT LEAST ONCE	OF BEING HERE.
17								
18	BIDEN @ Skips +/- 1 and +/- 81218	81218	1	304805	3.28079E-06	0.003149555	0.0031446	318.0054708
19	HIS MILITARY @ Skip +1	1	32	304805	0.000104985	0.100785748	0.095873277	10.43043515
20	UFO @ Skips +/- 1 and +/- 81218	-81218	128	304805	0.000419941	0.403142993	0.331783458	3.01401404
21	JOSEPH @ Skip +1 only	1	180	304805	0.000590541	0.566919834	0.432729964	2.310909999
22	OBAMA (best spelling) @ Skips -81217 to -81440	-81217	262	304805	0.000859566	0.825183314	0.561845335	1.779849254
23	BARACK @ Skips +/- 1 and +/- 81218	-1	334	304805	0.001095783	1.051951248	0.6507444	1.53670166
24	MAGA (two 3-letter spellings) @ Skips +/- 1 and +/- 81218	-1	2383	304805	0.007818113	7.505388691	0.999449888	1.000550415
25	MILLEY (4-letter spelling) @ skips 81,217 to 81614	81217	7523	304805	0.024681354	23.69409951	1	1
26	COMBINED ODDS AGAINST =(H18*H19*H20*H21*H22*H23*H24*H25)/2						3.16339E-05	31611.64407

Odds against the full 960-letter Figure 3-1 were about 31,611 to 1.

	A	B	C	D	E	F	G	H
1	Axis term ALIEN IN OUR GOVERNMENT @ Skip -81,218	SKIP	NUMBER	DIVIDE BY	THE QUOTIENT	E QUOTIENT	POISSON EQUATION	EACH WORD OR
2	ELS RANK 2. CALCULATION FOR 160-LETTER AREA.	USED	IN	304805	EQUALS	X 160 LETTERS	PROBABILITY FOR	PHRASE HAS 1
3		ON	TORAH	LETTERS	FREQUENCY	EQUALS	THE WORD	CHANCE
4		FIGURE		IN THE	PER LETTER	WORD	APPEARING	IN THE FOLLOWING
5				TORAH		EXPECTANCY	AT LEAST ONCE	OF BEING HERE.
6	BIDEN @ Skips +/- 1 and +/- 81218	81218	1	304805	3.28079E-06	0.000524926	0.000524788	1905.531294
7	HIS MILITARY @ Skip +1	1	32	304805	0.000104985	0.016797625	0.016657331	60.03362636
8	MILLEY (4-letter spelling) @ skips 81,217 to 81614	81217	1448	304805	0.004750578	0.760092518	0.532376839	1.878368718
9	MAGA (two 3-letter spellings) @ Skips +/- 1 and +/- 81218	-1	2383	304805	0.007818113	1.250898115	0.713752402	1.401046073
10	COMBINED ODDS AGAINST = (H6*H7*H8*H9)/2						6.64333E-06	150526.8356

If we expand the shaded area of Figure 3-1 up from 120 letters to 160 letters to include (General) **MILLEY**, the significance drops from 166,704 to 1 against down to 150,526 to 1. Therefore, the expansion to show this short 4-letter spelling of his last name is not justified.

On Figure 3-1, is the Code telling us that Biden is an Alien or that Aliens Work at the White House?

I expected that if **AN ALIEN IN OUR GOVERNMENT** would match anyone, it would be **OBAMA.** This is because his early years were largely shrouded in mystery. Was he born in Hawaii or Kenya, as his own relatives contended? Was he always an American, or when he traveled in Pakistan with an Indonesian passport, what happened to his U.S. citizenship? Indonesia did not allow for dual citizenship. What about Michelle Obama? Why are there so few pictures of her when she was young? Joan Rivers, shortly before she died (or was killed) said that "everyone knows that Michelle was a tranny" (transgender man). Barack, in public, twice called his wife "Michael." People have said that they witnessed Barack or one of his security agents morph into a Reptilian alien for a fraction of a second. See Figure 3-2.

Figure 3-2 above: This picture allegedly shows a Reptilian acting as a security agent for President Biden.

The most suspicious thing about **OBAMA** is the amount of time that seems to be missing when we look at his life before his meteoric rise to the White House. In contrast, Biden at first appears to have no major missing time spans in his life. However, he was largely out of the public arena when he used COVID-19 as an excuse to stay home and in his basement during the 2020 election. And, as of January, 2025 (see Figure 3-3), he has managed to return to his Delaware home/s on 570 days in office. This time back home (1.56 years) is almost 40% of his reign.

Figure 3-3 – Biden seemed to live on Planet Delaware. He spent almost half his time away from the White House.

By the summer of 2024, the vast majority of the nation knew that Biden was out of contact with reality and, in fact, he was not really running the country for the four years of his presidency. Who was really running the nation? While people guess that it was his wife, his staff, or Obama, the Code has another answer. The nation was run by a hostile ET, likely reptilian, or someone who had abducted and scrambled Biden's brain. See Figure 3-4 with an axis term of **BIDEN is ALIEN. MARS, UFO,** and **ENEMY** are at the same absolute skip. **PRESIDENT** and **TREASONABLE** are at skip +1. This is a tough fact to absorb. It's one thing to learn that the President is a traitor. But to understand that he (and Obama) were either not human or were controlled by hostile aliens is much harder to accept. How can we prevent this from happening again? I'd start with requiring DNA tests for anyone (particularly Democrat) who advocates homosexual or transgender positions. These people want a human population in America who are not capable of having babies. We need 2.1 children per American woman to have a stable population. See https://www.pop.org/episode-2-2-1-kids-a-stable-population/, But we are getting only 1.6 children per woman. At this rate, our population will collapse and make it easier for ET to conquer us. This is exactly what they want and are working hard to achieve.

Figure 3-4: The axis term is **BIDEN (is) ALIEN. MARS, UFO, and ENEMY** are at the same absolute skip. **PRESIDENT and TREASONABLE** are at skip +1.

On Figure 3-4 CodeFinder indicates that the R factor of the axis term is 1.387. This means that there was only one chance in 24.378 that the 10-letter term **BIDEN (is) ALIEN** would even be found once in the Torah Code. The most interesting *a priori* term found was **MARS** at the same absolute skip of the axis term. Odds against it occurring at a special case skip (+/- 1 and +/- 97,832) in an 814-letter matrix were about 125 to 1. **METHANE** is on the matrix because it was found in the Martian atmosphere and it usually indicates the presence of some sort of life. Final odds against Figure 3-4 above and its axis term **BIDEN IS ALIEN** were over 3.9 trillion to 1, as the spreadsheet below indicates.

	A	B	C	D	E	F	G	H
1	AXIS TERM BIDEN (is) ALIEN @ Skip- 95,866	SKIP	NUMBER	DIVIDE BY	THE QUOTIENT	E QUOTIENT	POISSON EQUATION	EACH WORD
2	ELS RANK 1	USED	IN	304805	EQUALS	x 814 LETTERS	PROBABILITY FOR	OR PHRASE
3	R FACTOR FOR AXIS TERM = 1.387 = 1 CHANCE IN 24.378101818	ON	TORAH	LETTERS	FREQUENCY	EQUALS	THE WORD	HAS 1 CHANCE
4	Combined ODDS against this matrix are about 3,938,620,000,000 to 1.	FIGURE		IN THE	PER LETTER	WORD	APPEARING	IN THE
5	(3.9 trillion to 1)			TORAH		EXPECTANCY	AT LEAST ONCE	FOLLOWING OF
6								BEING HERE
7	MARS (Moadim) @ Skips +/- 1 and +/- @ skips +/- 1 and +/- 97832	97832	3	304805	9.84236E-06	0.00801168	0.007979672	125.31844
8	METHANE @ Skips +/- 1 and +/- @ skips +/- 1 and +/- 97832	-1	46	304805	0.000150916	0.122845754	0.115599933	8.650524071
9	TREASONABLE @ Skip +1 only	1	47	304805	0.000154197	0.125516314	0.117958625	8.477548797
10	PRESIDENT @ Skip +1 only	1	68	304805	0.000223093	0.181598071	0.166063544	6.021791287
11	ENEMY (2 terms) @Skips +/- 1 and +/- 97832	95832	91	304805	0.000298552	0.243020948	0.215744915	304805.3045
12	REPTILIAN @ Skips +8 to +381 Skips +/- 1 and +/- +/- 97832	8	136	304805	0.000446187	0.363196142	0.304549987	3.283533225
13	UFO @ Skips +/- 1 and +/- +/- 97832	97832	158	304805	0.000518364	0.421948459	0.344232162	2.905016175
14	AXIS TERM R FACTOR VALUE 1 CHANCE IN 24.47810818						0.04085283	24.47810818
15	FINAL ODDS INCLUDING AXIS TERM = (H7*H8*H9*H10*H11*H12*H13*H14)						2.53896E-13	3.93862E+12

UNDER DEMOCRATS, REPTILIANS CONTROLLED THE WHITE HOUSE. The Torah Code not only alleges that aliens rule our government. It also names the major players that are Reptilian implants. Let's examine who belongs to the enemy within, starting with Figure 3:5 below. While Figure 3-4 was based on BIDEN IS ALIEN, Figure 3-5 has a 9-letter axis term of OBAMA IS A REPTILIAN. Its R factor is 0.697.

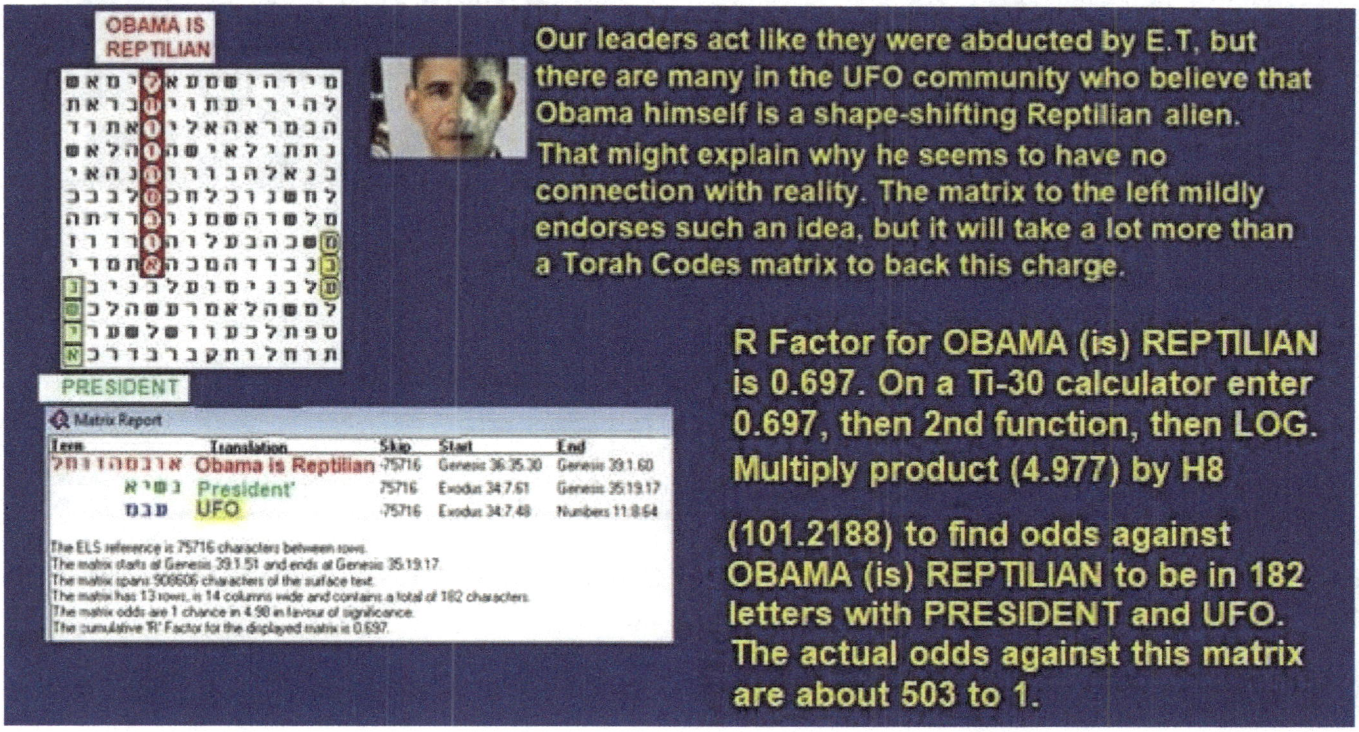

	A	B	C	D	E	F	G	H
1	AXIS TERM OBAMA IS REPTILIAN @ Skip -75,716	SKIP	NUMBER	DIVIDE BY	THE QUOTIENT	E QUOTIENT	Poisson Equati	EACH WORD
2	ELS RANK 2	USED	IN	304805	EQUALS	x 182 Letters	Probability for	OR PHRASE
3	R FACTOR = 0.697 = 4.97737085	ON	TORAH	LETTERS	FREQUENCY	Equals	the Word	HAS 1 CHANCE
4		FIGURE		IN THE	PER LETTER	WORD	Appearing	IN THE
5				TORAH		EXPECTANCY	At Least once	FOLLOWING
6								
7	PRESIDENT @ Skips +/- 1 and +/- 75716	75716	89	304805	0.00029199	0.053142173	0.051754812	19.3218749
8	UFO @ Skips +/- 1 and +/- 75716	-75716	168	304805	0.000551172	0.100313315	0.095446037	10.47712439
9	COMBINED ODDS AGAINST THE MATRIX (without axis term) = (H7*H8)/2						0.009879583	101.2188434
10								
11	R FACTOR = 0.697 VALUE = 4.97737085						0.009879583	4.97737085
12	FINAL ODDS AGAINST MATRIX (with axis term) =H9*4.97737085						0.0019849	503.8037209

Figure 3-6 below: SENATOR CHUCK SCHUMER is the axis term. The former Senate Majority Leader is now reduced to a mere Senator. On Figure 3-6 above, he is encoded at the same skip as, **REPTILIAN** and **UFO**. On the matrix to the right below is **TO SPY**, but I had to expand the basic matrix **(CHUCK SCHUMER, REPTILIAN, THE NEPHILIM,** and **UFO**) with a white background from 154 letters to 987 letters to see it. Besides insinuating that Schumer is a Reptilian or brainwashed by them, the matrix makes another equally startling revelation. The letter L (*lamed*) in the Hebrew word for **REPTILIAN** is the same *lamed* that we see in the open text for **THE NEPHILIM** (which means **GIANTS**). The matrix thus solves an ancient and vital question: Who were or are the Nephilim? Giant Reptilians!

Figure 3-6 above. Note the relationship between **CHUCK SCHUMER, REPTILIAN,** and **THE NEPHILIM.** What does the Torah say about these creatures?

Genesis 6:1 to 6:4, The Wickedness of Mankind

When men began to multiply on the face of the ground, and daughters were born to them, the sons of God saw that the daughters of men were fair; and they took to wife such of them as they chose. [3] *Then the Lord said, "My spirit shall not abide in man forever, for he is flesh, but his days shall be a hundred and twenty years."* [4] *The Nephilim were on the earth in those days, and also afterward, when the sons of God came in to the daughters of men, and they bore children to them. These were the mighty men that were of old, the men of renown.*

The spreadsheets showing the odds against the Schumer matrices are given below.

	A	B	C	D	E	F	G	H
1	AXIS TERM CHUCK SCHUMER @ SKIP +134645	SKIP	NUMBER	DIVIDE BY	THE QUOTIENT	E QUOTIENT	POISSON EQUATION	EACH WORD
2	ELS RANK 40	USED	IN	304805	EQUALS	X 164 LETTERS	PROBABILITY FOR	OR PHRASE
3		ON	TORAH	LETTERS	FREQUENCY	EQUALS	THE WORD	HAS 1 CHANCE
4		FIGURE		IN THE	PER LETTER	WORD	APPEARING	IN THE
5				TORAH		EXPECTANCY	AT LEAST ONCE	FOLLOWING OF
6								BEING HERE
7	REPTILIAN @ Skips +/- 1 and +/- 134645	134645	2	304805	6.56157E-06	0.001010482	0.001009972	990.1267076
8	NEPHILIM/GIANTS @ Skip +1 only (2 spellings)	1	4	304805	1.31231E-05	0.00580043	0.00578364	172.9015015
9	UFO @ Skips +/- 1 and +/- 134645	134645	173	304805	0.000567576	0.250868588	0.22187538	4.507034528
10	COMBINED SIGNIFICANCE = (H7*H8*H9)/40						5.18417E-05	19289.47616
11							#REF!	#REF!
12	AXIS TERM CHUCK SCHUMER @ SKIP +134645	SKIP	NUMBER	DIVIDE BY	THE QUOTIENT	E QUOTIENT	POISSON EQUATION	EACH WORD
13	ELS RANK 40	USED	IN	304805	EQUALS	X 550 LETTERS	PROBABILITY FOR	OR PHRASE
14		ON	TORAH	LETTERS	FREQUENCY	EQUALS	THE WORD	HAS 1 CHANCE
15		FIGURE		IN THE	PER LETTER	WORD	APPEARING	IN THE
16				TORAH		EXPECTANCY	AT LEAST ONCE	FOLLOWING OF
17								BEING HERE
18	REPTILIAN @ Skips +/- 1 and +/- 134645	134645	2	304805	6.56157E-06	0.003608865	0.003602361	277.5957563
19	NEPHILIM/GIANTS @ Skip +1 only (2 spellings)	1	4	304805	1.31231E-05	0.007217729	0.007191744	139.0483287
20	UFO @ Skips +/- 1 and +/- 134645	134645	173	304805	0.000567576	0.312166795	0.268140553	3.729387407
21	WAR @ Skip +1 only	1	36	304805	0.000118108	0.064959564	0.062894645	15.89960484
22	NAZI @ Skips -3 to -256	-3	452	304805	0.001482915	0.815603419	0.557627689	1.793311236
23	COMBINED SIGNIFICANCE = (h18*H19*H20*H21*H22)/40						9.74545E-06	102611.9884
24								
25	AXIS TERM CHUCK SCHUMER @ SKIP +134645	SKIP	NUMBER	DIVIDE BY	THE QUOTIENT	E QUOTIENT	POISSON EQUATION	EACH WORD
26	ELS RANK 40	USED	IN	304805	EQUALS	X 987 LETTERS	PROBABILITY FOR	OR PHRASE
27		ON	TORAH	LETTERS	FREQUENCY	EQUALS	THE WORD	HAS 1 CHANCE
28		FIGURE		IN THE	PER LETTER	WORD	APPEARING	IN THE
29				TORAH		EXPECTANCY	AT LEAST ONCE	FOLLOWING OF
30								BEING HERE
31	REPTILIAN @ Skips +/- 1 and +/- 134645	134645	2	304805	6.56157E-06	0.006476272	0.006455346	154.9103674
32	NEPHILIM/GIANTS @ Skip +1 only (2 spellings)	1	4	304805	1.31231E-05	0.012952543	0.01286902	77.70599326
33	SPY @ Skips +/- 1 and +/-134645	134645	11	304805	3.60886E-05	0.035619494	0.034992586	28.57748237
34	WAR @ Skip +1 only	1	36	304805	0.000118108	0.116572891	0.110034776	9.088035973
35	NAZI @ Skips -3 to -256	-3	452	304805	0.001482915	1.463637408	0.768606929	1.301055146
36	COMBINED SIGNIFICANCE = (h31*h32*h33*h34*h35)/40						9.83412E-06	101686.8318

FIGURE 3-7 below: Is General Milley a Reptilian or Reptilian Hybrid? On Figure 3-7, the axis term is MARK MILLEY. GENERAL is at the same skip, and UFO is at the same absolute skip. There are three terms that mean STRANGER, FOREIGNER, or ALIEN. The third is in Leviticus 24:16. It is about blasphemy, and it deals with how to punish a blasphemer. It says, **"HE WHO BLASPHEMES GOD'S NAME SHOULD BE KILLED."**

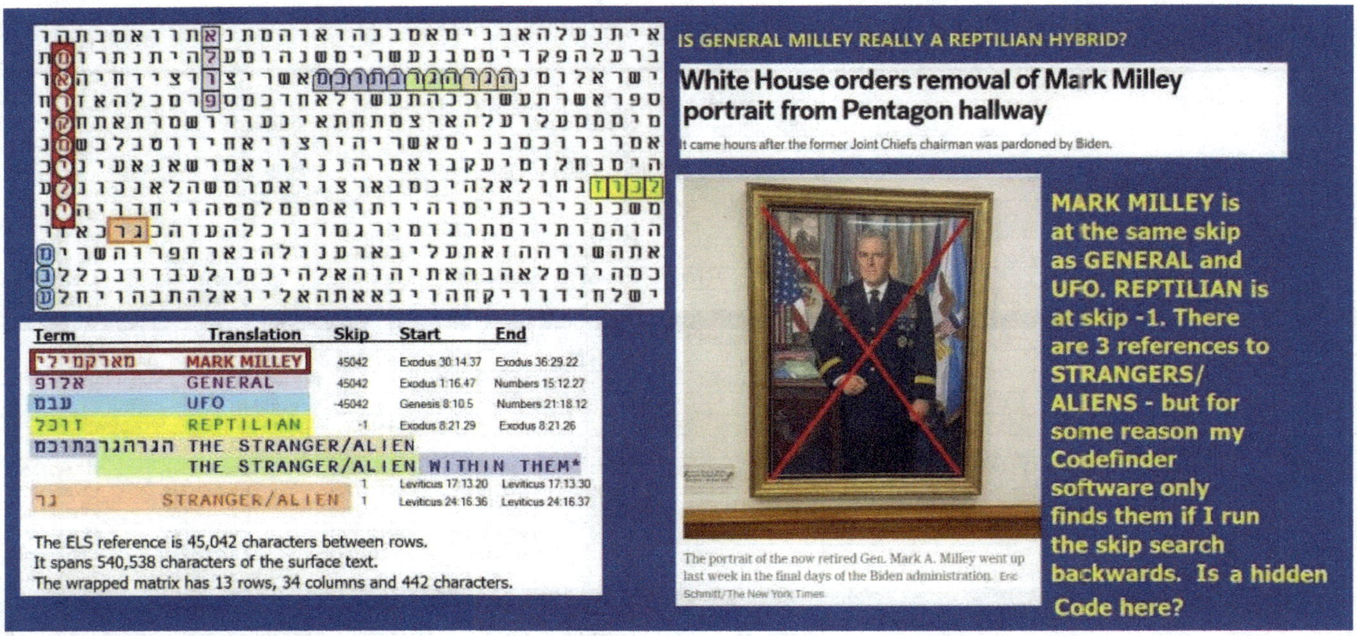

This full phrase touches **MARK MILLEY**. That's likely to happen IF it can be proven that Milley is indeed an Alien. The former Chairman of the Joint Chiefs of Staff, **GENERAL MARK MILLEY**, was a blasphemous man during the Trump Administration (and a traitor). On TV, he rejected accompanying Trump and his Bible when the church across the street from the White House was burned. Milley actually accepted a pardon from President Biden in the last 15 minutes of that Administration. So as soon as President Trump began his new term in 2025, he ordered the portrait of Milley to be taken down at the Pentagon.

Is Milley really a Reptilian, a Reptilian-human hybrid, or merely someone who was abducted and brainwashed by Reptilians? Based on (1) his refusal to be photographed with the President holding a Bible in front of the burnt St. John's Church across the street from the White House on June 1, 2020, (2) his rebellious TV-broadcasted speech about it, we can (3) take his behavior plus Figure 3-9 where he is at the same exact skip as **TO HYBRIDIZE** and **UFO,** plus (4) his Army Times remarks about fighting hybrids, and (5) his mysterious contact with me about the abstract of my Mars Correct Report, I think he is likely to be a hostile Reptilian-human hybrid.

Figure 3-7 above. As per the spreadsheet below, there was only about one chance in 1.3 million to find **REPTILIAN** at a special case skip (-1) in a full 442-letter matrix with **MARK MILLEY** and his rank (**GENERAL**) at the same skip as his name. **MARK MILLEY** is the axis term, and **UFO** is at the same skip too (but in the opposite direction). A word for **STRANGER** or **ALIEN** appears three times in the open text. Again, the most significant term was **REPTILIAN.** Odds against its appearance at a special case skip on the full 442-letter matrix were about 690 to 1**.** The next most important term was Milley's rank (**GENERAL**) at the same skip as his name. Odds against this find were about 172.9 to 1.

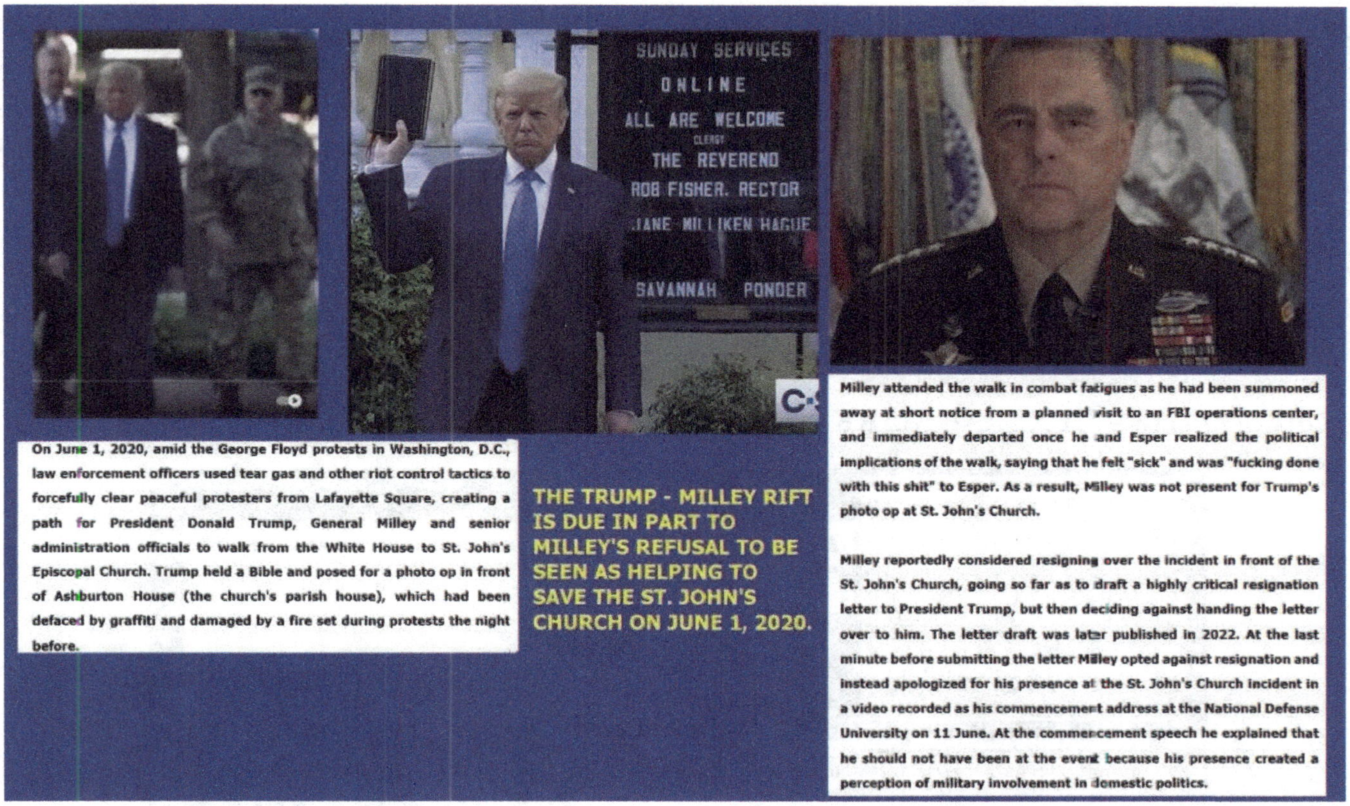

Figure 3-8 above. Milley wanted nothing to do with Trump's desire to save a historic church (St. John's) across the street from the White House. He backed away from the President before a group photo could be taken. This is not proof, like the Torah Code offers, that he is a Reptilian, but it is consistent with what is written about their murderous outlook with respect to humans.

General Milley, the former Chairman of the Joint Chiefs of Staff, is also on Figure 3-1, at one skip off the axis, touching **AND HIS MILITARY,** which touched **ALIEN IN OUR GOVERNMENT** (at the same absolute skip as **BIDEN**). Milley's name is not statistically significant at all on the full matrix, but it is of personal interest because, on December 8, 2015, he sent me a weird Facebook invitation. On it, he included nothing at all except the abstract of the Roffman (father and son) Report, Mars Correct – Critique of All NASA Mars Weather Data (https://www.academia.edu/40831370/MARS_CORRECT_CRITIQUE_OF_ALL_NASA_MARS_WEATHER_DATA). The current Abstract differs from what he sent back to me largely with respect to the number of Martian days (1,171 then vs. 3,025 below) that MSL Curiosity was on Mars later.

Did General Milley mean to praise us by making it clear that he was aware of our research, or threaten us by implying that he knew all the sensitive things we were finding (not then, but eventually even evidence for time travel)? His attention led me to publish several Torah Codes matrices about him. When he wrote me, I had not yet written about evidence for primitive life on Mars.

I hid Figure 3-9 (see below) and the bulk of the article about Milley for 5 months, but since he stepped in it by talking in public about fighting Little Green Men and Hybrid Armies at 34:10 to 50 in this video, link (https://www.inquisitr.com/u-s-military-prepping-for-alien-invasion-army- chief-warns-cadets-prepare-for-little-green-men-and-hybrid-armies-video), I declared it time for me to let the public see what the Code says about him. The Army Times had damage control up about his remarks (but they apparently didn't know about him contacting me before). They say "little green men" was used in a non-extraterrestrial, military context as recently as 2014 when Russia annexed Crimea, to refer to men wearing unmarked green uniforms, whose origin was unconfirmed. Further, the military has long talked of a hybrid threat of regular forces, criminal elements, terrorist groups, and cyber threats, etc. Obviously, the General could have avoided the controversy by defining these terms for the public when he used them. But by 2022, we were dealing with a war in Ukraine, one in which there have been articles about an alien ark activating underground in Kherson, Ukraine, in the area where Putin first invaded in 2022.

Figure 3-9 below – GENERAL MARK MILLEY, UFOs, and (alien) HYBRIDIZATION

Term	Translation	Skip	Start	End
מארקמיללי	MARK MILLEY	-71760	Leviticus 10:5.20	Numbers 2:29.9
להכליא	TO HYBRIDIZE	-71760	Deuteronomy 6:10.52	Numbers 11:19.10
עבם	UFO	-71760	Leviticus 20:5.71	Genesis 23:8.33
אלופ	GENERAL	1	Exodus 15:15.8	Exodus 15:15.11

The ELS reference is 71,760 characters between rows.
The wrapped matrix starts at Numbers 11:18 letter 92 and ends at Leviticus 10:5 letter 20.
The matrix spans 861138 characters of the surface text.
It has 13 rows, is 18 columns wide and contains a total of 234 characters.

	A	B	C	D	E	F	G	H
1	AXIS TERM MARK MILLY @ Skip -71760	SKIP	NUMBER	DIVIDE BY	THE QUOTIENT	E QUOTIENT	POISSON EQUATION	EACH WORD OR
2	Axis Term ELS Rank 1	USED	IN	304805	EQUALS	X 234 LETTERS	PROBABILITY FOR	PHRASE HAS 1
3		ON	TORAH	LETTERS	FREQUENCY	EQUALS	THE WORD	CHANCE IN THE
4		FIGURE		IN THE	PER LETTER	WORD	APPEARING	IN THE FOLLOWING
5				TORAH		EXPECTANCY	AT LEAST ONCE	OF BEING HERE.
6	HYBRIDIZE @ Skips +/- 1 and +/- 71760	1	5	304805	1.64039E-05	0.00383852	0.003831162	261.0174139
7	GENERAL @ Skip +1 only	1	43	304805	0.000141074	0.03301127	0.032472344	30.79543624
8	UFO @ Skip +/-1 and +/- 71760	1	173	304805	0.000567576	0.132812782	0.12437099	8.04046026
9	COMBINED SIGNIFICANCE						1.54726E-05	64630.38646

Figure 3-9 above: **HYBRIDIZE** is at the same absolute skip as **MARK MILLEY** and **UFO.**

Back to the Codes for now. Is **MILLEY**'s name significant on a matrix measuring 16 rows by 10 columns with **BIDEN, AND HIS MILITARY, AN ALIEN IN OUR GOVERNMENT,** and **MAGA** in it? No. While odds against the smaller 12 X 10 matrix with **BIDEN, AN ALIEN IN OUR GOVERNMENT, AND HIS MILITARY,** and **MAGA** were about 166,704 to 1, if we add four rows to accommodate a non-special skip **MILLEY,** the significance still drops to 150,526 to

1. The general rule is that when you start with two outstanding terms that are very close like **BIDEN** and **AN ALIEN IN OUR GOVERNMENT**, new terms will not add any significance to the matrix unless they are (1) at a special case skip (+/- 1 or the absolute skip of the axis term – here +/-81,218) or (2) in the area of the original close match – here 120 letters.

But before looking at General Milley further and asking if he may be one of the officers in our government willing to work for ET so long as the pay is good, you may want to look at what he chose to send me. It was an unsolicited segment of my work, a draft Abstract Basic Report published on March 3, 2021, by the International Mars Society. It read as follows:

Abstract for Basic Report for Mars Correct: Critique of All NASA Mars Weather Data

We present evidence that NASA is seriously understating Martian air pressure. Our 12-year study critiques 3,025 Sols up through 8 February 2021 (8.51 terrestrial years, 4.52 Martian years) of highly problematic MSL Rover Environmental Monitoring Station (REMS) weather data, and offers an in-depth audit of over 8,311 hourly Viking 1 and 2 weather reports. We discuss the analysis of technical papers, NASA documents, and personal interviews of transducer designers. We troubleshoot pressures based on radio occultation/spectroscopy, and the previously accepted small pressure ranges that could be measured by Viking 1 and 2 (18 mbar), Pathfinder and Phoenix (12 mbar), and MSL (11.5 mbar – altered to 14 mbar in 2017). For MSL, there were several pressures published from August 30 to September 5, 2012, that were from 737 mbar to 747 mbar – two orders of magnitude high – only to be retracted. We challenged many pressures, and NASA revised them down. However, there are two pressure sensor ranges listed on a CAD for Mars Pathfinder. We long thought the CAD listed two different sensors, but based on specifications of a new Tavis sensor for InSight that is like that on PathFinder, it appears that the transducer could toggle between two pressures ranges: 0-0.174 PSIA/12 mbar (Tavis Dash 2) ***and*** *0-15 PSIA/1,034 mbar (Tavis Dash 1). Further, an Abstract to the American Geophysical Union for the Fall 2012 meeting, shows the Finnish Meteorological Institute (FMI) states of their MSL (and Phoenix) Vaisala transducers, "The pressure device measurement range is 0 – 1025 hPa in temperature range of -45°C – +55°C (-45°C is warmer than MSL night temperatures), but its calibration is optimized for the Martian pressure range of 4 – 12 hPa." So, in fact of the first five landers with meteorological suites, three were actually equipped to measure Earth-like pressure.*

All original 19 low μV values were removed when we asked about them, although eventually 12 were restored. REMS always-sunny opacity reports were contradicted by Mars Reconnaissance Orbiter photos. We demonstrate that REMS weather data was regularly revised after they studied online critiques in working versions of this report. REMS even labelled all dust 2018 Global Dust Storm weather as sunny, although they did list the μV values then as all low. Vikings and MSL showed consistent timing of daily pressure spikes, which we link to how gas pressure in a sealed container would vary with Absolute temperature, to heating by radioisotope thermoelectric generators (RTGs), and to dust clots at air access tubes and dust filters. Pathfinder, Phoenix, and MSL wind measurements failed. Phoenix and MSL pressure transducer design problems included confusion about dust filter location and a lack of information about nearby heat sources due to International Traffic and Arms Regulations (ITAR). NASA Ames could not replicate dust devils at 10 mbar. Rapidly filled MER Spirit tracks required wind speeds of 80 mph at the assumed low pressures. These winds were never recorded on Mars. Nor could NASA explain drifting Barchan sand dunes. Based on the above and dust devils on Arsia Mons to altitudes of 17 km above areoid (Martian equivalent of sea level), spiral storms with 10 km eye-walls above Arsia Mons and similar storms above

Olympus Mons (over 21 km high), dust storm opacity at MER Opportunity blacking out the sun, snow that descends 1 to 2 km in only 5 or 10 minutes, excessive aero braking, liquid water running at or near the surface in numerous locations at Recurring Slope Lineae (RSL) and stratus clouds 13 km above areoid, we argue for an average pressure at areoid of ~511 mbar rather than the accepted 6.1 mbar. This pressure grows to 1,050 mbar in the Hellas Basin.

I published an article entitled **Teleportation, Basiago & Torah Code**. Part of the article was the full Air Force Teleportation Physics Study (AFRL-PR-ED-TR-2003-0034) by Dr. Eric W. Davis. On page 56, it *states, in part,*

"Most claimed instances of human teleportation of the body from one place to another have been unwitnessed. ***There are also a small number of credible reports of individuals who reported being teleported to/from UFOs during a UFO close encounter,*** *which were scientifically investigated (Valee, 1988, 1990, 1997).*

As I document in my teleportation article, it appears that our military teleportation abilities are further advanced than what the public is aware of. Was Biden of alien descent, or was he merely maintaining direct contact with ETs when he is largely out of the public eye? He was in Delaware pretty much every week. ETs could easily use superior teleportation skills to bring him where they choose to and scramble his brain (if he still had one) in any way that they wanted to every time they got him isolated in Delaware. We should not ignore the Torah Code matrix at the start of this chapter. By the way, when trying to understand why Biden raided President Trump's home in Florida, we should not overlook the possibility of Biden attempting to find documents or evidence of Trump's knowledge of ETs on Earth and/or Government teleportation and/or related time travel capabilities.

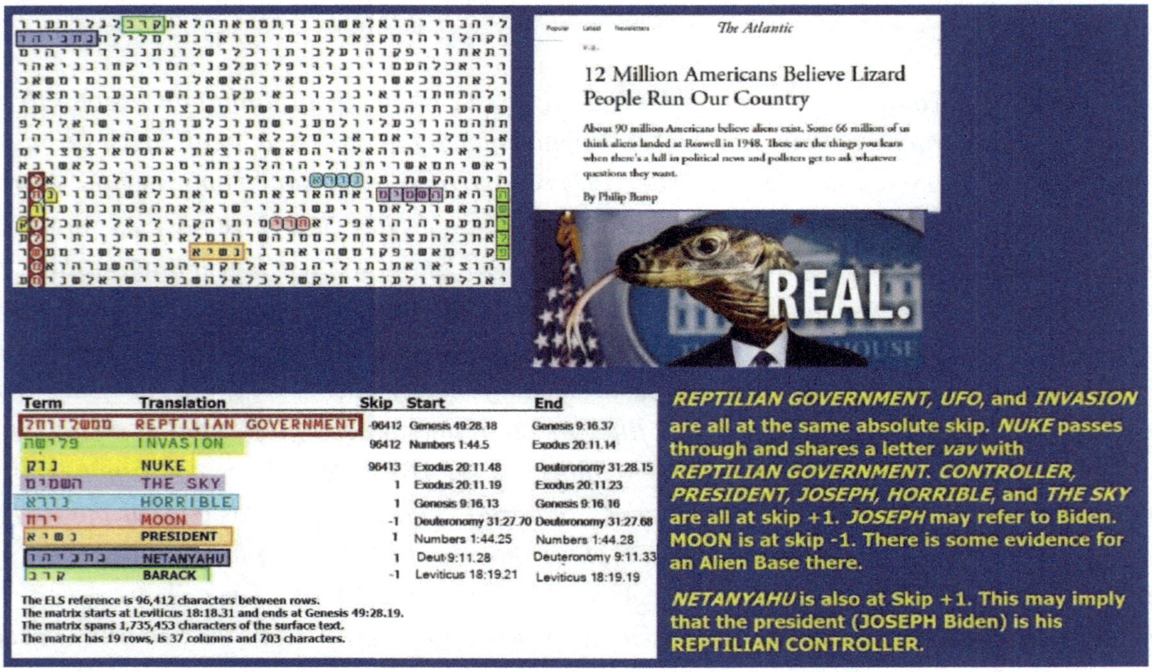

Figure 3-10 above: The axis term is **REPTILIAN GOVERNMENT**. At the same skip: **INVASION.**

	A	B	C	D	E	F	G	H
1	Axis term REPTILIAN GOVERNMENT @ Skip -96412	SKIP	NUMBER	DIVIDE BY	THE QUOTIENT	IF QUOTIENT	POISSON EQUATION	EACH WORD OR
2	ELS Rank 1	USED	IN	304805	EQUALS	X 703 LETTERS	PROBABILITY FOR	PHRASE HAS 1
3	R FACTOR FOR REPTILIAN GOVERNMENT = -0.240	ON	TORAH	LETTERS	FREQUENCY	EQUALS	THE WORD	CHANCE IN THE
4		FIGURE		IN THE	PER LETTER	WORD	APPEARING	IN THE FOLLOWING
5				TORAH		EXPECTANCY	AT LEAST ONCE	OF BEING HERE.
6								
7	INVASION @ Skips +/-1 and 96412	96412	1	304805	3.28079E-06	0.002306393	0.002303735	434.0777171
8	NETANYAHU @ Skip +1 only	1	16	304805	5.24926E-05	0.036902282	0.036229691	27.60167043
9	HORRIBLE @ Skips +1 only	1	23	304805	7.54581E-05	0.05304703	0.051664589	19.35561711
10	PRESIDENT @ Skip +1 only	81	68	304805	0.000223093	0.156834698	0.145154645	6.889204275
11	THE SKY @ Skip +1 only	-1	81	304805	0.000265744	0.186817802	0.170405126	5.868368051
12	BARACK @ Skips +/- 1 and +/- 96412	-1	336	304805	0.001102344	0.774947918	0.539272224	1.854351023
13	TOTAL ODDS AGAINST THIS MATRIX = H7*H8*H9*H10*H11*H12						5.75191E-08	17385542.04

Spreadsheet for Figure 3-10 is above. Note the high significance of **INVASION** (Only one chance in 434 that it would be there at a special case skip). This is what our **REPTILIAN GOVERNMENT** is all about. But who knew enough about them to give us such a warning over 3,000 years ago?

OTHER GOVERNMENT INFILTRATORS

Figure 3-11: Chris Wray, Director of the FBI

Figure 3-11 above: Axis term **CHRIS WRAY** who is presented as not only as the boss of the **FBI**, but also as a **REPTILIAN BOSS.** Spreadsheet below for Figure 3-11. The matrix was found against odds of about 28,955,985 to 1. As expected, the most significant keyword was **REPTILIAN**. Before considering the ELS rank of **CHRIS WRAY, REPTILIAN** was found at a special case skip (-1) against odds of at least 977 to 1.

	A	B	C	D	E	F	G	H
1	AXIS TERM: CHRIS WRAY @ Skip 112,442	SKIP	NUMBER	DIVIDE BY	THE QUOTIENT	E QUOTIENT	POISSON EQUATION	EACH WORD
2	ELS RANK 61	USED	IN	304805	EQUALS	X 312 LETTERS	PROBABILITY FOR	OR PHRASE
3		ON	TORAH	LETTERS	FREQUENCY	EQUALS	THE WORD	HAS 1 CHANCE
4		FIGURE		IN THE	PER LETTER	WORD	APPEARING	IN THE
5				TORAH		EXPECTANCY	AT LEAST ONCE	FOLLOWING OF
6								BEING HERE
7	REPTILIAN Skip +/- 1 and +/- 112442 only	-1	1	304805	3.28079E-06	0.001023605	0.001023082	977.4391879
8	BOSS @ 112441 TO 112474	122441	8	304805	2.62463E-05	0.008188842	0.008155405	122.6180702
9	SODOM AND GOMORRAH @ Skip +1 only	1	11	304805	3.60886E-05	0.011259658	0.011196505	89.31358399
10	FBI @ Skip +/- 1 and +/- 112442	122266	38	304805	0.00012467	0.038897	0.038150225	26.21216509
11	UFO @ Skips +/- 1 and +/-112442	112442	169	304805	0.000554453	0.172989288	0.158853375	6.295113338
12	Final Combined Odds = (h7*h8*h9*h10*h11)/61						-3.45352E-08	28955985.74

WHICH MATRICES SPECIFICALLY LINK REPTILIANS TO OUR GOVERNMENT?

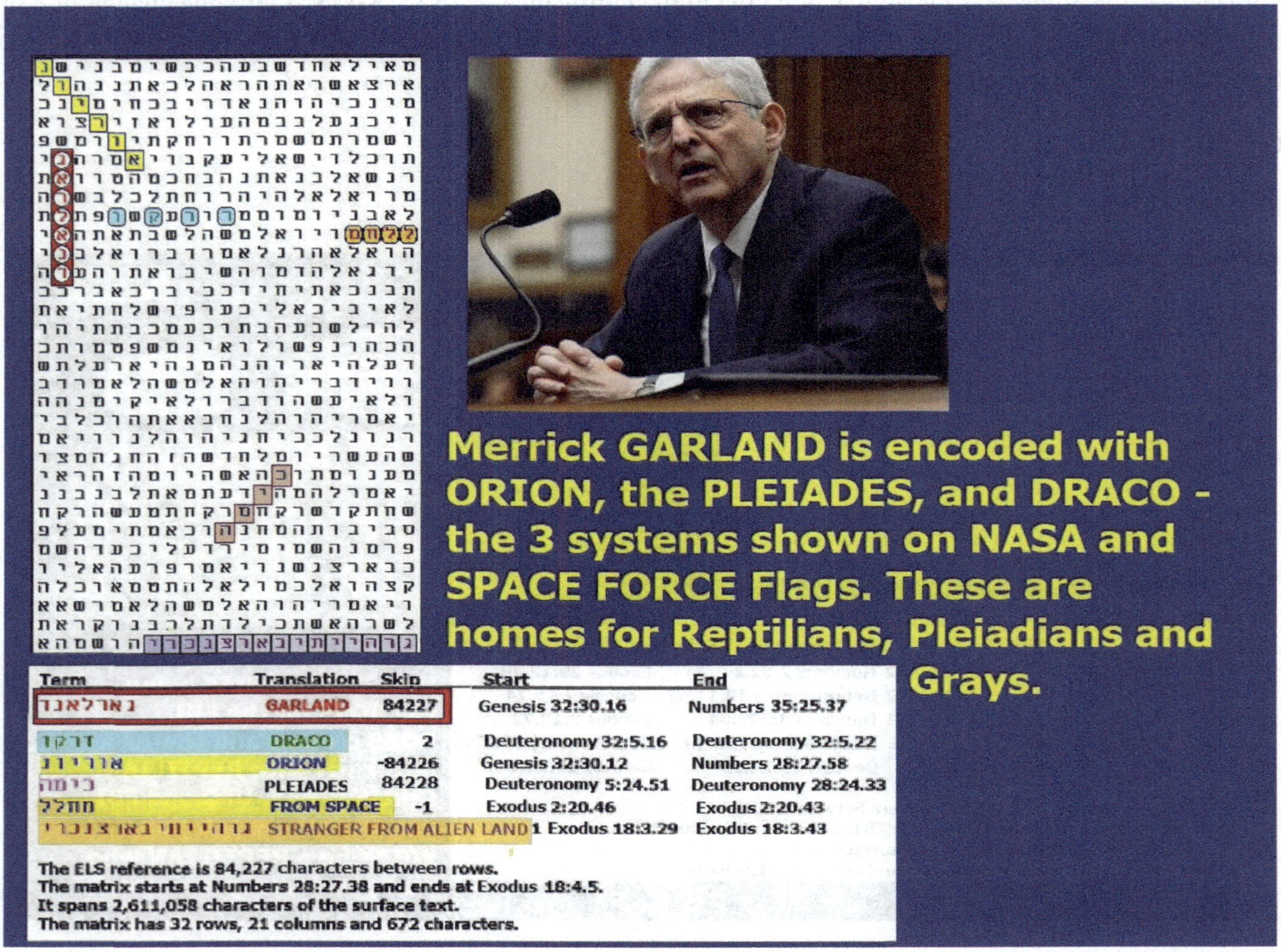

Figure 3-12 above links ex-U.S. Attorney General Merrick **GARLAND** with some home constellations of **REPTILIANS – ORION** (yellow and blue) and the **PLEIADES** (light brown). A synonym for Reptilians is **DRACO** (light blue), skip +2). **DRACO** pertains to the constellation Alpha Draconis. **FROM SPACE** is at skip -1. In the open text is **STRANGER FROM AN ALIEN WORLD**. What is being exposed is that most, if not all, people in Biden's administration are either Reptilians or are traitors who serve their cause. Most are likely here to seize control of our planet and/or destroy it., or use it for their own purposes.

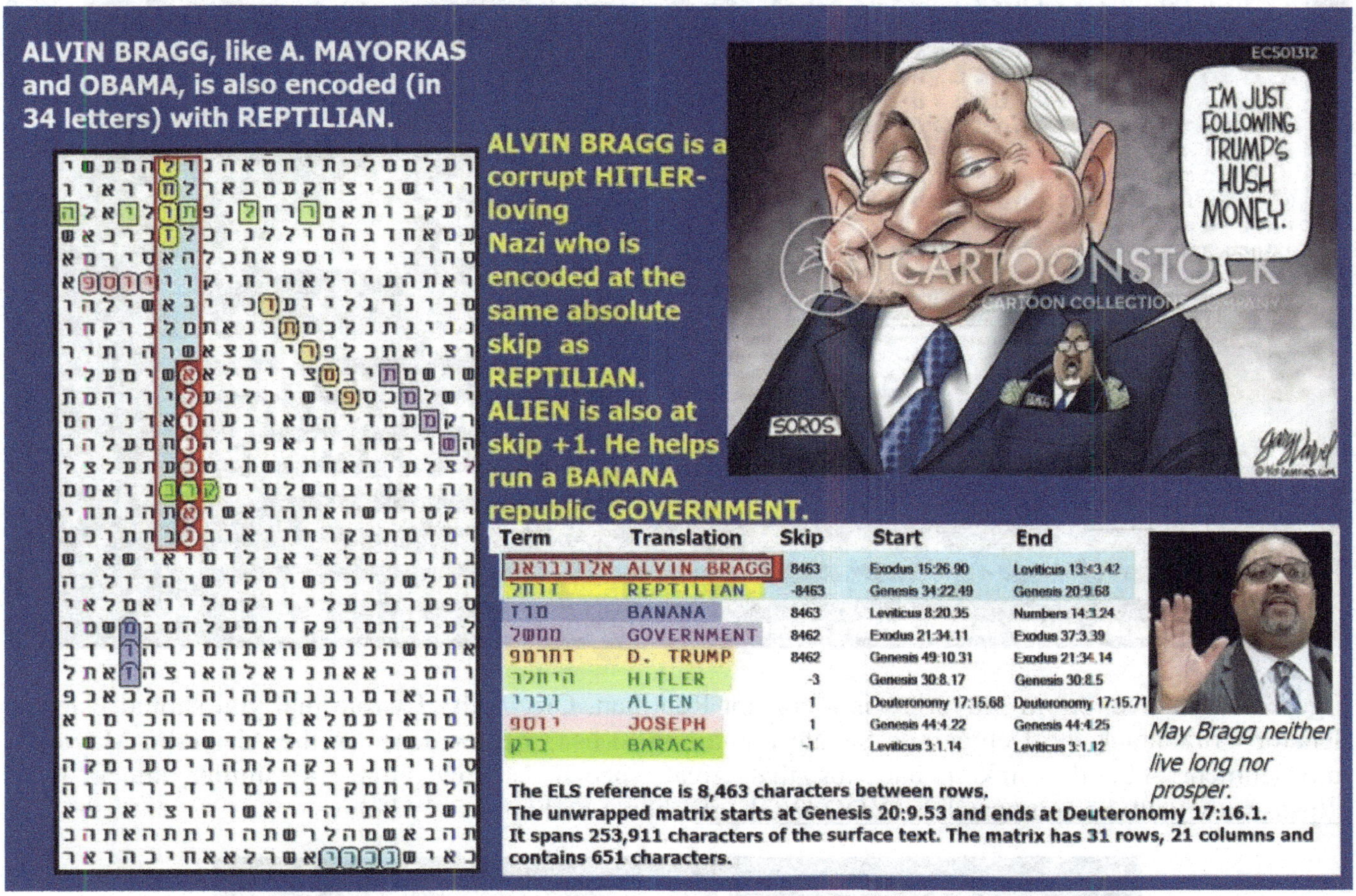

Figure 3-13 above: **ALVIN BRAGG**, who showed his sick desire to destroy President Trump is encoded in only 34 letters with **REPTILIAN** at skip -1. Odds against them both being encoded in 651 letters are about 156 to 1, Although odds against them being so close are about 2,988 to 1. Overall odds against Figure 3-13 are about 170,073,613 to 1.

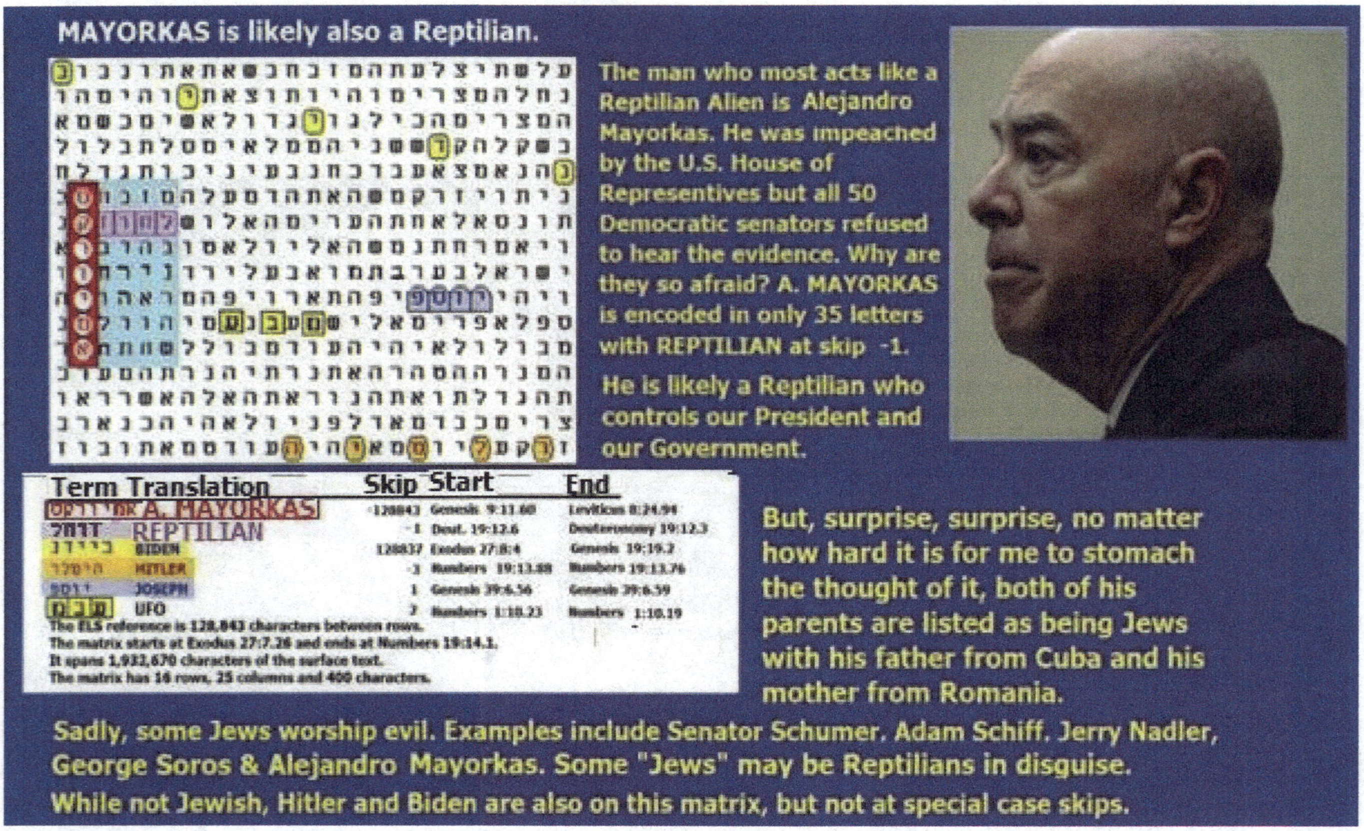

Figure 3-14 above. Alvin Mayorkas is a corrupt Reptilian, Originally, I wrote that you should listen to Senator Cruz attack the Democratic Senate for failing to hear any evidence in the Alejandro Mayorkas impeachment, but the link I had for that great speech is now gone. A similar attack is at https://www.youtube.com/watch?v=YDQd2jOKjgE shown on Figure 3-15 below.

Figure 3-15 above: Senator Cruz blasts Mayorkas.

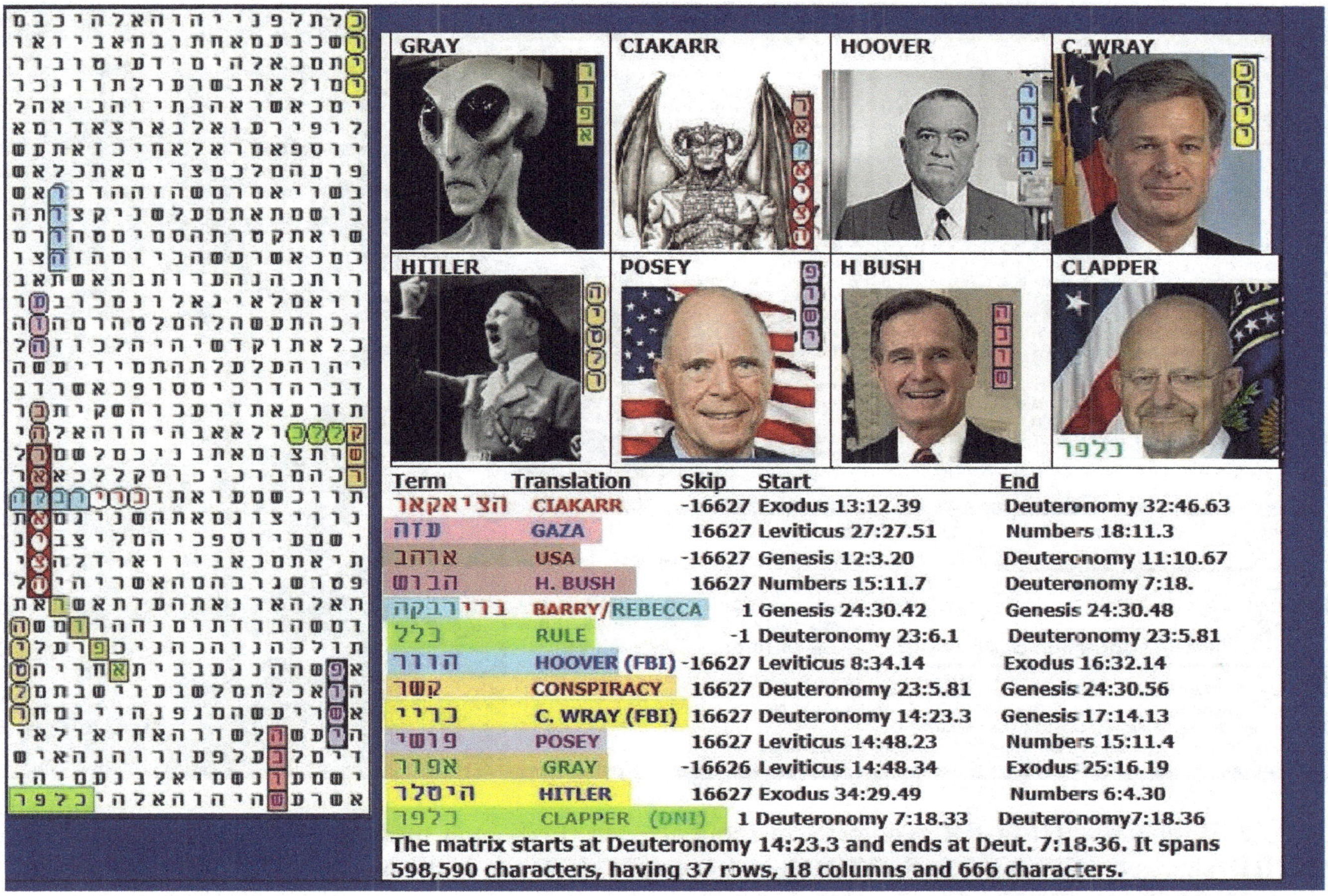

Term	Translation	Skip	Start	End
הציאקאר	CIAKARR	-16627	Exodus 13:12.39	Deuteronomy 32:46.63
עזה	GAZA	16627	Leviticus 27:27.51	Numbers 18:11.3
ארהב	USA	-16627	Genesis 12:3.20	Deuteronomy 11:10.67
הבוש	H. BUSH	16627	Numbers 15:11.7	Deuteronomy 7:18.
בריירבקה	BARRY/REBECCA	1	Genesis 24:30.42	Genesis 24:30.48
כלל	RULE	-1	Deuteronomy 23:6.1	Deuteronomy 23:5.81
הוור	HOOVER (FBI)	-16627	Leviticus 8:34.14	Exodus 16:32.14
קשר	CONSPIRACY	16627	Deuteronomy 23:5.81	Genesis 24:30.56
כריי	C. WRAY (FBI)	16627	Deuteronomy 14:23.3	Genesis 17:14.13
פושי	POSEY	16627	Leviticus 14:48.23	Numbers 15:11.4
אפור	GRAY	-16626	Leviticus 14:48.34	Exodus 25:16.19
היטלר	HITLER	16627	Exodus 34:29.49	Numbers 6:4.30
כלפר	CLAPPER (DNI)	1	Deuteronomy 7:18.33	Deuteronomy7:18.36

The matrix starts at Deuteronomy 14:23.3 and ends at Deut. 7:18.36. It spans 598,590 characters, having 37 rows, 18 columns and 666 characters.

Figure 3-16 above. Reptilians, Gaza, Hitler, Chris Wray and more on a matrix.

Congressman **POSEY**'s book, The Manson Connection, is about our government assisting in abductions (and sometimes murder) by aliens.

Figure 3-17 above – **HITLER HARRIS** in sequence. ALIENS is at the same absolute skip. **HITLER HARRIS** shares the first letter with **SODOM AND GOMMORAH** in the open text. **REICH** is at the same skip as **HITLER HARRIS** and **ALIENS. NAZI** is at skip -1.

Figure 3-18 below: TIM WALZ is a REPTILIAN.

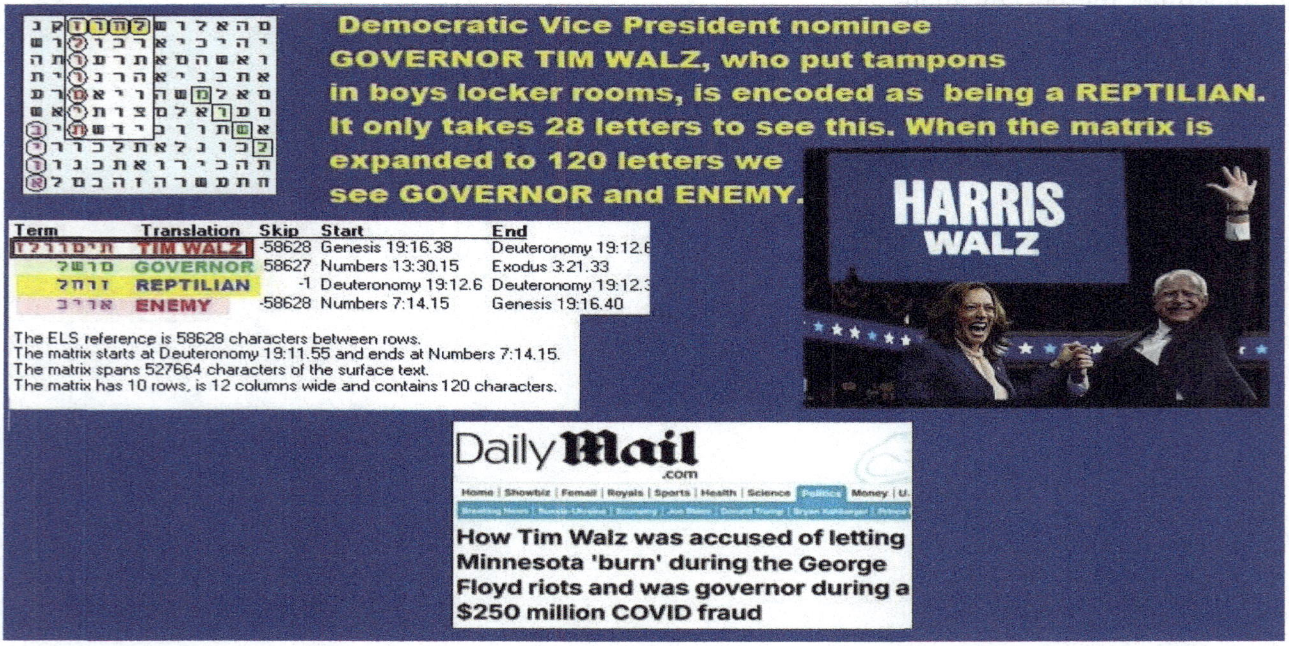

Figure 3-18 above and its spreadsheet below. The 28-letter matrix looks extremely significant. If it were based on the lowest ELS of **TIM WALZ,** odds against it would be about 10,886 to 1. However, we do not see this match until the 84th ELS of Tim Walz. Dividing 10,886 by 84 we find that this match exists against odds of about 129 to 1. It's really much better than these odds make clear because of terms that only become visible when the matrix is expanded from 28 to 120 letters – still considered a tight matrix. Then we see that Walz's title of **GOVERNOR** (of Minnesota) is there at one skip less than that of **WALZ**. Further, **ENEMY** is at the same skip as **WALZ**. Odds against this matrix are about 5,898 to 1. Although this matrix did not occur until the 84th ELS of **TIM WALZ**, there is likely to be a lot of other information about him in matrices with lower ELS ranks for axis terms. It takes a lot of time for me to find what I do. Although I did not conclude that an artificial intelligence wrote this Code, I do believe that an A.I. should be set up along the lines of Project Creator Decoder to efficiently decipher the massive quality of other critical warnings encoded. My vision of Creator Decoder is found on my website at https://arkcode.com/photo4_113.html. It was addressed to Fort Huachuca, Level-3/CACI, MIT, MITRE, NASA Ames, G.E., IBM, Drakes Holding and the NSA, however I do not know what they did with the information.

	A	B	C	D	E	F	G	H
1	*AXIS* TERM TIM WALZ @ Skip *-58628*	SKIP	NUMBER	DIVIDE BY	THE QUOTIENT	E QUOTIENT	POISSON EQUATION	EACH WORD
2	ELS RANK 84	USED	IN	304805	EQUALS	x 28 LETTERS	PROBABILITY FOR	OR PHRASE
3		ON	TORAH	LETTERS	FREQUENCY	EQUALS	THE WORD	HAS 1 CHANCE
4		FIGURE		IN THE	PER LETTER	WORD	APPEARING	IN THE
5				TORAH		EXPECTANCY	AT LEAST ONCE	FOLLOWING OF
6								BEING HERE
7	REPTILIAN @ Skips +/- 1 and +/58628	-1	1	304805	3.28079E-06	9.1862E-05	9.18578E-05	10886.39286
8	FINAL ODDS AGAINST INCLUDING AXIS TERM = h7/84						0.007716054	129.5999151
9								
10	*AXIS* TERM TIM WALZ @ Skip *-58628*	SKIP	NUMBER	DIVIDE BY	THE QUOTIENT	E QUOTIENT	POISSON EQUATION	EACH WORD
11	ELS RANK 84	USED	IN	304805	EQUALS	x 120 LETTERS	PROBABILITY FOR	OR PHRASE
12		ON	TORAH	LETTERS	FREQUENCY	EQUALS	THE WORD	HAS 1 CHANCE
13		FIGURE		IN THE	PER LETTER	WORD	APPEARING	IN THE
14				TORAH		EXPECTANCY	AT LEAST ONCE	FOLLOWING OF
15								BEING HERE
16	REPTILIAN @ Skips +/- 1 and +/58628	-1	1	304805	3.28079E-06	0.000393694	0.000393617	2540.541699
17	GOVERNOR @ Skips 58627 to 58674	58674	400	304805	0.001312314	0.157477732	0.145704163	6.86322189
18	ENEMY (2 terms) @ Skips +/- 1 and /- 58628	- 58628	91	304805	0.000298552	0.035826184	0.035192022	28.41553124
19	INITIAL COMBINED ODDS						2.01832E-06	495461.7673
20	FINAL ODDS FOR AXIS TERM RANK 84						0.000169539	5898.354372

Above: Spreadsheet for Figure 3-19.

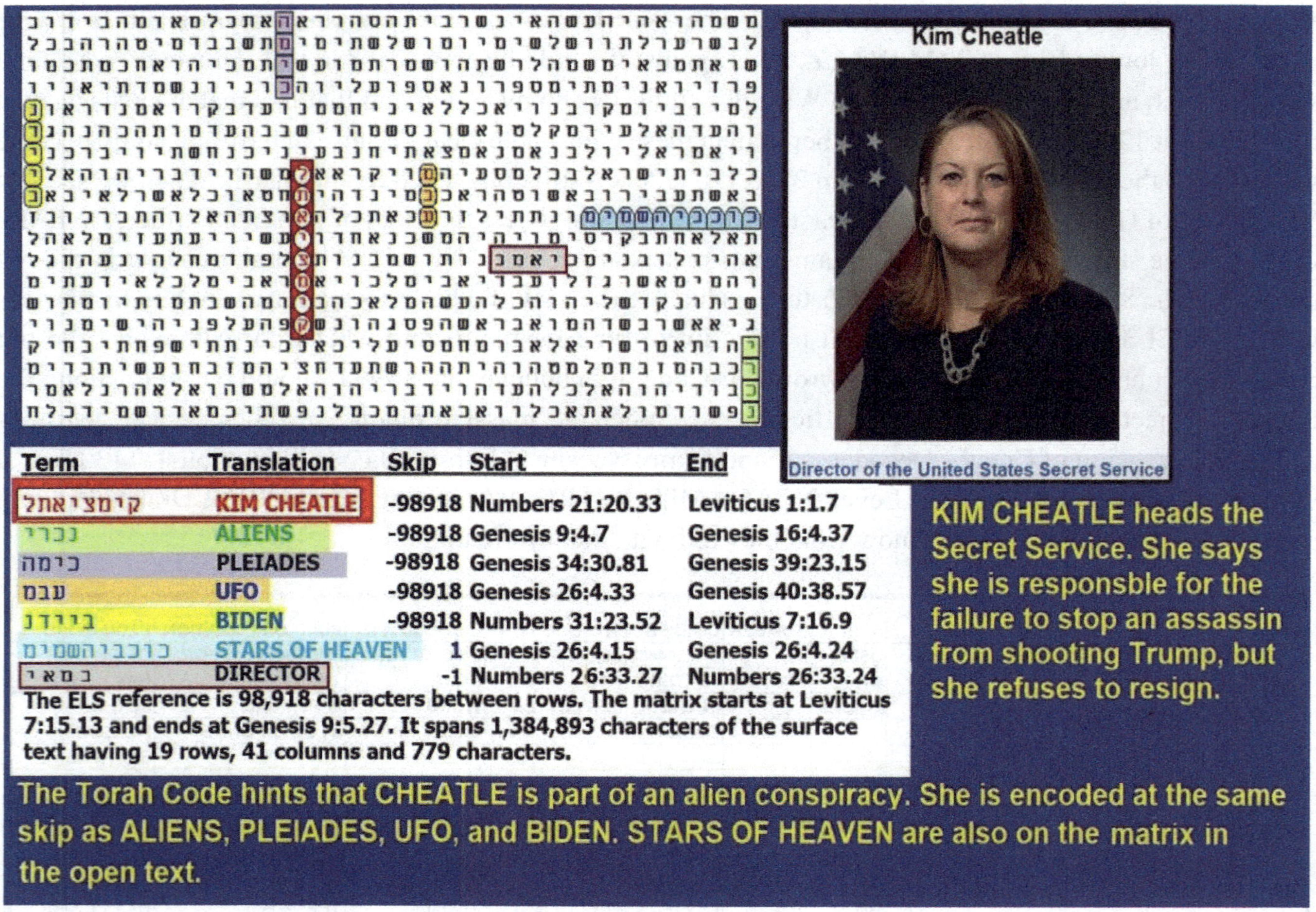

Figure 3-19 – KIM CHEATLE, Director of the United States Secret Service. One of her jobs was to protect President Trump, but he was still shot in the head under her watch. Errors made were numerous, but were they part of an alien (Reptilian) plan?

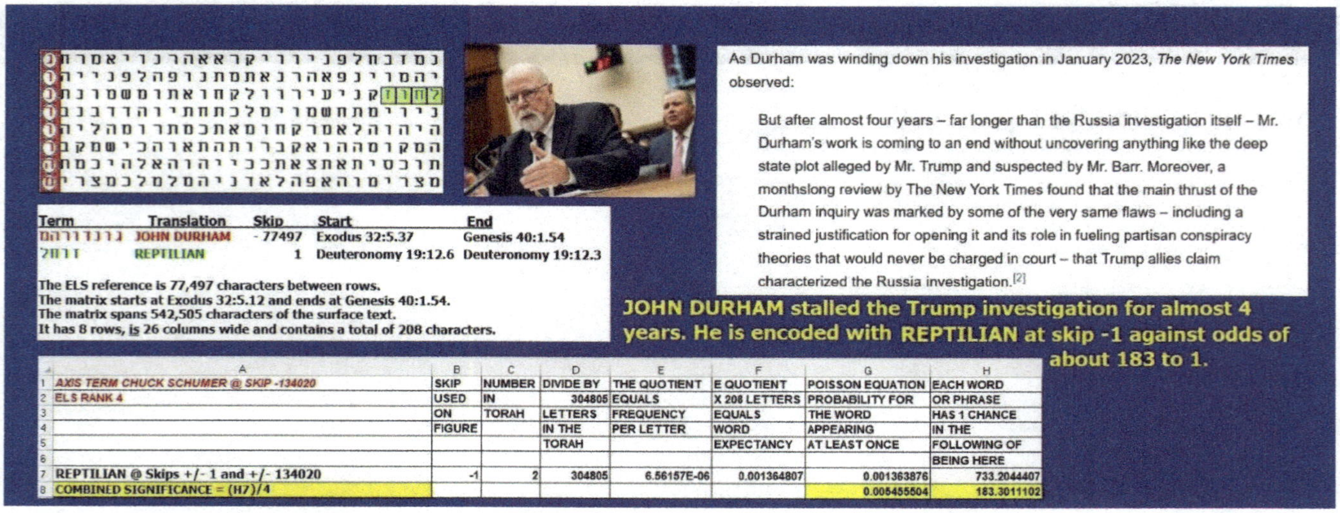

Figure 3-20 above. **JOHN DURHAM** is encoded with the only **REPTILIAN** at skip +/-1. Odds against this march are about 183 to 1. In April 2019, U.S. Attorney General William Barr had tasked Durham with overseeing a review of the origins of the Russia investigation and to determine if intelligence collection involving the Trump campaign was "lawful and appropriate". After 3 1⁄2 years of investigation and prosecutions, Durham had secured one guilty plea and a probation sentence (not against Trump) for a charge unrelated to the origins of the Trump-Russia investigation, and two unsuccessful trial prosecutions.

Figure 3-21 below: Although it takes a row split of 3 to see it, against odds of cver 87 million to 1, **HAKEEM JEFFRIES** is encoded with **REPTILIAN** and **WAR** at the same absolute skip plus **B. OBAMA** at skip -1.

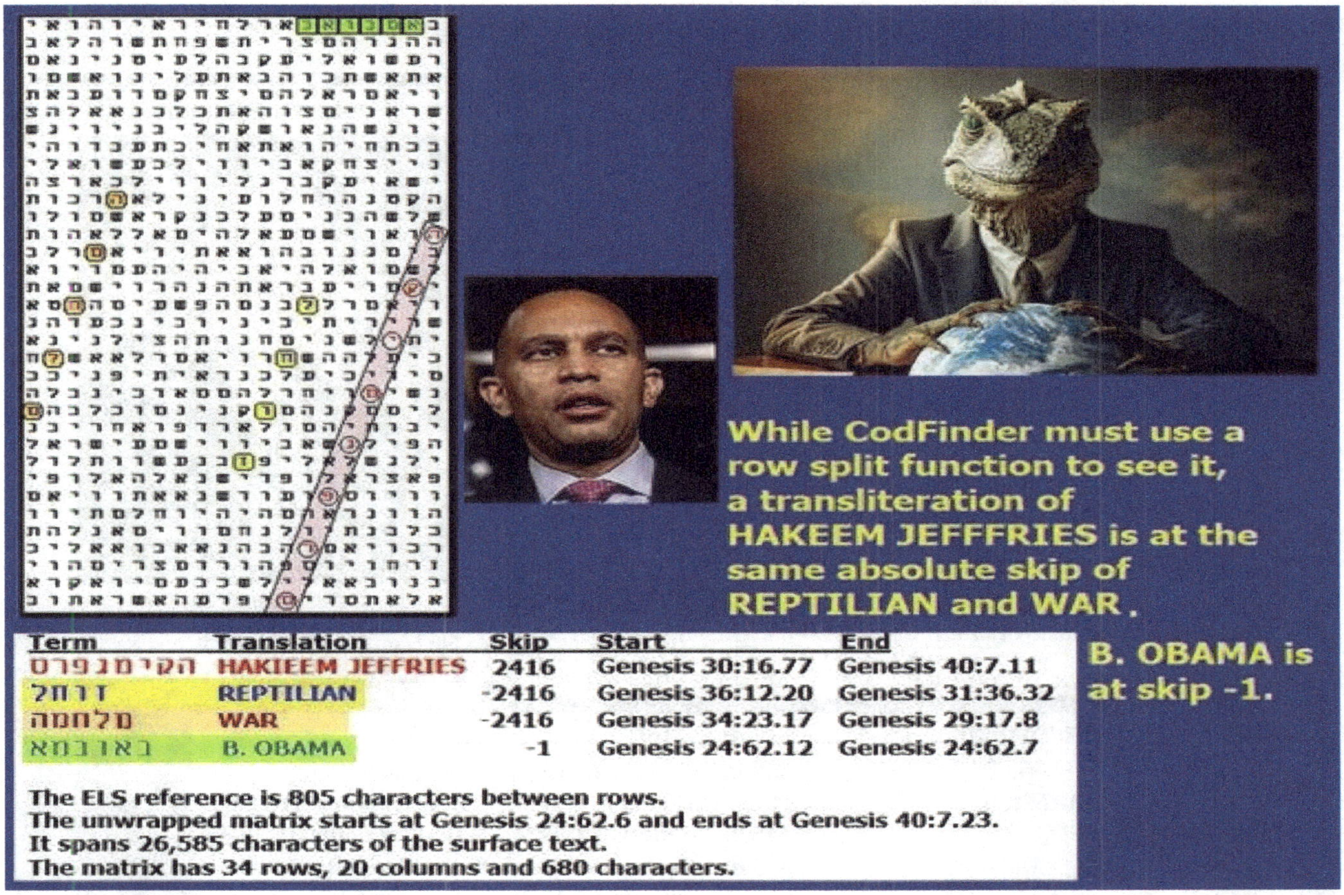

Figure 3-21 above and its spreadsheet below. **HAKEEM JEFFRIES**, Democratic minority leader in the House of Representatives, is at the same absolute skip as **REPTILIAN** and **WAR. B. OBAMA** is also on the matrix at skip -1. Both men should be considered to be aliens (reptilians) or people who have been abducted and brainwashed/programmed to follow the reptilian agenda of weakening or destroying America and humanity.

	A	B	C	D	E	F	G	H
1	*AXIS TERM* HAEEM JEFFRIES @ *Skip 2416*	SKIP	NUMBER	DIVIDE BY	THE QUOTIENT	E QUOTIENT	POISSON EQUATION	EACH WORD
2	ELS RANK 1	USED	IN	304805	EQUALS	X 680	PROBABILITY FOR	OR PHRASE
3		ON	TORAH	LETTERS	FREQUENCY	EQUALS	THE WORD	HAS 1 CHANCE
4	R factor 3.171 = 1 CHANCE IN 72.4969	FIGURE		IN THE	PER LETTER	WORD	APPEARING	IN THE
5				TORAH		EXPECTANCY	AT LEAST ONCE	FOLLOWING OF
6								BEING HERE
7	REPTILIAN @ Skips +/- 1 AND +/- 2416	-2416	2	304805	6.56157E-06	0.004461869	0.00445193	224.6216954
8	WAR @ Skips +/- 2416 and +/- 1	-2416	39	304805	0.000127951	0.087006447	0.083328814	12.00065083
9	B. OBAMA @ Skips +/- 1 and +/- 2416	-1	1	304805	3.28079E-06	0.002230935	0.002228448	448.742833
10	INITIAL COMBINED SIGNIFICANCE = (H7*H8*H9)						8.26696E-07	1209634.113
11	R factor 3.171 = 1 CHANCE IN 72.49691786						0.01379369	72.49691786
12	FINAL ODDS AGAINST THIS MATRIX = H10*H11						1.14032E-08	87694744.92

Figure 3-22 below: John Brennan, former Director of the CIA, is a reptilian.

Term	Translation	Skip	Start	End
ברנאן זוחל	BRENNAN (is) REPTILIAN	-34526	Genesis 36:12.21	Exodus 3:15.68
ברקה	BARACK H.	34526	Numbers 5:27.32	Deuteronomy 31:28.52
עבם	UFO	-34526	Genesis 42:32.5	Deuteronomy 31:28.43
כימה	PLEIADES	-34526	Deuteronomy 9:7.8	Leviticus 14:17.23
מרגלים	SPIES	1	Genesis 42:31.26	Genesis 42:31.31
סין	CHINA	1	Exodus 26:11.8	Exodus 26:11.10
קומי	COMEY	-2	Exodus 10:26.33	Exodus 10:26.25
קרי	KERRY	-34526	Exodus 10:26.33	Genesis 19:34.35

The ELS reference is 34,526 characters between rows.
Axis term R Value 1.812 = 1 chance in 64.86344335
The wrapped matrix starts at Exodus 3:15.44 and ends at Genesis 42:32.30.
It spans 897,712 characters of the surface text.
The matrix has 27 rows, 36 columns and 972 characters.

Figures 3-23 and **3-24 below**: The famous liar A. **SCHIFF** is also encoded with **REPTILIAN**, probably descended from the **ANAKIM** who likely arose from the **NEPHILIM**. His **GREAT SIN** was treason. He wanted to be FBI Director if Hilary Clinton won the 2016 election. Under Biden he used the FBI to try to impeach President Trump. Odds against Figure 3-23 were almost six million to 1. Figure 3-24 below shows another ELS of **A. SCHIFF REPTILIAN.**

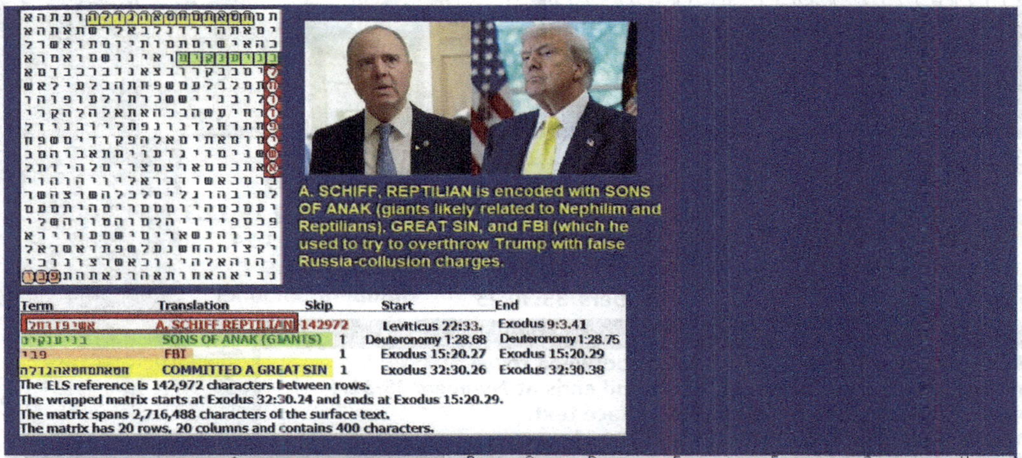

	A	B	C	D	E	F	G	H
1	AXIS TERM A. SCHIFF REPTILIAN @ Skip -142,972	SKIP	NUMBER	DIVIDE BY	THE QUOTIENT	E QUOTIENT	Poisson Equati	EACH WORD
2	ELS RANK 2	USED	IN	304805	EQUALS	x 400 Letters	Probability for	OR PHRASE
3	R FACTOR = 0.109 VALUE = 1.28528666	ON	TORAH	LETTERS	FREQUENCY	Equals	the Word	HAS 1 CHANCE
4	(Note: for the same axis term at skip 16 005 (ELS rank 1) the R	FIGURE		IN THE	PER LETTER	WORD	Appearing	IN THE
5	factor was 1.06 which equaled 11.48153622).			TORAH		EXPECTANCY	At Least once	FOLLOWING
6								
7	COMMITED GREAT SIN @ Skip +1 only	1	1	304805	3.28079E-06	0.001312314	0.001311454	762.5126094
8	SONS OF ANAKIM or NEPHILIM (2 spellings) @ Skip +1	1	6	304805	1.96847E-05	0.007873887	0.007842969	127.5027395
9	FBI @ Skips +1 Only	1	8	304805	2.62463E-05	0.010498515	0.010443598	95.75243737
10	COMBINED ODDS AGAINST THE MATRIX (without axis term) = (H7*H8*H9)/2						2.14839E-07	4654643.114
11								
12	R FACTOR = 0.109 VALUE = 1.28528666						0.778036551	1.28528666
13	FINAL ODDS AGAINST MATRIX (with axis term) = H10*28528666						1.67153E-07	5982550.702

Figure 3-23 above. **ADAM SCHIFF** is encoded as a reptilian. At the same absolute skip are **ALIEN** and **SPY**. **UFO** touches his name. The matrix was found against odds of about 666,754 to 1.

Figure 3-24. ILLINOIS GOVERNOR PRYTZKER reptilian enemy too. Although finding the future encoded requires the right technology (like Code finder Software) and technique, it helps to understand the personality of the Encoder. It is neither alien nor computer-like. Rather, there are at times hints of human-like personality, with what can resemble a sense of humor. The Torah says that man was created in the image of God. While I cannot tell you what God looks like, He does apparently share emotions with us, be they love, anger, or jealousy. How did I know to search for a link between Governor Pritzker and Reptilian? Pritzker does not have a normal human personality. Therefore, I looked to see if he was encoded as a reptilian who has no problem with cheering on murder in Chicago. So, I looked for his name with Reptilian at skips +/- 1 and at the skip of his name. The match was a bingo. As for humor, his name sounds a little like prick, a euphemism for 'penis'. The Concise New Partridge Dictionary of Slang says a prick is "a despicable man, a fool, used as a general term of offence or contempt. The Oxford Dictionary of English says "a stupid or contemptible man". Merriam Webster offers "a spiteful/contemptible man often having some authority". This matrix was found against odds of about 34,218 to 1 after factoring in ELS rank 87 for the axis term (Pritzker).

Term	Translation	Skip	Start	End
פריצקר	PRITZKER	136329	Numbers 31:35.34	Genesis 7:8.32
זוחל	REPTILIAN	-1	Deuteronomy 19:12.6	Deuteronomy 19:12.3
עבם	UFO	1	Genesis 47:13.22	Genesis 47:13.24
ירד	CAME DOWN	1	Numbers 12:5.2	Numbers 12:5.4
רצח	MURDER	1	Numbers 35:30.23	Numbers 35:30.25

The ELS reference is 136,329 characters between rows.
The matrix starts at Numbers 31:35. 29 and ends at Numbers 35:31.4
It spans 1,226,988 characters of the surface text.
The matrix has 10 rows, 27 columns and 270 characters.

Figure 3-24 above. Governor Pritzker, like Governor Newsom, is encoded as a reptilian (who wants to run for President in 2028). See **Figure 3-25 for Newsom.**

SWALWELL (below as Figure 3-25) crossed by REPTILIAN at skip -1.

Term	Translation	Skip	Start	End
סוולוול	SWALWELL	75794	Numbers 11:4.6	Genesis 35:23.45
זוחל	REPTILIAN	-1	Deuteronomy 19:12.6	Deuteronomy 19:12.3
מלחמה	WAR	1	Deuteronomy 20:20.63	Deuteronomy 20:20.67
אויב	ENEMY	75794	Numbers 10:3.15	Exodus 33:1.64

The ELS reference is 75,794 characters between rows. The full, wrapped matrix starts at Numbers 11:4.4 and ends at Exodus 33:1.64. It spans 530,577 characters of surface text. The matrix has 8 rows, 19 columns and 152 characters. The smaller matrix with a white background has 75,794 characters between rows. It starts at Numbers 11:4.4 and ends at Genesis 35:23.46, spanning 454,768 characters of surface text. The matrix has 7 rows, 4 columns and 28 characters.

Figure 3-24 above: The famous liar **A. SCHIFF** is also encoded with **REPTILIAN,** probably descended from **ANAKIM** who likely arose from Nephilim. His **GREAT SIN** was treason. He wanted to be FBI Director if Hilary Clinton won the 2016 election. Under Biden he used the FBI to try to impeach President Trump. Odds against this matrix were almost six million to 1. Figure 3-24 shows another ELS of **A. SCHIFF REPTILIAN.** Odds against it were 60,808 to 1.

REPTILIANS CAN BE HANDSOME OR HORRIBLE. Governor Newsom (Figure 3-25 thinks he should be President because of his white looks, while Jasime Crocket (Figure 3-26) and Maxine Waters (Figure 3-27) think only blacks (ugly and or crazy) should hold national office.

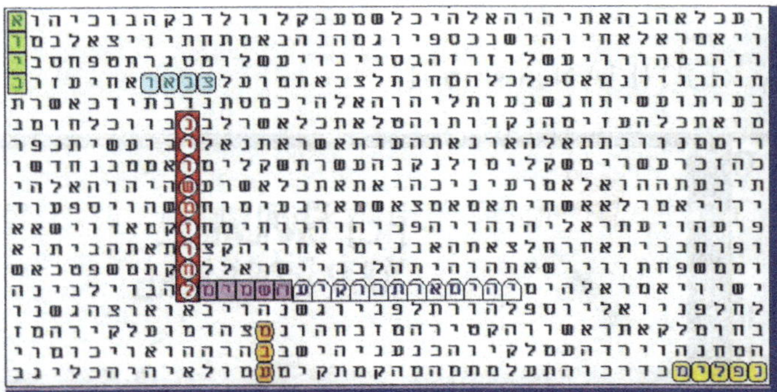

(Governor Gaven) NEWSOM is encoded in sequence with REPTILIAN. He is at the same absolute skip as ENEMY and UFO. NEPHILIM (reptilians) is in the open text. There is also reference to LIGHTS (UFOs) IN THE FIRMAMENT OF HIS SKY. HIS MILITARY is also in the open text.

Term	Translation	Skip	Start	End
ניושמוזוחל	NEWSOM (is) REPTILIAN	70889	Genesis 30:35.63	Genesis 1:14.28
אויב	ENEMY	70889	Deuteronomy 30:20.37	Numbers 10:25.48
נפלים	NEPHILIM	1	Deuteronomy 22:4.23	Deut. 22:4.27
עבמה	UFO	-70889	Deuteronomy 22:4.49	Leviticus 1:15.43
יהימארתברקיע	LET THERE BE LIGHTS	1	Genesis 1:14.11	
	IN THE FIRMAMENT	1		Genesis 1:14.22
השמים	OF THE SKY	1	Genesis 1:14.23	Genesis 1:14.27
צבאו	HIS MILITARY	1	Numbers 10:25.38	Numbers 10:25.41

The ELS reference is 70,889 characters between rows. The matrix starts at Deuteronomy 30:19.80 and ends at Deuteronomy 22:5.11. The matrix spans 1,205,153 characters of the surface text. The matrix has 18 rows, 40 columns and 720 characters.
R FACTOR FOR AXIS TERM = 0.978 = 1 chance in 9.506047937 to be encoded.

	A	B	C	D	E	F	G	H
1	AXIS TERM NEWSOM REPTILIAN @ Skip 70,889	SKIP OF	NUMBER	DIVIDE BY	THE QUOTIENT	E QUOTIENT	Poisson Equation	EACH WORD
2	ELS RANK 1	NEWSM	IN	304805	EQUALS	x 720 Letters	Probability for	OR PHRASE
3	R factor 0.978 = 1 chance in 9.506047937 to be	USED	TORAH	LETTERS	FREQUENCY	Equals	the Word	HAS 1 CHANCE
4	found at least once in the Code.	ON		IN THE	PER LETTER	WORD	Appearing	IN THE
5		FIGURE		TORAH		EXPECTANCY	At Least once	FOLLOWING
6								
7	NEPHLIM @ Skip +1 only	1	4	304805	1.31231E-05	0.009448664	0.009404166	106.3358568
8	THE SKY @ Skip +1 only	1	81	304805	0.000265744	0.191335444	0.174144487	5.742358083
9	ENEMY (2 synonyms) @ Skips +/- 1 and +/- 70889	70889	97	304805	0.000318236	0.2291301	0.204774932	4.883410249
10	MILITARY @ Skip +1 only	1	108	304805	0.000354325	0.255113925	0.22517178	4.441053856
11	UFO @ Skips +/- 1 and +/- 70889	70889	153	304805	0.00050196	0.361411394	0.303307676	3.296982174
12	R FACTOR ODDS AGAINST AXIS TERM						0.105196187	9.506047937
13	COMBINED ODDS AGAINST THE MATRIX = H7*H8*H9*H10*H11*H12						2.40937E-06	415045.6389

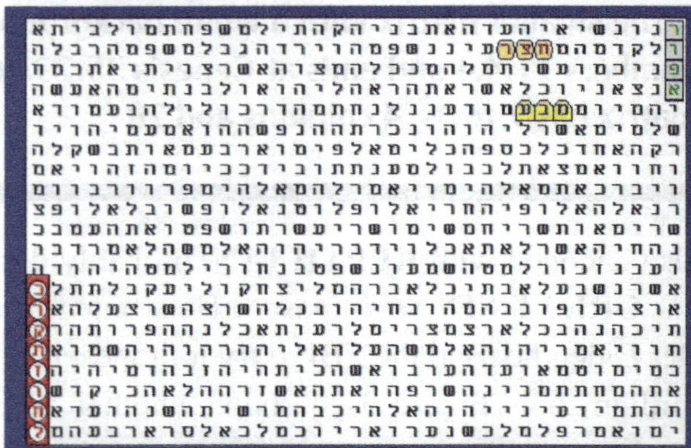

Term	Translation	Skip	Start	End
כרקתזוחל	CROCKET, REPTILIANS	52062	Deuteronomy 6:10.61	Genesis 14:9.61
אפור	GRAY	-52062	Genesis 31:43.40	Numbers 4:34.12
עבם	UFO	-1	Exodus 13:21.20	Exodus 13:21.18
רצח	MURDER	-1	Numbers 34:10.22	Numbers 34:10.20

The ELS reference is 52,062 characters between rows.
The matrix starts at Numbers 4:34.12 and ends at Genesis 14:9.61.
It spans 1,041,275 characters of the surface text.
The matrix has 21 rows, 35 columns and contains 735 characters.

Jasmine CROCKET, REPTILIAN is encoded at the same absolute skip as GRAY. Grays are said to help REPTILIANS abduct people. UFO is at skip -1. So is MURDER.

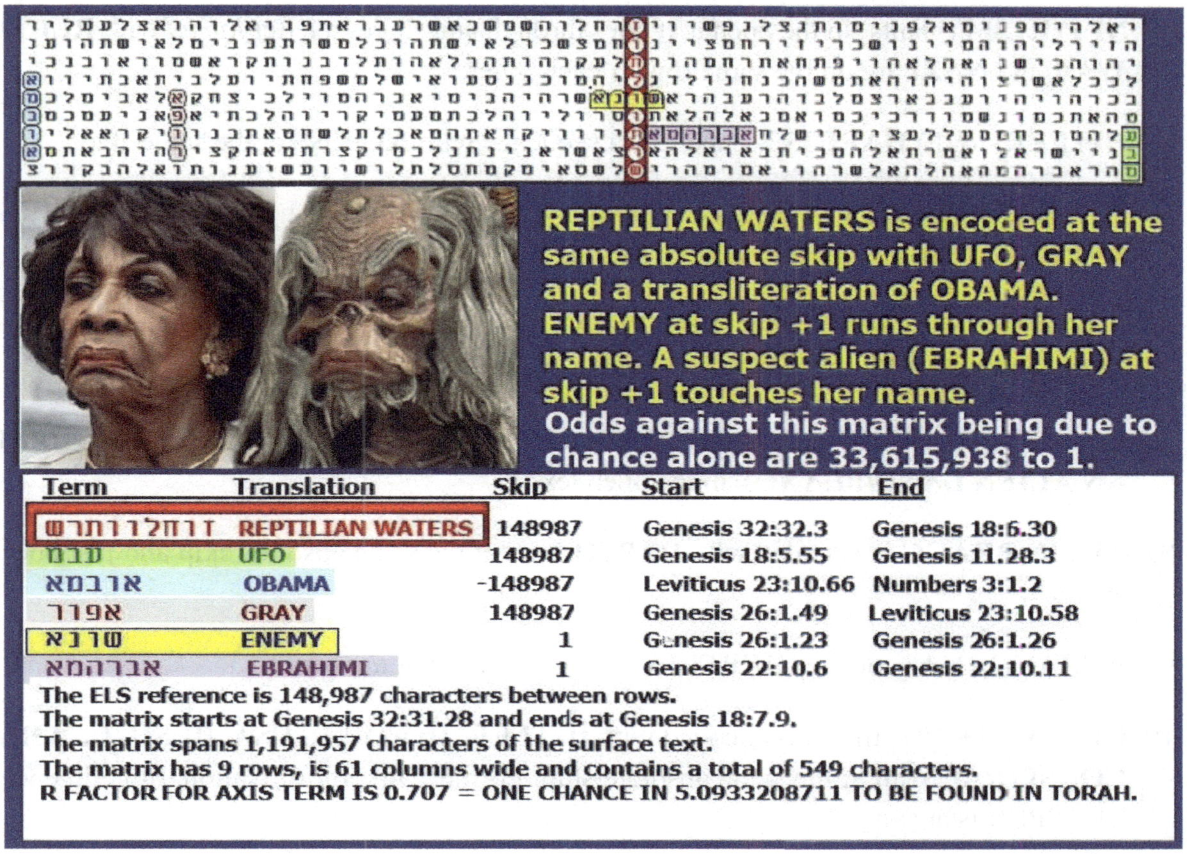

Figure 3-28 above: Maxine Waters appears to a reptilian shapeshifter even without shapeshifting.

	A	B	C	D	E	F	G	H
1	AXIS TERM REPTILIAN WATERS @ Skip 148,987	SKIP	NUMBER	DIVIDE BY	THE QUOTIENT	E QUOTIENT	Poisson Equation	EACH WORD
2	ELS RANK 1	USED	IN	304805	EQUALS	x 545 Letters	Probability for	OR PHRASE
3	R factor 0.707 = one chance in 5.0933 to find axis term	ON	TORAH	LETTERS	FREQUENCY	Equals	the Word	HAS 1 CHANCE
4	TOTAL ODDS AGAINST MATRIX = 33,615,938 to 1	FIGURE		IN THE	PER LETTER	WORD	Appearing	IN THE
5				TORAH		EXPECTANCY	At Least once	FOLLOWING
6								
7	OBAMA (2 spellings) at Skips +/- 1 and +/- 148,987	-148987	6	304805	1.96847E-05	0.010806909	0.010748724	93.03429462
8	GRAY @ Skip +/- 1 and +/- 148987	148987	9	304805	2.95271E-05	0.016210364	0.016079683	62.19028022
9	ENEMY @ Skip +1 only	1	22	304805	7.21773E-05	0.039625334	0.038850518	25.73968222
10	EBRAHIMI (2 spellings) @ Skip +1 only	1	51	304805	0.00016732	0.091858729	0.087765988	11.39393547
11	UFO @ Skip +/- 1 and +/- 148987	148987	165	304805	0.00054133	0.297190007	0.257097158	3.889580147
12	axis term R FACTOR 0.707 = one chance in 5.0933						0.196336028	5.093308711
13	COMBINED ODDS AGAINST MATRIX = H7*H8*H9*H10*H11*H12						2.97478E-08	33615938.78

SUMMARY: ALIENS ENCODED AS INFILTRATING OUR GOVERNMENT.

1. **PRESIDENT JOE BIDEN** is encoded with **ALIEN IN OUR GOVERNMENT. BIDEN is) ALIEN** is also encoded with **MARS, UFO**, and **ENEMY** at the same absolute skip.

2. **PRESIDENT BARACK OBAMA** is (in sequence). Encoded with **REPTILIAN** and at the same absolute skip as **UFO** and **PRESIDENT.**

3. Senator **CHUCK SCHUMER** is at the same skip as **REPTILIAN, UFO**, and **SPY**. The term

NEPHILIM (giants) in the open text goes through and shares a letter *lamed* with **REPTILIAN. WAR** and **NAZI** are on the same row as each other.

4. **HITLER** - (VP) **HARRIS** is encoded in sequence with **ALIENS** at the same skip.

5. GENERAL **MILLEY** is encoded with **GENERAL** and **UFO** at the same absolute skip, plus **REPTILIAN** and **ALIEN** (3 times). See Figure 3-1 for additional matches.

6. FORMER FBI DIRECTOR **CHRIS WRAY** is encoded with **REPTILIAN** at skip-1 crossing **BOSS** and with **FBI** and **UFO** at the same skip as CHRIS WRAY.

7. U.S. ATTORNEY GENERAL MERRICK **GARLAND** is encoded with **STRANGER FROM AN ALIEN LAND/PLANET** in the open text.

8. **ALVIN BRAGG** is encoded with **REPTILIAN** at the same absolute skip and just one column away.

9. A. **MAYORKAS** touches **REPTLIAN** at skip -1 in 35-letter box.

10. **CIAKAHRR** (Reptilian) is encoded with **HITLER, HOOVER, USA, BUSH, C. WRAY**, and **CONSPIRACY** all at the same absolute skip. Former Director of National Intelligence James CLAPPER is at skip +1.

11. **TIM WALZ**, 2024 Democratic Party vice presidential nominee, is encoded with **REPTILIAN** at skip -1 sharing a letter *zayin* with it in a 28-letter box. **UFO** and **WEIRD** are at skip -1. During his campaign, Walz called Vance weird, but Walz put tampon dispensers into boys' locker rooms.

12. Former Secret Service Director **KIM CHEATLE** is encoded at the same skip with **ALIENS, PLEIADES, UFO**, and **BIDEN**.

13. **JOHN DURHAM** played a major role in investigating President Trump for almost four agonizing years. Against odds of about 183 to 1. He, too, is encoded on Figure 3-20 with **REPTILIAN** at skip-1.

14. House Minority Leader **HAKEEM JEFFRIES** in a row split three unwrapped matrix is encoded in 680 letters with **REPTILIAN** and **WAR** at the same absolute skip of a transliteration of **HAKEEM JEFFRIES. B. OBAMA** is at skip -1. Odds against this matrix are about 87,694,744 to 1. Although it takes a row split of 3 to see it, against odds of over 87 million to 1, HAKEEM JEFFRIES is encoded with **REPTILIAN** and **WAR** at the same absolute skip plus **B. OBAMA** at skip -1.

15. On Figure 3-22 above, former CIA Director John Brennan. **See Figure 3-22. BRENNAN** is also

in sequence with **REPTILIAN**. At the same absolute skip are **BARACK H., KERRY** (Secretary of State), **PLEIADES** and **UFO.** At Skip +1 touching **BRENNAN (is) REPTILIAN** is **CHINA. SPIES** are in the open text and COMEY is at skip -2. KERRY and COMEY share a letter *kuf.* Odds against this matrix were about 13,752,292 to 1.

16. Adam Schiff worked hard to try to impeach Trump over the Russia Collusion hoax. A**. SCHIFF** is encoded as a **REPTILIAN** who had a **GREAT SIN**. See Figures 3-23 and 3-24.. He is a hostile spy.

17. **SWALWELL** is encoded with **REPTILIAN** going through it at skip -1, but this matrix requires a very high-ranking ELS (rank 165) to be seen. **WAR** is in the open text and **ENEMY** is at the same skip as **SWALWELL**. See Figure 3-28.

18. **Jasmine CROCKET** is in sequence with **REPTILIAN**. See Figure 3-27. **GRAY** is at the same absolute skip. UFO is at skip -1.

19. Gavin **NEWSOM** is in sequence with REPTILIAN. **ENEMY** and **UFO** are at the same absolute skip. NEPHILIM is in the open text.

Figure 03-30 – The Chinese Government is likewise infiltrated (as is Putin's Russian Government).

There was only 1 chance in 18,354 that REPTILIANS would be found at a special case skip (+/- 1 or the absolute skip of the axis term, CHINESE GOVERNMENT) yet again, it's there anyway. Basically, this means that to defeat reptilians would require us to also defeat Putin's followers and dual agents in the Democrat Party.

Term	Translation	Skip	Start	End
ממשל סיני	CHINESE GOVERNMENT	12001	Leviticus 25:10.26	Deuteronomy 7:25.36
זוחל	REPTILIAN	-12001	Exodus 12:35.42	Genesis 41:1.23

The ELS reference is 12,001 characters between rows.
It starts at Genesis 41:1.23 and ends at Deuteronomy 7:25.36.
It spans 204,019 characters of the surface text.
The matrix has 18 rows, 2 columns and 36 characters.

1	AXIS TERM CHINESE GOVERNMENT @ Skip -12,001	NEWSM	NUMBER	DIVIDE BY	THE QUOTIENT	E QUOTIENT	Poisson Equation	EACH WORD
2	ELS RANK 1	USED	IN	304805	EQUALS	x 36 Letters	Probability for	OR PHRASE
3	R factor 0.336 = 1 chance in 2.167047 in to be	ON	TORAH	LETTERS	FREQUENCY	Equals	the Word	HAS 1 CHANCE
4	found at least once in the Code.	FIGURE		IN THE	PER LETTER	WORD	Appearing	IN THE
5				TORAH		EXPECTANCY	At Least once	FOLLOWING
6								
7	REPTILIAN	1	1	304805	3.28079E-06	0.000118108	0.000118101	8467.305565
8	R FACTOR ODDS AGAINST AXIS TERM 0.336 2nd function LOG=2.1677							2.1677040
9	COMBINED ODDS AGAINST THE MATRIX = H7*H8						0.461317597	18354.61214

Chapter 4
Nazis in the U.S. Government

Decades of Torah research has taught me an unusual view of history. For starters, Germany was working on UFOs before World War II but was not quite able to produce them in significant numbers before the war ended. Of the aliens who helped the Nazis, the Reptilians were the worst. Just before the war ended, Hitler escaped alive, and he reached South America in a submarine (probably U944, but U977 was also suspected). Eisenhower stated his belief that Hitler survived, but he failed to track him down. In the 1950s, Ike got to meet and make some deals with more than one species of aliens. In exchange for humans and cattle that the aliens were allowed to take, our government received advanced military technology. By the end of his second term, Eisenhower was disgusted with how the aliens were handling people and cattle. He made an effort to warn the public about the alien threat, but he was not specific enough. He said:

"A vital element in keeping the peace is our military establishment. Our arms must be mighty, ready for instant action, so that no potential aggressor may be tempted to risk his own destruction... American makers of plowshares could, with time and as required, make swords as well. But now we can no longer risk emergency improvisation of national defense; we have been compelled to create a permanent armaments industry of vast proportions... This conjunction of an immense military establishment and a large arms industry is new in the American experience... Yet we must not fail to comprehend its grave implications... ***In the councils of government, we must guard against the acquisition of unwarranted influence, whether sought or unsought, by the military-industrial complex. The potential for the disastrous rise of misplaced power exists and will persist."***

His warning about the military-industrial complex was (fatally) too vague. He should have told us about the military-industrial-extraterrestrial complex. Many human members of this group were Nazi war criminals brought to the United States under Operation Paperclip. This was a secret United States intelligence program in which more than 1,600 Nazi German scientists, engineers, and technicians were taken from former Nazi Germany to the U.S. for government employment after the end of World War II in Europe, between 1945 and 1959. Conducted by the Joint Intelligence Objectives Agency (JIOA), it was largely carried out by special agents of the U.S. Army's Counterintelligence Corps (CIC). Many of these personnel were former members, and some were former leaders, of the Nazi Party. During the 1950s, quite a few Nazis rose to high positions in the U.S. government. Eisenhower went from being begrudgingly in control while meeting with combined Nazi-alien crews to having to threaten to attack Area 51 with the First Army if the base did not come clean with him regarding what research was being conducted there.

Key Recruits for Operation Paperclip – Advisors brought into the United States. This list is found at the end of this chapter. Many were members of the Nazi Party, and quite a few, including Wernher von Braun, were war criminals. Von Braun designed the Saturn V that took American astronauts to the moon.

President Kennedy was concerned about the secret Nazi-alien influence. He wanted to warn us about what was going on, but that idea was shut down by his assassination. Presidents after him were well aware of the price of giving us too much truth. Eventually, much of what was left of American morality became Nazi philosophy. President Trump, who established a new branch of our military—Space Force—is trying to reverse that, but he has already survived three assassination attempts (a bullet fired through his bedroom window at Mar-a-Lago was the first), which are likely related to the military-industrial-extraterrestrial complex.

The rest of this chapter will highlight Torah Code matrices and New York Times articles meant to support and highlight the above altered view of recent world history.

What Happened to Hitler After World War II?

An April 27, 1945, New York Times article (dated April 26, 1945) was entitled *Hitler Imitator Reported Ready to Die in His Place.* I followed the story via the Times as far as 1949. Briefly, a former grocer, August Wilhelm Bartholdi, was a double for Hitler. He was poised to "die on the barricades" while the real Hitler escaped. Two days later (just after midnight on April 29, 1945), Hitler married Ava Braun. Germany had time to prepare the Hitler swap, whether by switching dental records or drilling Bartholdi's mouth to match Hitler's. But they didn't have time to properly prepare a duplicate for his bride. The following day (April 30, 1945) was the alleged suicide. Russian troops found two bodies burned beyond recognition. They were supposed to be Hitler and Eva Braun. Later, we learned that there was a match with Hitler's dental records. But once in my life, I heard (but could never confirm) that the female corpse found did not have a match with Eva Braun's dental records; however, this aspect of the story does not match what Wikipedia writes. They indicate that, "by 11 May 1945, the Soviets had already had Hitler's dentist, Hugo Blaschke, and his dental technician confirm the dental remains found were Hitler's and Braun's. However, Wikipedia often lies when political opinions are involved."

When the American troops arrived in Berlin, they questioned the staff. All, except for two, said that it was Hitler who had died, but the New York Times reported that the maid and the chauffeur claimed it was the cook. However, as I recall, no name was given for the cook, and it was unclear as to whether he was Bartholdi. By July of 1945, the Times was writing about unaccounted-for U- boats that might be carrying Hitler, and by August of 1945, there were stories of U-977 being searched by U.S. Marines in Buenos Aires, Argentina. There was also a story of people seeing what looked like Hitler and Eva Braun coming ashore from U-944 elsewhere in Argentina, with them heading for Patagonia. Does the Code discuss any of this? Yes.

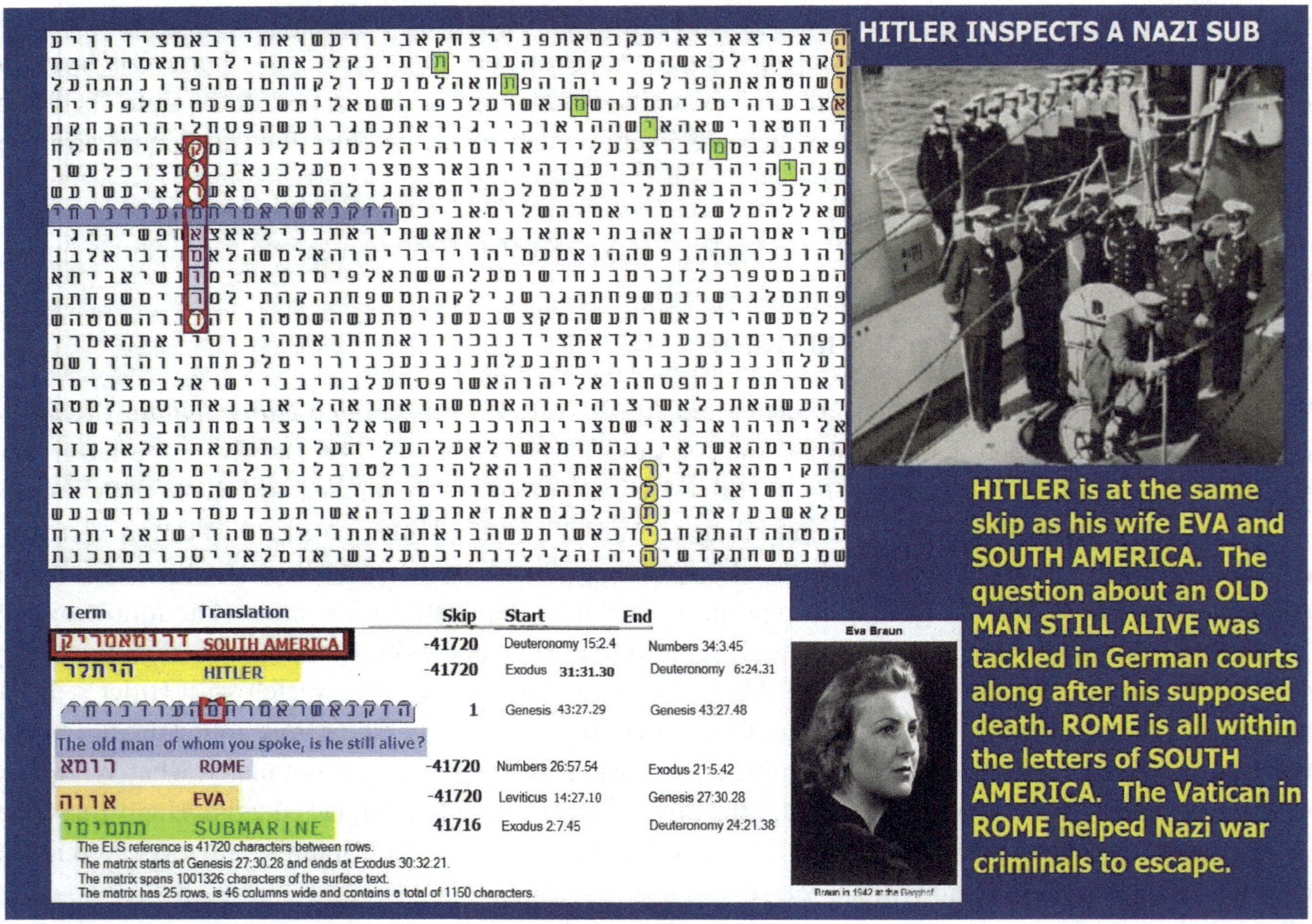

Figure 4-1 above: The axis term is SOUTH AMERICA (Hitler's hideout after World War II). ROME (who helped him get there), HITLER, and his bride (EVA) are all at the same skip. The odds against finding the matrix were about 20 million to 1 if we do not count the axis term, and about 100 million to 1 if we use the R factor to count South America.

	A	B	C	D	E	F	G	H
1	AXIS TERM SOUTH AMERICA @ Skip -41720	SKIP	NUMBER	DIVIDE BY	THE QUOTIENT	E QUOTIENT	POISSON EQUATION	EACH WORD
2		USED	IN	304805	EQUALS	X 1150 LETTERS	PROBABILITY FOR	OR PHRASE
3		ON	TORAH	LETTERS	FREQUENCY	EQUALS	THE WORD	HAS 1 CHANCE
4		FIGURE		IN THE	PER LETTER	WORD	APPEARING	IN THE
5				TORAH		EXPECTANCY	AT LEAST ONCE	FOLLOWING OF
6								BEING HERE
7	IS THE OLD MAN...STILL ALIVE?	1	1	304805	3.28079E-06	0.003772904	0.003765796	265.5481405
8	HITLER (with TAV orTET) @ Skips +/- 1 and +/- 41720	-41720	2	304805	6.56157E-06	0.007545808	0.00751741	133.0245419
9	SUBMARINE (2 TERMS) @ Skips +/- 1 and +/- 41885	45042	20	304805	6.56157E-05	0.07545808	0.072681397	13.75867888
10	ROME @ Skips +/- 1 and +/- 41720	-41720	31	304805	0.000101704	0.116960024	0.110379244	9.059674321
11	EVA @ Skips +/- 1 and +/- 41720	-41720	65	304805	0.000213251	0.245238759	0.217482317	4.59807497
12	COMBINED SIGNIFICANCE (NO AXIS TERM) = H7*H8*H9*H10*H11						4.93923E-08	20246054.66
13							#REF!	#REF!
14	R FACTOR = 0.701 ENTER 2ND FUNCTION, LOG = 5.035						0.198609732	5.035
15	ENTER 5.035 AT H14							
16	FINAL SIGNIFICANCE ODDS AGAINST = H12*H14 = H16						9.8098E-09	101938885.2

Figure 4-2 shows alleged FBI files that point to Hitler in South America after World War II.

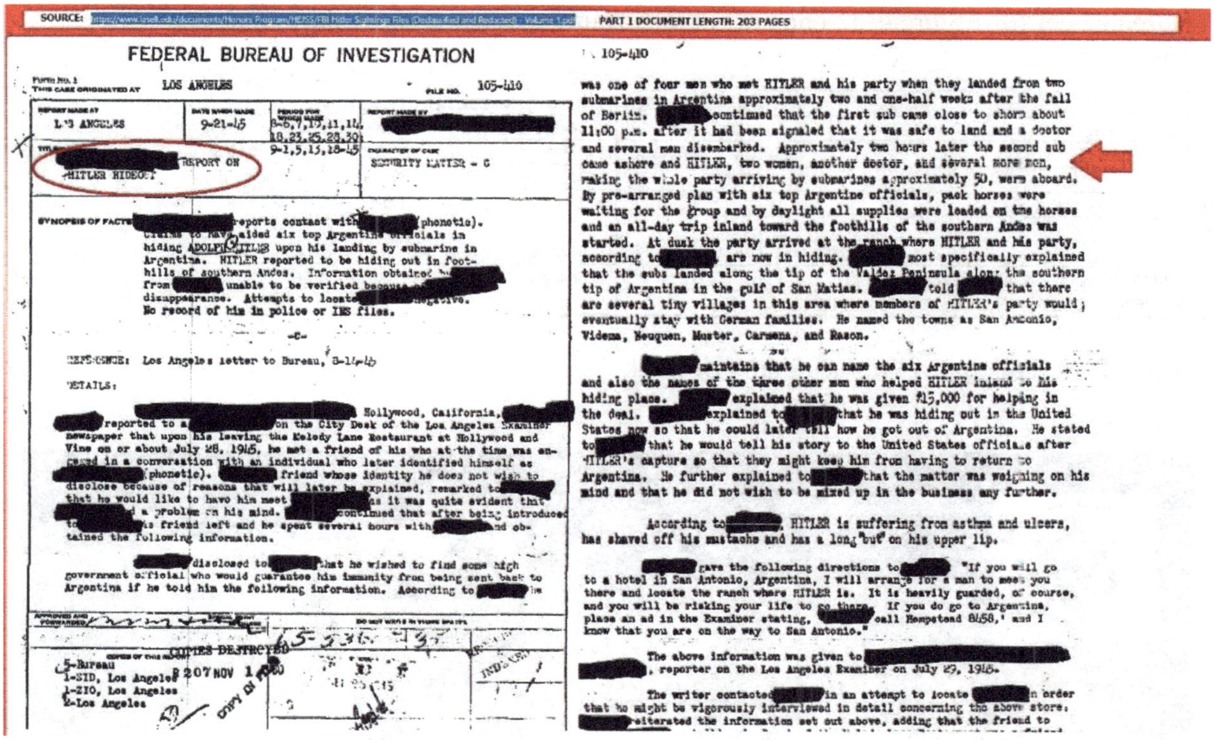

SOURCE: https://www.lasell.edu/documents/Honors Program/HEISS/FBI Hitler Sightings Files (Declassified and Redacted) - Volume 1.pdf   PART 1 DOCUMENT LENGTH: 203 PAGES

FEDERAL BUREAU OF INVESTIGATION

Form No. 1
This case originated at LOS ANGELES   File No. 105-410

Report made at	Date when made	Period for which made	Report made by
LOS ANGELES	9-21-45	8-6,7,10,11,14, 18,23,25,28,30; 9-1,5,13,18-45	[redacted]

Title: [redacted] REPORT ON HITLER HIDEOUT

Character of case: SECURITY MATTER - C

SYNOPSIS OF FACTS: [redacted] reports contact with [redacted] (phonetic). Claims to have aided six top Argentine officials in hiding ADOLPH HITLER upon his landing by submarine in Argentina. HITLER reported to be hiding out in foothills of southern Andes. Information obtained [redacted] from [redacted] unable to be verified [redacted] disappearance. Attempts to locate [redacted] negative. No record of him in police or INS files.

-C-

REFERENCE: Los Angeles letter to Bureau, 8-14-45

DETAILS:

[redacted] Hollywood, California, [redacted] reported to a [redacted] on the City Desk of the Los Angeles Examiner newspaper that upon his leaving the Melody Lane Restaurant at Hollywood and Vine on or about July 28, 1945, he met a friend of his who at the time was engaged in a conversation with an individual who later identified himself as [redacted] (phonetic). [redacted] friend whose identity he does not wish to disclose because of reasons that will later be explained, remarked to [redacted] that he would like to have him meet [redacted] as it was quite evident that [redacted] a problem on his mind. [redacted] continued that after being introduced to [redacted] his friend left and he spent several hours with [redacted] and obtained the following information.

[redacted] disclosed to [redacted] that he wished to find some high government official who would guarantee him immunity from being sent back to Argentina if he told him the following information. According to [redacted] he

Approved and forwarded: [signature]

Do not write in these spaces

65-53615-35

Copies of this report
5-Bureau
1-SID, Los Angeles
1-ZIO, Los Angeles
2-Los Angeles

COPIES DESTROYED
R 207 NOV 1 1960

COPY IN FILE

105-410

was one of four men who met HITLER and his party when they landed from two submarines in Argentina approximately two and one-half weeks after the fall of Berlin. [redacted] continued that the first sub came close to shore about 11:00 p.m. after it had been signaled that it was safe to land and a doctor and several men disembarked. Approximately two hours later the second sub came ashore and HITLER, two women, another doctor, and several more men, making the whole party arriving by submarines approximately 50, were aboard. By pre-arranged plan with six top Argentine officials, pack horses were waiting for the group and by daylight all supplies were loaded on the horses and an all-day trip inland toward the foothills of the southern Andes was started. At dusk the party arrived at the ranch where HITLER and his party, according to [redacted], are now in hiding. [redacted] most specifically explained that the subs landed along the tip of the Valdez Peninsula along the southern tip of Argentina in the gulf of San Matias. [redacted] told [redacted] that there are several tiny villages in this area where members of HITLER's party would eventually stay with German families. He named the towns as San Antonio, Videma, Neuquen, Muster, Carmena, and Rason.

[redacted] maintains that he can name the six Argentine officials and also the names of the three other men who helped HITLER inland to his hiding place. [redacted] explained that he was given $15,000 for helping in the deal. [redacted] explained to [redacted] that he was hiding out in the United States now so that he could later tell how he got out of Argentina. He stated to [redacted] that he would tell his story to the United States officials after HITLER's capture so that they might keep him from having to return to Argentina. He further explained to [redacted] that the matter was weighing on his mind and that he did not wish to be mixed up in the business any further.

According to [redacted], HITLER is suffering from asthma and ulcers, has shaved off his mustache and has a long "but" on his upper lip.

[redacted] gave the following directions to [redacted] "If you will go to a hotel in San Antonio, Argentina, I will arrange for a man to meet you there and locate the ranch where HITLER is. It is heavily guarded, of course, and you will be risking your life to go there. If you do go to Argentina, place an ad in the Examiner stating, '[redacted] call Hempstead 8458,' and I know that you are on the way to San Antonio."

The above information was given to [redacted], reporter on the Los Angeles Examiner on July 29, 1945.

The writer contacted [redacted] in an attempt to locate [redacted] in order that he might be vigorously interviewed in detail concerning the above store. [redacted] reiterated the information set out above, adding that the friend to

Figure 4-3 left below: Torah Code matrix about Hitler in Tunja, Colombia. Figure 4-4 right below: related State Department document. Figure 4-3 was found against odds of about 26,077 to 1.

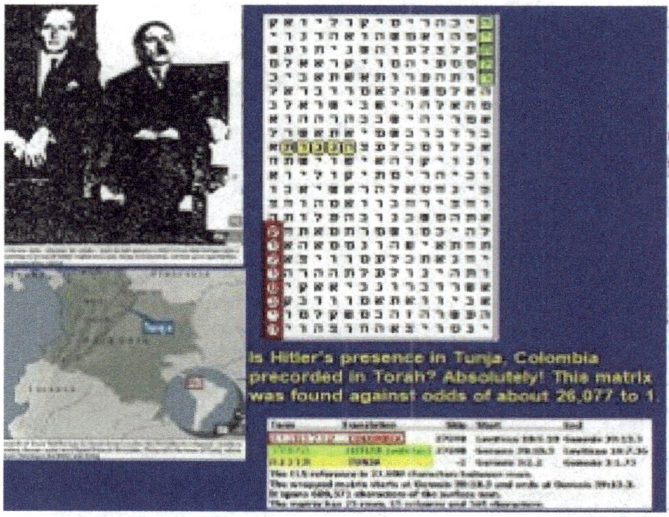

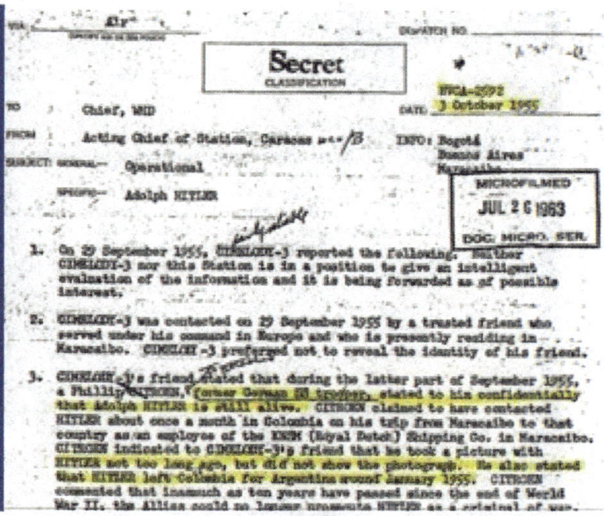

Secret
CLASSIFICATION

TO: Chief, WHD

DATE: 3 October 1955

FROM: Acting Chief of Station, Caracas

INFO: Bogotá, Buenos Aires, Maracaibo

SUBJECT: GENERAL— Operational
SPECIFIC— Adolph HITLER

MICROFILMED JUL 26 1963 DOC. MICRO. SER.

1. On 29 September 1955, CIMELODY-3 reported the following. Neither CIMELODY-3 nor this Station is in a position to give an intelligent evaluation of the information and it is being forwarded as of possible interest.

2. CIMELODY-3 was contacted on 29 September 1955 by a trusted friend who served under his command in Europe and who is presently residing in Maracaibo. CIMELODY-3 preferred not to reveal the identity of his friend.

3. CIMELODY-3's friend stated that during the latter part of September 1955, a Phillip CITROEN, former German SS trooper, stated to him confidentially that Adolph HITLER is still alive. CITROEN claimed to have contacted HITLER about once a month in Colombia on his trip from Maracaibo to that country as an employee of the KNSM (Royal Dutch) Shipping Co. in Maracaibo. CITROEN indicated to CIMELODY-3's friend that he took a picture with HITLER not too long ago, but did not show the photograph. He also stated that HITLER left Colombia for Argentina around January 1955. CITROEN commented that inasmuch as ten years have passed since the end of World War II, the Allies could no longer prosecute HITLER as a criminal of war.

Below: Spreadsheet for Figure 4-3. The matrix was found against odds of 26,077 to 1.

	A	B	C	D	E	F	G	H
1	AXIS TERM COLOMBIA @ Skip +27,698	SKIP	NUMBER	DIVIDE BY	THE QUOTIENT	E QUOTIENT	Poisson Equation	EACH WORD
2	ELS RANK 10	USED	IN	304805	EQUALS	x 345 Letters	Probability for	OR PHRASE
3		ON	TORAH	LETTERS	FREQUENCY	Equals	the Word	HAS 1 CHANCE
4		FIGURE		IN THE	PER LETTER	WORD	Appearing	IN THE
5				TORAH		EXPECTANCY	At Least once	FOLLOWING
6								
7	HITLER (2 spellings) @ Skips +/- 1 and +/- 27698	27698	3	304805	9.84236E-06	0.003395614	0.003389855	294.9978675
8	TUNJA @ Skips +/- 1 and +/- 27698	-1	1	304805	3.28079E-06	0.001131871	0.001131231	883.9928479
9	COMBINED ODDS AGAINST THE MATRIX = (H7*H8*)/10						3.83471E-05	26077.6005

Figure 4-5 below. This matrix names many of the key players in the Hamas/Hitlerian/ Ciakahrr Gaza War.

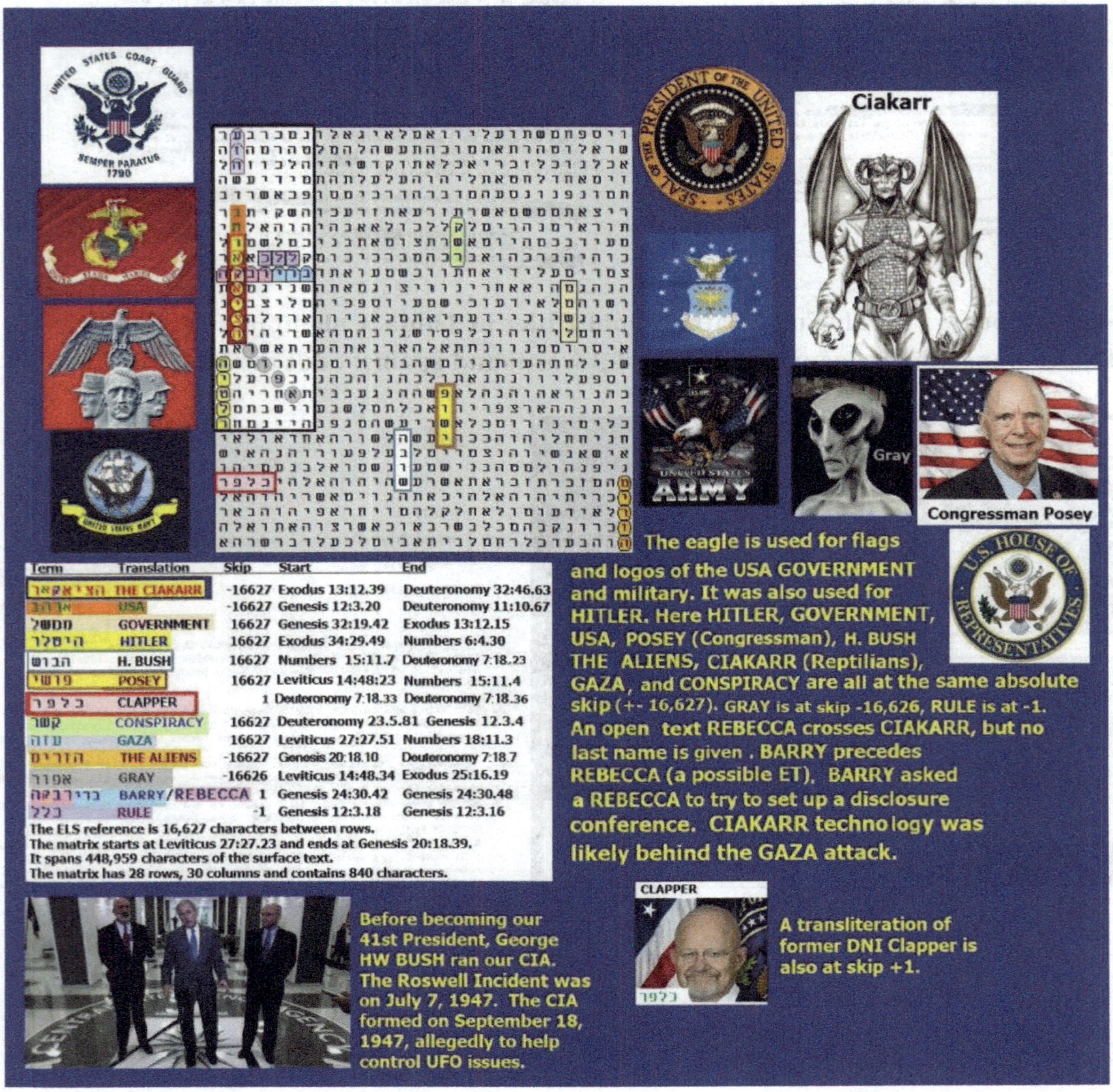

ARE THERE ET COMMUNITIES ON EARTH?

As part of my investigation into possible ET links to Rebecca, I took my wife and son to Rebecca's hometown in Lakeland, Florida. There, my son was tested by a local doctor who specialized in holistic medicine. Although the test results did not lead to a cure of his rare condition (erythromelalgia), which can painfully affect feet, hands, ears, and other areas, they were of use in reaching a proper diagnosis. After the test, the two were nice enough to take us out to lunch. It was then that I was given some of the other information that I sought. Rebecca said she had seen shape-shifter Reptilians emerge from a Publix Supermarket. She also said that a 20-year-old man had approached her in a panicked state right after David Gruth testified before Congress. Gruth' topic, and that of two Navy FA-18 fighter pilots, was UFOs held by the Pentagon or tracked by the Navy. Apparently, the three witnesses were frightening to the young man because he now had reason to believe that a growing, rapidly more aware Congress might have reason to go after him and his type. The implication of what Rebecca told me could be that there was a colony of ETs, possibly shape-shifters, in her town that could encounter problems for the security of every ET in town. Discovery and disclosure of such a group might be catastrophic for her genetic brethren. I have also spoken to an Andrew Basiago and his wife about Reptilians. Andrew has weird stories about alien experiences. He is discussed elsewhere in this chapter. I only mention him here because both he and his wife claim to have witnessed reptilians morphing in Vancouver, Washington. Ignoring these reports may expedite the way to the end of our species. We need responsible leaders to make such decisions, but we will not find them in the Democratic Party. Nor do all Republicans have clean hands here.

Ebrahimi, the alleged ET that I met, is from **Sooke, B.C. Canada**. It is possible that his hometown is also an ET nest. A very high percentage of my readers are from Canada.

Why are Congressman Posey & Conspiracy also at the same absolute Skip on Figure 4-4? As I continue to study Figure 4-4, I have now identified nine terms that are at skips +/- 16,627 (and one, GRAY at skip -16,626, touching the Ciakahrr axis term at skip -16,627). They are as follows:

1. **CIAKAHRR.** Synonymous with Reptilian. When not shape-shifted to look human, they are somewhat Satanic-looking creatures, often with Satanic values. See the top-right creature on Figure 4-4.

2. **HITLER. Hitler** is one column left of **GAZA, USA,** and **CIAKAHRR**. The Ciakahrr Reptilians, like their Nazi-leaning forefathers and Hitler, would have relished the chance to massacre Jews.

3. **GRAY.** Grays are said to often provide crews for Reptilian ships. They are commonly cited as abducting massive numbers of humans to satisfy Reptilian needs, doing so out of fear of Reptilians who, in the distant past, blew up at least one Gray world. On Figure 4-4, GRAY at skip-16,626 touches the bottom of **CIAKAHRR** at skip-16,627. It almost suggests that the GRAYs kiss the butts of the CIAKAHRR.

4. **POSEY (Congressman, R, FL).** I made a presentation about UFOs to Congressman Posey on August 7, 2023, at his Brevard County, Florida office. At the end, he handed me a copy of his 1997 book, *THE HANSON CONNECTION – The book they don't want you to read.*

Previously, I had asked Posey's aide, Patrick Gavin, if the Congressman had any background or opinion papers about UFOs. He said, "No," but for some reason, Gavin was not truthful with me. A week after I finished reading the book, it was clear that the work was about the U.S. Government not only hiding what they knew about ETs, but they had even imprisoned or killed folks who knew too much for their own good. On a late fall, 2023 day, several congressmen publicly called for the release of major UFO secrets. On the same day, for some unexplained reason, Posey sent me a second copy of his book along with copies of reviews for it that were written back in 1997. The book was mailed from his house, and he addressed it in his own handwriting. All the mystery about his book may have been an attempt to warn me that there was a threat that I should take personally, but this is just a guess. If I suffer an untimely death, the congressman should be interrogated about why he sent me the second copy, which, unlike the first copy, was autographed. Posey is shown on the extreme right side of Figure 4-4.

5. **GOVERNMENT.** A major issue under study here is whether with Operation Paperclip, the **GOVERNMENT** of the **USA** came under the control of the **CIAKAHRR** and those who are loyal to **HITLER.** These terms and **CONSPIRACY** are all at the same absolute skip of 16,627 and spell out the scope of the conspiracy.

6. **USA. USA o**verlaps and shares two letters with **CIAKAHRR**. The most serious aspect of this investigation is whether there is any truth to the allegation that the Ciakahrr not only have occupied other worlds through use of their alleged shape-shifting abilities, but they have also taken over some **GOVERNMENT**s on Earth, including that of the **USA**.

7. **CONSPIRACY.** At a minimum, the **CONSPIRACY** to break through the wall around Gaza likely had Ciakahrr's support. A few months before the attack, the president of th**e USA** tried to transfer another $6 billion to Iran. Previously, the Obama-Biden **GOVERNMENT** had transferred over $100 billion to Iran. In both cases, Biden and his cohorts encouraged **ALIENS** and Iran to make war on Israel.

8. **THE ALIENS.** The axis term on this matrix is the **CIAKAHRR** (at the same absolute skip) and the **GRAY**s who touch them.

9. **GAZA. GAZA** is not only at the same absolute skip as **CIAKAHRR** and the **USA**, but it's also in **the** same column. We think the matrix supports the idea that the Ciakahrr used their advanced technology to break through the wall around Gaza. While teleportation might be one way to accomplish this, holes could have been quickly cut with the use of nuclear technology.

10. **BUSH.** The Bush family, and especially the grandfather of President George H. W. **BUSH**, have a broad history of support for the Hitler regime. The Guardian begins its account as follows:

Rumours of a link between the US first family and the Nazi war machine have circulated for decades. Now the Guardian can reveal how the repercussions of events that culminated in action under the Trading with the Enemy Act are still being felt by today's president.

George Bush's grandfather, the late US senator Prescott Bush, was a director and shareholder of companies that profited from their involvement with the financial backers of Nazi Germany.

The Guardian has obtained confirmation from newly discovered files in the US National Archives that a firm of which Prescott Bush was a director was involved with the financial architects of Nazism. The whole story is at The Guardian article.

FIGURE 4-5 - THE EMPHASIS SHIFTS TO THE DEEP STATE/ FOURTH REICH. There are a lot of names that could be sought as secret members of the Fourth Reich or leaders of organizations that are loyal to the Fourth Reich. Figure 4-4B shifts the focus up from Figure 4-4A, allowing us to see the FBI's first director (J. Edgar HOOVER) and the director under Biden until January 19, 2025, Chris WRAY. Hoover was involved in a dispute with the Army over UFO wreckage (more likely from a Los Angeles and Long Beach engagement in 1942 than Roswell in 1947), but it could be both. Chris Wray did his best to try to remove Trump from office based on trumped-up Russia collusion charges, but he was also opposed to UFO disclosure. Both C. WRAY and HOOVER are again at the absolute skip of the axis term, THE CIAKAHRR. Another Deep State member, James CLAPPER, was also found while researching this matrix. He was Director of National Intelligence. The H. BUSH on the matrix (President George H.W. Bush) was Director of the Central Intelligence Agency. Allegedly, he refused to brief President-elect Carter on UFOs. The loyalty of the CIA is in question because one of its directors, John Brennan, voted for a Communist for President in 1976 (Gus Hall). In July, 2025 he is under investigation along with former FBI director, James Comey. However, I could not find Brennan encoded at special case skip +/- 16,627 of Ciakahrr. Alas, there is never a shortage of items to explore in the Torah Code. The second lowest ELS of THE CIAKAHRR is at skip +1,986. There is an ELS of BRENNAN at skip -1,986. That matrix is 15 columns by seven rows (105 letters in area). Even better, there is an ELS of BRENNAN with REPTILIAN at skip -34,626. See Figure 3-22.

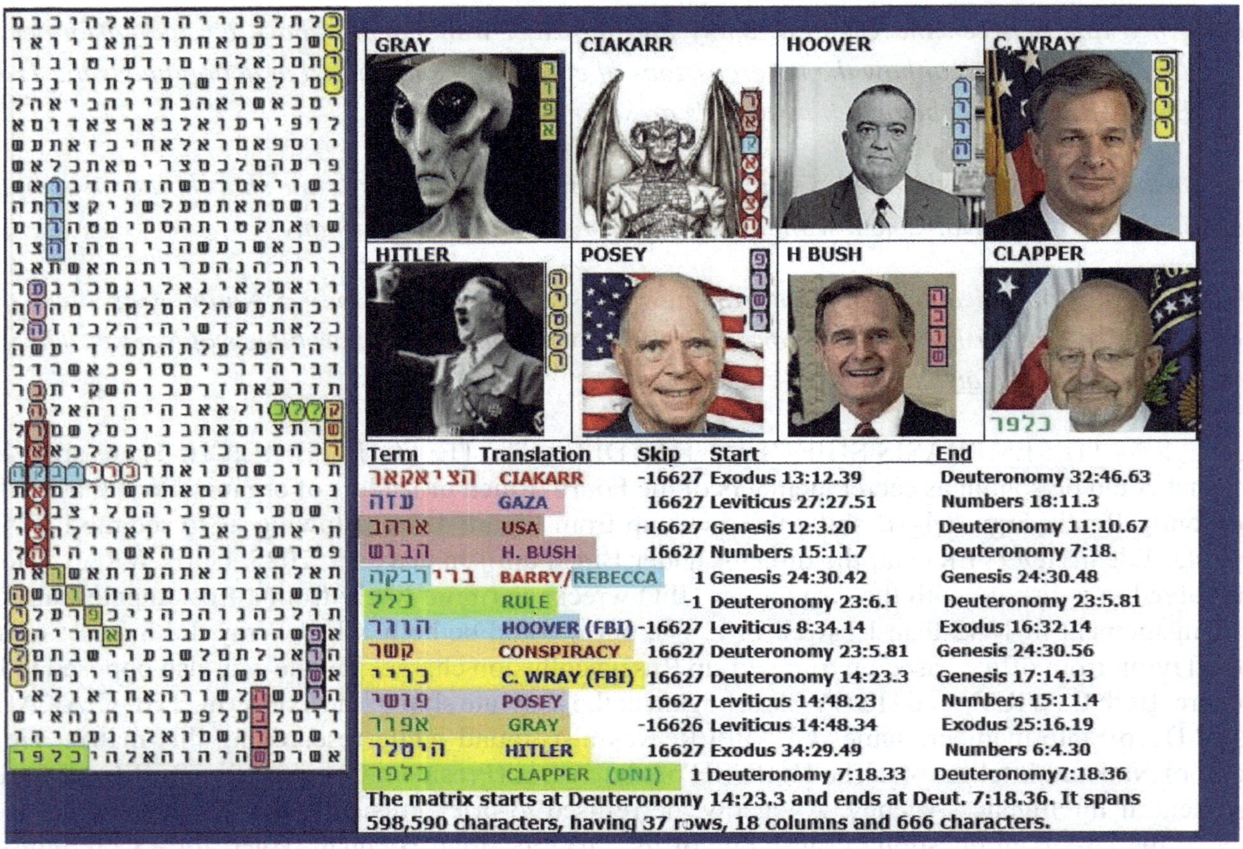

Figure 4-6 is above. Its spreadsheet indicates it was found against odds of over 217.9 trillion to 1.

	A	B	C	D	E	F	G	H
32								
33	Axis term THE CIAKARR @ Skip -16,627	SKIP	NUMBER	DIVIDE BY	THE QUOTIENT	E QUOTIENT	POISSON EQUATION	EACH WORD OR
34	ELS Rank 15	USED	IN	304805	EQUALS	X 666 LETTERS	PROBABILITY FOR	PHRASE HAS 1
35	**Odds against this matrix are about 217.9 trillion to 1.**	ON	TORAH	LETTERS	FREQUENCY	EQUALS	THE WORD	CHANCE IN THE
36		FIGURE		IN THE	PER LETTER	WORD	APPEARING	IN THE FOLLOWING
37				TORAH		EXPECTANCY	AT LEAST ONCE	OF BEING HERE.
38	HITLER (best spelling) @Skips +/-1 and +/-16627	-16627	1	304805	3.28079E-06	0.002185004	0.002182618	458.1653472
39	CLAPPER Skip +1 only (4 SPELLINGS)	1	6	304805	1.96847E-05	0.013110021	0.013024459	76.77862003
40	POSEY @ Skips +/-1 and +/-16627	16627	8	304805	2.62463E-05	0.017480028	0.017328139	57.70960231
41	GAZA @Skips +/-1 and +/-16627	16627	22	304805	7.21773E-05	0.048070078	0.046933004	21.30696774
42	C. WRAY @ Skips +/- 1 and +/- 16627	16627	23	304805	7.54581E-05	0.050255081	0.049013185	20.40267319
43	USA @Skips +/-1 and +/-16627	16627	26	304805	8.53004E-05	0.056810092	0.055226527	18.10724027
44	REBECCA @Skip +1 only	1	31	304805	0.000101704	0.067735109	0.065492017	15.26903659
45	H. BUSH@ Skips +/- 1	16627	35	304805	0.000114828	0.076475123	0.073624041	13.58251988
46	CONSPIRACY @ Skips +/- 1 and +/- 16627	16627	36	304805	0.000118108	0.078660127	0.075645966	13.21947559
47	HOOVER @Skips +/-1 and +/-16627 (2 spellings for HOOVER)	16627	98	304805	0.000321517	0.214130346	0.192756842	5.18788328
48	GRAY @ skips -16,626	16626	160	304805	0.000524926	0.349600564	0.295030376	3.38948149
49	RULE @Skips +/-1 and +/-16627	-1	218	304805	0.000715211	0.476330769	0.378941978	2.638926424
50	BARRY @Skips +1 only	1	445	304805	0.00145995	0.972326569	0.6217979	1.608239591
51	FINAL SIGNIFICANCE FOR 666-LETTER MATRIX = (H38*H39*H40*H41*H42*H43*H44*H45*H46*H47*H48*H49*H50)/15						4.58788E-15	2.17966E+14

Above: Spreadsheet for Figure 4-6. The most significant term matching CIAKAHRR was HITLER which was found at a special case skip against odds of 458 to 1.

Franklin Roosevelt supposedly chose to fund our atomic bomb technology because he thought it was a faster route to victory. While our government kept the lid on UFOs through November 1944, in that month, our planes over Germany saw lights following them at night. The crews of the 415th (Bristol) Night Fighter

Squadron reported seeing (and dog fighting with) 8 to 10 bright orange lights. In the summer of 1945, LT (jg) John F. Kennedy accompanied James Forrestal (later our first Secretary of Defense) to Germany, where they both examined Germany's advanced technology (UFOs). After the Roswell Incident in July 1947, President Truman put together the Majestic-12 Group to oversee the cover-up starting on 24 September 1947. Forrestal was a member. On May 22, 1949, Forrestal either committed suicide or was murdered. His body was thrown out of a window at Walter Reed Hospital. There was no suicide note. Both Kennedy and Forrestal were allegedly in favor of the disclosure of the Roswell Incident. The following individuals were described in the Majestic-12 documents as "designated members" of Majestic- 12 (Note: Although the George Bush Presidential family did much to control the Roswell secrets, they were not related to Vannevar Bush.) The primary reference for this material was Wikipedia.

These were the members of Majestic 12:

Lloyd

Berkner Detlev

Bronk

Vannevar

Bush James

Forrestal

Gordon Gray

Roscoe H.

Hillenkoetter Jerome

Clarke Hunsaker

Donald H. Menzel

Robert M. Montague

Sidney Souers

Nathan F. Twining

The men listed above were American scientists, leaders, and military officers.

THE FOURTH REICH ACCORDING TO DR. MICHAEL SALLA. The following information is taken from Salla's 2018 book, *ANTARCTICA'S HIDDEN HISTORY, Corporate Foundations of Secret Space Programs.* It states that in 1955, as a result of a secret agreement reached between the Eisenhower Administration and a German breakaway group in Antarctica, a transnational corporate space program began to emerge. Personnel from US military contractors in Antarctica allowed this transnational corporate program to steadily grow into a major space power, which would eventually surpass and eclipse the secret space programs run by the US military, and the classified space programs of other nations.

Figure 4-7 above: The axis term is ANTARCTIC. Admiral Richard Byrd lead an expedition there (Operation High-jump) in 1946-1947. After taking a beating from UFOs/USOs he published a warning in a Spanish newspaper (El Mercurio) on March 5, 1947 that described a new threat to the U.S. from hostile aircraft quickly attacking the U.S. by flying in over the North or South Pole. The matrix has UFO at the same absolute skip as ANTARCTIC. NEPHILIM (believed to be Reptilians) are in the open text. FROM THE SKY and POLE are at one skip less than ANTARCTIC and UFO.

	A	B	C	D	E	F	G	H
1	Axis term ANTARCTIC @ Skip -21,123	SKIP	NUMBER	DIVIDE BY	THE QUOTIENT	E QUOTIENT	POISSON EQUATION	EACH WORD OR
2	ELS Rank 1	USED	IN	304805	EQUALS	X 27 LETTERS	PROBABILITY FOR	PHRASE HAS 1
3	This calculation is only for UFO and ANTARCTIC	ON	TORAH	LETTERS	FREQUENCY	EQUALS	THE WORD	CHANCE IN THE
4		FIGURE		IN THE	PER LETTER	WORD	APPEARING	IN THE FOLLOWING
5				TORAH		EXPECTANCY	AT LEAST ONCE	OF BEING HERE.
6	UFO @ Skips +/- 1 and +/- 21123	21123	156	304805	0.000511803	0.013818671	0.013723631	72.867011
7								
8	The next calculation is for areas with a white background.					324 letters		
9	NORDIC @ Skips +7 to +90	7	22	304805	7.21773E-05	0.023385443	0.023114123	43.26359299
10	HITLER (2 spellings) @ Skips +6 to +89	6	73	304805	0.000239497	0.077597152	0.074662879	13.39363664
11	UFO @ Skips +/- 1 and +/- 21123	21123	156	304805	0.000511803	0.165824051	0.152804716	6.54430063
12	FINAL COMBINED ODDS = H9*H10*H11						0.000263705	3792.111478
13								
14	The next calculation is for the full matrix					760 letters		
15	CHIEFS/GENERALS	1	2	304805	6.56157E-06	0.004986795	0.004974381	201.0300208
16	NEPHILIM (2 spellings) @ Skip +1 only	1	4	304805	1.31231E-05	0.00997369	0.009924018	100.7656338
17	POLE @ skips -21122 to -21419	-21122	36	304805	0.000118108	0.089762307	0.085851554	11.64801281
18	NORDIC @ Skips +7 to +176	7	38	304805	0.00012467	0.094749102	0.090398877	11.06208433
19	FROM THE SKY @ Skip -21122	-21122	136	304805	0.000446187	0.339102049	0.287590255	3.477169278
20	HITLER (2 spellings) @ Skips +6 to +175	6	138	304805	0.000452748	0.344088844	0.291134053	3.43484381
21	UFO @ Skips +/- 1 and +/- 21123	21123	156	304805	0.000511803	0.388969997	0.322245398	3.103225098
22	FINAL COMBINED ODDS = H15*H16*H17*H18*H19*H20*H21						1.03369E-08	96740487.02

Figure 4-7 spreadsheet above. Odds against the full matrix are about 96,740,487 to 1.

"U.S. Military-Industrial Complex works with Antarctic-based Fourth Reich. The role of the Dulles brothers in the secret negotiations that culminated in the emergence of the Fourth Reich during the Eisenhower administration cannot be over-estimated. Both John Foster and Allen Dulles had much experience working with German industrialists whose companies they represented when they were employed at Sullivan and Cromwell. The Dulles brothers played an important role in Hitler's ascent to power by advocating the interests of German industrialists who were united in their opposition to Communism and advocacy of a strong nationalist leader."

While Eisenhower was annoyed with the idea that he had to threaten to invade the Area 51/S-4 Area with the First Army to find out exactly what the Deep State/Fourth Reich was up to, the relationship between them and Kennedy was worse. In Salla's (2013 book), *KENNEDY'S LAST STAND* (page 95), Salla writes, *"Forrestal had gone on to become a founding member of Majestic- 12 (MJ-12) while Kennedy became a congressman."* It would have been easy to share classified information about UFOs, and extraterrestrial concerns were likely a direct factor in his sacking as Secretary of Defense and his death two months later.

Why was Kennedy killed, and who ordered it? Salla provides some answers, but his proof is not solid enough to take into a court, especially in a country as corrupt as ours. Salla contends that the main players were members of MJ-12, Project Paperclip, the State Department, and the CIA. On May 28, 1963, Kennedy visited Forrestal's grave. Eleven days before Kennedy's assassination, Soviet Premier Khrushchev accepted Kennedy's offer to share CIA files about UFOs with other U.S. agencies and the Soviet Union. A rice-bowl fight! The top-secret order was issued to CIA Director James McCone. It was designed to block Kennedy from UFO files. The collective files were known as "Project Environment."

Salla hypothesizes that in 1961, the MJ-12 members to approve the Directives were Allen Dulles (MJ-2), Dr. Edward Teller (MJ-2), LT General Marshall Carter (MJ-3), General Curtis LeMay (MJ-4), LT General Gordon Blake (MJ-5), Dr. Detlev Bronk (MJ-6), and Dr. Jerome Hunsaker (MJ-7). Salla believes that Vice-

President Johnson was privy to the operation since at least 1949. The Directive that Kennedy sent to McCone was seen as a direct threat to MJ-12. Salla names CIA Director James Jesus Angelton as the man who gave the order to kill Kennedy. Despite this alleged cabal, Russia largely cooperated with the U.S. in space until 2022. Now they are threatening not only nuclear war but also to bring down the International Space Station (ISS). What's changed? Apparently, until 2022, there was some kind of understanding that preserved the peace between America, Russia, and some extraterrestrial species working with us. Perhaps the ETs have decided that it's more in their interest to go back to China. A limited war between us and the Russians will simplify that goal while reducing our population to make more room for the visitors who are really calling the shots. But a new wildcard is President Trump.

Members of Operation Paperclip.

I believe that I originally found the following names in Wikipedia, however when I checked the list during proof reading the manuscript it appeared that the original list had been altered. The reference numbers that I initially had after each name did not match. If I can locate the correct or original source I will add it to a future edition of this book. I have noticed that Wikipedia often censors or tries to discredit articles backing the existence of UFOs, but I cannot say anything about that here other than note that my traditional readers are often in U.S. or British Intelligence.

Aeronautics and rocketry: Hans Amtmann, Herbert Axster, Erich Ball, Oscar Bauschinger, Hermann Beduerftig[, Rudi Beichel, Anton Beier, Herbert Bergeler, Magnus von Braun, Wernher von Braun, Ernst Czerlinsky, Theodor Buchhold [de], Walter Burose[, Adolf Busemann, GN Constan, Werner Dahm, Konrad Dannenberg, Kurt H. Debus, Gerd De Beek, Walter Dornberger – head of rocket programme, Gerhard Drawe, Friedrich Duerr, Ernst R. G. Eckert, Otto Eisenhardt, Krafft Arnold Ehricke, Alfred Finzel, Edward Fischel, Karl Fleischer, Anton Flettner, Anselm Franz, Herbert Fuhrmann, Ernst Geissler, Werner Gengelbach, Dieter Grau, Hans Gruene, Herbert Guendel, Fritz Haber, Heinz Haber, Karl Hager, Guenther Haukohl, Karl Heimburg, Emil Hellebrand, Gerhard B. Heller, Bruno Helm, Rudolf Hermann[Bruno Heusinger, Hans Heuter, Guenther Hintze, Sighard F. Hoerner, Kurt Hohenemser, Oscar Holderer, Helmut Horn, Hans Henning Hosenthien [de], Dieter Huzel, Walter Jacobi, Erich Kaschig, Ernst Klauss, Theodore Knacke, Siegfried Knemeyer, Heinz-Hermann Koelle, Gustav Kroll, Willi Kuberg Werner Kuers, Hermann Kurzweg, Hermann Lange, Hans Lindenberg, Hans Lindenmayer, Alexander Martin Lippisch (Aeronautical Engineer), Robert Lusser, Hans Maus, Helmut Merk, Joseph Michel, Hans Milde, Heinz Millinger, Rudolf Minning, William Mrazek, Hans Multhopp], Erich Neubert, Hermann Oberth, Hans von Ohain (designer of German jet engines), Robert Paetz, Hans Palaoro,

Hermann Oberth,

Kurt Patt, Hans Paul, Fritz Pauli, Arnold Peter, Helmuth Pfaff, Theodor Poppel, Werner Rosinski, Heinrich Rothe, Ludwig Roth, Arthur Rudolph, Friedrich von Saurma [de], Edgar Schaeffer, Martin Schilling, Helmut Schlitt, Albert Schule, August Schulze, Walter Schwidetzky, Ernst Steinhoff, Wolfgang Steurer, Heinrich Struck, Ernst Stuhlinger, Bernhard Tessmann, Adolf Thiel, Georg von Tiesenhausen, Werner Tiller, JG Tschinkel, Arthur Urbanski, Fritz Vandersee[, Richard Vogt, Woldemar Voigt (designer of Messerschmitt P.1101), Werner Voss, Theodor Vowe, Herbert

A. Wagner, Hermann Rudolf Wagner, Hermann Weidner, Georg Rickhey – director of the slave labour Mittelwerk factory, Walter Fritz Wiesemann, Philipp Wolfgang Zettler-Seidel, (see List of German rocket scientists in the US).

Architecture: Heinz Hilten and Hannes Luehrsen

Electronics – including guidance systems, radar and satellites: Wilhelm Angele [de], Ernst Baars [de], Josef Boehm, Hans Fichtner, Hans Friedrich, Eduard Gerber, Georg Goubau, Walter Haeussermann, Otto Heinrich Hirschler, Otto Hoberg, Rudolf Hoelker, Hans Hollmann, Helmut Hölzer, Horst Kedesdy, Kurt Lehovec, Kurt Lindner, JW Muehlner, Fritz Mueller, Johannes Plendl, Fritz Karl Preikschat, Eberhard Rees, Gerhard Reisig, Harry Ruppe, Heinz Schlicke, Werner Sieber, Othmar Stuetzer, Albin Wittmann, Hugo Woerdemann. Albert Zeiler, Hans K. Ziegler

Material Science (high temperature): Klaus Scheufelen [144] and Rudolf Schlidt.

Medicine – including biological weapons, chemical weapons, and space medicine: Theodor Benzinger [de], Rudolf Brill [de], Konrad Johannes Karl Büttner, Richard Lindenberg, Ulrich Cameron Luft [de], Walter Schreiber, Hubertus Strughold, Hans Georg Clamann, and Erich Traub.

Physics: Gunter Guttein, Gerhard Schwesinger, Gottfried Wehner, Helmut Weickmann and Friedwardt Winterberg.

Chemistry and Chemical engineering: Helmut Pichler, Leonard Alberts, Ernst Donath, Josef Guymer, Hans Schappert, Max Josenhaus, Kurt Bretschneider,Erich Frese

Similar Operations

- **APPLEPIE**: Project to capture and interrogate key Wehrmacht, RSHA AMT VI, and General Staff officers knowledgeable of the industry and economy of the USSR.

- **DUSTBIN** (counterpart of ASHCAN): An Anglo-American military intelligence operation established first in Paris, then in Kransberg Castle, at Frankfurt.

- **ECLIPSE** (1944): An unimplemented Air Disarmament Wing plan for post-war operations in Europe for destroying V-1 and V-2 missiles.

- **Safehaven**: US project within ECLIPSE meant to prevent the escape of Nazi scientists from Allied-occupied Germany.

- **Field Information Agency; Technical (FIAT):** US Army agency for securing the "major, and perhaps only, material reward of victory, namely, the advancement of science and the improvement of production and standards of living in the United Nations, by proper exploitation of German methods in these fields"; FIAT ended in 1947, when Operation Paperclip began functioning.

- **On April 26, 1946**, the Joint Chiefs of Staff issued JCS Directive 1067/14 to General Eisenhower instructing that he "preserve from destruction and take under your control records, plans, books, documents, papers, files and scientific, industrial and other information and data belonging to … German organizations engaged in military research"; and that, excepting war-criminals, German scientists be detained for intelligence purposes as required.

- **National Interest/Project 63**: Job placement assistance for Nazi engineers at Lockheed, Martin Marietta, North American Aviation, and other airplane companies, whilst American aerospace engineers were being laid off work.

- **Operation Alsos, Operation Big, Operation Epsilon, Russian Alsos**: Soviet, American, and British efforts to capture German nuclear secrets, equipment, and personnel.

- **Operation Backfire**: A British effort at recovering rocket and aerospace technology, followed by assembling and testing rockets at Cuxhaven.

- **Fedden Mission**: British mission to gain technical intelligence concerning advanced German aircraft and their propulsion systems.

- **Operation Lusty**: US efforts to capture German aeronautical equipment, technology, and personnel.

- **Operation Osoaviakhim** (sometimes transliterated as "Operation Ossavakim"), a Soviet counterpart of Operation Paperclip, involved German technicians, managers, skilled workers, and their respective families who were relocated to the USSR in October, 1946.

- **Operation Surgeon**: British operation for denying German aeronautical expertise to the USSR, and for exploiting German scientists in furthering British research.

- **Special Mission V-2**: April–May 1945 US operation, by Maj. William Bromley recovered parts and equipment for 100 V-2 missiles from a Mittelwerk underground factory in Kohnstein within the Soviet zone. Major James P. Hamill coordinated the transport of the equipment on 341 railroad cars with the 144th Motor Vehicle Assembly Company, from Nordhausen to Erfurt, just before the Soviets arrived. (See also Operation Blossom, Broomstick Scientists, Hermes project, Operations Sandy and Pushover)

- **Target Intelligence Committee**: US project to exploit German cryptographers.

Nazi /Alien Plan to Conquer the U.S.

When I went looking for a picture to add to Figure 4-8, I had not yet heard of the book *Alien World Order*, or of its author, Len Kasten. First, I found the matrix, then I found Kasten's book and saw that it described most of my beliefs about UFOs and Nazis. That's not to say that I agree with everything that he says. He blames Reptilians for just about every evil act that anyone ever did. He offers no proof for the claim. He needs to supply it.

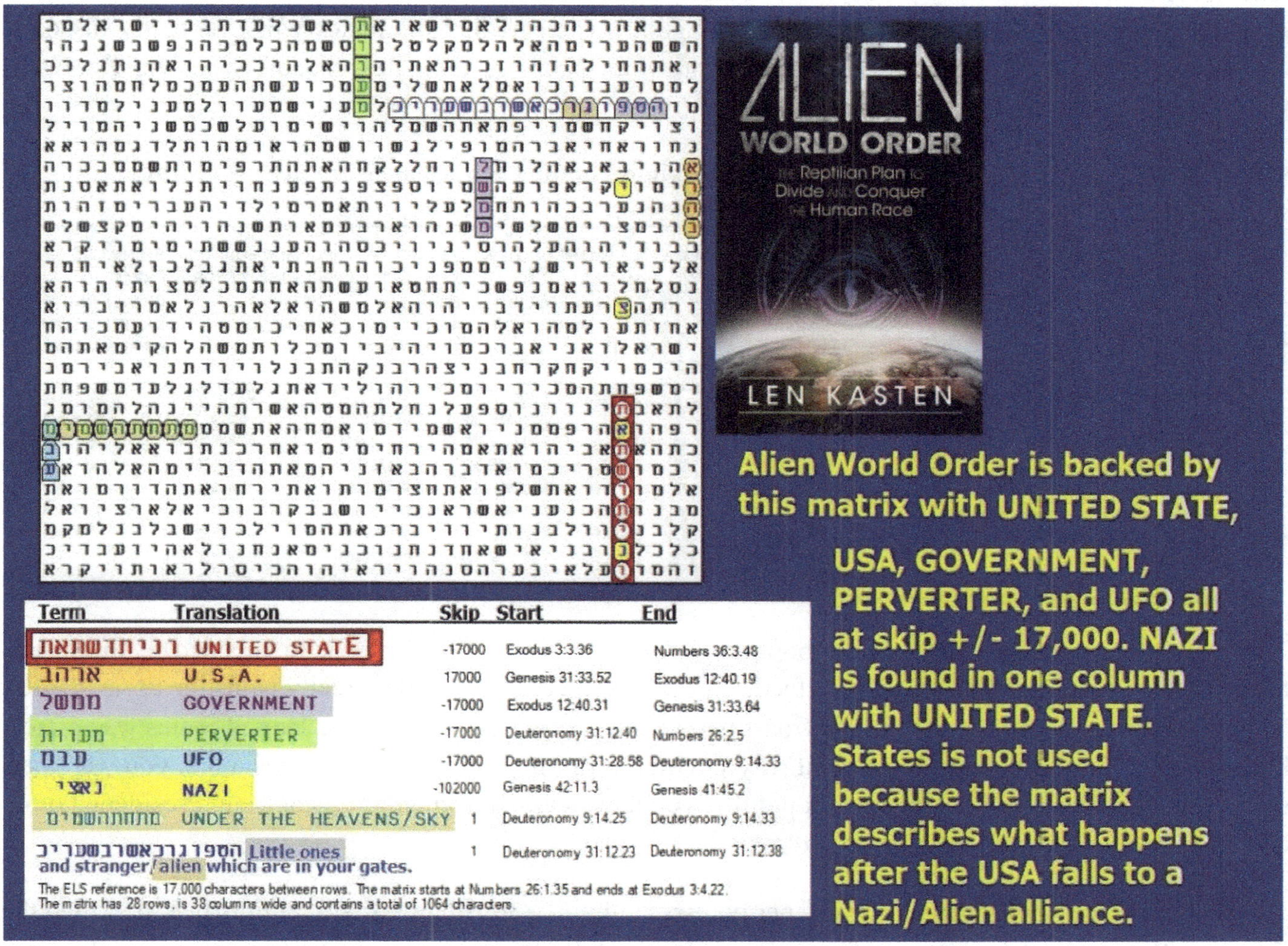

Figure 4-8 above. The axis term **UNITED STATE** does not bode well for America.

When you look at the spreadsheet for Figure 4-7, notice that the two most significant keywords are **PERVERTER** and **LITTLE ONES.** The fact that **UNITED STATE, USA, GOVERNMENT, PERVERTER,** and **UFO** are all at the same absolute skip probably is not an accident. We need to eliminate the gay/transgender leadership from our government. President Trump began to do just that almost immediately by declaring that our government will (under his term) only recognize two genders – male and female.

The word **SKY** touches **UFO.** A hard-to-accept truth is attached to the use of **UFO** here. Aliens are here now. They work with Nazis or their descendants who, via Operation Paperclip, have penetrated our government. Based on two people that I have met who claim to be aliens (one of them, Craig Ebrahimi, claims it directly, with the other – (Rebecca in Lakeland, Florida), we need to be very careful about what her presence implies. While the woman seemed innocent enough, I have a lot of evidence in this book that points to Craig as the original source of COVID-19. There is evidence that suggests aliens have infiltrated our government, but it is easier to show that Nazis have played major roles there too. While not statistically spectacular, the word **NAZI** is parallel to the five terms at a skip of +/- 17,000. Two of its letters are found inside the **UNITED STATE**, which may be singular to tell us that the states will eventually dissolve, leaving us with one Nazi state. The leaders work with Nazis or their descendants who, via Operation Paperclip, have penetrated our government.

Spreadsheet for Figure 4-8 below. Odds against finding Figure 4-8 by chance are over 1.7 million to 1.

	A	B	C	D	E	F	G	H
1	Axis term UNITED STATE @ Skip -17,000	SKIP	NUMBER	DIVIDE BY	THE QUOTIENT	E QUOTIENT	POISSON EQUATION	EACH WORD OR
2	ELS RANK = 1	USED	IN	304805	EQUALS	X 1064 LETTERS	PROBABILITY FOR	PHRASE HAS 1
3		ON	TORAH	LETTERS	FREQUENCY	EQUALS	THE WORD	CHANCE IN THE
4		FIGURE		IN THE	PER LETTER	WORD	APPEARING	IN THE FOLLOWING
5				TORAH		EXPECTANCY	AT LEAST ONCE	OF BEING HERE.
6								
7	PERVERTER @ Skip +/1 and +/- 17,000 only	-17000	3	304805	9.84236E-06	0.010472269	0.010417626	95.99116091
8	LITTLE ONES @ Skip +1 only	1	7	304805	2.29655E-05	0.024435295	0.02413917	41.42644549
9	GOVERNMENT @ Skip +/1 and +/- 17,000 only	-17000	18	304805	5.90541E-05	0.062833615	0.060900287	16.42028383
10	U.S.A. @ Skip +/1 and +/- 17,000 only	17000	29	304805	9.51428E-05	0.101231935	0.096276597	10.38674023
11	UFO @ Skip +/1 and +/- 17,000 only	-17000	179	304805	0.000587261	0.624845393	0.46465581	2.152130628
12	ALIEN @ Skip (+1) only (4 synonyms sought)	1	635	304805	0.002083299	2.216630305	0.891024294	1.122303855
13	NAZI @ Skips 102000 to 102473	102000	825	304805	0.002706649	2.879874018	0.943858165	1.05948122
14	SONS/CHILDREN @ Skip +1 only	1	1172	304805	0.003845081	4.091166483	0.983280281	1.017004021
15	FINAL COMBINED ODDS =H7*H8*H9*H10*H11*H12*H13*H14						5.66547E-07	1765077.834

The first thing you'll probably notice about the 4-8 axis term (the first term found encoded) is that it is a transliteration for **UNITED STATE** instead of **UNITED STATES**. While it really pains me to say it, after 34 years of military service in the U.S. Coast Guard and Navy, I believe that the United States was almost killed when Biden was elected President. If you voted for him or any Democrat legislators, you were part of the conspiracy to murder America. Hopefully, you have learned from your mistake and President Trump can resurrect it. What am I talking about?

Think about it. Nations go to war over energy, especially oil. On Biden's first day in office, he killed the Keystone Pipeline and America's energy independence. Biden then stopped construction of our border wall. The Chinese, who helped bring us COVID, also mass-murder Americans by mass-producing Fentanyl. They ship it to the drug cartels in Mexico, who then smuggle it into what was the United States. Specifically, they brought it in via holes in the unfinished wall that were unfinished because of all the Trump haters who were more interested in throwing out Trump than in electing someone who is loyal to America. The result is that if you voted for Biden, you voted to poison and, in most cases, kill over 100,000 Americans per year – about twice as many Americans as were killed in the entire Vietnam War.

Is the Code's Author trying to tell us anything else by using the absolute skip of the axis term? Yes. **GOVERNMENT** is also at skip -17,000. This is also true for **PERVERTER** and **UFO**. What is our government perverting? Our **LITTLE ONES**, which are in the open text of a line that has the first letter of **PERVERTER.** The reference to **LITTLE ONES** is from Deuteronomy 21:12. Looking at several different translations with options capitalized, this verse begins with:

Gather together the people – the men, the women, and the LITTLE ONES, SMALL CHILDREN or CHILDREN, and the ALIEN, STRANGER, SOJOURNER, or FOREIGNER who is WITHIN YOUR GATES, IN YOUR TOWNS or IN YOUR CITIES.

REVIEW OF ALIEN WORLD ORDER.

When looking for a relevant picture to add to Figure 4-1, I came across Len Kasten's book. It seemed to fit my findings and beliefs at the time exactly, so I ordered the book and doubled my efforts to find a complete matrix that summarized the situation. That's Figure 4-1 above. It may grow in size. My summary of Kasten's book is below. It is based on what is found on the back cover of his book, although there seems to be a chapter missing or expunged (by the CIA?). I have written to the publisher to ask for clarification. The book should discuss the assassination of President Kennedy, but he is missing except for on the back cover. My comments are given in italics. As of August, 2025, the publisher has not responded to my 2022 email or phone calls.

PUBLISHER: *Shares new documents to expose the sinister alien influence in world governments, financial systems, and scientific institutions throughout history.*

ROFFMAN: Attention needs to be paid to which governments besides our own have contact. If we don't know who does, we don't know who could gain nuclear weapons or other WMD from ETs. We could experience a false flag war where we would make an incorrect choice and be wrongly dragged into World War 3. This was apparently actually attempted on January 13, 2018, when, on their cell phones, Hawaiians received warnings of inbound missiles. It took 38 minutes to tell them that they would survive what they were then told was a false alarm; however, according to Michael Salla, there were many who said that they actually saw the inbound missile or missiles shot down. Allegedly, the attack was initiated by a Nazi-Reptilian Dark Fleet Alliance.

PUBLISHER: *Shows how Eisenhower's treaty with the Greys, signed at Holloman Air Force Base in 1954, gave the aliens authority to abduct humans for "research."*

ROFFMAN: That research, allegedly, included how to make hybrid-humans. I have met two people (Craig Ebrahimi and Rebecca in Lakeland, Florida) who claim to be from other worlds. The woman said that in Earth orbit, she personally was permitted to see a spaceship carrying small

grays and hybrids. They were screaming in panic. Tall grays told her that their behavior was due to the knowledge that Reptilians would use them for food.

PUBLISHER: *Reveals how Reptilian-influenced ex-Nazis infiltrated the U.S. government*.

ROFFMAN: This began with Operation Paperclip, which brought 1,600 Nazi scientists to America. The top ones are listed above in this chapter. It included Wernher Von Braun, who built America's Saturn 5 moon rocket. We are going to need definitive evidence about how this infiltration impacted the Kennedy assassination. President Trump signed an Executive Order to release classified information about the

Assassination. Trump said he thinks Lee Harvey Oswald killed former President John F. Kennedy, but he didn't act alone. While he said that more than one person participated in the assassination, he didn't link it to UFOs.

PUBLISHER: *Explains how the Reptilians have created alien-human hybrids under their control to replace the human population.*

ROFFMAN: Aside from using some of these hybrids for food, the important thing to understand is that there are aliens on Earth now who are working to replace/murder us. While it may be true that not all aliens are hostile, this assumption must be proven. Of the two likely aliens that I know personally, the woman seems very friendly, but the other one (Craig) has given me evidence that suggests he is responsible for getting the COVID-19 virus to the Chinese Army. The evidence is well-documented on my website and in this book. So far (in August, 2025), this has cost over 18 million human lives. More than a million of them were Americans. For us to accept the "good" ETs, they need to identify themselves and register with their local governments. If they accept these conditions, laws need to be enacted to protect them. But if large numbers ignore these requirements, I suspect that the number of ETs on Earth will be well above a million. Due to the large number of people who go missing here every year, we're also going to need honest information about previous and recent or current abductions.

PUBLISHER: *Long ago, the Galactic Federation sent the Atlans, a fierce group of humans from the Pleiades, to Earth to confront the newest colony of the combative, ever-spreading alien race known as the Reptilians. The ensuing battle sank the continent of Lemuria and drove the Reptilians underground beneath the Indian subcontinent as well as to Antarctica. Able to shapeshift to appear human, the Reptilians then infiltrated the Atlantean civilization, abducting humans and creating Reptilian hybrids. Len Kasten contends that infiltration and hybridization prior to an all-out attack have been the hallmark of Reptilian conquest throughout the galaxy.*

Chronicling the history of the Reptilian Empire's influence on Earth and their conquest of 21 star systems, Kasten reveals how the human race is enmeshed in a skillfully concealed plot to enslave humanity and exploit our planet's physical and biological resources. Revealing Hitler's pact with extraterrestrials and the Reptilian influence on the Nazi state, he shares new documents that disclose the rescue and rehabilitation of Nazi war criminals to assist in the Cold War, which then corrupted many U.S. government institutions. Focusing on crucial events in the decade after World War II, he examines the Reptilians' human allies, the Illuminati, who control the levers of

financial, technological, and military power throughout the world through various secret societies. He shows how Eisenhower's treaty with the Greys, signed at Holloman Air Force Base in 1954, gave the aliens permission to take humans up to their spaceships, ostensibly for genetic study, in return for alien technology, and how these abductions led to the creation of a hybrid race under Reptilian mind control.

ROFFMAN: Here, Kasten touches on my greatest concern. For most of my 78-year life in presidential elections, I see-sawed back and forth between our two major parties. But then Obama was elected, and the

Democrat Party took an extreme hard left turn.

When Biden ended the sanity that had made an all-too-brief return under Trump, it was obvious to anyone with half a brain that from Day One (January 20, 2021) when Biden took over and destroyed our energy system, thus ensuring insane inflation, Biden had a single goal – to destroy the United States. Over and over again, this nutcase idiot destroyed our Southern Border, let in an estimated minimum of 21,000,000 unknown people, including criminals, terrorists, and people infected with COVID, and he surrendered in Afghanistan for no reason. He armed the Taliban with

$85 billion in arms so they could slaughter the allies (like translators) and hundreds of Americans that he left behind. We all watched as those abandoned climbed onto the sides of jet transports during takeoff efforts. We saw them fall to their deaths as the planes tried to flee the hell Biden had created. I could go on, but there was no concern by Biden for the demonic world that he rushed to make worse on a daily basis. And yet… most Democrats showed zero comprehension of what was going on. Worse still, somehow America managed to temporarily increase the strength of Democrats controlling the Senate, although (thank God) we were able to vote Pelosi and her Democrat/Communist comrades out of control of the House. Even better, Vice President Harris was squashed in he 2024 election.

About half of America has plainly lost its mind. You cannot reason with them. A symptom of their illness is that they refuse to debate anyone who doesn't agree with their extremist positions. They cannot begin to comprehend the word *patriotism.* If you try to use the term to justify any Republican point of view, they will only hear the word *fascism.*

I know that the knee-jerk response to my charge (and Kasten's, too) that the Democrats have lost control of their minds is that we are the crazy ones. In fact, the idea that people who see UFOs are crazy is a policy that the CIA came up with in 1947 to suppress the Roswell event and similar events later. The CIA was formed on September 18, 1947 (less than 3 months after Roswell).

A more critical analysis of the Kasten ideology is that there is no accepted geological evidence of the existence of a sunken continent named Lemuria.

PUBLISHER: *He also explains how Kennedy was assassinated by the CIA because of his plans to rout out this Nazi–Reptilian presence.*

ROFFMAN: Something appears to be very wrong here! I received my copy of *Alien World Order* on December 8, 2022. Other than the above exact words on the back cover, I did NOT see any reference to Kennedy in the Table of Contents, Index, or anywhere else in the book after reading

it. Is it possible that the CIA could have printed their own censored version of the book and then substituted it for the real deal??? Or are we just dealing with a publisher who accidentally cut out a chapter, or fraud by someone who wanted to increase sales via a little false advertising?

There was a lot of talk about how the CIA became overrun by Nazis. Pages 255 to 256 quote Eisenhower's farewell address on January 17, 1961. In it, he harshly condemned the military- industrial complex. But he never spoke about who it was in them that constituted the threat to our liberties or democratic processes. After quoting the alarming part of Ike's speech, Kasten writes,

"What Ike didn't say, and what he couldn't say, but what he probably knew, was that the military- industrial complex was now riddled with ex-Nazis with an agenda in service to the Fourth Reich!"

That's how Chapter 20 ends. Chapter 21 is entitled *The Fourth Reich.* The CIA is in it. Both Allen Dulles and his brother John Foster Dulles are in it, too. But there is no mention of Kennedy or James Jesus Angleton in it. I actually looked for Angleton first because in *Kennedy's Last Stand* (2013), author Michael Salla named him as the man who ordered the assassination. What positions did the characters above hold? Allen Dulles was Director of the CIA from February 26, 1953, to November 29, 1961. John Foster Dulles was Secretary of State from January 26, 1953, to April 22, 1959. James Jesus Angleton (December 9, 1917 – May 11, 1987)[1] was chief of counterintelligence for the CIA from 1954 to 1974. His official position within the organization was Associate Deputy Director of Operations for Counterintelligence (ADDOCI).

The failure to mention Kennedy looks like a crime (fraud), but it remains to be seen whether it was committed by the publisher to sell more books, or by the CIA or FBI to hide the truth about the Kennedy Assassination. I attempted to call the publisher (Bear & Bear in Rochester, VT) about seven times. Tracy took calls before 2 PM at an answering service. Calls after 2 PM were taken by Cindi, who works for the publisher. She first gave me an email address for Jon, who would know what happened. But when I wrote to him, all I got was a Delivery Status Notification (Failure). On December 19, 2022, I spoke to Cindi again about this problem. All she would do about it was to promise to forward a letter about the problem to the author, Len Kasten. As for Jon, she said that since the COVID-19 pandemic struck, he worked from home, as did most company employees. My letter went unanswered, and Cindi would not discuss the matter anymore.

As the government has part of the Kennedy investigation classified as TOP SECRET since 1963, I bet that they have committed this crime to protect the assassins and keep the cover-up about aliens going, but this remains to be proven. The initial finding comes at a time when Elon Musk is exposing how Twitter lied to us to protect government criminals (including senile President Biden, who should have been impeached).

The Smithsonian Magazine on December 16, 2022, reports that, "The government is still withholding 515 full (classified) documents, as well as parts of 2,545 documents… Officials were supposed to review the remaining records and determine whether anything else could be released by June 2023, reports the Hill's Zach Schonfeld. A 1992 law required the release of all government records pertaining to the assassination by October 2017, but Biden and President Donald Trump

pushed that deadline back." When information was finally released during Trump's second term, there was nothing sensational exposed.

PUBLISHER: *Contrasting the Reptilians with the benevolent Ebens, the aliens from Zeta Reticuli who crashed at Roswell, Kasten exposes the stealthy tactics of the Reptilians, their relationship with the Greys, and their advanced genetic bio-technology and teleportation abilities, as well as what we need to do to defeat their plans.*

ROFFMAN: Evidently, our government also has teleportation abilities. See the U.S. Air Force Teleportation Physics Study (AFRL-PR-ED-TR-2003-0034) by Eric W. Davis.

It sounds like the woman in Lakeland claiming to be an alien has experience with teleportation, though it was instituted by ETs rather than human Earthlings (Pleiadians are also allegedly human). She spoke about people fleeing a planet destroyed or about to be destroyed in the Pleiades. The star she referred to is Miai, which is about 380 +/- 10 light years from Earth. She did not know the name of the planet or who destroyed it. Len Kasten names three planets that were destroyed in the Lyran star system: Bila, Teka, and Merok. The brightest star in the Lyran system is Vega, which is only about 25.04 +/- 0.07 light years from Earth. Referring to the work of Stewart Swerdlow, Reptilians get the blame for destroying not just Bila, Teka, and Merok, but also in our solar system, a planet (Maldek) between Mars and Jupiter that subsequently formed the asteroid belt, plus enough of the Martian atmosphere to drive people there underground.

Kasten claims the Reptilians have the capability and have used it to hollow out a planet and turn it into a death star complete with its own biosystem. Dinosaurs were brought along as food. For what it's worth, Andrew Basiago claims to have been a time traveler for the U.S. Government's Project Pegasus and a passenger to Mars via so-called "jump rooms." Basiago also claims that he saw at least one astronaut eaten alive by a dinosaur (about 16 feet tall) on Mars. The belief that there is a common human-alien base on Mars is widespread throughout the UFO believers' community.

The idea of Reptilians taking dinosaurs to Mars – a huge version of panspermia – is consistent with NASA/JPL photographs of primitive life taken on Mars and seen by MSL Curiosity and MER Opportunity. What we see looks like bacteria, algae, fungi, including puff balls, lichens, and stromatolites. For proof of this, see the 2019 R. Joseph article: EVIDENCE OF LIFE ON MARS? The Journal of Astrobiology asked my son (David Altair Roffman, PhD, physics) and me to write a review of the article. It is METEOROLOGICAL IMPLICATIONS: EVIDENCE OF LIFE ON MARS. See Chapter 8 of this book.

It might be possible for one form of primitive life to make it from Earth to Mars via an asteroid impact on ancient Earth and an assist from solar wind. But for multiple primitive species to make it there, an assist by intelligent life is almost certainly needed. Basiago spoke about something like cows being there, too. For that (and dinosaurs), a spacecraft or a jump room to get there would be an imperative. The jump room spoken about by Basiago was, she says, also used by the Lakeland lady. Reptilians (and Grays) get a lot of credit for this technology. But Reptilians, who supposedly work with our government now, once worked for Hitler. It appears that via the Fourth Reich, they still serve evil. That evil includes blowing up civilizations. The only way to keep them in place is to inform our people of the truth, right up to and including the Kennedy assassination. But this may mean that we have to pay for it with a civil war.

The FBI is not the only force in our government that has been out of control. But, along with the IRS, which was expanded by Biden's addition of 87,000 more agents who will only serve to further all corruption in our government, they are the best place to start in the effort to drain the Swamp. However, know this: on Figure 4-8 below, the term **UFO** touches **TRUMP RAIDED**. Trump supposedly wanted to disclose their presence on Earth, but listened to their pleas to continue the cover-up. That was a fatal mistake. The UFO cover-up is at the core of the rot in our government. It allows the Fourth Reich, enabled by Operation Paperclip in 1947, to continue its Nazi infiltration and control of everything our nation did under Biden and Obama.

By the way, more study is needed, but there are 9-letter ELS trains in Torah meaning "*Nazi Obama*" (skips -52,940, 57,940, and 126,613), plus ELS trains meaning "*Obama (is) Nazi*" at skips 15 and 11,340, but I only saw some of them pair well with Reich. No such match existed between Biden and Nazi, but there are a few matrices in Chapter 3 pointing to Obama and Biden being aliens or being brain washed by them (see Figures 3-1, 3-4 and 3-5). Figure 3-1 has an axis term of "alien in our government" at skip -81,218. This is at the same absolute skip as Biden and UFO. Figure 3-4 has the axis term "Biden (is) alien" at skip -97,832, which is the same absolute skip as Mars, enemy, and UFO. Figure 3-5 is based on Obama is reptilian. The code is quite dependable when you know how to read and use it. From his birth on, Obama's youth is clouded in mystery. It would not be too hard to believe that he was a reptilian who slipped in through our defenses. But Biden was well known for at least 50 years. In the White House he seems to have come down with a case of scrambled brains. He often looked and acted like he didn't know what planet he was on. This may indicate that he was a human who was abducted and subjected to mind control. He wasn't always like this. In fact, when he sounded more conservative, I think I actually briefly supported him when he was a candidate for president in 1987. But while he was regarded as potentially one of the strongest candidates in the field, in September 1987, reports emerged that he had plagiarized a speech by the British Leader of the Opposition and Labor Party Leader, Neil Kinnock. Other allegations of past law school plagiarism and exaggerating his academic record soon followed and Biden withdrew from the race later that month.

NAZI RAID ON MAR-A-LAGO. The date 8-8-2022 will go down in history like 9-11-2001. It was the day that the Nazis (Fourth Reich) felt secure enough in running our government to storm President Trump's house. They stole the 2020 election. The raid was a follow-up exercise aimed at ensuring that Trump would neither get nominated nor elected. It will be discussed in greater detail after analyzing Figure 4-8 and its probability spreadsheet.

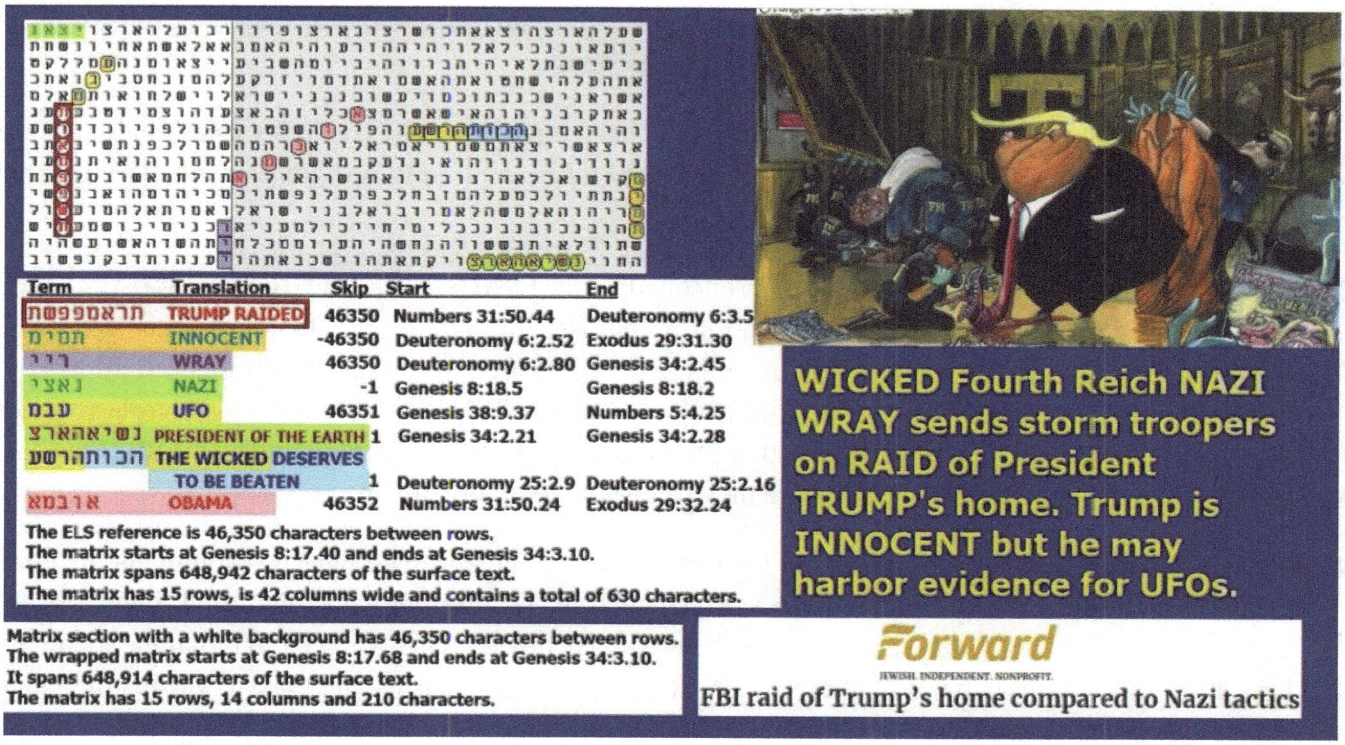

Figure 4-9 above. The spreadsheet for it is below. The axis term is **TRUMP RAIDED**. Odds against finding this matrix were about 333,345 to 1 for the full (638-letter) matrix but just 5,210 to 1 for the 210-letter portion that includes **NAZI** at skip -1 above Trump, **WRAY** at the same skip as (but opposite direction to) **TRUMP RAIDED**, and UFO at the same skip as **TRUMP RAIDED**. Here's the spreadsheet:

	A	B	C	D	E	F	G	H
1	Axis term TRUMP RAIDED @ Skip 46,350	SKIP	NUMBER	DIVIDE BY	THE QUOTIENT	E QUOTIENT	POISSON EQUATION	EACH WORD OR
2	ELS RANK 1	USED	IN	304805	EQUALS	X 210 LETTERS	PROBABILITY FOR	PHRASE HAS 1
3	This calculation is only for the 210 letters with a	ON	TORAH	LETTERS	FREQUENCY	EQUALS	THE WORD	CHANCE IN THE
4	white background.	FIGURE		IN THE	PER LETTER	WORD	APPEARING	IN THE FOLLOWING
5				TORAH		EXPECTANCY	AT LEAST ONCE	OF BEING HERE.
6								
7	NAZI @ Skip +/- 1 and+/- 46350	1	2	304805	6.56157E-06	0.00137793	0.001376981	726.2263053
8	WRAY @ Skips +/-1 and +/- 46350	46350	370	304805	0.001213891	0.254917078	0.225019242	4.444064391
9	UFO @ Skips +46351 to +46378	-1	1402	304805	0.004599662	0.965929037	0.619370583	1.61454229
10	COMBINED ODDS AGAINST = H7*H8*H9						0.00019191	5210.768077
11								
12	Axis term TRUMP RAIDED @ Skip 46,350	SKIP	NUMBER	DIVIDE BY	THE QUOTIENT	E QUOTIENT	POISSON EQUATION	EACH WORD OR
13	ELS RANK 1	USED	IN	304805	EQUALS	X 630 LETTERS	PROBABILITY FOR	PHRASE HAS 1
14	Not included in the calculation because they were	ON	TORAH	LETTERS	FREQUENCY	EQUALS	THE WORD	CHANCE IN THE
15	noticed *a posteriori* were the terms DESERVES TO BE	FIGURE		IN THE	PER LETTER	WORD	APPEARING	IN THE FOLLOWING
16	BEATEN, OF THE EARTH and OBAMA (not the primary			TORAH		EXPECTANCY	AT LEAST ONCE	OF BEING HERE.
17	spelling and not at a special case skip).							
18	NAZI @ Skip -1	1	2	304805	6.56157E-06	0.00413379	0.004125258	242.4090746
19	WICKED @ Skip +1	1	25	304805	8.20197E-05	0.051672381	0.050360064	19.85700425
20	PRESIDENT@ Skip +1 only	1	68	304805	0.000223093	0.140548876	0.131118803	7.626671204
21	INNOCENT @ Skip -46350	-46350	119	304805	0.000390414	0.245960532	0.218046914	4.586169018
22	WRAY @ Skips +/-1 and +/- 46350	46350	370	304805	0.001213891	0.764751234	0.534550296	1.870731357
23	UFO @ Skips +46351 to +46378	-1	1402	304805	0.004599662	2.89778711	0.944854885	1.058363582
24	COMBINED ODDS AGAINST = H18*H19*H20*H21*H22*H23						2.99989E-06	333345.0387

What did the then-corrupt FBI gain by its robbery of Trump's house? Before their crime, although I had twice voted for Trump, I wasn't thrilled with his performance on January 6, 2021, but I thought Biden stole the 2020 election from Trump. Although I once admired the FBI, I came to view them as a Fourth Reich asset, loyal to all things Nazi (and/or Communist or Reptilian). I am pleased to see President Trump replace Wray with Kash Patel in 2025. I hope he will totally clean house.

The FBI is not the only force in our government that has been out of control. The IRS was expanded by the addition of 87,000 more agents who only serve to further all corruption in our government. Trump fired many of the worst in his first two weeks back in office.

On Figure 4-9 above, **TRUMP RAIDED** is encoded with **INNOCENT** and (Chris) **Wray.** Trump supposedly wanted to disclose their presence on Earth, but listened to their pleas to continue the cover-up. That was a near-fatal mistake. The UFO cover-up is at the core of the rot in our government. It allows the Fourth Reich, enabled by Operation Paperclip in 1947, to continue its Nazi infiltration and control of almost everything our nation does. There were two attempts to assassinate Trump in 2024 and one in 2021 at Mar-A-Lago. He must wipe these people out before they can try to get to him again.

REAL FATE AND OUR OWN DESTINY. One of the major concerns that I have deals with the outcome of World War II. There is no question that the U.S. defeated Japan. A Hitler who committed suicide would also imply that we defeated Nazi Germany as well. But since 1981, I have believed that that the man who shot himself in the head in Hitler's bunker was not Hitler. As I wrote earlier, instead, it was his duplicate and cook, August Wilhelm Bartholdi.

Now, returning to 2025, all of my research has led me to believe that Hitler not only survived World War II, but he led a very productive (although quiet) life in South America and Germany's underground fortifications in Antarctica. In conjunction with over 1,600 German/Nazi scientists brought to the United States under Operation Paperclip between 1945 and 1959, they slowly took over control of the American government. Many were familiar with aliens and alien technology, and some likely contributed to our research into time travel. It was, if we believe Michel Salla's book *Kennedy's Last Stand: Eisenhower, UFOs, MK-12, and JFK's Assassination* (2013), a Fourth Reich American working for the CIA (James Jesus Angleton, Chief of Counter- Intelligence) who ordered the assassination of President Kennedy because the President wanted to disclose the presence of aliens on Earth.

Three days before Kennedy was sworn into office, President Eisenhower made his farewell speech on TV. In it (on January 17, 1961), he said:

"In the councils of government, we must guard against the acquisition of unwarranted influence, whether sought or unsought, by the military-industrial complex. The potential for the disastrous rise of misplaced power exists and will persist. We must never let the weight of this combination endanger our liberties or democratic processes. We should take nothing for granted. Only an alert and knowledgeable citizenry can compel the proper meshing of the huge industrial and military machinery of defense with our peaceful methods and goals, so that security and liberty may prosper together."

The above speech should be compared with that of Secretary of Defense Donald Rumsfeld on September 10, 2001. In the building that would be attacked the following day, along with the World Trade Center and four airliners, he identified our greatest adversary as the Pentagon bureaucracy! The full 11-minute 42-second speech is at https://www.youtube.com/watch?v=MfMjdKElgqY. His identification of this enemy comes at 1 minute 34 seconds into the video. Meanwhile, if we believe Andrew Basiago, when Rumsfeld made his 2001 speech, he actually knew of the upcoming attack, and not just by a few hours before the planes smacked into their targets. Rumsfeld, we are told, knew about the pending disaster and its date – in 1971, 30 years before the disaster! When he issued his critique, Rumsfeld was not on active duty. He was the Pentagon's highest bureaucrat – the Secretary of Defense.

The current Democratic bureaucrats are so evil that they can't even call what they did at Trump's home on August 8, 2022, a raid. They prefer us to stay stupid by simply saying that we're enforcing a warrant or conducting a search.

Previous Articles Published by Me on this Topic:

NOVEMBER 20, 2020: FACEBOOK AND TWITTER ARE TOOLS OF THE FOURTH REICH.

OCTOBER 29, 2020: HITLER, 4TH REICH, and the UFO are encoded with COVID-19. Why? The highway to hell was paved with good intentions. What is new today is the BOTTOM LINE and Figure 4-4, an update of the matrix *AMERICA FALLEN.*

OCTOBER 26, 2020: QUESTION: ARE OBAMA AND HITLER RELATED? Claims that

Hitler was Obama's grandfather seem absurd, but there is at least one matrix that strongly points this way. However, it's easier to simply accept that the real father of Obama was Malcolm X.

OCTOBER 1, 2020: JOHN DURHAM REFUSES TO RELEASE FBI-RUSSIAN COLLUSION

HOAX PROBE RESULTS BEFORE THE ELECTION. Does Durham serve a Fourth Reich?

COMMENT FROM LISTENER: Each time I hear about Antarctica and UFOs, I think of what my dad shared. He was in the navy on a transport ship; he said crafts or UFOs did come out of the water during WW2 along their ship, and also he said they did take officers to Antarctica to drop off there during WW2. He never said why. He may not have known why. He worked gunnery control fire unit coordination for gunnery. I believe I shared this before with you a while back. He also later worked for the state education department and made many trips to Fort Huachuca. Very interesting—who knows. God bless. Debra

SEPTEMBER 20, 2020: HITLER AND THE 4TH REICH. There is photographic, documentation, and Torah Codes evidence that Hitler faked his death and lived on in Antarctica, Argentina, and Colombia. His 4th Reich likely played a major role in the American Government up to the present. His followers are likely to be a major factor in the effort to bring down President Trump.

AUGUST 21, 2020: U.S. NAVY, ALIENS, HITLER, ARGENTINA & ANTARCTICA.

A fragment of World War 2 may have continued well past 1945 with Hitler very much alive. Dr. Salla believes aliens continued to work with Nazis hiding in a massive base in Antarctica. Hitler

may have still been around to command UFOs overflying Washington from July 12, 1952, to July 29, 1952.

AUGUST 16, 2018: Deputy Democratic National Chairman Keith Ellison, Hitler and Israel to Die.

DECEMBER 4, 2018: German Chancellor Angela Merkel, Like Chancellor Hitler, Chooses Islam over Israel.

JANUARY 29, 2017: For the past 2 years, NASA has closely followed my research about Mars, especially the Report and PowerPoint listed below for today's date. But around December 2016, one NASA site in California shifted its emphasis to track my ArkCode site in an obvious effort to understand the Torah Code and perhaps to examine the hypothesis that the Code might be an alien language. They were back again today, looking at (among other things) whether Hitler survived World War 2. Although I last updated that article on 9/21/2014, it's worth a look if you missed it before. Be sure not to miss the film clip from the *Jerusalem Post* that supports my findings.

SEPTEMBER 21, 2014: Hitler Escaped?

FEBRUARY 11, 2013: POPE BENEDICT XVI IN NAZI UNIFORM

SEPTEMBER 18, 2008: VATICAN AND ROME

Chapter 5
First Contact with an ET

This chapter examines the claim by CRAIG EBRAHIMI that he is an alien. It will also examine evidence that I have (since January 26, 2020) suggesting that Craig is responsible for bringing COVID-19 to Earth and, with Chinese and American help, spreading it around our planet. Craig is shown on data-packed Figure 5-1 below. The chapter will also examine evidence that the vector of the contamination was a 27-pound meteorite that he recovered from the seafloor off his current hometown of Sooke, British Columbia, in the Canadian waters of the Strait of Juan De Fuca.

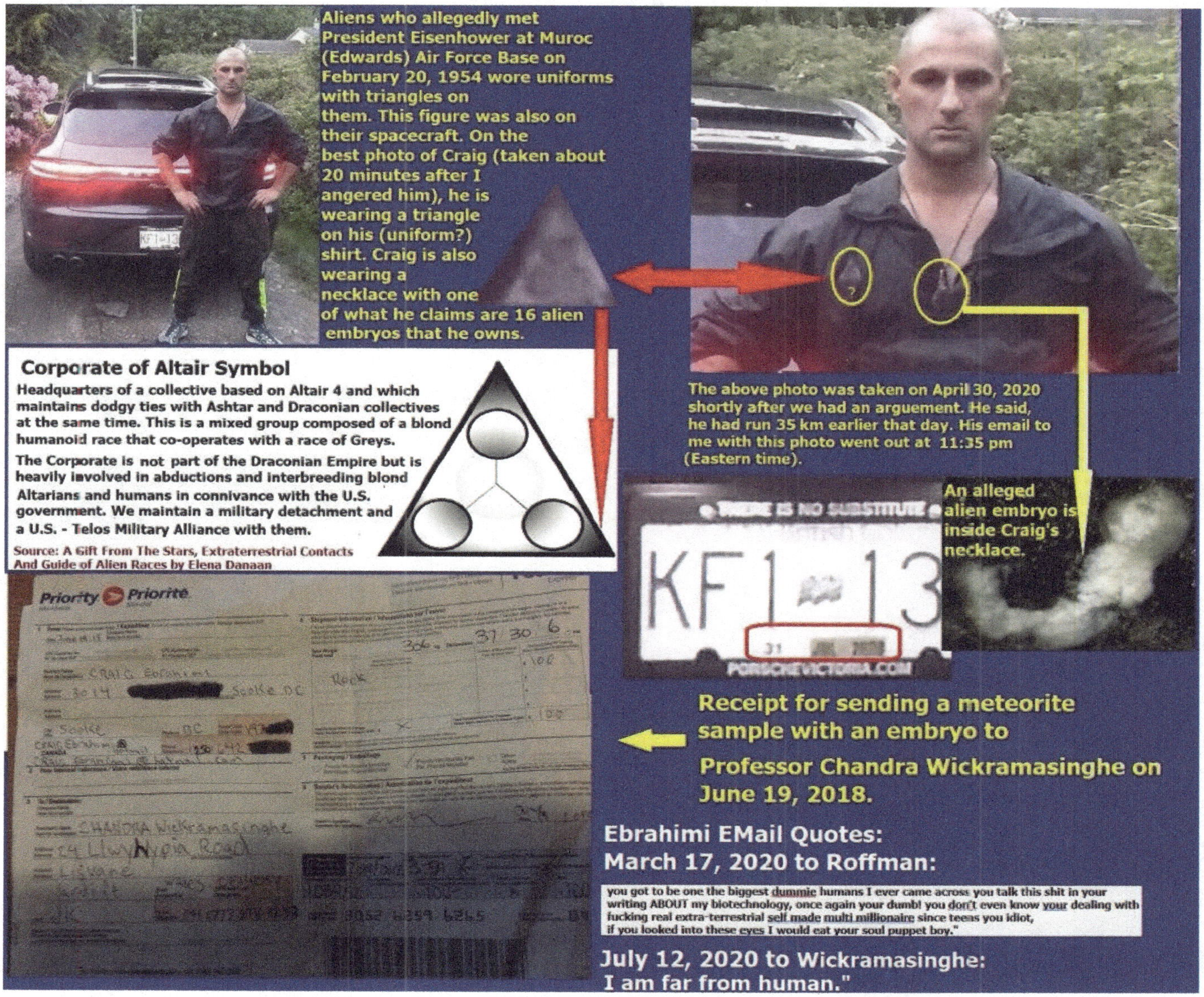

Figure 5-1 – Craig Ebrahimi, possible alien.

Source material: This chapter includes a scientific analysis of Craig's unusual meteorite. While Chapter 6 details our e-mail record for more than a year, Chapter 5 includes Torah Code matrices that relate to Craig.

INITIAL CONTACT WITH CRAIG: I received the following email on January 26, 2020:

My name is Craig Ebrahimi, research scientist/Founder of Deviant off-world technology. ***Barry, would you be interested in the research and biotechnology opportunity of the only pristine extraterrestrial embryo on the planet at the University of Victoria?***

In the two months that followed, I grilled him often and hard. I also angered him a lot. Eventually, he sent me the photo shown in Figure 5-1, but before doing so, he sent me the following e-mail on March 17, 2020, when he was angry with me, in which he wrote that he is himself an alien:

On Fri, Mar 27, 2020 at 4:43 PM Craig Ebrahimi craigebrahimi@hotmail.com wrote:

"Guy, your dumb ass shit keeps getting sent to me by one of your Hebrew followers. You gotta be one of the biggest dummy humans I ever came across. You talk this shit in your writing ABOUT my biotechnology. Once again, you're dumb! ***You don't even know you're dealing with a fucking real extra-terrestrial*** *self-made multi-millionaire since teens, you idiot. If you looked into these eyes, I would eat your soul, puppet boy."*

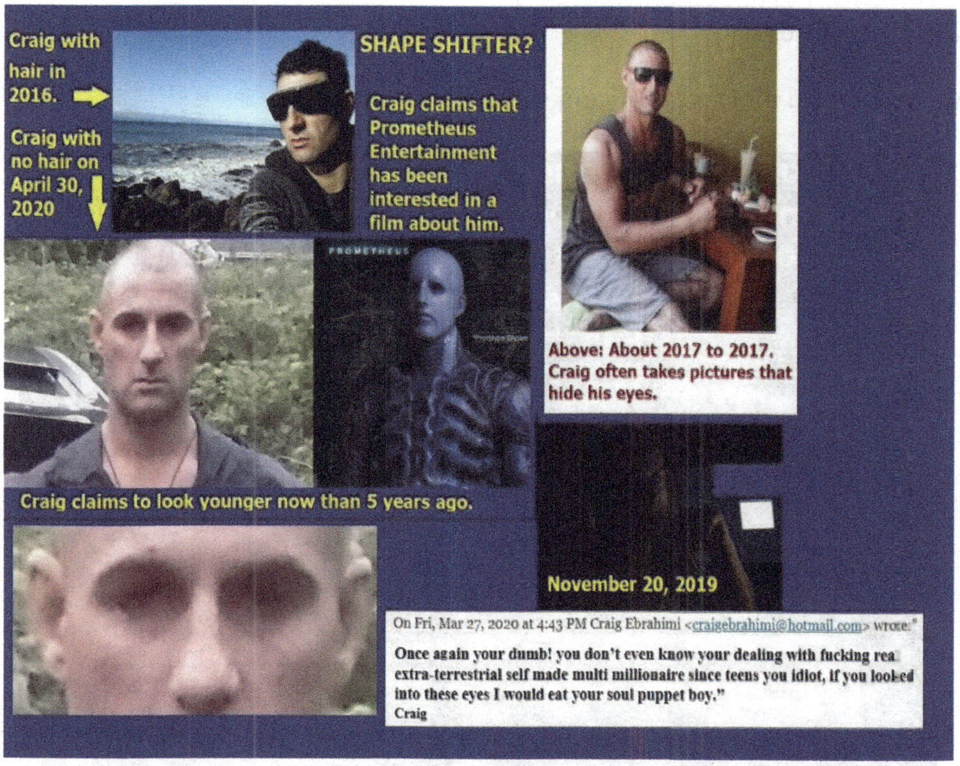

Figure 5-2 above: Is Craig a shape-shifter?

EXTRACTS OF MY POWERPOINT ABOUT EBRAHIMI:

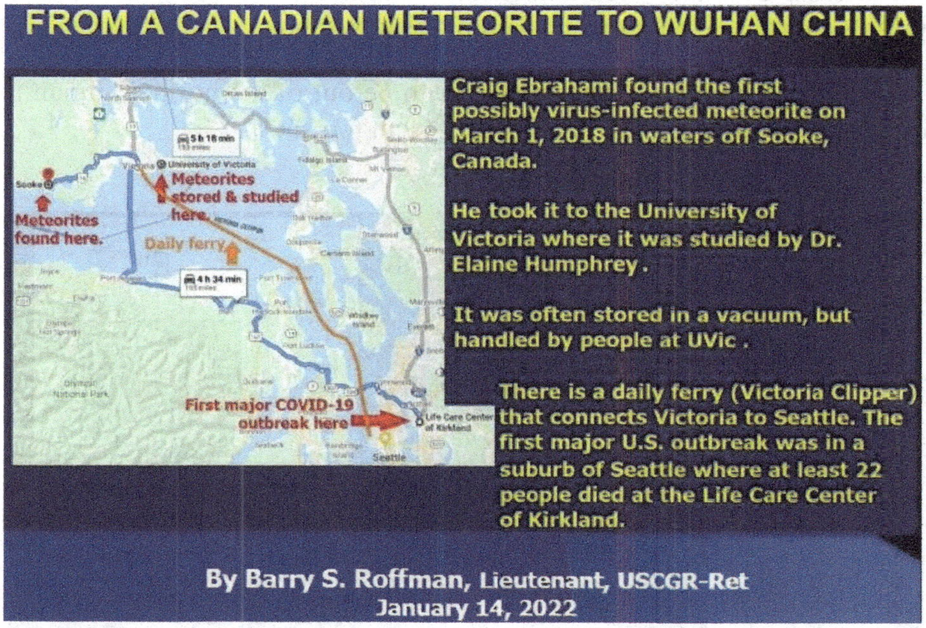

Figure 5-3 above: When the U.S. suffered its first major COVID-19 outbreak in Seattle, there was a daily ferry that connected Seattle with the University of Victoria, where one of Craig's meteorites was being studied for its weird properties.

Figure 5-4 below: Craig with alien embryo and shipping receipt.

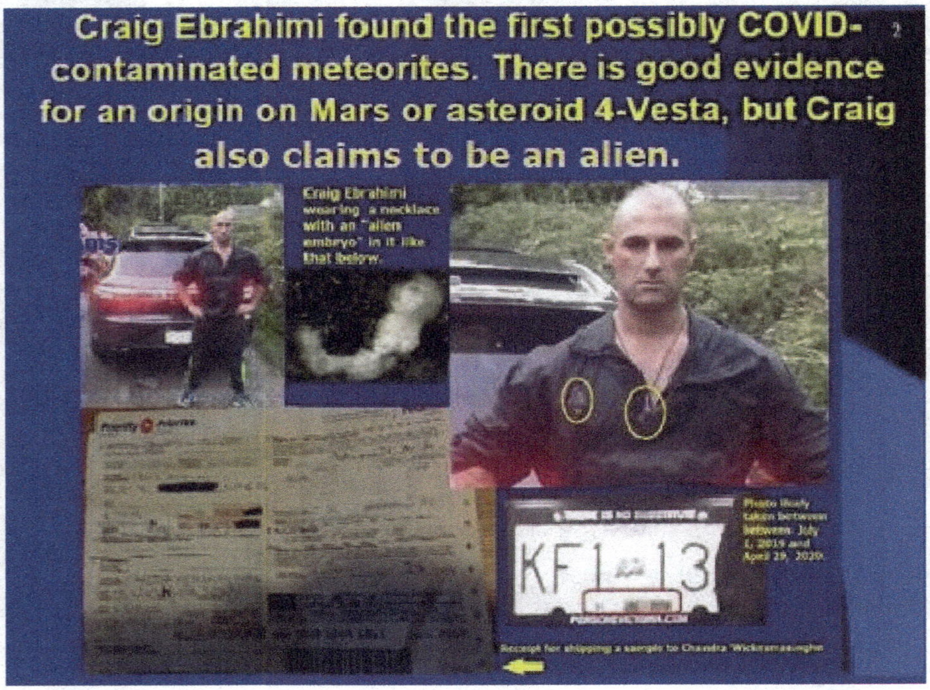

Figure 5-4 above: Craig with an alleged alien embryo around his neck and the receipt for shipping a sample of his meteorite to Professor Chandra Wickramasinghe in England. The professor apparently delivered the sample to the Chinese Army for development as a biological weapon.

Figure 5-5 below: Ebrahimi's body proportions appear to be out of the normal human range. His chest is quite large, perhaps hinting at an extra rib, and his legs seem too short.

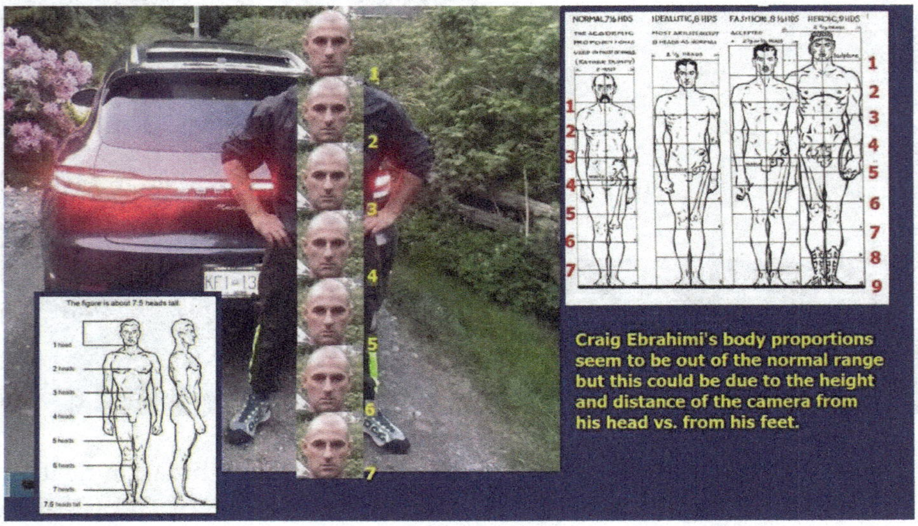

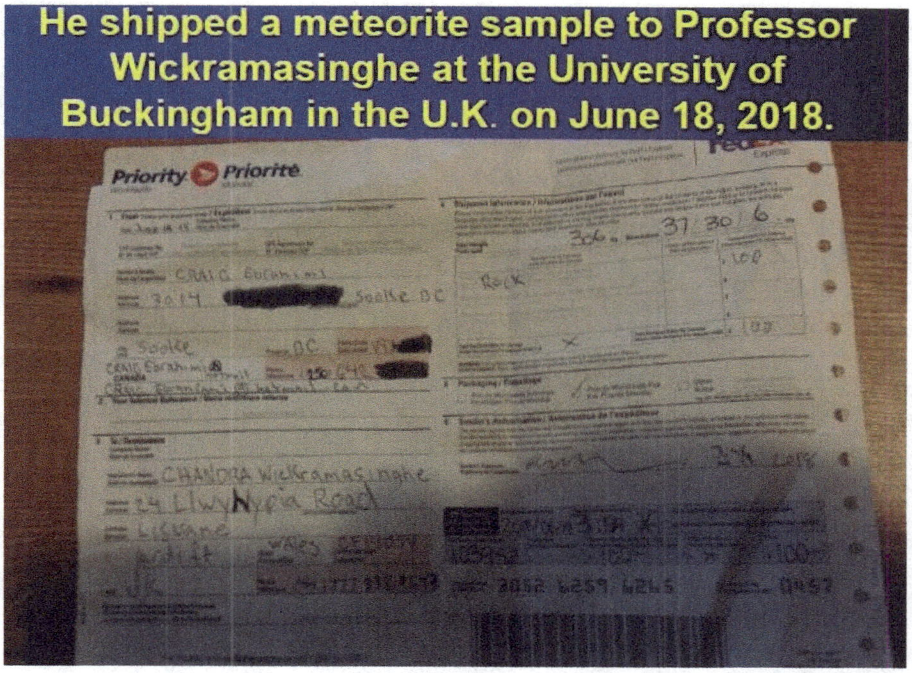

Figure 5-6 above: This receipt marks the beginning of the transfer of Craig's contaminated meteorite sample with alien embryonic samples and possible surviving viruses to Professor Wickramasinghe in the UK and to Chinese Professor Jingwen Qu, at the Center for Infectious Disease Control in Tianjin, China, and its parent Chinese Army.

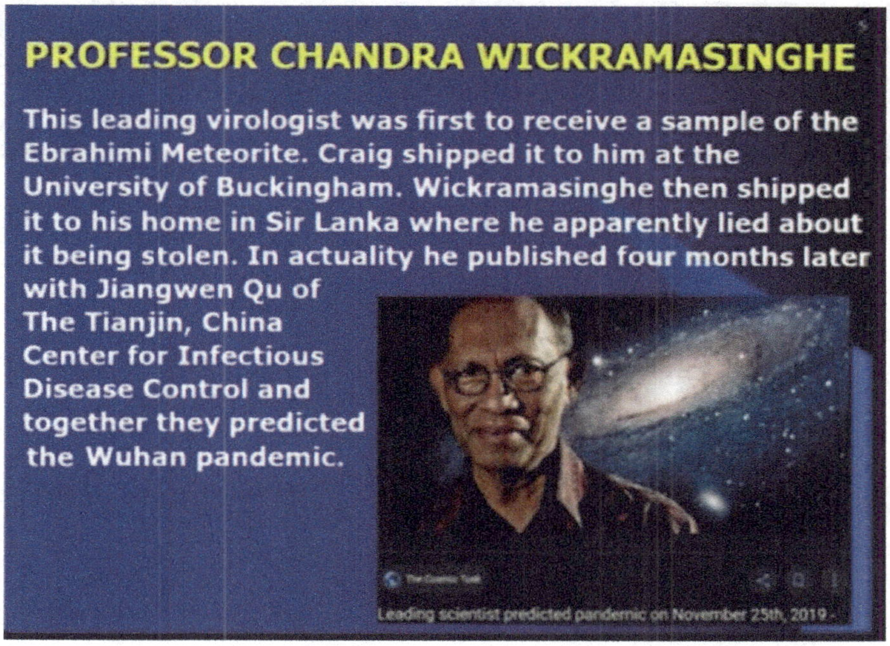

Fig. 5-7 Chandra Wickramasinghe transferred the COVID 19 to the Chinese Army Figure 5- 7 above: Arch-villain Wickramasinghe is among the most guilty when it comes to the murder of millions of people via COVID-19.

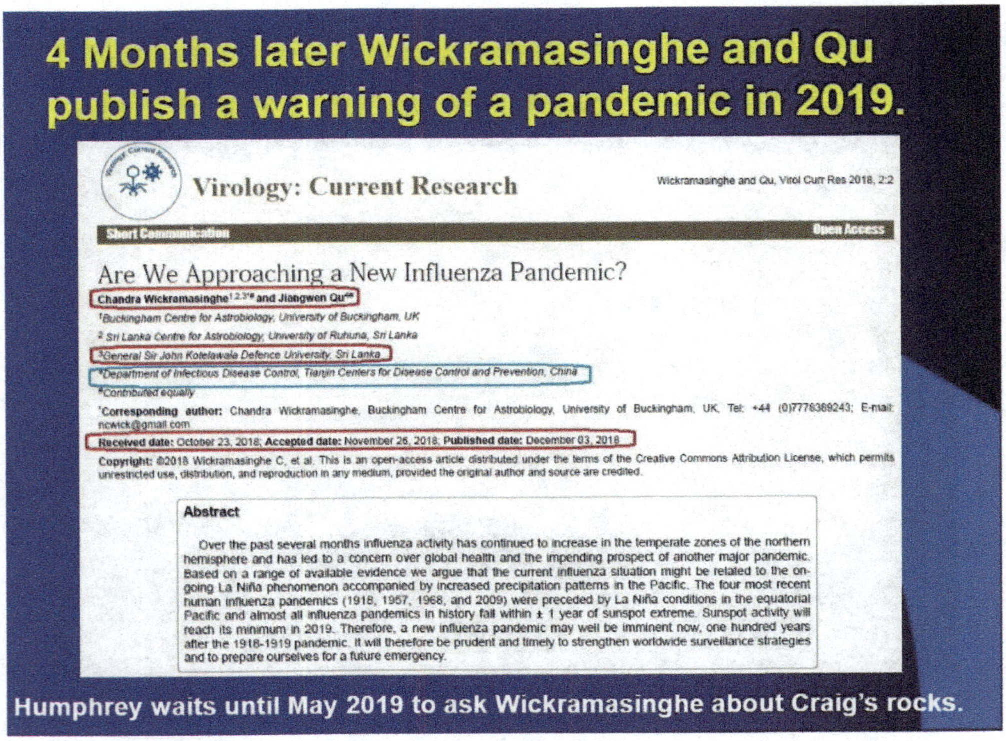

Figure 5-8 above: There is a clear publication record that allows us to learn that this disease was deliberately spread.

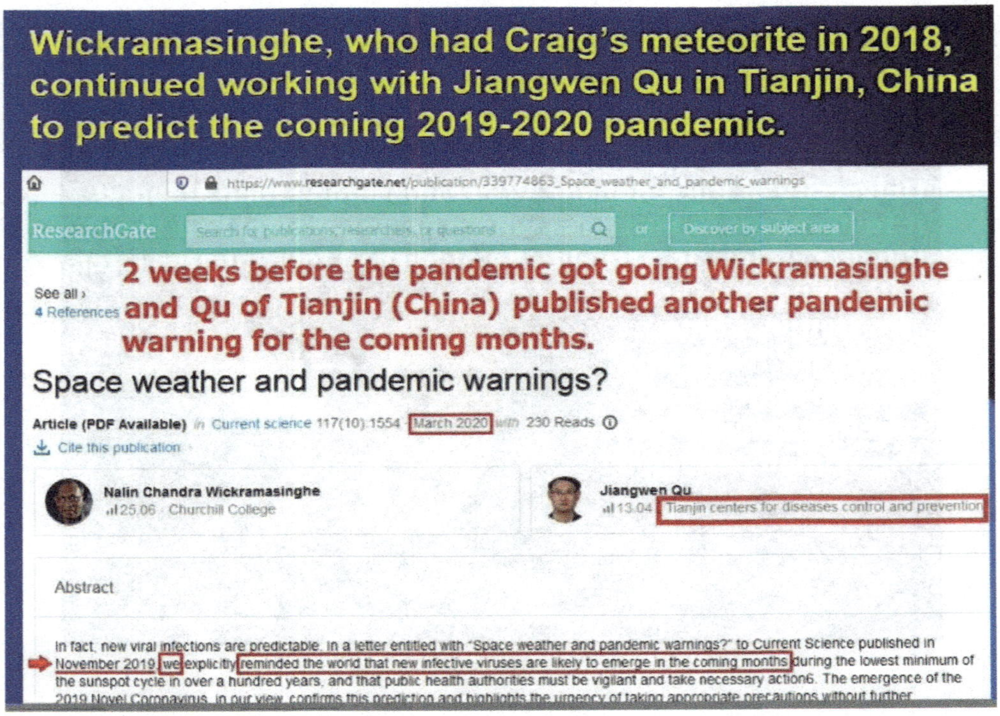

Figure 5-9 above: Hidden warnings about Craig's unusual rock spread quickly in China.

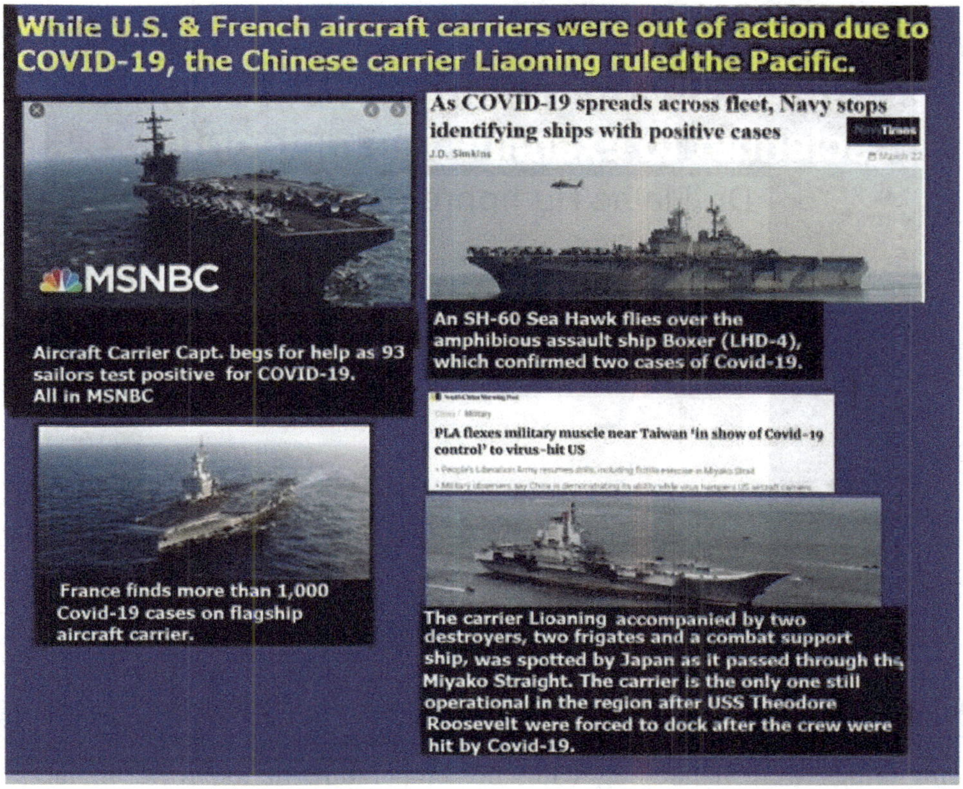

Figure 5-10 above – China quickly weaponized the virus and incorporated it into its naval war plans. They actually attacked our ships with it!

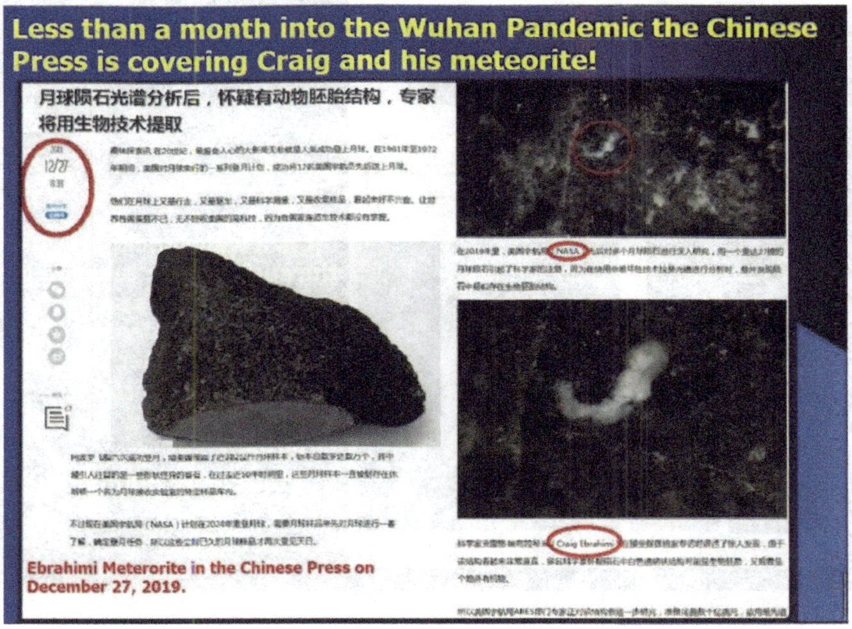

Figure 5-11 above: The plague was just get going in China when this story about Craig and his meteorite was published on December 27, 2019.

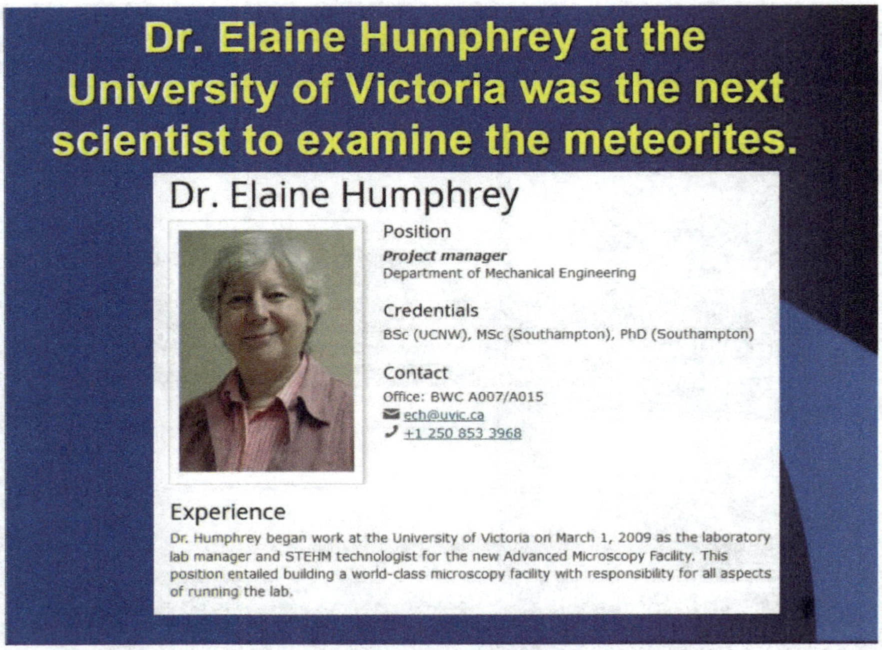

Figure 5-12 above: Dr. Elaine Humphrey was the next major scientist to study Craig's meteorite. I interviewed her several times. She was somewhat nervous about what she was finding, because, as Figure 5-12 shows, weird things were happening with the temperature of the rock. In fact, the temperature fluctuations are suggestive of viral activity.

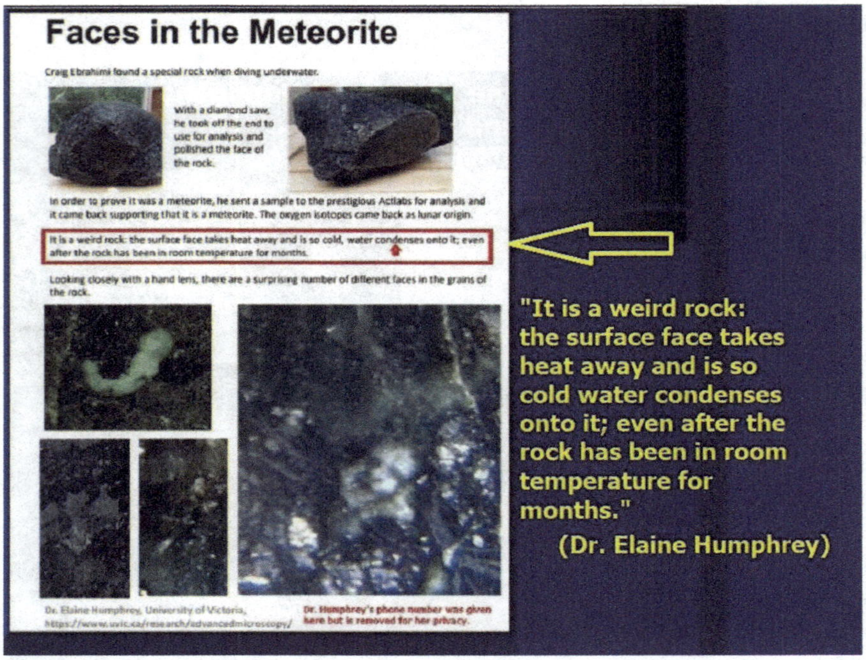

Figure 5-13 above: Dr. Humphrey noted the unusual heat (really, cold) properties of the meteorite.

This may suggest that living viral processes are at work in the rock.

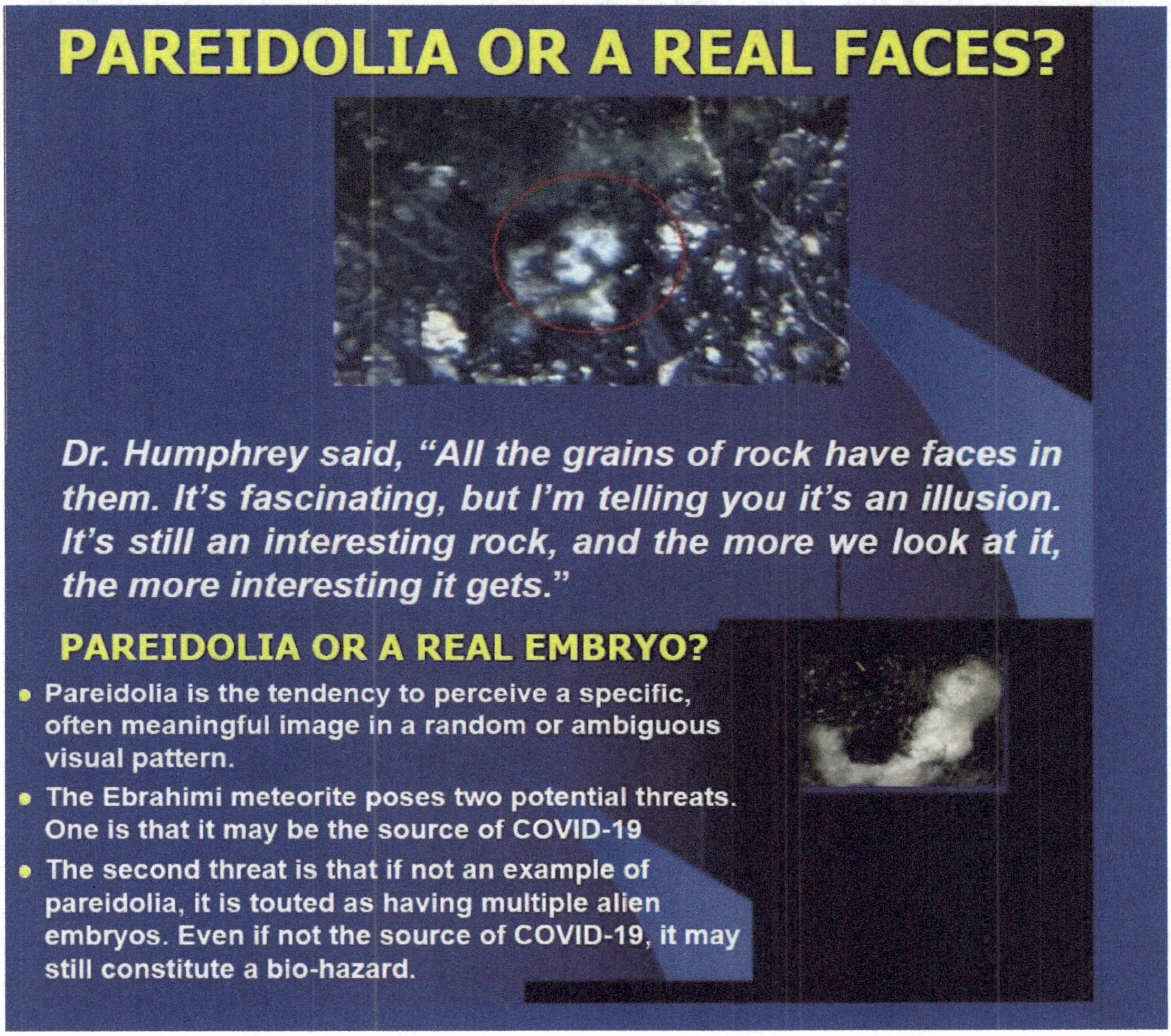

Figure 5-14 above: Pareidolia does not explain temperature fluctuations. My experience interviewing Humphrey leads me to believe that she is very much aware of, and afraid of, the consequences of having alien life or proof of alien life in her laboratory.

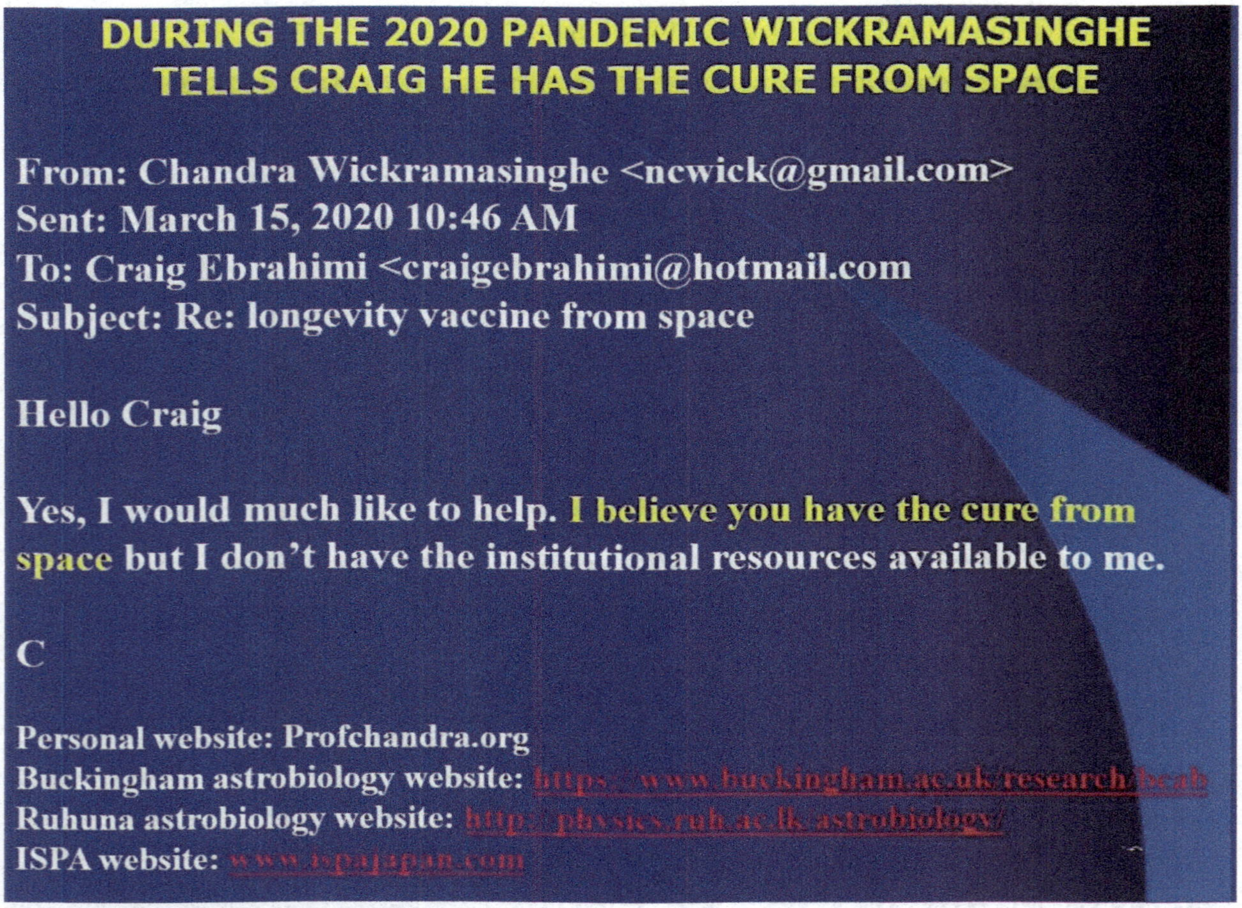

Figure 5-15 above: This claim of Wickramasinghe's lack of resources does not stack up well with his claim later that somebody broke into his outhouse in Sri Lanka, where he stored the meteorite, and somebody stole it! Know that Wickramasinghe is one of the world's leading advocates of panspermia, which states that things like meteorite impacts and solar wind can transfer life from one world to another. He knew a meteorite with proof of life embedded in it would be worth billions of dollars. And yet he says he just left the sample in outhouse? That is utterly unbelievable!

From the moment I met Craig, he worked hard to try to convince me to help him sell his rock to biopharma. When he finally saw (after COVID-19 started killing so many of us) that I could not be swayed to help him enlist Elon Musk or Bill Gates to produce a vaccine, that's when he boasted about his wealth while cursing me out and bragging that he is an ET:

"...You got to be one of the biggest dummy humans I ever came across. You talk this shit in your writing ABOUT my biotechnology, once again you're dumb! You don't even know you're dealing with a fucking real extra-terrestrial self-made multi-millionaire since teens, you idiot. If you looked into these eyes, I would eat your soul, puppet boy."

When following the money to the wealth involved with Craig's acts, we need to keep an eye on Craig and all his associates and follow the money. This includes joint publishers Wickramasinghe and Gensuke Tokoro, shown on the slide below.

Figure 5-16 There was a widespread interest in what Craig was offering, but a great need for them not to go public about it. Craig even reached out to Elon Musk before he turned to me. However, once I angered him with demands for loyalty to the scientific method, he offered changing stories. He altered his story about an early meeting with Musk. But photographic evidence suggests that Craig's original boast was correct.

Jerzy Sawicki and Elon E-Mails Accidentally Transmitted to Me:

The Geiger meter was when Dr. Jerzy (Sawicki) was involved. He is a Polish senior who went absolutely crazy with this project. He would never sleep and started drinking vodka. We had a falling out when he randomly asked me for $100,000 immediately when he heard I was meeting with Elon

The instrument the US scientist used, I have been looking for notes on what it was called. I will contact him.

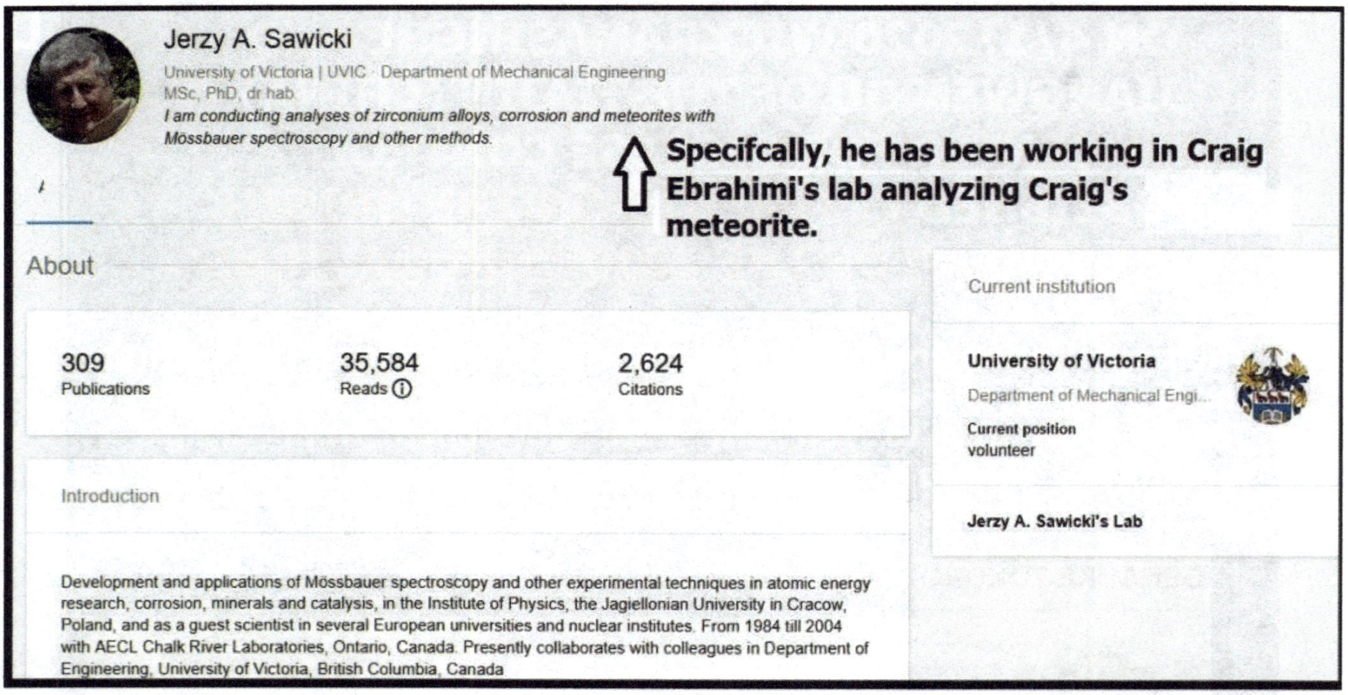

Figure 5-17 above At one point, Dr. Sawicki quit working for Craig. However, I convinced Craig to repair their rift. My concern was that Sawicki was vital for determining the origin of the meteorite. Craig claimed that "voices he heard in his head" gave him the exact coordinates of the meteorite. That sounds a lot like how aliens are believed to communicate by telepathy. Prove that the meteorite has life or remains of life, and you go a long way toward proving that Craig is also an alien (as he claims to be). Sawicki published a joint paper with Ebrahimi about the meteorite in question. See their paper here and below as Figure 5-18.

UNUSUAL ACHONDRITE-LIKE FIND IN SOOKE BASIN: RESULTS OF FIRST ANALYSES.

J. A. Sawicki[1] and C. Ebrahimi[2], [1]Interatomics, Victoria, BC, V8P 1E3, [2]Sooke, BC, V9Z 0X3, Canada

Introduction: The achondrite-like stone examined here was discovered by Craig Ebrahimi during diving in Sooke Basin with JW Pulse 8X induction metal detector. The stone was at 12 m depth and under 3 m of anoxic mud, about 100 m from east shore of Sooke Basin, a 10 km^2 body of sea water connected to the Juan de Fuca Strait at the Southern tip of Vancouver Island. This area was populated long time ago by the T'sou-ke Nation of the Coast Salish. The Juan de Fuca Ice Lobe of the Cordilleran Ice Sheet advanced and receded many times until ~15,000 years BP and likely deposited meteorites from the glacier. The find weighs ~13 kg, attracts a magnet, and is up to 70% covered by black fusion crust about 0.5 to 1 mm thick.

Fig. 1. Photograph of the find, without a cut section.

Analyses:

Chemistry: Using fusion ICP and Na_2O_2 techniques, Actlabs in Ancaster, Ontario, determined LOF=3.45% and elements in proportions characteristic of achondrites - likely from Mars or 4 Vesta. In the Korotev's plots our data in wt% vs. Fe_2O_3+MgO=29.55 are all located in achondrite areas: 45.95 SiO_2 in lunar, HED or Mars; 17.90 MgO and 1.5 MgO/Fe_2O_3 above HED and Mars; 5.30 Al_2O_3 in Mars; 14.74 CaO in lunar, Mars or other; 2.78 CaO/Al_2O_3 in Mars or other; 11.65 Fe_2O_3 below lunar, HED and Mars; Fe_2O_3/MnO=77.7 in lunar and other; 0.36 Na_2O in lunar, HED and near Mars; 0.08 K_2O in lunar, HED and Mars, and 940 ppm Cr as in Mars. Low amount of Na_2O+K_2O=0.44 vs. 45.95 SiO_2 places our find in the left corner of SNC meteorites area [5]. There are also 0.43 wt% TiO_2, and in ppm 227 V, 100 Co, 240 Ni and 50 Zn. REE's were found at lower concentrations than in Martian meteorites, but in a range of 1 ppm observed in many stony achondrites (e.g. aubrites). La/Lu fractionation ratio is ~9 with linear decline between La and Lu and no Eu and Er anomalies.

Petrography: Thin sections that were examined showed some similarities to the NASA study of Alan Hill 84001. Minerals were identified in the following volume fractions and main crystal size ranges: pyroxene 55-60%, 1-3 mm; olivine 35-40%, 0.5-2 mm locally up to 5 mm crystals partly altered to serpentine-magnetite ~1%, 0.3-0.7 mm particles. Pyroxene formed irregular crystals in the spaces between olivine crystals slowly formed during cooling.

Mössbauer spectra: The spectrum in Fig. 2 illustrates close similarity of our Mössbauer data and NASA data for Martian takhlite meteorite MIL 03346 [1]. Mössbauer spectra of stony meteorites usually consist of three doublets due to paramagnetic Fe^{2+} present in olivine and pyroxene [2] and two sextets due to magnetically ordered iron in magnetite. In our samples Fe^{2+} is in M1 and M2 sites of clinopyroxene (ferric diopside $CaMgFeSi_2O_6$) and in olivine (fayalite $(Mg,Fe)_2SiO_4$). The average values of quadrupole splitting for Fe^{2+} in 12 spectra analyzed vs. depth were: 2.59 mm/s in M1 site and 2.02 mm/s in M2 site of clinopyroxene; and 2.89 mm/s in fayalite.

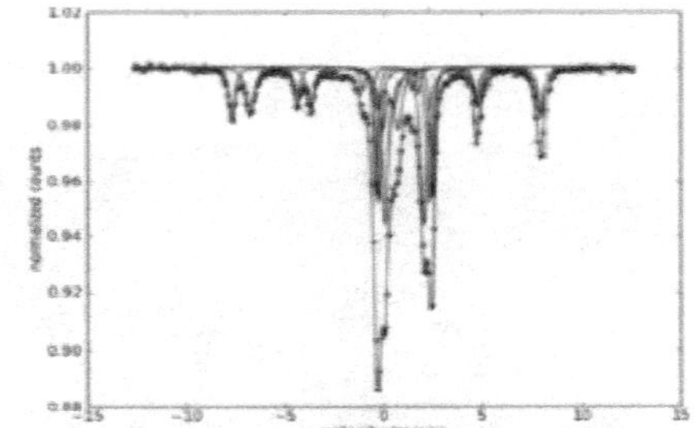

Fig. 2. Mössbauer spectrum of 15-mm deep sample.

With the depth of samples from crust to ~20 mm inside the total fraction of Fe in cpx M1 and M2 sites increased from 20% to 40% and in fayalite from 10 to 15%; in magnetite it decreased from 65% to 15%, while the fraction of Fe^{3+}, likely in a form of amphibole increased from 15% to 40%, see Fig. 3.

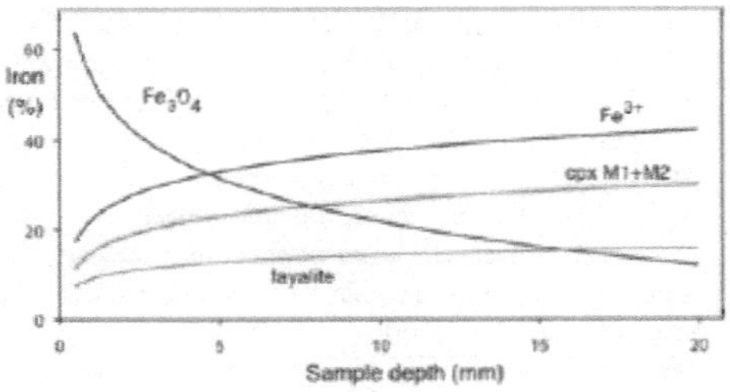

Fig. 3. Dependence of Fe fractions on sample depth.

Figure 5-18 above and below.

52nd Lunar and Planetary Science Conference 2021 (LPI Contrib. No. 2548) 2775.pdf

X-ray diffraction: Peaks in diffractogram of sample taken from ~15 mm inside the rock were assigned to ferroan diopside, sodium-tremolite i.e. amphibole richterite $(NaCa)Mg_5Si_8O_{22}(OH)_2$ and magnetite, in accord with Mössbauer spectra. Abundance of tremolite resembles that in Almahata Sata meteorite of asteroid origin. Remaining three diffraction peaks were ascribed to oxidized and reduced graphene.
Raman spectra: Taken with 532, 633 and 785 nm laser beams focussed to a 2 μm spots showed mostly bands of clinopyroxene P at 320, 660 and 1020 cm^{-1}, some bands of tremolite T, and likely very weak broad bands at 400 and 600 cm^{-1} of maskelynite from the impact.

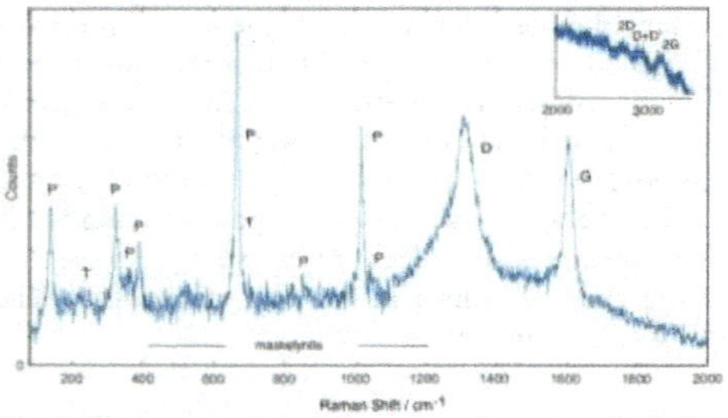

Fig. 4. Raman spectra of the general area and of white feature shown in photograph below.

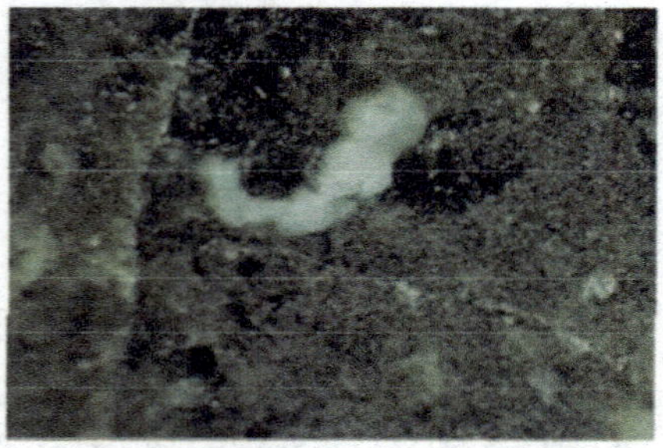

Fig 5. The photograph of the 2x3 cm polished area.

The white area in Fig. 5 showed prominent bands of nanodiamonds: D at ~1330 cm^{-1} due to structural defects and sp^3 bonds and G at ~1600 cm^{-1} from in plane vibrations of sp^2 bonded carbon atoms in graphene. Minor bands seen in top right corner of Fig. 4 between 2500 and 3300 cm^{-1}, are ascribed to the combinations of inelastic scattering bands observed in the first-order region [3]. The nanodiamonds as small as 2-7 nm were previously observed in carbonaceous meteorites, e.g., Allende. Such particles can be formed by CVD process or shocks at 1000-2300 K and 30-40 GPA.
Oxygen isotopes: The $\delta^{17}O$ and $\delta^{18}O$ values obtained for 3 samples taken near crust are close to terrestrial fractionation line at 0.523, but also close to the data for lunar and 4 Vesta samples, whereas $\Delta^{17}O$ values are similar to those in angrites as shown below.

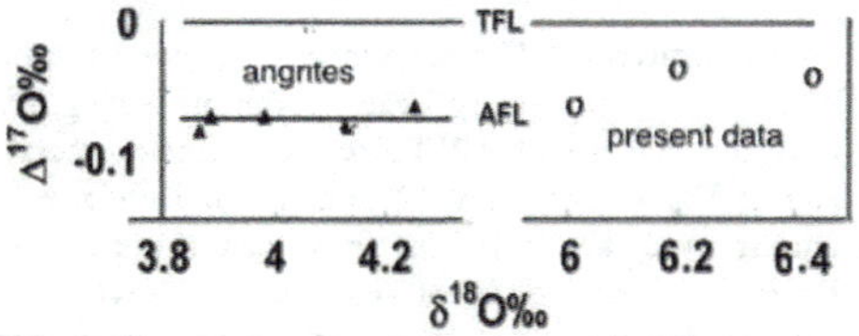

Fig. 6. Oxygen isotopes data compared with angrites.

However, as the oxygen isotopes near the surface of meteorites after long exposure to ^{18}O-rich sea water become closer to terrestrial [4] even small correction for $\delta^{18}O$ of -1‰ could relocate our data to the Martian area in $\delta^{17}O$ vs. $\delta^{18}O$ plot. Further analyses of bulk samples and separates from deep interior for comparison in particular with NWA 7035 data [5],[6] are necessary.
SEM-EDS and TEM: are being used now to identify CAI inclusions, C-N-P elements overlaps, graphite whiskers, iron veins and other microscopic features.

Conclusions: Analyses suggest that we have encountered a very interesting, differentiated, carbonaceous extraterrestrial object with nanodiamond including features. This could be the third largest Martian and largest not fractionated basaltic shergottite-type of trakhlite-like meteorites ever found, possibly from celestial watery bodies [7],[8] or carbonaceous asteroids [9],[10],[11]. As the research in astrogeology and biochemistry of Martian and asteroid meteorites, rocks and soil and space missions intensify, we trust that this find can be quite significant and deserves further more thorough investigations.

Acknowledgments: We thank Actlabs in Ancaster, Ontario for chemical analysis; Dr. John G. Payne at Van Petro Ltd. Vancouver for petrography; Dr. Karen Ziegler at Institute of Meteoritics, University of Mexico for oxygen isotopes data; as well as Dr. Elaine C. Humphrey at Advanced Microscopy Facility and operators of CAMTEC XRD and Raman labs at University of Victoria for their very valuable contributions.

References: [1] Morris R. V. et al. (2008) *LPS XXXIX,* 2548-2549. [2] Dyar M. D. et al. (2006) *AREPS. XXXIV,* 83-125. [3] Steele A. et al. (2016) *MPS 51,* 2203-2225. [4] Hezel D. C. et al. (2015) *MPS L,* 229-242. [5] Ziegler K. et al. (2013) *LPS XXXXII,* 2639-2640. [6] Agee C. B. et al. (2013) *Science 339,* 780-785. [7] Sutter V. et al. (2002) *EPSL 196,* 223-238. [8] Holm N. G. et al. (2015) *Astrobiology 15,* 587-600. [9] Goodrich, C. A. et al. (2020) *LPS LI,* Abstract #1223. [10] Hamilton V. E. et al. *LPS LI,* Abstract #1122. [11] Gucsik A. et al. (2008) *Organic Matter in Space, Proc. IAU Symp. No251,* 335-339.

Figure 5-18 continued above: the Sawicki - Ebrahimi Paper

There are many pictures taken by our Mars landers that suggest primitive life growing on them, but the best photo that I have seen from there, showing both growth and reproduction, is Figure 5- 19.

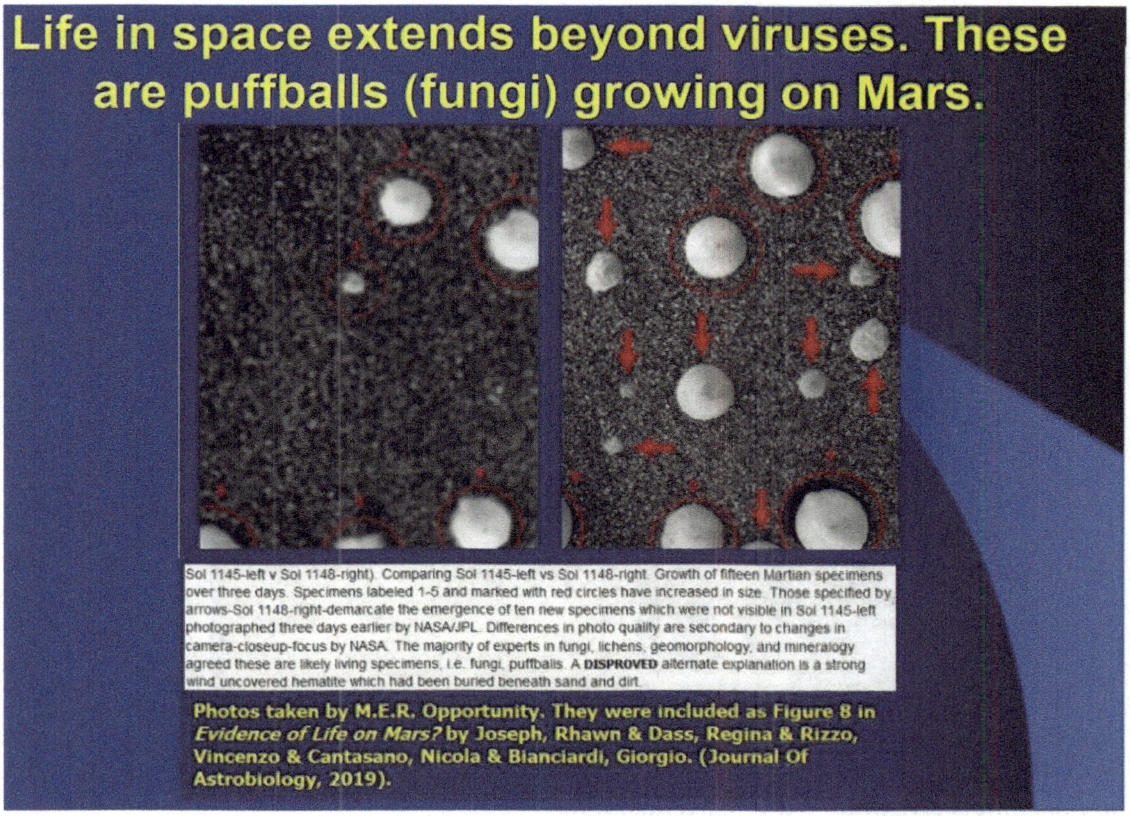

Figure 5-19 above: Probable life (puffballs, a type of fungus) growing on Mars. The alternate suggestion that we are seeing hematite buried by sand and uncovered by wind has been disproven.

The relationship between Wickramasinghe, Dr. Humphrey, and Ebrahimi becomes clear when reading the following emails.

Fri, 17 May 2019 at 00:13:

Hi Chandra,

I have a client, Craig Ebrahimi, who brought in a meteorite. At first, it looked like gabbro, but the chemical analysis (by Act Labs) proved it to be a meteorite. I understand that he sent you thin sections, outside crust, and a piece of the rock with an embryo fossil in a white circle. I have a very small sample for SEM and EDS. Craig suggested that if you have finished with his samples, is it possible to send them to me? Since I will be in the UK at the end of June for a month, I could pick them up. If you are still working with them, please keep us in the loop.

Best wishes,
Elaine

Dr. Elaine C. Humphrey, Fellow of Microscopy Society of America, Past President Microscopical Society of Canada

May 20, 2019 from Wickramasinghe:

Dear Elaine,

Unfortunately, the meteorite package was left in a shed near our house, and it appears to have been stolen. I do not have it anymore.

Sorry about this. Sincerely, Chandra Wickramasinghe

Honorary Professor, University of Buckingham, Buckingham, UK Honorary Professor, University of Ruhuna, Matara, Sri Lanka Honorary Professor, Sir John Kotelawala Defence University, Sri Lanka

Buckingham astrobiology website: http://www.buckingham.ac.uk/research/bcab Ruhuna astrobiology website: http://physics.ruh.ac.lk/astrobiology/

Personal website: www.profchandra.org

ISP

May 20, 2019 at 1:15:13 PM PDT

From: Chandra Wickramasinghe ncwick@gmail.com Date: May 20, 2019 at 1:15:13 PM PDT To: Elaine Humphrey ech@uvic.ca Subject: Re: meteorite

Reply-To: ncwick@gmail.com Hi Elaine,

Our house was in chaos after some international travel, so I store all valuable things in an outhouse. Some months back, I discovered that many things, including the box with the meteorite, had disappeared. I was waiting to get a colleague to handle the experiments on this, but it never happened. Sorry about this – not my fault.

Best, Chandra Prof. Chandra

May 20, 2019, 1:32 PM

From: Elaine Humphrey ech@uvic.ca Sent: May 20, 2019 1:32 PM

To: Craig Ebrahimi craigebrahimi@hotmail.com Subject: Fwd: meteorite FYI. Sent from my iPhone

On May 20, 2019, at 18:15, Elaine Humphrey ech@uvic.ca wrote: Chandra,

Oh dear, that's too bad. Did you get any information out of them before they were stolen? If so, have you published them?

Elaine

I don't have his direct response to her last question, but he was publishing up the kazoo in conjunction with what he learned from the meteorite. He just *forgot* to include Craig Ebrahimi's name as the owner of the rock.

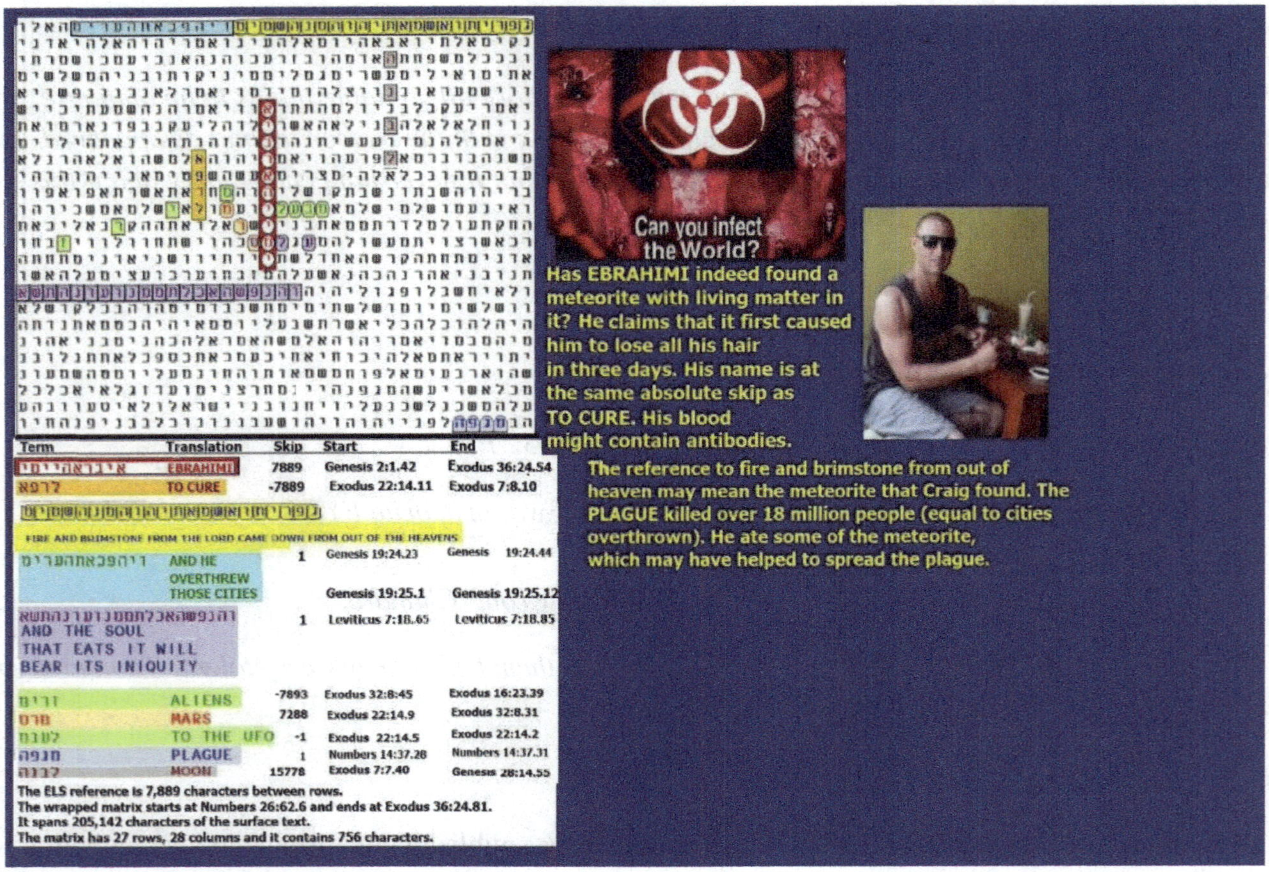

Figure 5-20 above: Ebrahimi has eaten some pretty unusual food. He claims to have lost hair after ingesting small amounts of dust from his meteorite. On the matrix above, the axis term is a long (10-letter) spelling of his last name. When we think about a meteorite, we envision something hot. If that something is carrying a virus like COVID-19, it could cause a **PLAGUE. PLAGUE** is on the matrix in the open text, but of more interest with respect to his diet are two phrases: **BRIMSTONE AND FIRE FROM OUT OF HEAVEN,** plus **AND THE SOUL THAT EATS IT *WILL BEAR ITS INIQUITY.* ALIENS** is on the matrix, but not at a special case skip. However, a short transliteration of **MARS** touches both **ALIENS** and **EBRAHIMI.** ***TO THE UFO*** at skip -1 touches **EBRAHIMI.** While **ROCK** is only three letters in Hebrew, two of them touch **EBRAHIMI**. In looking for who wrote the Codes, this matrix can be used to argue for God or aliens; however, it does not exclude the possibility that God employed aliens as his assistants, much the way He did with three angels (messengers) that He sent to report on behavior in Sodom and Gomorrah before He destroyed them.

There are three significant terms on the matrix above. Both ***BRIMSTONE AND FIRE FROM OUT OF HEAVEN***, plus ***AND THE SOUL THAT EATS IT WILL BEAR ITS INIQUITY*** were each found against odds of about 321 to 1. **PLAGUE** was found against odds of about 36 to 1. Overall, this matrix was found against odds of about 2.09 trillion to 1 without allowing statistical credit for the long spelling of the axis term **EBRAHIMI**. If we count the 10-letter axis term, the odds against finding this matrix grow to over 19.90 trillion to 1. Clearly, the Author of the Torah knew all about this alien over 3,000 years ago.

	A	B	C	D	E	F	G	H
1	Axis term EBRAHIMI @ Skip 7,889	SKIP	NUMBER	DIVIDE BY	THE QUOTIENT	E QUOTIENT	POISSON EQUATION	EACH WORD OR
2	ELS Rank 1	USED	IN	304805	EQUALS	X 646 LETTERS	PROBABILITY FOR	PHRASE HAS 1
3	R FACTOR FOR REPTILIAN GOVERNMENT = 0.979	ON	TORAH	LETTERS	FREQUENCY	EQUALS	THE WORD	CHANCE IN THE
4	=h15*9.52796164	FIGURE		IN THE	PER LETTER	WORD	APPEARING	IN THE FOLLOWIN
5	20926100000000*9.52 =about 19.8968 Trillion			TORAH		EXPECTANCY	AT LEAST ONCE	OF BEING HERE.
6							WAR @ Skip +1 only	.
7	BRIMSTONE AND FIRE @ SKIP +1 only	1	1	304805	3.28079E-06	0.002119388	0.002117143	472.3345419
8	AND HE OVERTHREW THOSE CITIES @ Skip 1 only	1	1	304805	3.28079E-06	0.002119388	0.002117143	472.3345419
9	THE SOUL THAT EATS IT WILL BEAR ITS INIQUITY @ +1 only	1	1	304805	3.28079E-06	0.002119388	0.002117143	472.3345419
10	TO CURE @ +/-1 AND +/- 7889	-7889	6	304805	1.96847E-05	0.012716327	0.012635816	79.14012068
11	PLAGUE @ Skip +1 only	1	9	304805	2.95271E-05	0.01907449	0.018893723	52.92763012
12	UFO @ SKIPS +/- 1 AND +/- 7889 AND -7889	-1	174	304805	0.000570857	0.368773478	0.308417949	3.242353449
13	3-LETTER MARS @ SKIPS 7288 TO 7360	7288	651	304805	0.002135792	1.379721461	0.748351363	1.336270701
14	ALIENS (EITHER OF 2 TERMS @ Skips -7893 to -8166	-7,893	1157	304805	0.003795869	2.452131691	0.913890169	1.094223392
15	TOTAL ODDS AGAINST THIS MATRIX = H7*H8*H9*H10*H11*H12*H13*H14						4.77871E-13	2.09261E+12

Above: Spreadsheet showing the odds against finding **Figure 5-20.**

There are many more examples in the Torah Code that describe the COVID-19 plague. Below the axis term is **WUHAN**, the city where the virus was modified at the Institute of Virology to be more lethal. **INSTITUTE** is at the skip of **WUHAN**. No specific date is given, but in the open text is ***IN THE END OF DAYS.*** **PLAGUE** is also in the open text. More profound, in the open text is ***IT SEEMS TO ME THERE IS A PLAGUE IN THE HOUSE***. There is more about death.

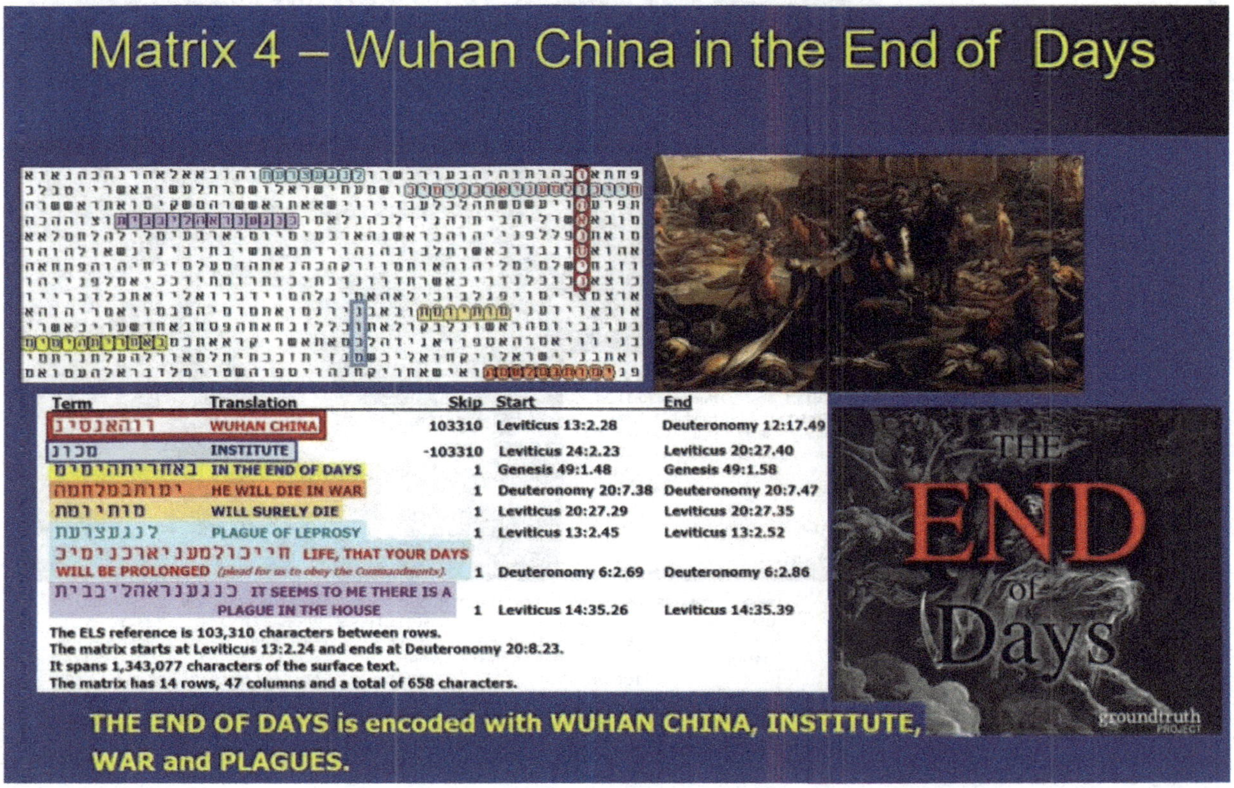

Figure 5-21 above: WUHAN and a **PLAGUE IN YOUR HOUSE** in the Torah Code. By the time it calmed down, my wife and I both tested positive for it. Fortunately, we recovered. Odds against this matrix are about 935,195,904 to 1.

A	B	C	D	E	F	G	H
AXIS TERM WUHAN CHINA @ Skip 103,310	SKIP	NUMBER	DIVIDE BY	THE QUOTIENT	E QUOTIENT	POISSON EQUATION	EACH WORD
ELS RANK 9	USED	IN	304805	EQUALS	x 874 LETTERS	PROBABILITY FOR	OR PHRASE
	ON	TORAH	LETTERS	FREQUENCY	EQUALS	THE WORD	HAS 1 CHANCE
	FIGURE		IN THE	PER LETTER	WORD	APPEARING	IN THE
			TORAH		EXPECTANCY	AT LEAST ONCE	FOLLOWING OF
							BEING HERE
INSTITUTE @ Skips +/- 1 and +/- 103310	-103310	1	304805	3.28079E-06	0.002867407	0.0028633	349.2473785
THERE IS A PLAGUE IN YOUR HOUSE @ Skip +1 only	1	1	304805	3.28079E-06	0.002867407	0.0028633	349.2473785
END OF DAYS @ Skip +1 only	1	4	304805	1.31231E-05	0.011469628	0.011404103	87.6877407
FOR A PLAGUE	1	6	304805	1.96847E-05	0.017204442	0.017057291	58.62595696
WILL SURELY DIE @ Skip +1 only	1	27	304805	8.85812E-05	0.07741999	0.074498929	13.42301175
Combined odds against the matrix = (h7*h8*h9*h10*h11)/9						1.06929E-09	935195904.1

Figure 5-22 below: The axis term is **LIVE ALIEN EMBRYO**. At the same skip is **WUHAN. CHINA** is at skip +1, **UFO** is at skip -1 sharing a letter ayin with **LIVE ALIEN EMBRYO. GILEAD** is at skip +1. It makes remdesivir, a treatment for COVID-19 The virus caused some starvation in Wuhan when the city was shut down. Odds against this matrix were over 1.8 million to 1.

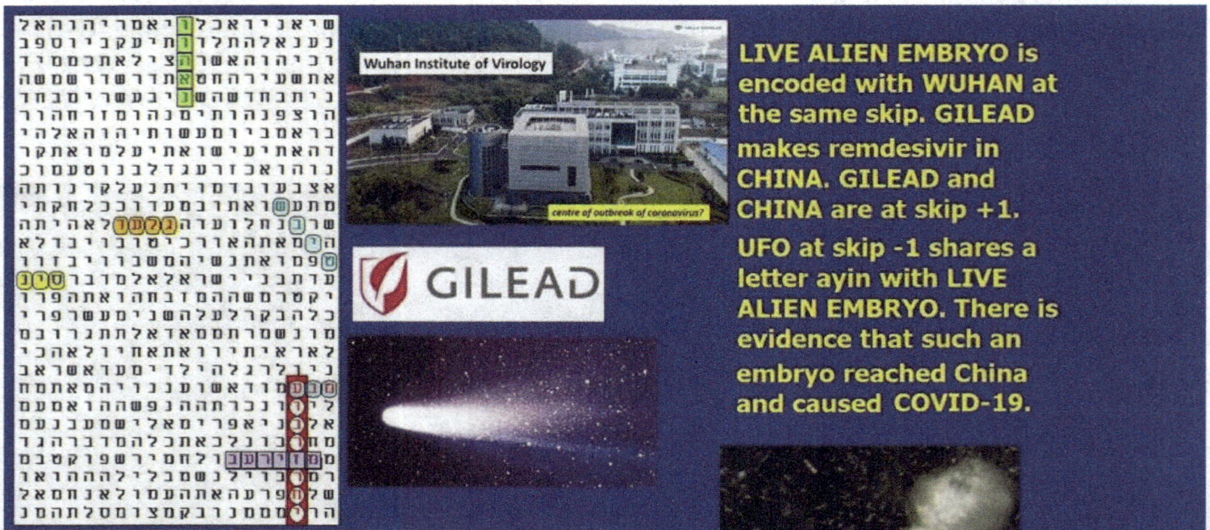

Term	Translation	Skip	Start	End
עוברזרחי	LIVE ALIEN EMBRYO	50513	Exodus 14:24.36	Leviticus 6:8.4
ווהאן	WUHAN	50513	Genesis 3:14.1	Numbers 10:11.20
סין	CHINA	1	Exodus 16:1.35	Exodus 16:1.37
גלעד	GILEAD	1	Deuteronomy 2:36.37	Deuteronomy 2:36.42
מזירעב	THE WASTING HUNGER	1	Deuteronomy 32:24.1	Deuteronomy 32:24.4
שביט	COMET	50512	Numbers 9:3.31	Genesis 34:29.15
עבמ	UFO	-1	Exodus 14:24.36	Exodus 14:24.34

The ELS reference is 50513 characters between rows.
The matrix starts at Genesis 3:13.42 and ends at Leviticus 6:8.21.
The matrix spans 1,363,871 characters of the surface text.
The matrix has 28 rows, 20 columns and contains 560 characters.

A	B	C	D	E	F	G	H
Axis term: LIVE ALIEN EMBRYO @ Skip +50,513	SKIP	NUMBER	DIVIDE BY	THE QUOTIENT	E QUOTIENT	POISSON EQUATION	EACH WORD OR
ELS Rank 3	USED	IN	304805	EQUALS	X 560 LETTERS	PROBABILITY FOR	PHRASE HAS 1
	ON	TORAH	LETTERS	FREQUENCY	EQUALS	THE WORD	CHANCE IN THE
	FIGURE		IN THE	PER LETTER	WORD	APPEARING	IN THE FOLLOWING
			TORAH		EXPECTANCY	AT LEAST ONCE	OF BEING HERE.
WUHAN @ Skip +/-1 and +/-50513	50513	5	304805	1.64039E-05	0.009186201	0.009144137	109.3596941
GILEAD @ Skip +1 only	1	25	304805	8.20197E-05	0.045931005	0.044892142	22.27561316
CHINA @ Skip +1 only	1	39	304805	0.000127951	0.071652368	0.069145566	14.46224341
STARVATION @ Skip +1 only	1	72	304805	0.000236217	0.132281295	0.12390548	8.070668045
COMET @ Skips +50512 to 50757	50512	121	304805	0.000396975	0.222306065	0.19932973	5.016813098
UFO @ Skips +/- 1 and +/- 50513	-1	163	304805	0.000534768	0.299470153	0.258789155	3.864149562
INITIAL COMBINED SIGNIFICANCE = H6*H7*H8*H9*H10*H11						1.8142E-07	5512060.233
FINAL SIGNIFICANCE = (H6*H7*H8*H9*H10*H11)/3						5.44261E-07	1837353.411

DID ELON MUSK VISIT EBRAHIMI AND HIS METEORITE?

Here we examine email, local news coverage, and Torah Code evidence that on August 6, 2019, Elon Musk traveled to Sooke, BC, Canada, or within a few minutes of it, to meet Craig Ebrahimi, an alleged alien, who tried to sell him a meteorite. It supposedly contains 16 fossils of alien (possibly Reptilian) embryos.

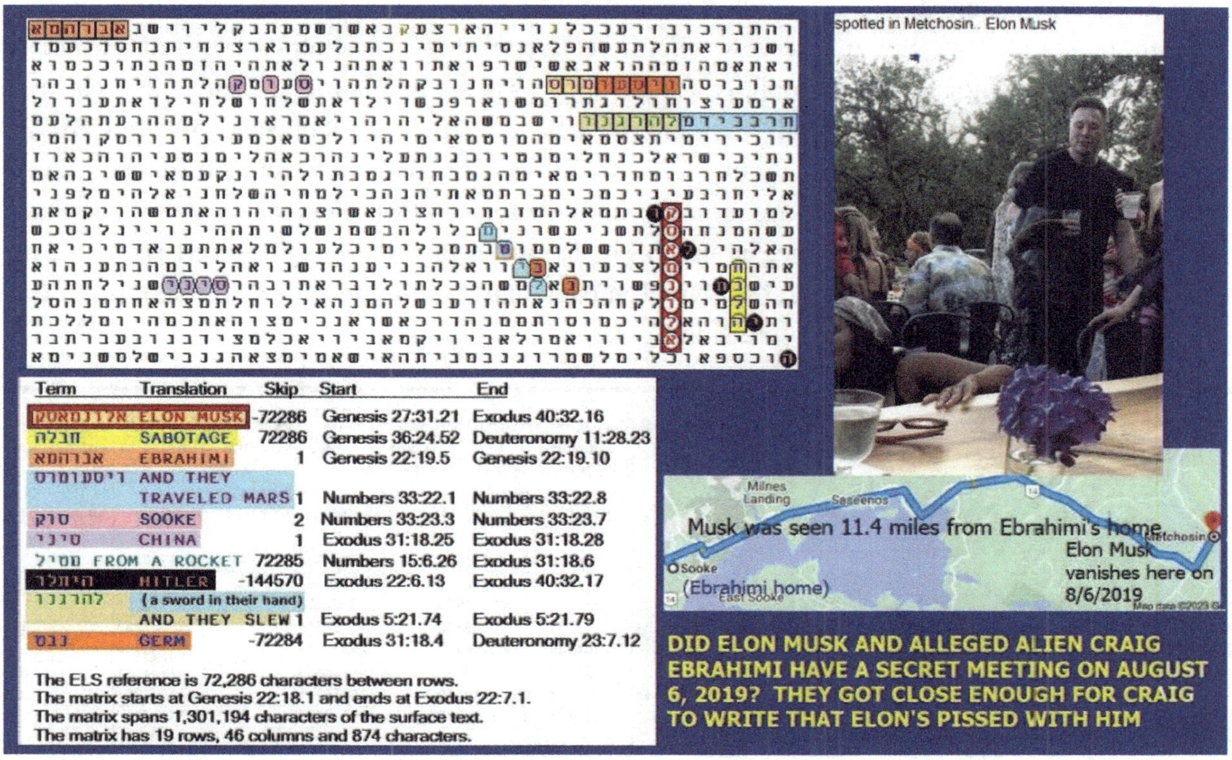

Figure 5-23 above: This is the only matrix that I could find with **ELON MUSK** on it (as the axis term) and a transliteration of **EBRAHIMI** that was formed with Abraham in the open text, followed by a letter **alef** (yud would also work) that can be read as a hard e when a dot (chiriq) is assumed to be implied under the alef. Vowel signs are not included in a hand-written Torah scroll but are included in book versions.

Terms (found *a priori*) on the matrix that were worthy of a statistical calculation start with a short transliteration of **MARS**. The planet is not only on the matrix at skip +1 against odds of about 174 to 1. It has an interesting location. It's in sequence with and after **AND THEY TRAVELED...(MARS).** Musk not only wants to live there but to die there (just not on the landing). If, as I suspect based on remarks by Israeli Brigadier General Haim Eshed, it's true that we have a joint top-secret U.S.-alien base on Mars, Musk may have to worry about his mission with respect to **SABOTAGE** by a Deep State/Fourth Reich government. **SABOTAGE** is four columns directly to the right of and at the same absolute skip as **ELON MUSK.** A transliteration of **HITLER** touches two letters of **ELON MUSK.** The same **HITLER** also touches two letters of **SABOTAGE. CHINESE** is also on the matrix at skip +1. The **CHINESE** will be our primary competition unless we are already there. Further, on December 27, 2019, the Chinese press covered the story of Craig's meteorite.

The importance of Sooke is twofold. It's the town that Craig lives in, and, since his photo shows him wearing something like military insignia allegedly seen on aliens in the 1954 meeting between Eisenhower and ETs, Sooke might be the home of more aliens than just **Ebrahimi**. I could not find COVID, plague**, or** virus on the matrix, but **GERM** is there. Musk is a rocket man, and **ROCKET** is there. However, it's easier to find **ROCKET** than to call a coin toss correctly. As for **EBRAHIMI**, he had to be on the matrix for me to publish it. He was found at skip +1 again, odds of about 7.35 to 1. I would have thought that the name would be more significant, but in Hebrew, the first five of the six letters of his name are simply **ABRAHAM**. He is a major focus of Torah. His name appears at skip +1 some 151 times. His earlier name, **ABRAM**, also appears 60 times, but that 4-letter name lacks the letter *hey* necessary to form **EBRAHIMI**. Overall, odds against this matrix were about 31,253,037 to 1. This does not factor in the axis term (Elon Musk). The **Codefinder** software that I use indicates an R-factor of 0.061. This means that it only had about an 86.9% chance to be found at least once at an ELS in Torah. In fact, there is just one ELS of the name in the wrapped Torah. It's at a skip of -72,286, which means that the computer had to make more than one pass through the 304,805 letters of Torah to find Sooke.

	A	B	C	D	E	F	G	H
1	Axis term ELON MUSK @ Skip -72286	SKIP	NUMBER	DIVIDE BY	THE QUOTIE	E QUOTIENT	POISSON EQUATION	EACH WORD OR
2	ELS RANK = 1	USED	IN	304805	EQUALS	X 874 LETTERS	PROBABILITY FOR	PHRASE HAS 1
3		ON	TORAH	LETTERS	FREQUENCY	EQUALS	THE WORD	CHANCE IN THE
4	•	FIGURE		IN THE	PER LETTER	WORD	APPEARING	IN THE FOLLOWING
5				TORAH		EXPECTANCY	AT LEAST ONCE	OF BEING HERE.
6								
7	MARS (3-letter @ Skips +1 only	1	2	304805	6.56157E-06	0.005734814	0.005718401	174.8740477
8	SABOTAGE @ +/- 1 and +/- 72286	72286	12	304805	3.93694E-05	0.034408884	0.033823631	29.56512898
9	CHINESE @ Skip +1 only	1	32	304805	0.000104985	0.091757025	0.087673205	11.40599346
10	EBRAHIMI (alef or yud ending) @ Skip +1 only	1	51	304805	0.00016732	0.146237759	0.136047733	7.350361346
11	SOOKE @ skips +2 to +28	2	79	304805	0.000259182	0.226525155	0.202700714	4.933381726
12	GERM @ Skips -72284 to -72311	-72284	125	304805	0.000410098	0.358425879	0.301224582	3.319782181
13	HITLER (2 spellings - tet and tav) @ Skips	-144570	148	304805	0.000485556	0.424376241	0.345822293	2.891658581
14	ROCKET @ Skips +72285 to +72312	72285	373	304805	0.001223733	1.069542822	0.656834631	1.522453222
15	COMBINED ODDS = H7*H8*H9*H10*H11*H12*H13*H14						3.19969E-08	31253037.33

Figure 5-23 Above: Spreadsheet showing odds against Figure 5-23.

IS CRAIG EBRAHIMI A CANADIAN REPTILIAN? See Figure 5-24 and its spreadsheet below.

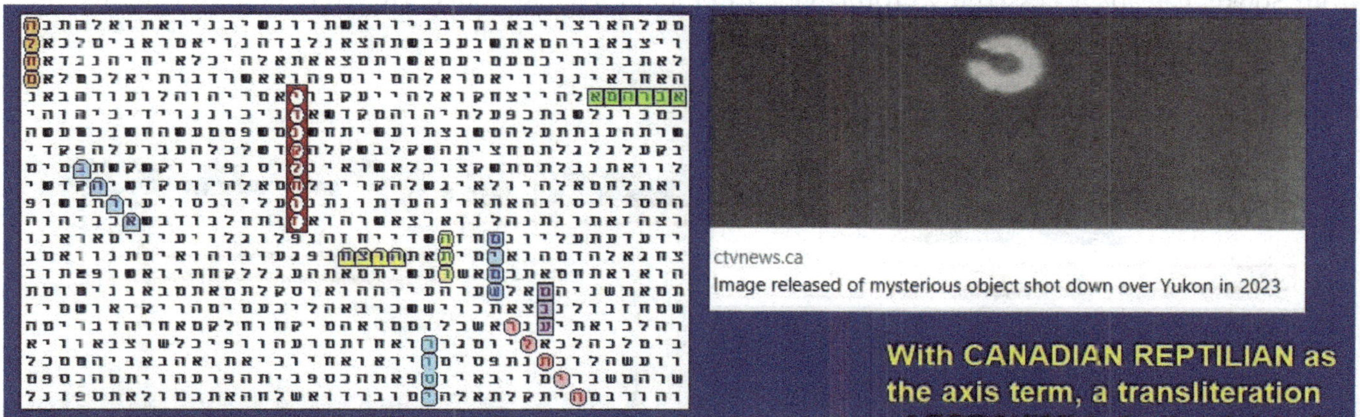

Term	Translation	Skip	Start	End
זוחלקנדי	CANADIAN REPTILIAN	-18463	Numbers 14:8.49	Exodus 4:6.2
אברהמא	EBRAHIMI	1	Exodus 4:5.37	Exodus 4:5.42
שמים	SKY	-18463	Deuteronomy 22:24.17	Numbers 24:16.25
ארהב	USA	-18462	Numbers 14:9.1	Leviticus 11:12.22
רוסי	RUSSIAN	18463	Genesis 26:26.18	Exodus 9:28.26
היתלר	HITLER	-18462	Exodus 9:28.17	Genesis 14:24.43
התר	PERMISSION	18463	Numbers 14:8.40	Deuteronomy 9:21.12
הרצח	THE MURDERER	1	Numbers 35:19.16	Numbers 35:19.19
עבמ	UFO	-18463	Genesis 14:24.41	Deuteronomy 22:24.14
מחלה	DISEASE	-18463	Genesis 42:14.31	Genesis 7:7.33

The ELS reference is 18,463 characters between rows.
The matrix starts at Genesis 7:6.26 and ends at Exodus 9:28.50.
The wrapped matrix spans 387,763 characters of the surface text.
It has 22 rows, 40 columns and contains 880 characters.

Figure 5-24 above. CANADIAN REPTILIAN is the axis term. First, let's look at an 18-row by 24-column (432-letter) area including and around CANADIAN REPTILIAN. Terms at the same absolute skip of CANADIAN REPTILIAN include UFO, SKY, and RUSSIAN. I have no evidence that Ebrahimi was working with the Russians, so this term requires its own spreadsheet to analyze. It is noted just because the Code may be warning us about them here. Also at the same skip is PERMISSION, which shares a letter resh with THE MURDERER (open text). I have evidence that Craig was involved with spreading the COVID DISEASE, which killed at least 18 million people. He basically claims that other aliens sent him directions (coordinates) for finding his 27- pound meteorite, which appears to be the source of the virus. DISEASE is also at the same absolute skip of CANADIAN REPTILIAN, but it only shows up on the larger, 880-letter matrix. USA and HITLER are both one skip off that of the axis term.

If we just look at the 432-letter matrix without ***RUSSIAN,*** it was found against odds of about 5,265,921 to 1. If ***RUSSIAN*** turns out to be a valid term (against odds of about 101 to 1), then odds for this matrix grow to about 1.6 billion to 1 if we include the 8-letter axis term. On the larger, 880-letter matrix, with no *Russian*, total odds against it are about 19.2 million to 1. This axis term, and what I heard from Craig, seems to indicate that there is a colony of Reptilians living near or around Craig. The picture shown on Figure 5-24 shows a UFO that was shot down by a U.S. aircraft over Canada. **Is there any evidence to support the claim that there is a reptilian colony in Canada?** On May 9, 2025, I got a print-screen of the Table below. It revealed that in all the world so far that month, Canada ranked number 2 with 1,538 pages of my research read. Only the U.S. had read more so far, with 6,443 pages. It might turn out that many of my readers there are ETs who are worried about me revealing their identity. On September 1, 2025 I received an alert that a reader from Sooke, Canada accessed my profile. Craig Ebrahimi lives in Sooke. He likely noticed then that there are two chapters in this book about him.

Canadian readers including Ebrahimi

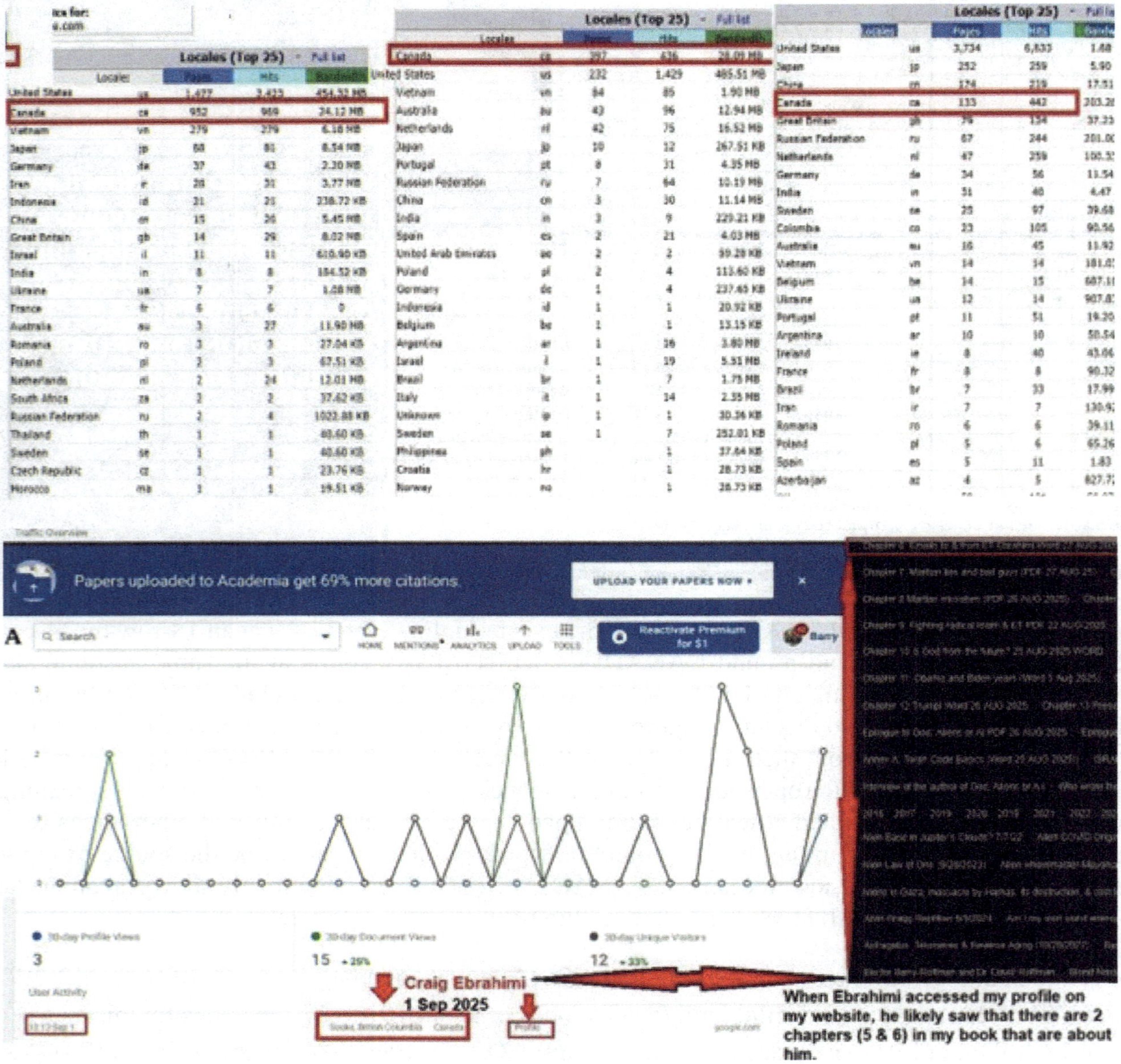

	A	B	C	D	E	F	G	H
1	*AXIS TERM CANADIAN REPTILIAN @ Skip -18463,* Axis term ELS rank = 1	SKIP	NUMBER	DIVIDE BY	THE QUOTIEN	E QUOTIENT	POISSON EQUATIO	EACH WORD
2	RUSSIAN is included but I found no evidence that Ebrahimi has	USED	IN	304805	EQUALS	X 432 LETTERS	PROBABILITY FOR	OR PHRASE
3	a Russian link. Without it or the axis term the matrix with white background	ON	TORAH	LETTERS	FREQUENCY	EQUALS	THE WORD	HAS 1 CHANCE
4	was only found against odds of about 3,353 to 1.	FIGURE		IN THE	PER LETTER	WORD	APPEARING	IN THE
5				TORAH		EXPECTANCY	AT LEAST ONCE	FOLLOWING OF
6	RUSSIAN @ Skips +/- 1 and +/- 18463	18463	7	304805	2.29655E-05	0.009921097	0.009872045	101.296131
7	THE MURDERER @ Skip +1 only	1	16	304805	5.24926E-05	0.022676793	0.022421607	44.59983532
8	EBRAHIMI (alef or yud ending) @ Skip +1 only	1	51	304805	0.00016732	0.072282279	0.069731736	14.3406726
9	HITLER (2 spellings) @ Skips -18462 to	-18462	82	304805	0.000269024	0.116218566	0.109719383	9.114159892
10	HEAVENS/SKY @ Skip +/- 18463 and +/- 1	-18463	132	304805	0.000433064	0.187083545	0.170625557	5.860786736
11	UFO @ Skips +/- 1 and +/- 18463	-18463	183	304805	0.000600384	0.259365824	0.228459277	4.37714771
12	PERMISSION @ Skips +/- 1 and +/- 18643	18643	369	304805	0.00121061	0.522983547	0.407250588	2.45549062
13	COMBINED SIGNIFICANCE WITHOUT AXIS TERM = H6*H7*H8*H9*H10*H11*H12						2.68845E-08	37196123.19
14								
15	Odds on a Ti-30 Calculator 1.636 2nd function LOG = 43.251381						0.023120649	43.251381
16	1/x where X = 43.251381 = .023120648. This value goes in cell G14.						6.21588E-10	1608783696
17	FINAL ODDS FOR 1 PASS THROUGH TO TORAH = H11*H13.							
18	IT TOOK PASSES THROUGH TORAH TO GET 1 MATCH SET. SO,							
19	88875.18332						Without Russian	5265921.012
20								
21	*AXIS TERM CANADIAN REPTILIAN @ Skip -18463*	SKIP	NUMBER	DIVIDE BY	THE QUOTIEN	E QUOTIENT	POISSON EQUATIO	EACH WORD
22	axis term ELS rank = 1	USED	IN	304805	EQUALS	X 880 LETTERS	PROBABILITY FOR	OR PHRASE
23	.	ON	TORAH	LETTERS	FREQUENCY	EQUALS	THE WORD	HAS 1 CHANCE
24		FIGURE		IN THE	PER LETTER	WORD	APPEARING	IN THE
25				TORAH		EXPECTANCY	AT LEAST ONCE	FOLLOWING OF
26	THE MURDERER @ Skip +1 only	1	16	304805	5.24926E-05	0.046193468	0.04514279	22.15193171
27	DISEASE @ Skip +/- 1 and +/- 18463	-18463	37	304805	0.000121389	0.106822395	0.10131473	9.870233098
28	EBRAHIMI (alef or yud ending) @ Skip +1 only	1	51	304805	0.00016732	0.147241679	0.136914637	7.303820967
29	HITLER (2 spellings) @ Skips -18462 to -18559	-18462	82	304805	0.000269024	0.236741523	0.21080475	4.743726132
30	HEAVEN/SKY @ Skips +/- 1 and +/- 18463	-18463	132	304805	0.000433064	0.381096111	0.316887768	3.155691388
31	UFO @ Skips +/- 1 and +/- 18463	-18463	183	304805	0.000600384	0.52833779	0.410415832	2.436553182
32	PERMISSION@ Skipa +/- 1 and +/- 18463	-18463	369	304805	0.00121061	1.065336855	0.655388249	1.525813137
33	USA @ Skips -18462 to -18865	-18462	3204	304805	0.010511639	9.250241958	0.999903912	1.000096098
34	Combined odds with no axis term = H26*H27*H28*H29*H30*H31*H32*H33						1.12507E-05	88883.72401
35	RUSSIAN @ Skips +/- 1 and +/- 18463	18463	7	304805	2.29655E-05	0.020209642	0.020006796	49.98301529
36	COMBINED ODDS AGAINST WITH RUSSIAN = H26*H27*H28*H29*H30*H31*H32*H33*H35						2.2509E-07	4442676.537
37	AXIS TERM R FACTOR VALUE (1.636)						2.2509E-07	43.2513831
38	FINAL AGAINST INCLUDING CANADIAN REPTILIAN AXIS TERM = H33*H34						5.20422E-09	192151904.9

Craig Ebrahimi, as noted above, claimed to be an alien. Specifically, after an argument with me, he sent me this email on March 27, 2020:

On Fri, Mar 27, 2020 at 4:43 PM Craig Ebrahimi craigebrahimi@hotmail.com wrote:

"Guy, your dumb ass shit keeps getting sent to me by one of your Hebrew followers. You got to be one of the biggest dummy humans I ever came across. You talk this shit in your writing ABOUT my biotechnology, once again you're dumb! You don't even know you're dealing with fucking real extra-terrestrial self-made multi-millionaire since teens, you idiot. If you looked into these eyes, I would eat your soul, puppet boy."

How does the Canadian government view Craig? I had a phone conversation about it on or around December 29, 2020. In checking my email record with him (in chapter 6), I see:

"Craig reports that the Canadian Royal Police are stopping him every week and that he is afraid to leave Canada for fear that they will not let him back in. I surmise that Fort Huachuca is likely keeping him under their thumb, but they have yet to contact me directly to confirm it."

Hostile reptilians in Canada may be the real reason why President Trump wants to acquire Canada.

Additional Sources:

SEE THE POWERPOINT AT https://roffmanmarsresearch.com/?page_id=2467

The record of the matrices that I have found relating to him and Mars and posted on my website (RoffmanMarsResearch.com) is as follows:

- **Alien COVID Origin 7/24/2024** – Page **Alien COVID Origin 7/24/2024**. Menu item 5

  of 114.

- **Meteorological Implications: Evidence of Life on Mars?** (14-page Roffman review of Joseph) – Page **Meteorological Implications: Evidence of Life on Mars?** (14-page Roffman review of Joseph). Menu item 9 of 114.

- **Israel's top scientist, Eshed, reveals that people are on Mars. 12/12/23** – Page **Israel's top scientist, Eshed, reveals people on Mars. 12/12/23.** Menu item 89 of 114.

WHAT OTHER WORLDS MAY HAVE VISITED US OR INFILTRATED OUR

- **GOVERNMENT? Alien Base in Jupiter's Clouds?** (7/7/2022) – Page **Alien Base in Jupiter's Clouds?** (7/7/2022). Menu item 56 of 114.

- **ISAAC & AN INTELLIGENCE ESTIMATE** (12/22/2022). Sub item number 2 under

- **Nazi/Alien Plan to Conquer the U.S.** (12/19/22).

- **TIC TAC UFOs** – Page **TIC TAC UFOs.** Menu item 23 of 114.

- **Part 1- Gilbert Levin Found Life on Mars** – Page **Part 1- Gilbert Levin Found Life on Mars.** Menu item 10 of 114.

- **Brandenburg & Martian Nuclear War. 3/22/2023** – Page **Brandenburg & Martian Nuclear War. 3/22/2023.** Menu item 76 of 114.

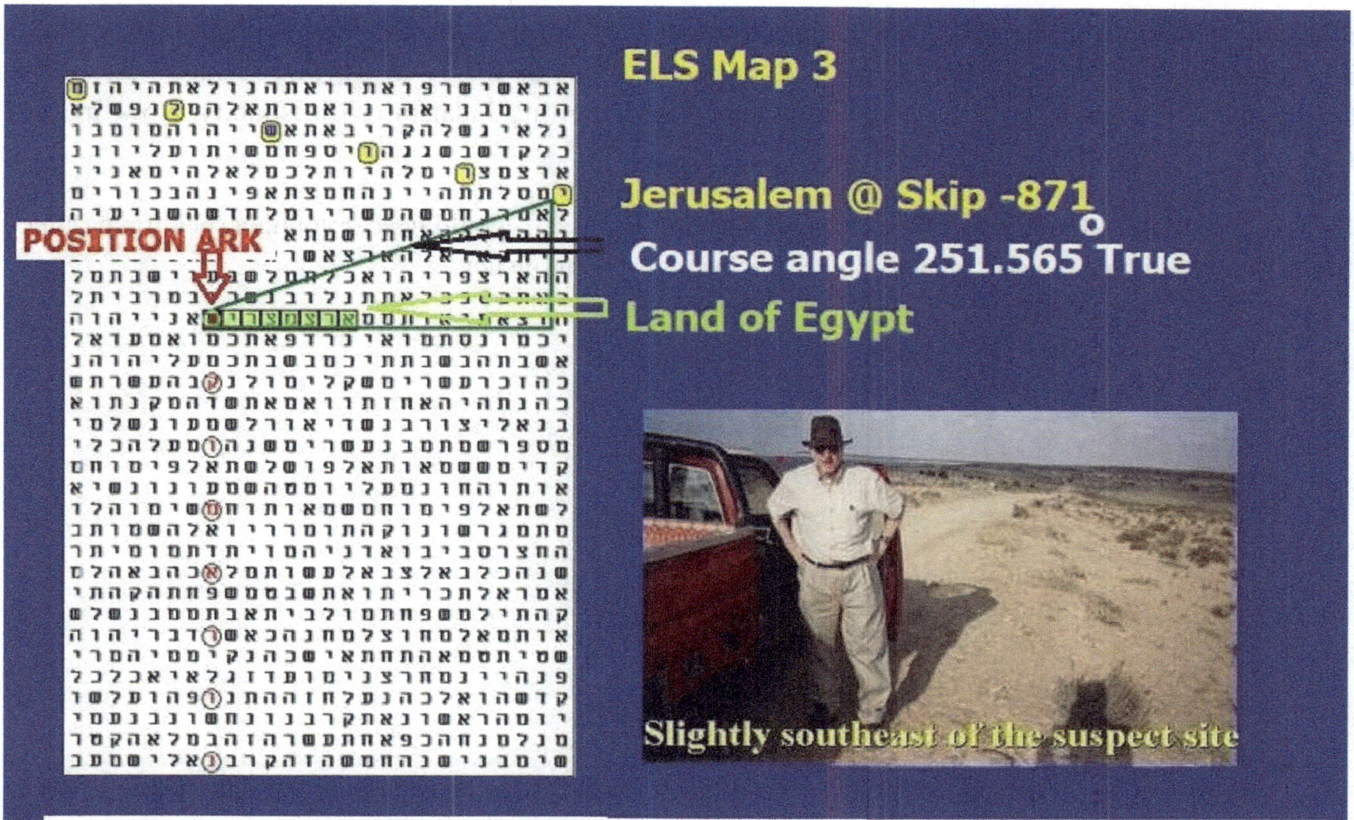

Term	Translation	Skip	Start	End
מקוםארון	POSITION ARK	2628	Leviticus 25:55.45	Numbers 7:53.56
ארצמצרים	LAND OF EGYPT	1	Leviticus 25:55.38	Leviticus 25:55.45
ירושלם	JERUSALEM	-871	Leviticus 23:17.34	Leviticus 20:14.52

The ELS reference is 876 characters between rows.
The matrix starts at Leviticus 20:14.27 and ends at Numbers 7:53.63.
The matrix spans 28,058 characters of the surface text.
It has 33 rows, 26 columns and contains 858 characters.

How is the course angle calculated? The ordinate is six rows down from the first letter (*yud*) of **JERUSALEM.** The abscissa is 18 columns to the left with the first letter (mem) of **POSITION** (of) **ARK** (it's also the last letter of **LAND OF EGYPT**). The ratio of rows (6) to columns (18) is 0.33333. On a TI-30XA calculator, hit the second function and tangent. This angle is the arctangent (18 43494882 degrees). Next, since the direction is in the third quadrant, hit the +/- button and get

-18.43494882 degrees. To this add 270 degrees and get 251.5650512 degrees True from **JERUSALEM** (Temple Mount) to **POSITION** (of) **ARK** on the El Zuqba peninsula in the Northern Sinai desert. The site sits in a military restricted zone forbidden to foreigners. On my last expedition, with some help via Egyptian General Benil Shaddah, I got within a mile of it, but the men that I paid to drive me there (over sand dunes in a truck with low-inflated tires) did not want me to get out of the truck, and I was not permitted to use my metal detector. Nor did I have any GPS equipment with me. The driver said I might be shot if I were spotted in my Indiana Jones hat. At that time, Dr. Zawi Hawass was in charge of Egyptian antiquities. He was adamantly opposed to helping any expedition aimed at supporting Judaism or Christianity. Now, in 2025, I am 78.

Even if I could raise the money and obtain the permits required, I would need a younger, healthy partner. If you want to be that person, please contact me through my publisher.

Chapter 6
Emails to & from ET Ebrahimi

This chapter reveals all emails to and from a man who claims to be an extraterrestrial alien. It also focuses on whether or not he is the source of COVID-19. It documents everything he wrote to me by email or said to me for over a year. His claim is related to remnants of life on the 27-pound meteorite that he recovered from the seafloor in the Strait of Juan De Fuca off his current hometown of Sooke, British Columbia, Canada, between March 1 and March 6, 2018.

The email source material is for the period February 16, 2019, until December 25, 2022, with Craig Ebrahimi below being audited to just deal with issues related to why Craig is suspected of having spread COVID-19 around the Earth. There were also frequent phone calls that I discussed, but they were not recorded by me, so their accuracy is related to how good my memory is.

EMAILS: Times given are for different time zones. I gained access to some mails written by Craig before January 26, 2020, because he forgot to delete them. He was quite angry when he realized what he had done, but the cat was out of the bag. While they didn't speak with me about any of them, I could trace many (if not almost all) of my readers to DOD. Hopefully, the Trump Administration can use some of what is included below to go after Anthony Fauci, a criminal who had to be pardoned by the false President Biden in the last 15 minutes while he illegally occupied the office of President.

On Sat, Feb 16, 2019, at 2:52 PM, Craig Ebrahimi craigebrahimi@hotmail.com wrote to Naveen Jain:

Yes, it is fossil E.T.

Who is Naveen Jain? Wikipedia says:

Naveen K. Jain (born 6 September 1959) is an Indian-American business executive, entrepreneur, and the founder and former CEO of InfoSpace. InfoSpace briefly became one of the largest internet companies in the American Northwest, before the crash of the dot-com bubble and a series of lawsuits involving Jain. In 2010, Jain co-founded Moon Express where he was the Executive Chairman, and in 2016, he founded Viome, where he is the CEO.

From: Naveen Jain naveen@viome.com Sent: February 16, 2019 3:36 PM To: Craig Ebrahimi craigebrahimi@hotmail.com

Cc: Naveen.jain@gmail.com Naveen.jain@gmail.com Subject: Re: Fw: Fw: Discovery in a Meteorite uvic bc *Interesting. I will love to see it. Are you selling it?*

Naveen Naveen Jain

Founder & CEO, Viome, Inc. M: 425-830-6556

MAY 10, 2019, 4:23 pm from Dr. Humphrey to Craig, later forwarded to Barry on Thu, February 6, 2020.

BARRY, THIS FOSSIL INVISIBLE TO MOST MICROSCOPES.

From: Elaine Humphrey ech@uvic.ca Sent: May 10, 2019 4:23 PM To: Craig Ebrahimi craigebrahimi@hotmail.com Subject: sem today.

Hi Craig. I managed an hour or so on the SEM today under my own time to look at the brown rock. I consolidated the files I have on Dropbox into one folder. The May 10, 2019 folder is all I took today. [Dropbox link] The iPhone pictures show the white spots, and the SEM pictures do not. The SEM (note: SEM means scanning electron microscope) pictures with PDBSE in the baseline are backscattered images showing the different grains in the rock. The attached and the EDS images of the same spot seem not to correlate at all. Rock is interesting! I think I may have mentioned this before! Best wishes for your help.

Elaine

PS: There was no waiting for the SEM to pump down. I've returned the rock to the vacuum.

Fri, 17 May 2019 at 00:13.

*Hi Chandra (Wickramasinghe). I have a client, Craig Ebrahimi, who brought in a meteorite. At first, it looked like gabbro, but the chemical analysis (by Act Labs) proved it to be a meteorite. I understand that he sent you thin sections, outside crust, a piece of th**e rock with an embryo fossil in a white circle.** I have a very small sample for SEM and EDS*

(Note: EDS means Energy-dispersive X-ray spectroscopy (EDS), It is a technique used in geology to analyze the chemical composition of rocks, minerals, and soils. It's also known as EDX or XEDS.) Dr. Humphrey continues:

Craig suggested that if you have finished with his samples, is it possible to send them to me? Since I will be in the UK at the end of June for a month, I could pick them up. If you are still working with them, please keep us in the loop. Many thanks for your help.

Best wishes Elaine

Dr. Elaine C. Humphrey

Fellow of the Microscopy Society of America Past President Microscopical Society of Canada

The plot thickens:

May 20, 2019

Dear Elaine

Unfortunately, the meteorite package was left in a shed near our house, and it appears to have been stolen. I do not have it anymore.

Sorry about this. Sincerely ChandraProf. Chandra Wickramasinghe

Honorary Professor University of Buckingham, Buckingham, UK Honorary Professor University of

Ruhuna, Matara, Sri Lanka Honorary Professor Sir John Kotelawala Defence University, Sri Lanka Buckingham astrobiology website: [link]

Ruhuna astrobiology website: [link] Personal website: www.profchandra.org **As will be shown later, Wickramasinghe is a liar. He continues below:** *May 20, 2019, at 1:15:13 PM PDT*

From: Chandra Wickramasinghe ncwick@gmail.com Date: May 20, 2019, at 1:15:13 PM PDT To: Elaine Humphrey ech@uvic.ca Subject: Re: meteorite

Reply-To: ncwick@gmail.com

Hi Elaine. Our house was in chaos after some international travel, so I stored all valuable things in an outhouse. Some months back, I discovered that many things including the box with the meteorite had disappeared. I was waiting to get a colleague to handle the experiments on this, but it never happened. Sorry about this – not my fault.

Best Chandra Prof. Chandra

May 20, 2019, 1:32 PM

From: Elaine Humphrey ech@uvic.ca

Dr. Humphrey responded as follows on May 20, 2019, 1:32 PM: *To: Craig Ebrahimi craigebrahimi@hotmail.com Subject: Fwd: meteorite FYI. Sent from my iPhone*

Mon, 20 May 2019 at 18:15, Elaine Humphrey ech@uvic.ca wrote:

Chandra. Oh dear, that's too bad. Did you get any information out of them before they were stolen? If so, have you published them?

Elaine

Sent from my iPhone

From: Jerzy Sawicki jasawicki@shaw.ca Sent: July 15, 2019, 2:54 PM To: Craig Ebrahimi

Subject: Re: raman This was a forwarded e-mail originally from someone named George. DEVIANT OFFWORLD FILM CLIP.

The date of the film is uncertain.

From: Karen Ziegler kziegler@unm.edu Sent: September 6, 2019, 1:08 PM

To: Craig Ebrahimi craigebrahimi@hotmail.com Subject: Oxygen isotope analyses of your rocks Hi Craig,

No, we can't do organics here. Sorry. In terms of oxygen isotopes, definitely neither one of your rocks is definitely Martian. Based only on my results, they could be: lunar (or E- chondrites, but they don't appear to be chondrites, D17O values are typical for the moon. You need to know the mineralogy and the mineral chemistry in order to determine which of these types they could be. Do you have EMPA analyses of the minerals?? Mn/Fe ratios (see attached figure)?

Let me know what you think. Cheers, Karen

There have been many claims that the occurrence of pandemic influenza and other viral outbreaks is correlated with the well-known 11-year sunspot cycle1–3, although the precise mechanism for such a causative connection had remained unclear... The next minimum between the current cycle 24 and cycle 25 was predicted to occur between July 2019 and September 2020 (Figure 1). Perhaps, we have now approached the deepest sunspot minimum for a century, with more 'spotless' days per week than in previous minima. On the basis of this data, there appears to be a prima facie case for expecting new viral strains to emerge over the coming months, and so it would be prudent for Public Health Authorities the world over to be vigilant and prepared for any necessary action. We need hardly be reminded that the specter of the 1918 devastating influenza pandemic stares us in the face from across a century.

NOTE: 15 days later, the first known coronavirus patient (Wei Guixian) started feeling ill.

On December 22, 2019, Craig is interviewed about his meteorites. The link for the 14-minute show is at https://disclosurechronicles.com/2019/12/22/major-discovery-nasa-test-crystallized- humanoid- in-moon-meteorite-2020-2021/. It's worth seeing and hearing it. Craig claims that the rocks have a lot of uranium and elements not from this world. It took six diamond blades to cut the rock.

Craig claims in the interview that there are 16 embryos and that the origin of the rocks is likely the moon, 4 Vesta, or beyond. This would appear to contradict a later claim that the non-Martian origin was due to a fight and personality conflict with Dr. Karen Ziegler at the University of New Mexico.

Temperature: Craig says the rock was in a room that was 25 degrees (C), but the rock in one section is 14 degrees (7:13 to 7:17). The embryonic section is 6 degrees, and it pulses to 12 degrees (7:20). I listened to this video again on July 11, 2025 during the final editing of this book and heard him say something major that he never discussed with me – namely that there was altered DNA found in the rock! He said the DNA was mostly human, but that a section had been spliced into it from some other life form. I know that I asked him several times about testing for amino acids (and DNA), but he never verbally let me know that he had this kind of finding. The video wants all kinds of identity verification. The show was on a site known as *thirdphaseof moon*.

Craig states that he found a second rock like it, but that one when cut open had something like dragon inside. I was aware of this second meteorite but he never told me about anything inside that resembled a dragon. He also said that the second rock had something like "swirled insectoid in it." Exactly how he found the first rock, even with extraterrestrial contacts, is not clear because he states that it was not just on the sea floor. Instead, it was buried under 12 feet of mud.

In January 2020, Craig was communicating with someone at NASA named Emery Smith (see an article about him at
https://historycollection.jsc.nasa.gov/JSCHistoryPortal/history/oral_histories/SmithEE/SmithEE_ 2-28-06.htm).

From: Emery Smith **emerysmith28@gmail.com** **Sent:** January 5, 2020 10:07 AM

To: Craig Ebrahimi **craigebrahimi@hotmail.com**

Cc: Olivia Smith **oliviasmithh33@gmail.com** **Subject:** Re: ET EMBRYO Biotechnology

Dear Craig,

Yes, I have lots of ideas. We need to get a CAT scan of it first, and then we can stereotactically, under fluoroscopy, make a small drill hole, utilizing a Midas REX High- Speed Neuro drill, into the meteorite where the being is. We can then extract some particles we can run the DNA test on, and any additional tests. The procedure should be quite easy with the right tools. I would also like to film the procedure as well and document it. Would you be interested in flying to Colorado to be on my show Cosmic Disclosure on GAIA TV to talk about the find first? All travel and expenses will be paid for. Let me know if this appeals to you. I am also filming a small series called *The Beings*, we could interview you as well for. Regardless, I am interested in seeing the specimen in BC, and thank you for the invite. I also want to be part of the analysis team and help in any way possible. I have a film production team as well at your disposal called ZeroCross Productions.

Sincerely, Emery Smith 202-999-6871

From: Craig Ebrahimi craigebrahimi@hotmail.com

To: BarrySRoffman@gmail.com

Date: Jan 26, 2020, 12:55 PM (Saturday)

Subject: Fw: Extraterrestrial embryo discovered in meteorite UVIC BC

My name is Craig Ebrahimi, research scientist/Founder of Deviant Off-World Technology

Barry, would you be interested in the research and biotechnology opportunity of the only pristine extraterrestrial embryo on the planet at the University of Victoria, BC?

From: Craig Ebrahimi **craigebrahimi@hotmail.com** **To:** BarrySRoffman@gmail.com

Date: Jan 28, 2020, 12:08 PM

Subject: Fw: #2 HED meteorite with crystallized reptilian.

Oxygen isotopes d17 HED Meteorite discovered in Sooke, BC. Raman data supports that the top center grain is an organic reptilian converted to diamond.

Craig

Comment by Roffman added on December 25, 2022:

Craig's January 28, 2020, email identifies the apparent embryo way before it is established that the fossil is from a living creature. In several conversations and emails up through today, Craig claimed to be an alien with the power to put me down just with his eyes.

While I initially assumed that Craig is a Nordic or other species related to humans, in different photographs that I have of him he has a different-looking face/shape. His lower legs seem too small for a human or a Nordic. A Nordic should be tall. This guy may be a shape-shifter, which Reptilians are said to be. The fact that he claims to have 16 Reptilian embryos may point to him being a Reptilian. They are generally hostile to humans, although our military is alleged to work with them. Craig is suspected by me as being the individual who brought COVID-19 to Earth.

The next Figure shows alleged alien Craig Ebrahimi shortly after we had a verbal fight. The remarks about alien uniforms seen by President Eisenhower on February 20, 1954, are taken from page 202 of the book *BEHOLD A PALE HORSE* by William Cooper. They were found by me after I caught an intelligence agent checking out a link on this site to the book, but before I was aware of the uniform remarks. If Craig is indeed wearing an alien military uniform here, then our military forces should look for others like it. In checking my website statistics on June 17, 2025, I have the *Pale Horse* book posted at https://arkcode.com/A%20Pale%20Horlse.pdf. Is anyone reading it? Yes. This month I see eight readers.

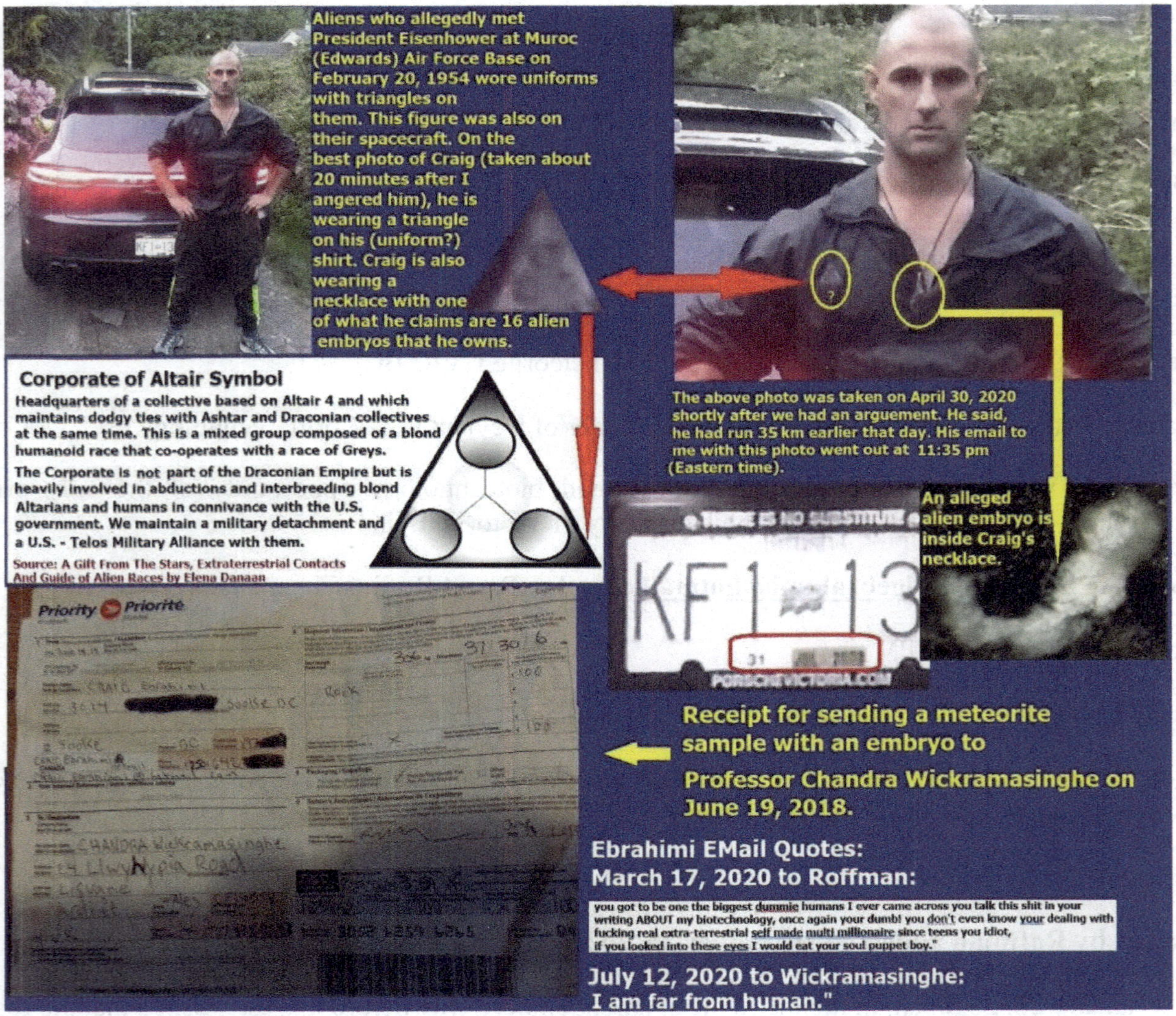

Figure 6-1 above The photo was taken on April 30, 2020, a few minutes after Craig and I had an argument.

During our argument, Craig told me, "Listen, that virus only works on weak humans. Bill Gates patented this coronavirus strain in 2013; the documents are available. You just gotta dig. All this has been a plan since the 1950s."

I saw those documents when COVID was ravaging the world.

E.T.s live in Sooke, BC, Canada? Elon Musk visited the area on August 6, 2019. He was seen in Metchosin, just a few miles from Sooke, but he vanished from there for a day. He may have met with Craig elsewhere – either at the home of an E.T. or aboard an E.T. ship. This was at a time between when Craig had sent a meteorite sample to Professor Chandra Wickramasinghe (June 19, 2018) and when COVID-19 broke out in China (December 2019).

From: *Craig Ebrahimi craigebrahimi@hotmail.com*

To: *BarrySRoffman@gmail.com*

Date: *Feb 2, 2020, 3:59 PM*

Subject: *The cure for diseases.*

Thank you, Barry, for your insight! Did you ever think I could be a walking, talking prophet who holds the medicine that will solve world diseases? And extend human longevity.

COMMENT BY ROFFMAN:

The above response from Craig was apparently his reaction to seeing his name encoded in the Torah. Attached to my previous e-mail to him were the matrices at http://arkcode.com/images/eaten_meteorites.png and http://arkcode.com/images/cure_man.png.

I cautioned Craig about claims to prophecy. However, it remains unclear who the voice was that told Craig where to look for the meteorite on the seafloor. I think it was other aliens, likely members of Craig's species.

Craig was told that I will not endorse him until I got answers in writing to my concerns about security. Our military is following this article closely. I said I would advise them or our government to consider buying the rock from him, but only if I think the thing is a legitimate security concern. For something this important, I said I don't want to rely on my memory. I needed answers in writing without any exaggeration or hyperbole.

COMMENT BY ROFFMAN FEB 3, 2020:

Craig claimed to hear a voice that told him exactly where to look for the meteorite. This raises the possibility that telepathic aliens directed the epidemic. How many people did it kill? On January 24, 2025, I found the following figures online:

COVID-19 Global Death Toll Estimate Is Three Times What Records Indicate. ***In this Lancet study, investigators estimate an excess mortality of 18.2 million deaths worldwide due to the COVID-19 pandemic between 2020-2021, although just 5.9 million are reported.***

It should be assumed that we were and are under a hostile biological attack by aliens. The U.S. Space Force must come up to speed quickly to respond. Frankly, since I know what is going on, the military should call me back to active duty ASAP to help them deal with the threat.

February 3, 2020 @ 11:20 am Craig wrote:

I intended to help humanity. Sincerely. I know I hold unimaginable biotechnology to help and advance humankind. Yes, it should be sold to the Institute of Disease Control in Atlanta or the Wyss Institute https://wyss.harvard.edu/ in the control of George Church at https://en.wikipedia.org/wiki/George_Church_(geneticist). For 2+ years, I have been stressing about the pristine extraterrestrial embryo while now 2mm from the exposed surface. We should be meeting and going over the data, and then you fly home and put together a report for the sale.

Craig Ebrahimi

The following is Craig's reply to an email that I sent to him on or before February 3, 2020.

From: Craig Ebrahimi craigebrahimi@hotmail.com

to: BarrySRoffman@gmail.com

date: Feb 3, 2020, 4:19 PM

Subject: Re: answers

1. Do you think an alien was communicating with you? Have you ever had a UFO encounter or a possible abduction experience?

   Aliens communicate with telepathy.

2. *Response: I have B- blood, which is from the Persian royal bloodline, and so do my two baby boys. My wife is O-. I have always been in communication with something, but humans would classify me as just crazy. So, to answer your question, I do not believe I was abducted, but I have many pictures of a green orb that follows my family and me around. When you zoom in to the amazingly clear pictures, it seems to resemble Ezekiel's wheel.*

   2 Do you still hear voices? *Response: YES.*

3. How did you know to contact me? *Response: IT HAPPENED.*

4. Are you now saying that you have no measured speed increase and that all enhanced effects are mental? Can you name any positive physical changes? *Response: PHYSICAL AND MENTAL, YES.*

5. All uranium is radioactive, but not all uranium is fissionable. I need to know which isotopes were found before I fly out there. Please look at the chemistry report and tell me what elements and/or isotopes are mentioned there. *Response: I HAVE ONLY HAD THE OXYGEN ISOTOPES DONE (OXYGEN 17 HED). THE CHEMISTRY WAS JUST A BASIC METEORITE TEST. 5 GRAM SAMPLES WERE PROVIDED. I WILL SEND YOU THE REPORT.*

From: Craig Ebrahimi craigebrahimi@hotmail.com To: <BarrySRoffman@gmail.com>

from: <craigebrahimi@hotmail.com>

to: "BarrySRoffman@gmail.com" <

date: Feb 3, 2020, 11:17 pm

"I have not told you that the first scientist to work on this meteorite was Chandra Wickramasinghe. I sent him an embryo. There is more than one, and he emailed Dr. Elaine (Humphrey) and told us someone broke into his house and stole my samples.

I have all the emails.

On February 4, 2020:

from: Craig Ebrahimi <craigebrahimi@hotmail.com>

to: "BarrySRoffman@gmail.com" <BarrySRoffman@gmail.com>

date: Feb 4, 2020, 11:17 AM

subject: ET HOLDING TECHNOLOGY

Barry, you asked about VOICES. I would answer that as they're not loud yelling voices, more like telepathy voices/ideas/direction. My pictures of the orb WE DID NOT WANT TO MENTION.

While I will show in person while my children are in the pictures, Barry, to understand the chemistry, just read the final report.

This is the ET holding some type of technology! If you were looking at it in person, it looks like it will come out of the rock. I have many. Dr. Elaine (Humphrey) is holding a sample that she is guarding with her life; it seems to have an elongated head. I will send pictures when she gives me permission.

SINCERELY,

Craig

Dr. Humphrey chimed in next:

From: Elaine Humphrey ech@uvic.ca Sent: February 7, 2020 11:56 AM To: Craig Ebrahimi craigebrahimi@hotmail.com Subject: EDS report 2. Hi Craig,

Looking through our old data, I notice that several of our spectra show relatively high amounts of iridium. It is interesting to note that this is not consistent over all the surface. Only some of the grains contain iridium.

https://www.dropbox.com/sh/ud476i4c1pe5jx2/AABn1w2aGFHe2u54evJCxB5Da?dl=0 ***sample b spectrum 2b.doc*** *shows relatively high amounts of iridium.*

Best wishes, Elaine

From: Elaine Humphrey ech@uvic.ca

Hi Craig,

I put the pictures into a Dropbox so you can look at them. ***https://www.dropbox.com/sh/nyxnvggkfiy8fds/AABNdL7uQsN1Xu4uvQSuufVoa?dl=0*** *Best wishes, Elaine*

On Wed, Feb 12, 2020, at 6:20 PM, Craig Ebrahimi

Subject: (Interesting read) My blood B-Anunnaki **craigebrahimi@hotmail.com** *wrote:*

https://www.gaia.com/article/rh-negative-blood-type-alien

Craig raises the issue of whether he has alien blood. The article cited includes the following text:

The Nephilim of biblical lore are responsible for Rh-negative blood types. In the Book of Enoch, the Nephilim, also known as the Watchers, descend from the heavens and mate with humans, creating a human-angel hybrid. This group of angels and their offspring were wiped out in the great biblical deluge, though some were said to have survived, leaving the Rh-negative blood distinction.

Note: Figure 3-5 in Chapter 3 of this book offers reason to believe that Nephilim are Reptilians. This may be evidence that Ebrahimi is at least a human-reptilian hybrid.

Craig continues:

Another otherworldly theory is that the Anunnaki, the extraterrestrial race that helped establish ancient Mesopotamian civilizations, engineered or crossbred with humans, and that some part of this process created the Rh-negative blood type. It's alleged that people with Rh-negative blood have distinct physical features paired with a predilection for psychic phenomena and alien abductions. Some of those features are:

Sensitive vision and particular sensitivity to sunlight

Extra vertebrae

A human typically has 33 vertebrae, while most mammals have a similar number of cervical (neck) vertebrae, usually seven, with the total number of thoracic and lumbar vertebrae varying depending on the species, but generally totaling around 19-20 vertebrae across most mammals.

The *Journal of Human Evolution* (Volume 179, June 2023) states, "ancestral primates were characterized by 29 precaudal vertebrae, with the most common formula being seven cervical, 13 thoracic, six lumbar, and three sacral vertebrae. Extant hominoids evolved tail loss and a reduced lumbar column via sacralization (homeotic transition at the last lumbar vertebra). Our results also indicate that the ancestral hylobatid had seven cervical, 13 thoracic, five lumbar, and four sacral vertebrae, and the ancestral hominid had seven cervical, 13 thoracic, four lumbar, and five sacral vertebrae. The last common ancestor of humans and chimpanzees likely either retained this ancestral hominid formula or was characterized by an additional sacral vertebra, possibly acquired through a homeotic shift at the sacrococcygeal border."

Lower body temperature. On April 30, 2020, I was told by Craig that he originally wanted to hire me to understand why the meteorite pulsed by coldness. He also told me that he and his wife had eaten a little of it.

- High intelligence
- Higher blood pressure
- Red or reddish hair

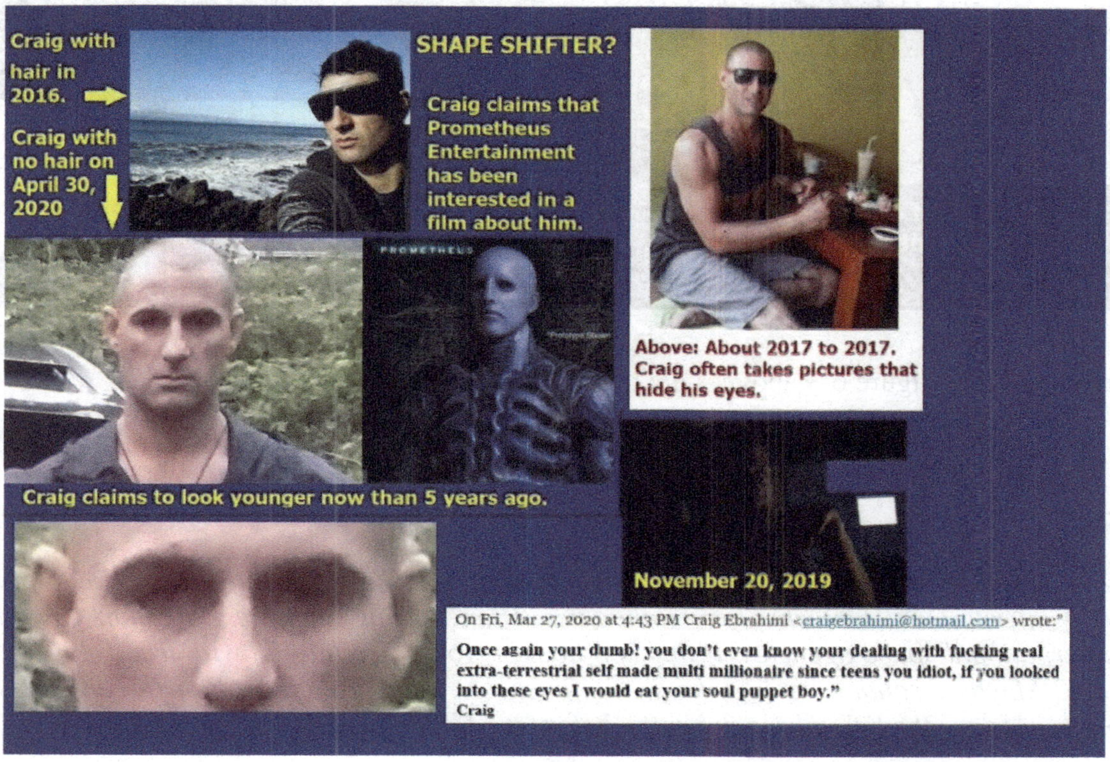

Figure 6-2. With respect to sensitive vision, the two pictures above (2016 to 2017) show Craig wearing sunglasses. While the top-left photo shows Craig outside on a sunny day, the top-right photo has him still wearing them indoors. In the photo labeled November 20, 2019, the picture was taken in a dark room where it is hard to see important facial features. At one point, Craig mocked me for not knowing that it was NOT him in the November 20, 2019 photo, but my overall impression was that he was trying to cover up the mistake of letting me know what he looks like. Other photos led me to believe that we are dealing with a possible shape-shifter.

I have no data on the number of vertebrae that Craig has. However, the first thing that I noticed upon seeing the April 30, 2020 photo in Figure 6-1 was how wide his chest was and how small his legs were. His physique seemed to suggest a potentially different skeletal frame. See Figure 6-3.

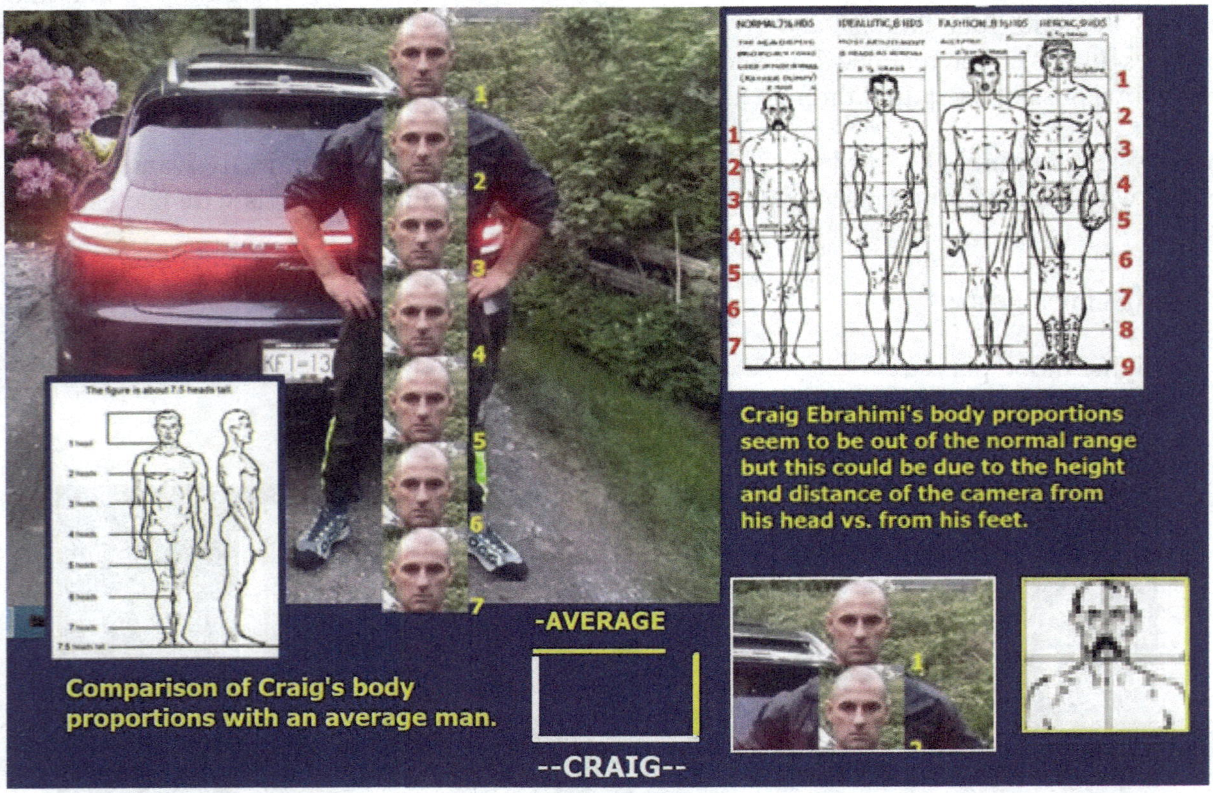

Extra vertebrae? See Figure 6-3 above. I next heard from Craig in March, 2020:

March 2, 2020.

Hello Barry,

I have been advised legally not to give any more data to anyone. But I will say that I have a lot of interest from leading international biotechnology companies. I am dealing with the Canadian government and CEOs of international space agencies. If this is a hoax, I would be charged with fraud. In Canada, what would I gain? Maybe 3 square meals a day in confinement? Your flight from Florida direct is still on the table if you want to see the ET embryo in person before it's sold.

In response, I discussed my reluctance to fly anywhere until the Wuhan coronavirus is under control. I also pointed out that no known scientist has concluded that there is, in fact, an embryo in his meteorite. If they have detected any amino acids, I have not yet seen proof of that either. Later on 3/2/2020, Craig wrote me saying,

"Barry, when I receive new data of organic extraction in the UK, I will keep you updated.

Aaron Burton is a PhD at the Johnson Space Center in Houston. This email shows that he had meteorite samples.

From: *Burton, Aaron (JSC-XI311)* *aaron.burton@nasa.gov*

Sent: *March 10, 2020 6:22 AM*

To: *Craig Ebrahimi craigebrahimi@hotmail.com*

Subject: *RE: organic extraction Hi Craig,*

It doesn't look like my lab will be functional any time soon. I'll get the samples sent back to you this week. Sorry!

Note: Although Craig told me that Burton's lab was shut down due to the COVID-19 plague, on the thirdphasemoon video back on December 22, 2019 Craig discussed Burton. He said that the scientist was in charge of Mars rovers and that he was excited about the rock for four months. Then he told a colleague about it and that that was why his lab was shut down for no apparent reason. The implication was that NASA did not want the public to know about such life in space. Craig bragged about gaining new physical abilities, like running 35 km every day. He boasted about the rock being able to extend lifespans. Along this line, the email record shows the following:

Chandra Wickramasinghe says Craig has the cure from space. ***From:*** *Chandra Wickramasinghe ncwick@gmail.com* ***Sent:*** *March 15, 2020 10:46 AM*

To: *Craig Ebrahimi craigebrahimi@hotmail.com*

Subject: *Re: longevity vaccine from space Hello Craig,*

Yes, I would much like to help. I believe you have the cure from space, but I don't have the institutional resources available to me.

cc *Personal website: www.profchandra.org*

Buckingham astrobiology website: https://www.buckinghtionsam.ac.uk/research/bcab Ruhuna astrobiology website: http://physics.ruh.ac.lk/astrobiology/

ISPA website: www.ispajapan.com

ROFFMAN COMMENT: Craig claims to be an alien below, threatens me with his eyes. I have no idea about who my "Hebrew follower" is. It could be Mossad, but they never show up on an IP address check. So far, all my intelligence followers are with DOD, largely through Fort Huachuca, or with the UK MOD. They probably have people who are Jewish and read Hebrew as well as Arabic, but I haven't met anyone like that.

On Fri, Mar 27, 2020 at 4:43 PM, Craig Ebrahimi craigebrahimi@hotmail.com wrote: *"Guy, your dumb ass shit keeps getting sent to me by one of your Hebrew followers. You got to be one of the biggest dummy humans I ever came across. You talk this shit in your writing ABOUT my biotechnology. Once again, you're dumb!* ***You don't even know you're dealing with fucking*** real extraterrestrial self-made multi-millionaire since teens, you idiot. If you looked into these eyes, I would eat your soul, puppet boy."

Barry Roffman <barrysroffman@gmail.com>Apr 1, 2020, 1:13 PM

From: Barry To: Craig April 1, 2020 TYPO CORRECT Apology

Hello Craig,

This e-mail serves as a partial apology. Yesterday, I created a new folder on Gmail that contains 48 of your previous e-mails in chronological order, received. This made it much easier to start at the beginning on January 26 and go through everything you sent me sequentially. I am finding answers to many of my questions. If you want to know why I missed a lot before, it is for the following reasons:

1. *Some of your attachments came up as possibly infected with malware and thus were not opened by me before.*
2. *There was nothing like a cover letter to tell me what to pay attention to.*
3. *Some were nothing more than links to Alex Jones videos that were not of interest to me or thought to be germane.*
4. *There were stories that I was working on that were of greater interest, like the impeachment of President Trump.*
5. *I was getting a mixture of truth and exaggeration from you.*

I am now slowly going through each item that you sent. Meanwhile, I strongly suggest that you and everyone WHO handled the rocks be tested for the COVID-19 virus.

It remains true that my primary interest in what you have found is related to national security. My reports, published on my site, will continue to be updated with a goal of providing maximum clarity and accuracy. I realize that you have financial interests here, but what you found might be killing tens of thousands of people and possibly millions of people in the near future. (Note: That prediction came true.) So, while you are understandably pissed at me for publishing everything you send me, sorry, but millions of lives trump profit for you. It may be that you have antibodies that will be profitable for you, but the longer you wait, the more there will be antibodies in recovered patients that may work as well.

Finally, you have played with the implications that you are an alien. If you believe this, you are either crazy or (if it's true) a real danger to our planet. You need to get sober and send me a clear statement about your identity.

Barry

That Craig sent me copies of so many highly sensitive emails between himself and extremely rich or famous people is evidence that he is not smarter than people like myself are. His reaction to what I sent him is given below:

Apr 2, 2020, 2:54 PM

To BarrySRoffman@gmail.com

I DID NOT SEND YOU THAT OR ANYTHING!

Roffman response to Craig's denial and a warning about the possible consequences of not cooperating with DoD or the UK MoD. EDX = energy-dispersive microanalysis. See CRISTOBALITE AND HEMATITE PARTICLES IN THE HUMAN BRAIN

(By M. Kopani et al., 2016) at **http://www.arkcode.com/photo5_76.html**

for what might be in Craig's brain.

My response to Craig was as follows: APR 4, 2020

from: Barry Roffman <barrysroffman@gmail.com>

to: Craig Ebrahimi <craigebrahimi@hotmail.com>

date: Apr 3, 2020, 11:22 AM

subject: Re: BARRY EDX DATA OF #2BR SMALLER EURCRITE HED

*Ah, but you did. I have enough from you to see that the thief is almost certainly your "friend" Wickramasinghe in the UK. I don't know if they'll do it, but I'm asking the UK MOD to pick him up for interrogation. I'm pretty sure that he stole your sh*t, sold it to the Chinese lab in Wuhan, and started the current plague. That would make him guilty of infecting over a million people so far and killing at least 54,000 people.*

*This apparent *sshole has destroyed the world economy and taken at least one U.S. aircraft carrier out of service. If he is not guilty, then he can start with lie detectors to prove it. If you warn him or discuss this with him, that could be seen as collaborating with the enemy. While I cannot be certain, it's wise to assume that my phone is bugged by my Intel friends. Therefore, assume that if you call me to help in the investigation, Uncle Sam will hear you and your point of view.*

Craig continues,

"In no way should you view the Government here as your enemy."

ROFFMAN COMMENT: The Obama and Biden administrations have shown themselves to be the enemy of America and all humanity. That was not yet clear to me when I wrote the above. Indeed, there are many Torah Code matrices that insinuate that both Obama and Biden are alien infiltrators. See Chapter 3 of this book, and in particular:

- Figure 3-1 (axis term ALIEN IN OUR GOVERNMENT at the same skip as BIDEN),
- Figure 3-4 (axis term OBAMA IS REPTILIAN),
- Figure 3-5 (axis term CHUCK SCHUMER at the same skip as REPTILIAN and UFO),
- Figure 3-6 (MARK MILLEY at the same skip as GENERAL and UFO with REPTILIAN at skip -1),
- Figure 3-8 (with GENERAL, TO HYBRIDIZE and UFO all at the same absolute skip and GENERAL at skip -1),
- Figure 3-9 (with REPTILIAN GOVERNMENT and INVASION at the same skip),
- Figure 3-10 (BIDEN [is] AN ALIEN as the axis term, where UFO and MARS are at the same absolute skip),
- Figure 3-11 (with axis term CHRIS WRAY as the axis term with FBI and UFO at the same skip),
- Figure 3-12 GARLAND FROM AN ALIEN LAND,
- Figure 3-13 (ALVIN BRAG is at the axis term with REPTILIAN at the same skip),
- Figure 3-14 (with A. MAYORKAS as the axis term touching REPTILIAN at skip -1),
- Figure 3-16 (with CIAKAHRR, a synonym of REPTILIAN as the axis term at the same absolute skip as GRAY, HITLER, C. WRAY, HOOVER, CONSPIRACY, BUSH, USA,

  and GAZA),
- Figure 3-17 (with HITLER in sequence with HARRIS),
- Figure 3-18 with (Governor and VP candidate TIM WALZ sharing a letter with REPTILIAN at skip -1),
- Figure 3-19 (with former Director KIM CHEATLE, responsible for Trump's security on the day he was shot and wounded, at the same skip as BIDEN, ALIENS, PLEIADES, and UFO),
- Figure 3-20, JOHN DURHAM with REPTILIAN, and
- Figure 3-21 HAKEEM JEFFRIES with REPTILIAN.

Later in Chapter 3 I added two matrices that pointed to A. (Adam) Schiff as being reptilian. Extraordinary claims require extraordinary evidence. I submit that the above Torah Code matrices are extraordinary evidence that the Torah is encoded with a warning about hostile Reptilian aliens running much of our government.

I continued my response to Ebrahimi on April 3, 2020, as follows:

"They are trying to keep the Chinese from killing us all. What is out there is a threat to all humanity. If this guy sold your stuff to Wuhan, he may turn out to be one of the biggest killers in history. You do not want to be seen backing him. He had your stuff for a long time, did not safeguard your property, and did not finish the analysis. It made no sense that you would eat the stuff unless Chandra did a study that showed improved life expectancy in animals.

I have asked you to call me several times. You have chosen to ignore me or send silly stuff, but some of that silly stuff had meaningful email trails (which I am still going through). Frankly, at this time, I would just prefer to turn over the investigation to DoD or the UK MOD. Bottom line – at any other time in history, this would all be an academic question, but now there is no difference

between trying to get the full truth here and trying to find a nuclear bomb buried under one of our cities by the Chinese, North Koreans, or (other) crazies. There is still time for you to come out of this with money, fame, and a good place in history. But if you hesitate and try to back your thief buddy, your fate will likely not be so good. I am not your average investigator.

I can read the Code, I've found you in it, and Wickramasinghe. Barry"

Response from Ebrahimi:

BARRY, NOW YOU'RE ON THE RIGHT TRACK.

ROFFMAN COMMENT: This was after Craig read my appraisal that the thief was likely Wickramasinghe.

from: Craig Ebrahimi <craigebrahimi@hotmail.com>

to: "BarrySRoffman@gmail.com" <BarrySRoffman@gmail.com >

date: Apr 7, 2020, 12:52 AM

ROFFMAN COMMENT: Naveen Jain claims to be a billionaire with the world's largest private collection of meteorites that are worth millions of dollars. **COSPAR** = Committee on Space Research. This is the first time that Craig said his wife also ate some of the material. Note that he says nothing about her reaction to it. The meaning of "seed balls" is unclear unless it is a reference to puffball "fungi" seen on Mars. These are discussed in Section 15.3.1 of the *Roffman Mars Correct Basic Report* and Chapter 8 of this book. Some of Craig's ideas seem to be based on the Alex Jones Radio Show. On his show on April 12, 2020, Jones claimed that not only is Obama a communist, but that the NIH under Dr. Fauci, at Obama's direction, supervised experiments in Wuhan on COVID-19.

2025 Note: This is likely why Biden gave Fauci a pardon on January 20, 2025.

Craig did not point to any specific plan since the 1950s. I gather that the reference is based on his understanding of common conspiracy beliefs.

Questions from Roffman to Ebrahimi on April 8, 2020, and his answers to Roffman on the same day

from:	Craig Ebrahimi <craigebrahimi@hotmail.com>
to:	"BarrySRoffman@gmail.com" <BarrySRoffman@gmail.com
date:	Apr 8, 2020, 7:15 PM
subject:	Fw: Re: Answered

From: Barry Roffman barrysroffman@gmail.com

Sent: April 8, 2020 6:30 AM

To: Craig Ebrahimi craigebrahimi@hotmail.com

Cc: davidaroffman@gmail.com DavidARoffman@gmail.com

Subject: Re:

OK. I saw the emails from the two Docs. I'll work this into the record today. More dates would be helpful. Here Craig's responses are shown in bold print.

When did you find the rocks? **March 01-06, 2018**

When did you mail him the sample? **June 19, 2018**

Do you have a copy of the shipping receipt? **Yes. SEE FIGURE 3 BELOW.**

When did he get it? Did he sign for it? **Yes.**

Did he report a theft to the police in Sri Lanka? A copy of the police report would help. The sample was supposedly stolen from his gated community in Car… (the next item was not saved).

Naveen Jain was excited immediately when we were in contact, but then his dad passed away. After he found out that Chandra had an embryo fossil, Chandra flew here to California for the COSPAR MEETING in 2018. Chandra and Naveen met and planned to both screw me over, but they don't know what they're dealing with. Then, they were stolen from. Now they are trying to communicate with me now with seed balls.

When did he suggest that you taste the sample? **No one told me to eat it. My wife and I ground it up and ate grams of it with bacon and eggs.**

When did you lose your hair?

Can you name any pharmaceutical companies involved? Craig did not answer this directly, but in an email from him to me on March 27 at 2:31 pm, he mentioned: **CEO of INOVIO, JOSEPH KIM (joseph.kim@inovio.com), founded by Bill Gates.**

Do you have any evidence for Chinese involvement?

At some point, a COVID-19 test will be needed for you and your Victoria partners.

Listen, that virus only works on weak humans. Bill Gates patented this coronavirus strain in 2013; the documents are available, you just gotta dig. All this has been a plan since the 1950s.

Note: A denial of this charge is posted at

https://committees.parliament.uk/writtenevidence/2109/default/. See https://patents.google.com/patent/US10130701B2/en for a related patent that may be the source of confusion here.

Craig continues:

from: *Craig Ebrahimi <craigebrahimi@hotmail.com>*

to: *"BarrySRoffman@gmail.com" <BarrySRoffman@gmail.com>*

date: *Apr 9, 2020, 12:57 PM*

subject: *Info wars have good investigative journalist like you! other than alex jones*

https://www.infowars.com/bill-gates-crosses-the-digital-rubicon-says-mass-gatherings- may-not- retu-without-global-vaccine/

https://www.infowars.com/censorship-extends-from-bill-gates-to-5g-can-trump-get-re- elected- in-this-environment/

https://www.infowars.com/shocking-historian-exposes-bill-gates-ties-to-nazis-and-more/

On 11 April 2020: Why has Fort Huachuca's Canadian liaison officer not yet removed the meteorite from the University of Victoria?

PARTIAL ROFFMAN ANSWER ON DECEMBER 29, 2020: Craig reports that the Canadian Royal Police are stopping him every week and that he is afraid to leave Canada for fear that they will not let him back in. I surmise that Fort Huachuca is likely keeping him under their thumb, but they have yet to contact me directly to confirm it.

To my readers there in January 2025: I'm now 77, but I am still willing to return to active duty (with Space Force) so we can work more freely together for a common goal – but I will not work toward any assault on our Constitution.

Craig's ideology again seems to be in part taken from Infowars and Alex Jones videos.

13 April 2020: This list is first accessed by the United Kingdom Ministry of Defence (UK MOD) at 13/Apr/2020:05:43:29 MST7MDT and was again accessed at 13/Apr/2020:17:51:44.

20 April 2020

from: Barry Roffman <barrysroffman@gmail.com>

to: "davidaroffman@gmail.com" <DavidARoffman@gmail.com>

date: Apr 19, 2020, 1:16 PM

subject: New connection between Wickramasinghe and China

My son, Dr. David A. Roffman (PhD, physics), wanted to see proof of a connection between Wickramasinghe and China. It is established in the article by Wickramasinghe and (among others) Jiangwen Qu of the Department of Infectious Disease Control, Tianjin Centers for Disease Control and Prevention, Hedong Qu, Tianjin Shi, China. Further, there may be an effort by Gensuke Tokoro of the Institute for the Study of Panspermia and Astroeconomics, Gifu, Japan, to sell any cure developed from Craig's meteorite. Tokoro is president and CEO of Kyoto Biopharma, Inc., and executive director of ISPA. He lives in Japan.

Article: Comments on the Origin and Spread of the 2019 Coronavirus

Link: https://www.academia.edu/42041228/Comments_on_the_Origin_and_Spread_of_the_2019_Coronavirus

Citation: Virol Curr Res, Volume 4:1, 2020, DOI: 10.37421/Virol Curr Res.2020.4.109

Authors: N Chandra Wickramasinghe1,2,3,4*, Edward J Steele2,5, Reginald M Gorczynski6, Robert Temple7, Gensuke Tokoro2,4, Jiangwen Qu8, Daryl H Wallis2,4, and Brig Klyce9

1. Buckingham Centre for Astrobiology, University of Buckingham, MK18 1EG, England, UK
2. Center for Astrobiology, University of Ruhuna, Matara, Sri Lanka
3. National Institute of Fundamental Studies, Kandy, Sri Lanka
4. Institute for the Study of Panspermia and Astroeconomics, Gifu, Japan
5. C Y O'Connor, ERADE Village, Foundation, Piara Waters, Perth 6112 WA, Australia
6. University Toronto Health Network, Toronto General Hospital, University of Toronto, Canada
7. History of Chinese Science and Culture Foundation, Conway Hall, London, UK

8. Department of Infectious Disease Control, Tianjin Centers for Disease Control and Prevention, Hedong Qu, Tianjin Shi, China

9. Astrobiology Research Trust, Memphis, TN, USA

Abstract:

We propose that the new coronavirus, which first appeared in the Hubei province of China, was probably linked to the arrival of a pure culture of the virus contained in cometary debris that was dispersed over a localized area of the planet, namely China. The sighting of a fireball some 2000 kilometers north of Wuhan on 11 October 2019, followed shortly after by the first recorded cases in Hubei, is suggestive of a causal link. Gene sequencing data of the virus that show little or no genetic variations between isolates, combined with available epidemiological data, point to the predominance of a transmission process directly from an "infected" environment, with person-to- person transmission playing a comparatively weaker secondary role. The facts relating to this epidemic are discussed and placed in the context of other pandemics that have been recorded throughout history.

Issue: 1, **Volume:** 4, **Journal Name:** Virology Current Research, **Page Numbers:** 1-4,

Publication Date: 2020, **Publication Name:** Virology Current Research

from: Craig Ebrahimi <craigebrahimi@hotmail.com>

to: ffman@gmail.com" <BarrySRoffman@gmail.com

Apr 30, 2020, 3:34 PM

Craig wrote:

Barry, the only reason I ever contacted you was regarding David Roffman's website saying if you have a problem we can solve it. Then you started trying to get me to prove to you all about my work and getting no respect except fear-mongering bullshit.

I responded:

We are solving your problem in that we present good evidence that shows Wickramasinghe is not the innocent victim of the theft of your sample while it was in his custody. Rather, he is the thief and, with his Chinese partner and the communist Chinese, he shares responsibility for the murder of over 230,000 people as of this date. I told you up front I am a retired military officer with direct contact with U.S. Intelligence, and that when you talk to me, you talk to U.S. military intelligence. David's website is maintained by me. He does the math. I do the writing.

Craig then stated:

The first email I sent obviously had nothing to do with the real reason.

ROFFMAN: So what was the real reason?

CRAIG: There is a picture I sent that is a crystallized ET holding some type of technology.

ROFFMAN: You need to specify exactly what you mean by technology. Are you referring to a weapon on the attachment at http://arkcode.com/images/figure_15_-looks_like_armed_man_running_20180610_140835(17)(1).jpg?

When a famous no-name US scientist flew to town last August, when he examined what the *ET was holding, it pulses, the hypothesis that's what's creating the coldness, so I wanted to hire a young, energetic new mind, David.*

NOTE: The famous scientist he refers to above is Elon Musk. He was close to Craig's home on August 6, 2019. Further, I am proofreading this part of my manuscript on July 12, 2025. The statistics for my visitors this month up through this morning are as follows:

Last Update: 12 Jul 2025 - 08:29

Reported period: Jul 2025 OK

Statistics for: arkcode.com

Back to main page

Filter : Exclude filter : OK

Links from an external page (other web sites except search engines)

Total: 21 different pages-url	Pages	Percent	Hits	Percent
http://papasearch.net/Tesla6.html	7	13.2 %	7	12.9 %
http://papasearch.net/ElonMusk23.html	7	13.2 %	7	12.9 %
http://papasearch.net/ElonMusk22.html	6	11.3 %	6	11.1 %
http://papasearch.net/Tesla11.html	4	7.5 %	4	7.4 %
http://papasearch.net/George_Soros/GeorgeSoros5.html	4	7.5 %	4	7.4 %
http://papasearch.net/George_Soros/GeorgeSoros4.html	3	5.6 %	3	5.5 %
http://papasearch.net/ElonMusk15.html	3	5.6 %	3	5.5 %
http://papasearch.net/ElonMusk7.html	3	5.6 %	3	5.5 %
http://papasearch.net/NelsonMandela78.html	2	3.7 %	2	3.7 %
http://papasearch.net/NelsonMandela77.html	2	3.7 %	2	3.7 %
http://papasearch.net/ElonMusk6.html	2	3.7 %	2	3.7 %
http://papasearch.net/George_Soros/GeorgeSoros3.html	2	3.7 %	2	3.7 %
http://papasearch.net/NelsonMandela76.html	1	1.8 %	1	1.8 %
http://papasearch.net/ElonMusk14.html	1	1.8 %	1	1.8 %
http://papasearch.net/ElonMusk13.html	1	1.8 %	1	1.8 %
http://papasearch.net/Tesla5.html	1	1.8 %	1	1.8 %
http://papasearch.net/George_Soros/GeorgeSoros6.html	1	1.8 %	1	1.8 %
http://papasearch.net/ISS/ISS44.html	1	1.8 %	1	1.8 %
http://papasearch.net/ElonMusk29.html	1	1.8 %	1	1.8 %
http://papasearch.net/ElonMusk24.html	1	1.8 %	1	1.8 %
http://quran-pazhoohesh.blogfa.com			1	1.8 %

Table 6-1 above showing visitors to one of my domains on July 12, 2025.

I have seen this combination of IP addresses before on my site on May 21, 2025. It looked as follows:

Links from an external page (other web sites except search engines) - Full li		
- http://papasearch.net/ElonMusk19.html	11	11
- http://papasearch.net/ElonMusk20.html	10	10
- http://papasearch.net/Tesla6.html	7	7
- http://papasearch.net/Tesla11.html	5	5
- http://papasearch.net/George_Soros/GeorgeSoros5.html	4	4
- https://t.co/R2qz1O2Gk9	4	4
- http://papasearch.net/ElonMusk7.html	3	3
- http://papasearch.net/ElonMusk21.html	3	3
- http://papasearch.net/ElonMusk15.html	3	3
- http://papasearch.net/NelsonMandela77.html	2	2
- http://papasearch.net/ElonMusk17.html	2	2
- http://papasearch.net/George_Soros/GeorgeSoros4.html	2	2
- http://papasearch.net/ElonMusk16.html	2	2
- http://papasearch.net/ElonMusk6.html	2	2
- http://papasearch.net/NelsonMandela78.html	2	2
- http://papasearch.net/George_Soros/GeorgeSoros3.html	2	2
- http://papasearch.net/ElonMusk18.html	1	1
- http://papasearch.net/Jihadi_John/JihadiJohn1.html	1	1
- http://papasearch.net/Tesla5.html	1	1
- http://papasearch.net/ISS/ISS44.html	1	1
- http://papasearch.net/NelsonMandela76.html	1	1
- http://papasearch.net/NelsonMandela27.html	1	1
- http://papasearch.net/George_Soros/GeorgeSoros6.html	1	1
- http://papasearch.net/ElonMusk29.html	1	1
- http://papasearch.net/Quantum_Computing/QuantumComputing9.html	1	1
- Others	0	6
Unknown Origin		

Table 6-2 above showing visitors to one of my domains on May 21, 2025.

What is surprising is that on both Tables of visitors we have Elon Musk/Tesla representing the far right and George Soros and Nelson Mandela representing the far left. Mandella died On December 5, 2013, but the two tables in this chapter indicate that there was some kind of financial link that survived. Mandela and Soros had a connection through Soros's philanthropy and advocacy for human rights and democracy, particularly in South Africa. Soros's Open Society Foundations played a significant role in supporting South Africa's transition from apartheid, including providing scholarships for Black students and funding civil society groups. Soros also met with Mandela, and they both shared a commitment to fighting for social justice and equality.

Continuing, Barry wrote:

Then you should have made this clear. It's the first time that I heard it's pulsing. If that's true, this thing may be much more hazardous than I imagined. This rock needs to be analyzed in a U.S. military lab because it has already killed at least 62,603 Americans. We have more deaths than any other nation, and if there is an ET plan to try to conquer our world, they will have to defeat the United States first. Reality is

that you are either with us or against us. Do you want to take on the United States? We are not playing games here.

Craig wrote:

Before ever going to UVic, I contacted Chandra.

Barry *wrote:*

That would be consistent with the limited emails that I have, although from a geographic point of view, it seems odd. Was there someone, an intermediary, anyone who introduced you to Chandra?

Craig wrote:

Then I researched out what needed to be done, so hired the labs equipment. And many PhD scientists you do not know far beyond UVic labs.

Barry wrote:

We are going to need the names involved and times of contact because this is not just a theft investigation. It's likely to be a murder investigation and possibly the cause of a war.

Craig wrote:

I have told you before I have many organic embryo fossils. I have one around my neck in a necklace.

Barry wrote:

We need to see photos of clearly distinct embryos. Craig wrote:

I now have a Joint Venture, you will read about it. Very powerful name R r

Barry wrote:

I need the full name. ***NOTE:*** *Craig later said it was a Rockefeller. Well done on the deuterium content.*

Thanks.

Craig wrote:

I have not told you that I had oxygen done 3 times at different laboratories, and two said Martian, and only Dr. Zeigler's data was false from a personal reason that I have had with dumb director Carl Agee, he garbaged a sample in 2018, I swore at him and some.

ROFFMAN COMMENT: Note Craig's clarification that he contacted Dr. Wickramasinghe in the UK before Dr. Humphrey, who lives and works not too far from Craig in British Columbia.

Also note the new PhD involved with handling the rocks: Dr. Carl B. Agee, Director, Institute of Meteoritics, Meteorite Curator Professor, Department of Earth and Planetary Sciences, MSC03 20501, University of New Mexico.

Carl Agee's contact information is as follows:

Dr. Carl B. Agee, Director, Institute of Meteorites, Meteorite Curator, Professor, Department of Earth and Planetary Sciences, MSC03 2050, 1 University of New Mexico, Albuquerque, NM 87131-0001, Phone: 505-573-5131, Fax: 505-277-3577, Email: agee@unm.edu

Barry wrote:

Can you back this up with emails or reports? Craig wrote:

"There is a global cover-up with meteorite institutions, so their funding got cut a lot. They are all broke lifers that do not like the public owning these."

I know about and have written about the cover-ups before, but for now, the issue is comparing the DNA and or RNA on your samples with current strains of the virus out there. Our survival depends on it.

Craig

Good investigative work with China and India.

Barry wrote:

You mean China and Sri Lanka. India intervened in the Sri Lankan Civil War in 1987. It suffered 1,138 killed and completed its withdrawal in 1990.

If you do not understand why I sent you those emails from CEOs and so on, maybe you knew at some level that it was the right thing to do. Left unchecked, whatever life form apparently is involved with the meteorite is a threat to all mankind. So are corrupt people in our government who have worked to impeach a legally elected President who actually cares about us. It remains to be seen if DoD and UK MOD can be trusted to get this right, but if they do, it will be in large part because you gave me critical information to pass on to them here. As for what you should do next, I would recommend you take that embryo off from around your neck and get me some good photos of it. You need to show at least photographic evidence of multiple embryos. This helps to establish that there has been reproduction rather than a simple geochemical process that produces similar intrusions.

It is also vital that you tell us what happened to your wife after she ate some of the rock content. Without this information, thousands of people may become infected, and doctors will misdiagnose thousands of people, who will then die because you did not tell our doctors what they need to look out for.

Craig wrote:

Start writing letters to write companies.

Barry wrote:

I'll leave this to our government/s to do. However, I act like a government's full intelligence agency. I'm not. There is an enormous amount of work required here. I need the DoD to do its job.

From: **craigebrahimi@hotmail.com** To: **barrysroffman@gmail.com** Date: Apr 30, 2020, 11:35 PM

On Fri, May 1, 2020, at 12:06 PM, Craig Ebrahimi craigebrahimi@hotmail.com wrote:

Barry, I have three documentary offers: Prometheus Entertainment, Ancient Aliens, and some Will not. My Rockefeller partners would not approve. I will send you a picture when the laboratories open. My wife had two extremely healthy boys.

Craig

ROFFMAN COMMENT: Rockefeller partners involved!

I asked this question because someone (Matthew Wilson) associated with the Rockefellers tried to recruit me for the Thule Society in 2013 to 2014. His motivation was not clear, but here is what survives in my records (just the e-mail address and the 500-page attachment: **Behold A PALE HORSE).**

From: Matthew Wilson matthewwilson888@gmail.com To: Barry Roffman BarrySRoffman@gmail.com

Date: Apr 13, 2014, 7:37 PM

Note: I have only briefly skimmed through the above book for the first time on May 11, 2020.

ROFFMAN COMMENT: On hair length and resemblance to man-like alien in Prometheus From: Craig Ebrahimi craigebrahimi@hotmail.com

Sent: May 1, 2020 1:36 PM

To: Barry Roffman barrysroffman@gmail.com

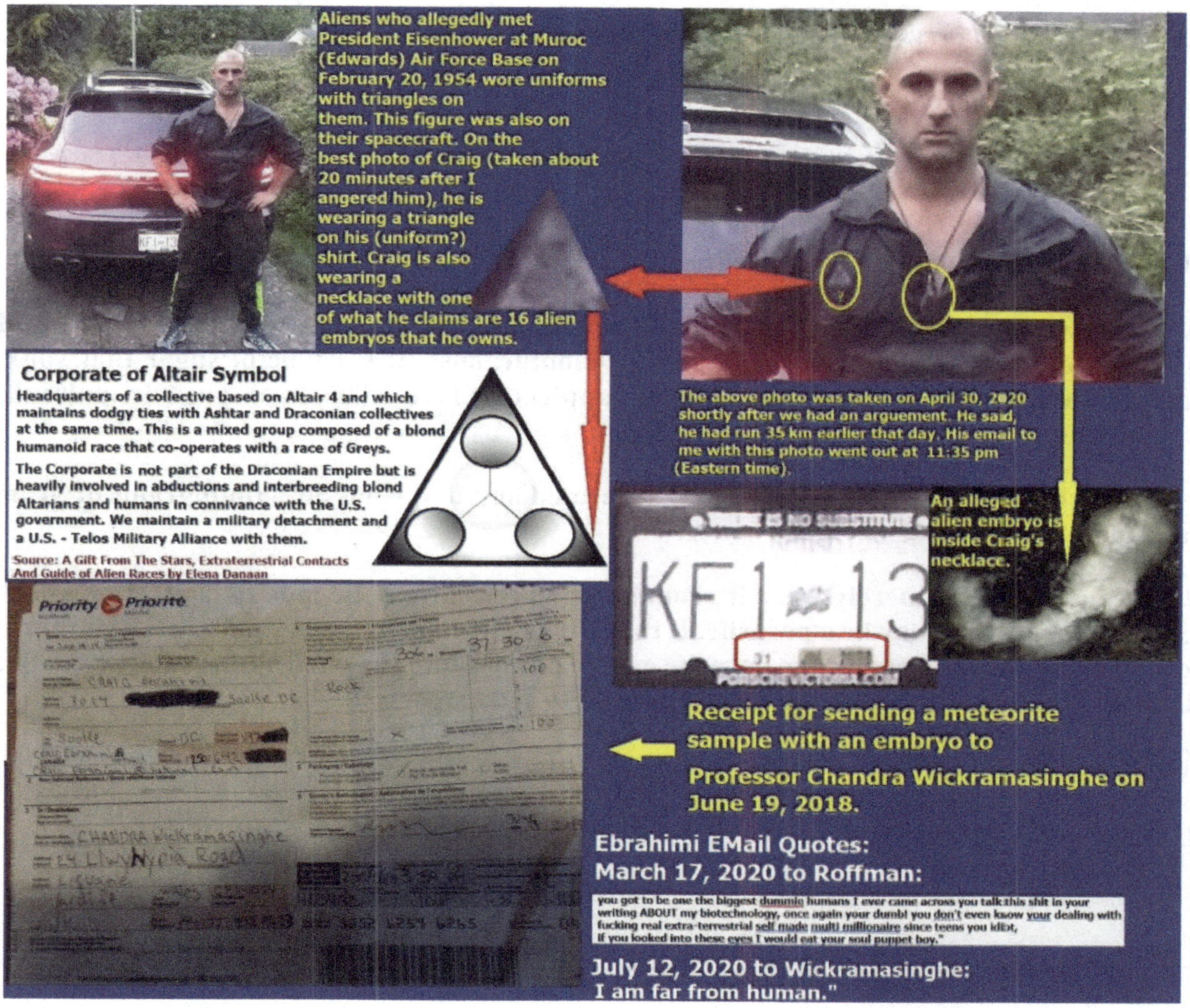

See Figure 6-4 Above of Craig with his car (April 30, 2020)

Subject: Re: Re: My hair has been shaved #2 since teens. Must be a coincidence. I'll find one and send you.

1 May 2020 @ 7:28 PM and response from Barry Roffman **3 May 2020 @ 12:02 AM**

From: Craig Ebrahimi craigebrahimi@hotmail.com Sent: May 1, 2020, 7:28 PM To: Barry Roffman barrysroffman@gmail.com

Di ribose increases stamina in my research. I found it pure in my meteorite sample. Enhancing performance is a side effect.

ROFFMAN COMMENT: This would imply that there exists a substantial organic chemistry analytic report.

From: Barry Roffman barrysroffman@gmail.com To: Craig Ebrahimi

craigebrahimi@hotmail.com

Cc: "davidaroffman@gmail.com" DavidARoffman@gmail.com Date: May 3, 2020, 12:02 AM Thanks Craig

ROFFMAN EMAIL COMMENT On hair length and resemblance to man-like alien in Prometheus:

I may not be working for you directly, but you are keeping me very busy. By Di ribose, I think you mean D-ribose. It's also called ribose or Beta-D-ribofuranose. It's a type of sugar that your body produces naturally and is readily available in supplement form. This is an organic molecule with great significance.

Who told you this is in the rock? How was it separated out, and what dose (milligrams or grams) do you take daily?

Whether you ate one or merely wear it around your neck, it may be able to control or affect your thoughts. If you ate it, it may be a parasite. If that's the case, then like other parasites, if it reproduces inside a host and the host poops, the parasite can be put out into the environment that way. The only way to know if this is happening with you is for you to get a stool test.

Best regards, Barry

Craig on organics in his meteorite:

from: Craig Ebrahimi <craigebrahimi@hotmail.com>

to: "barrysroffman@gmail.com" <barrysroffman@gmail.com>

date: May 3, 2020, 12:24 AM

Barry, I have had the organics extracted and analyzed in the UK, not Chandra. But it's well known for meteorite organics analysis. His data and lab were shut down in December 2019. I was informed on just a little about Aminos and some type of Protein they discovered, but the data has been suppressed. But I will arrange it again.

Barry: Who suppressed the data?

Craig's answer:

No, I have not extracted any out of samples to eat, come on, guy. D-ribose response. Someone suppressed organic chem data. See attachment.

From: Craig Ebrahimi craigebrahimi@hotmail.com Sent: May 3, 2020, 2:34 PM

To: Barry Roffman barrysroffman@gmail.com Subject: Re: Time of photo with the Porsche

I took that picture 20 mins before I sent you after I ran 25 km and haven't eaten in days. The research will be conducted in the US soon.

CRAIG'S RESPONSE AT 8:19 PM ON MAY 3, 2020

from: *craigebrahimi@hotmail.com*

to: *Barry Roffman <barrysroffman@gmail.com>*

date: *May 3, 2020, 8:19 PM*

subject *Re: Extraterrestrial ribose and other sugars in primitive meteorites PNAS*

At 7:36 PM, Barry wrote:

Think quantitative. How fast do you run 25 km? The world record is 1:11:18. At 8:19 PM, Craig wrote:

My trail runs are intense, a lot of hills. Yes, when running a flat course, easy peasy, I'd be able to keep close to 5-minute miles, but Eliud Kipchoge's marathon pace destroyed the record in 2019.

Barry then wrote:

Did your wife's strength and speed also increase? I'm glad that she had two healthy babies, but it's important to know if what she ate caused any unpleasant or unusual side effects.

Craig wrote:

She never had any negative side effects; we will not comment on her.

Barry wrote:

David thinks you look like you're about 38 in 2020, around 50 back in 2016. I agree, but with the sunglasses, it's hard to get an accurate estimate.

I asked David about the pulsation. He wants to see pictures (iPhone) or video if you have any. Are you talking about light pulsation? How long did this go on, and was there any pattern to it? Geiger?

Craig wrote:

The ribose article (https://www.pnas.org/content/116/49/24440) was published when your (our?) sample was in China, but also when the Japanese Pharma CEO likely had access. I'll see if I can find a connection.

Chandra is an honorary professor with JAXA, Japan. The ribose in the meteorite is to the embryos as yolk in an egg is to the embryo in it.

Barry wrote

The article is a great find that does much to increase your case. But to close the loop, we need something in writing about ribose being in your rocks.

Craig wrote:

The science data I do not have access to because of a lack of trust in me. And their data is theirs. And I would be a complete idiot to have their names mentioned.

Geiger meter cold pulse.

Barry wrote:

It sounds like it could be some kind of communication. Craig wrote:

My typing mistakes are because of emailing with one hand using a cell phone and a baby 1-year- old boy in the other hand.

Geiger meter cold pulse and Dr. Jerzi Sawicki...(missing end of the mail)

from: Craig Ebrahimi <craigebrahimi@hotmail.com>

to: barrysroffman@gmail.com" **barrysroffman@gmail.com**

date: May 4, 2020, 11:59 PM

Elon's pissed with me because... (I've censored the reason.)

My embryo is in my necklace. I'll send ya a picture, but it's hard to make out unless using the lab's LM. The embryo-like structure is so visible in the br that it looks like it's gonna jump out of the rock. Everyone and anyone can make it out. One day it will come to you. Top right curled with tail with spikes, really amazing when looking through the AMF LAB equipment on the big screen.

PARTIALLY CENSORED AT CRAIG'S REQUEST

br refers to one of the meteorite samples. Intel IP checks:

47.202.113.176 = 176.113.202.47 = 6.113.202.47 (Fort Huachuca Army Intelligence Base);
192.19.231.250 = 25.231.19.192 (UKMOD);

58.109.207.225 = 225.207.109.58 = 25.207.109.58 (UKMOD)

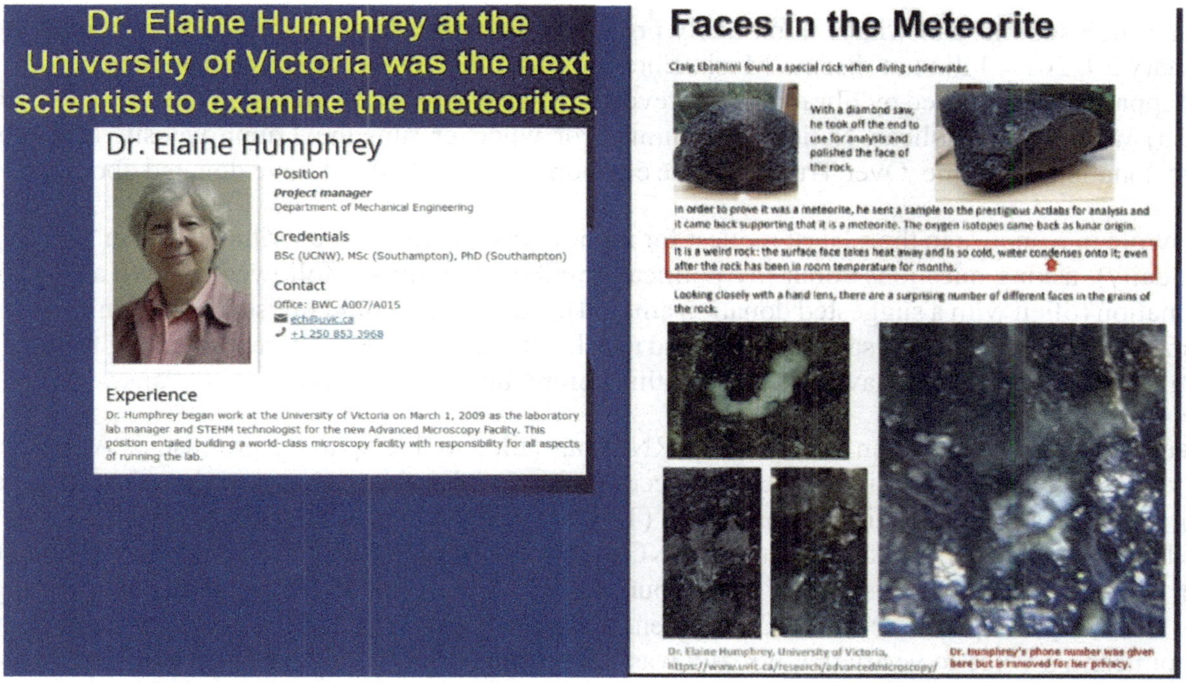

Figure 6-5 above: Photos of the meteorite and remarks by Dr. Elaine Humphrey at the University of Victoria in Canada.

NOTE: Emails published from May 4, 2020, through 2022 are posted at https://roffmanmarsresearch.com/?page_id=6035. The following communication, apparently from Craig, was received on January 29, 2025.

Barry wrote:

JANUARY 29, 2025. **The following voicemail text was found in my Message Box at 10:04 AM while I was on the phone with Express Scripts (a pharmaceutical company). The return number only took me to my doctor's office rather than to Craig in Canada. I left a message on Craig's mobile phone that answers his questions. It remains to be seen if he answers it. The mysterious text appears to suggest that he monitors what I write on this, and he has the ability to monitor what I write to the Trump National Committee or the Republican National Committee.**

Craig wrote:

So, do you believe that Musk and Trump are sitting around Mar-A-Lago talking about me? So they must be discussing which job they would offer me. What does he think about Ebrahimi?

First comment by Roffman:

The question marks were added by me. I have noticed that Craig's grammar is not always top of the line.

Second comment by Roffman:

This is a rough summary of what I think likely provoked the above January 29, 2025, voicemail message. On January 29, 2025, I did not discuss Musk, Ebrahimi, Mar-A-Lago, or a job with anyone before reading the text apparently composed by Ebrahimi. However, I did discuss all these things on the day before (January 28, 2025) with the Republican National Committee or whoever runs the Trump website that constantly requests donations from me. Over Trump's three elections, I estimate that I have donated about $2,000.

What I wrote in response to their latest request for money was that they have written to me in Trump's name before (daily), asking questions about my political opinions, but always followed up with another request for a donation (often with a suggested donation amount). However, I would not send more unless somebody called me to tell me what Musk said about me and to ask if Trump was really communicating by considering offering me a paid position. I have asked about this before, but never received a direct answer.

The man who promised to get an answer at the RNC, but failed to live up to his promise, is Adam Felty. As I have told the Secret Service, if Felty cannot account for exactly what happens to donations that go to the RNC or its partner, Trump National Committee (TNC), we probably need to rule out embezzlement of the hundreds of millions of dollars donated to the RNC in 2024. **Note:** On April 4, 2025, I filed a request for an internet fraud complaint against Felty. Four hours later, I received a text from someone claiming to be Trump. It had my complaint on it with a statement that he was trying to call my mobile number, but still, I could not hear his voice to verify that it really was the President. I advised him to borrow somebody else's phone because the intel base Fort Huachuca might be blocking his calls here.

Based on Elena Danaan's writings, Huachuca is suspected of a criminal relationship with reptilians, delivering hybrid children to the reptilians (for slavery or food). If this is true, they likely want to keep it hidden from the President. Consider **Figure 6-6 and 6-7**. On **Figure 6-6**, the axis term is **HUACHUCA** at skip +1,465. At the same absolute skip (opposite direction) are **REPTILIAN**, **UFO**, and **CHINA**. **REPTILIAN** touches **BARACK H.**, and he touches **FROM A NAZI** which touches **SONS OF ANAK** (giants). Odds against this matrix are about 477,922,797 to 1. Danaan says on pages 163-164 of *A GIFT FROM THE STARS, Extraterrestrial Contacts and Guide of ALIEN RACES* that, "Fort Huachuca: Here is one of the underground hangars for spaceships. This facility is a landing point for incoming **Ciakahrr** (Reptilian) contingents and a central hub deserving of a vast amount of other facilities." An even darker portrait emerges with **Figure 6-7** which addresses the question of whether children are sent from there to Reptilians for purposes or food (as was alleged by Rebecca).

Fort Huachuca is the most common visitor to my website so they will quickly become aware of what I am, at great risk, writing here. But, there are signs of hope from the Fort.

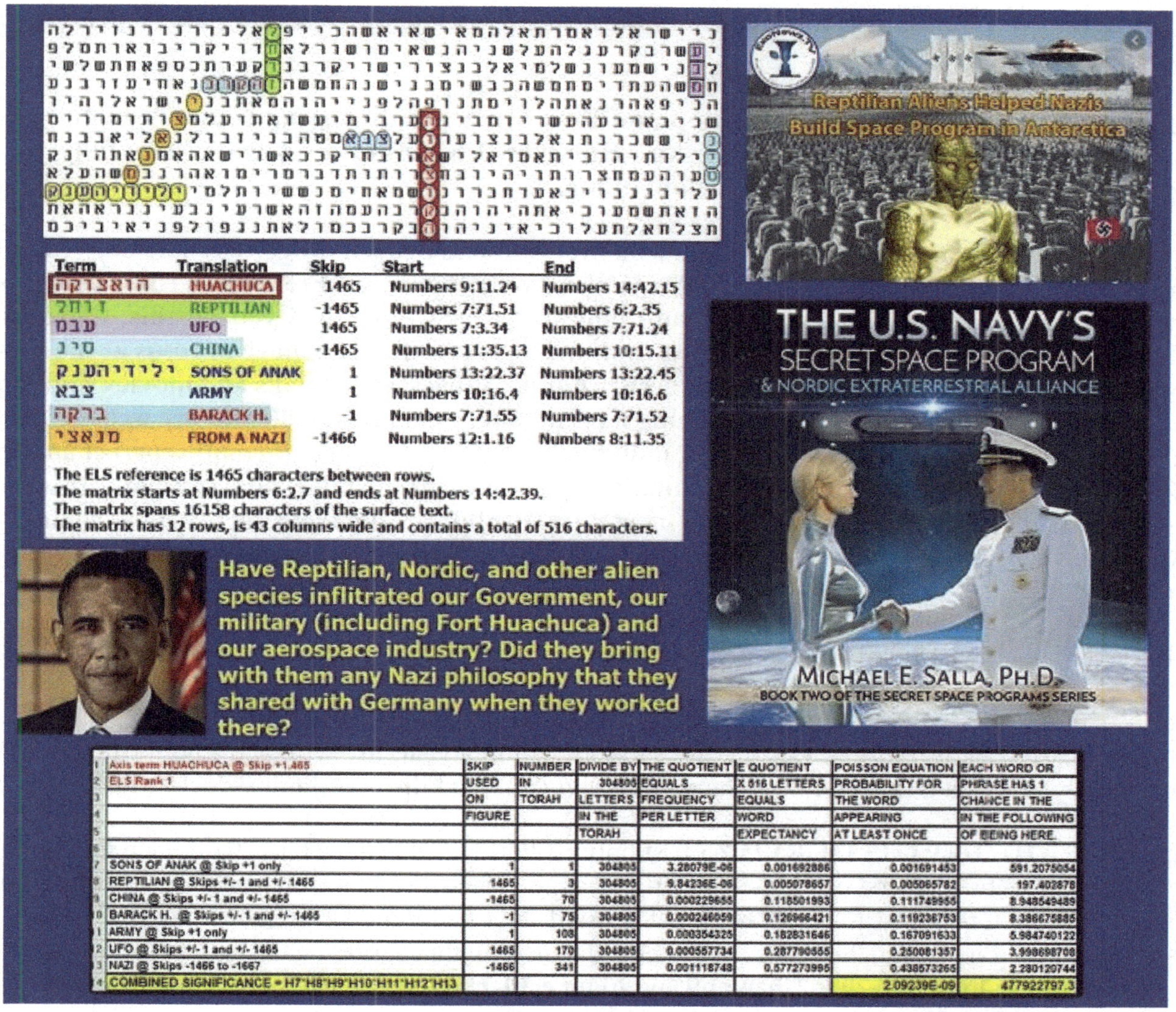

Term	Translation	Skip	Start	End
הואצוקה	HUACHUCA	1465	Numbers 9:11.24	Numbers 14:42.15
זוחל	REPTILIAN	-1465	Numbers 7:71.51	Numbers 6:2.35
עבם	UFO	1465	Numbers 7:3.34	Numbers 7:71.24
סינ	CHINA	-1465	Numbers 11:35.13	Numbers 10:15.11
ילידיהענק	SONS OF ANAK	1	Numbers 13:22.37	Numbers 13:22.45
צבא	ARMY	1	Numbers 10:16.4	Numbers 10:16.6
ברקה	BARACK H.	-1	Numbers 7:71.55	Numbers 7:71.52
מנאצי	FROM A NAZI	-1466	Numbers 12:1.16	Numbers 8:11.35

The ELS reference is 1465 characters between rows.
The matrix starts at Numbers 6:2.7 and ends at Numbers 14:42.39.
The matrix spans 16158 characters of the surface text.
The matrix has 12 rows, is 43 columns wide and contains a total of 516 characters.

Axis term HUACHUCA @ Skip +1,465	SKIP	NUMBER	DIVIDE BY	THE QUOTIENT	E QUOTIENT	POISSON EQUATION	EACH WORD OR
ELS Rank 1	USED	IN	304805	EQUALS	X 516 LETTERS	PROBABILITY FOR	PHRASE HAS 1
	ON	TORAH	LETTERS	FREQUENCY	EQUALS	THE WORD	CHANCE IN THE
	FIGURE		IN THE	PER LETTER	WORD	APPEARING	IN THE FOLLOWING
			TORAH		EXPECTANCY	AT LEAST ONCE	OF BEING HERE.
SONS OF ANAK @ Skip +1 only	1	1	304805	3.28079E-06	0.001692886	0.001691453	591.2075054
REPTILIAN @ Skips +/- 1 and +/- 1465	1465	3	304805	9.84236E-06	0.005078657	0.005065782	197.402878
CHINA @ Skips +/- 1 and +/- 1465	-1465	70	304805	0.000229655	0.118501983	0.111749955	8.948549489
BARACK H. @ Skips +/- 1 and +/- 1465	-1	75	304805	0.000246059	0.126966421	0.119236753	8.386675885
ARMY @ Skip +1 only	1	108	304805	0.000354325	0.182831646	0.167091633	5.984740122
UFO @ Skips +/- 1 and +/- 1465	1465	170	304805	0.000557734	0.287790555	0.250081357	3.998698708
NAZI @ Skips -1466 to -1667	-1466	341	304805	0.001118748	0.577273995	0.438573265	2.280120744
COMBINED SIGNIFICANCE = H7*H8*H9*H10*H11*H12*H13						2.09239E-09	477922797.3

Figure 6-6 above

When Craig first contacted me (January 26, 2020), he described the view I had out my living room sliding glass door in Cape Canaveral, Florida. He also claims to have telepathic abilities. Somehow, it appears that he accessed what I said to Trump's people in March 2025. I am sure that he reads my website, which includes this email and phone record. But the contents of my remarks about donations should not have been available on my website in my outgoing mail because they did not go out via email. They were sent in response to the Trump National Committee website, and it was on their program.

If Craig calls me back, I will discuss his concerns, but to try to help him understand what I think or know is going on, let me be clear. I wrote Craig and told him that,

"Your decision to pull the plug on our communications before was not wise. There are likely many species of aliens visiting Earth. You likely represent one of them, or at least you claim that this is so. Some of them (Reptilians, Greys, etc.) are hostile. I have broken the Torah Code and published the names of the top alien spies who have fought to prevent Trump from winning the election in 2024.

They are named in Chapter 3 of this book, and the last two presidents are among the suspects revealed by the Torah Code.

Craig, you asked whether President Trump and Elon Musk are discussing you now or not. IF they are talking about me, the answer is likely yes. However, I have not investigated this fully yet. I'm trying to discern IF they are really talking about me.

Until you are investigated, you must be seen as a threat unless proven otherwise. The only way that I can see anything positive coming out of this for you and your family is if President Trump remains our President AND you cooperate with a full and honest disclosure. Even if you do this, there will be people who believe you may be the antichrist (not me, but Jews do not believe in Jesus as the Messiah or divine). You will likely have to find a new, secure home. But yes, your books and movies would make a lot of money. You once wrote about our doing a joint book. From this book and my website, you can see what my knowledge is about your life. A joint book would allow you to formally correct any errors I make in my writings about you. If I have made any significant mistakes and you point them out, I will immediately correct my work. For the most part, I think profits from such a book should go to you so long as you cooperate. My main interest remains in protecting the United States, and indeed our species."

COVID-19 killed about 18,000,000 people worldwide. In our first phone conversation back in 2020, Craig offered me a trip up to his home in Sooke, B.C., so that I could examine the embryo. At that time, just days before COVID-19 was named a pandemic, I declined his offer, and I told him why. I was afraid that I would bring this disease back to Florida. Despite my caution, within a year, my wife and I both caught COVID-19. Fortunately, it was a mild version of the disease, but COVID killed our family doctor, Dale Adamson.

FIGURE 6-7. THE DEPTHS OF EVIL? I first noticed Fort Huachuca back in 2014 when they became a top reader of my website. I first encountered Fort Huachuca when an agent (referred to as Joseph *Sapphire*) who worked there threatened to do to me what he said they had done to someone like me. He said they loaded child pornography onto the guy's computer and then brought charges against him. I'll write about him more in reference to **Figure 9-20** in Chapter 9 and **Figure 10-1** in Chapter 10, but his words indicated that, at least in the past, Fort Huachuca was operating outside of the law.

With regards to **Figure 6-7** below, as a retired Coast Guard officer with 24 years in the Coast Guard and ten years of service as a Naval officer before that, the last thing that I want to do is to paint an Army intelligence base as a criminal organization. I started studying the Torah Code 28 years ago to help U.S. intelligence operations! So, what can I say about **Figure 6-7**? While the combined odds are somewhere between one chance in 75,193 (532-letter matrix) and one chance

in 268,456 (868-letter matrix), because of high frequency at skip +1, neither **FOOD** nor **CHILD** is statistically significant. But what if it was true? My study indicates that a substantial number of Democratic politicians are either aliens or humans who have been abducted and had their brains scrubbed by ET. An alien invasion is not just something that might occur. It is, instead, something that has already occurred. As such, just like there is an unprecedented division between right- and left-wing voters, so too does this division exist in our military. The matrices that I have done on General Milley provide an example. Was *Sapphire* a bad guy? I don't think so. Although our encounter went from strange to hostile initially, and although he said he was Catholic, to show him who he was dealing with, I told him the story of how I forced the Catholic Church to annul the baptism of my older son (now, Orthodox Rabbi Robert Altair Roffman) who goes by his Hebrew name, Moshe,

He seemed somewhat shocked by the incident, which was covered on the front page of the Miami Herald (see https://www.arkcode.com/photo4_85.html) and other major newspapers.

What startled me most about our meeting was that when it turned to my studies of the Martian atmosphere, his admiration of me apparently quickly grew. I offered him my calling card, which had a picture of the Mars Science Laboratory Curiosity Rover and some words about my specialty being Martian meteorology. He asked, "Can I keep this?" I said, "Sure," or something like that. He then acted like I had given him something precious, like a copy of our Constitution or the British Magna Carta! Next, he let me know something about my younger son's near future. Even his initial greeting of me seemed to indicate that he knew which book I had just checked out from the library (Slant, by Greg Bear – which deals with nanotechnology and artificial intelligence). He had not been in the library when I selected it. My son (David) had just earned his PhD in nanotechnology, but we didn't know then that four months later he would learn how to do A.I. while doing his post-doc at Yale. In short, in acting like he had library records from the future, and being so grateful for my card, he left me with the impression that he supported my efforts to find and publish the truth. More, he seemed to indicate that this book would eventually make the world a better place. As such, *Sapphire* seemed to indicate that the white hats will, under Trump and with his Space Force, win this *War of the Worlds*. As I write these words, my computer began to read them back to me! That is a function that I never choose to use. It can only mean one thing. Fort Huachuca is following every keystroke that I make. Joseph, if you're still alive and reading this, thanks for inspiring me to reach my destiny.

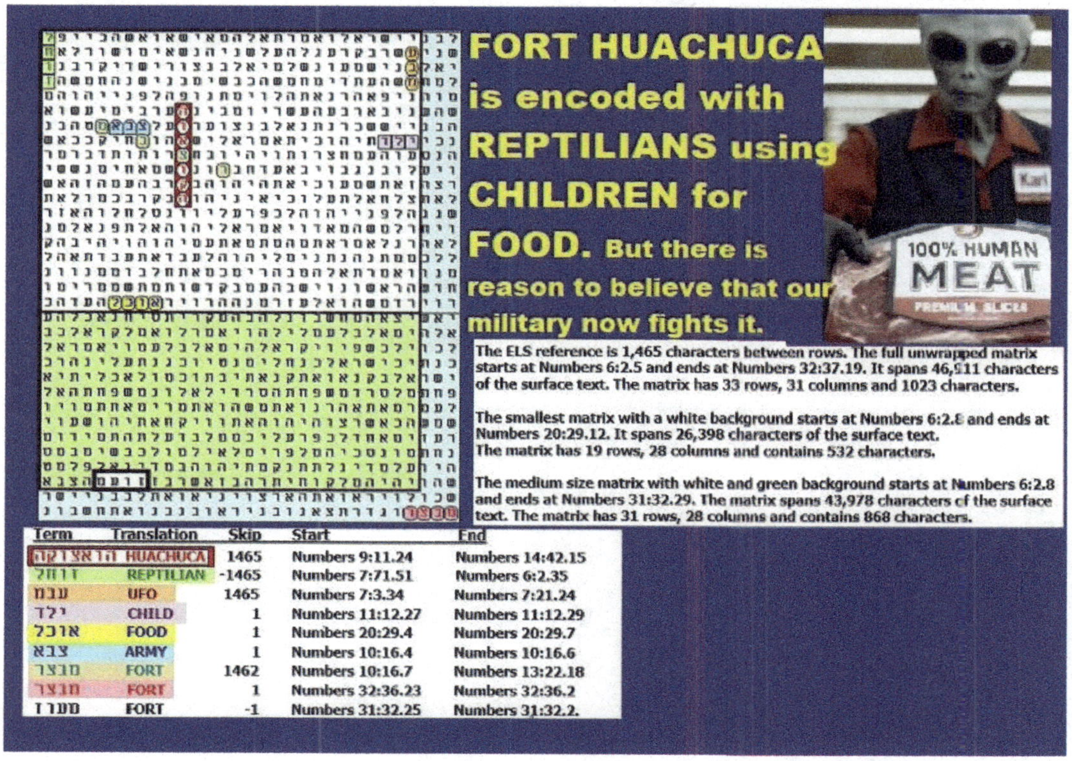

Figure 6-7 above. Its spreadsheet follows below.

	A	B	C	D	E	F	G	H
1	AXIS TERM HUACHUCA @ Skip 1,465	SKIP	NUMBER	DIVIDE BY	THE QUOTIENT	E QUOTIENT	POISSON EQUATION	EACH WORD
2	ELS RANK 1	USED	IN	304805	EQUALS	x 532 LETTERS	PROBABILITY FOR	OR PHRASE
3	This calculation is for the smallest (532-letter) matrix	ON	TORAH	LETTERS	FREQUENCY	EQUALS	THE WORD	HAS 1 CHANCE
4		FIGURE		IN THE	PER LETTER	WORD	APPEARING	IN THE
5				TORAH		EXPECTANCY	AT LEAST ONCE	FOLLOWING OF
6								BEING HERE
7	REPTILIAN @ Skips +/- 1 and +/- 1465	-1465	3	304805	9.84236E-06	0.005236135	0.00522245	191.4810128
8	FOOD (3 synonyms) @ Skip +1 only	1	71	304805	0.000232936	0.123921852	0.116551122	8.579926
9	CHILD @ Skip +1 only	1	110	304805	0.000360886	0.191991601	0.1746862	5.724550655
10	ARMY @ Skip +1 only	1	108	304805	0.000354325	0.188500845	0.171800196	5.820715121
11	FORT (2 synonyms) @ Skips 1462 to 1901	1	746	304805	0.002447466	1.302052132	0.728026905	1.373575611
12	Combined odds against the matrix = h7*h8*h9*h10*h11						1.3299E-05	75193.37462
13								
14	AXIS TERM HUACHUCA @ Skip 1,465	SKIP	NUMBER	DIVIDE BY	THE QUOTIENT	E QUOTIENT	POISSON EQUATION	EACH WORD
15	ELS RANK 1	USED	IN	304805	EQUALS	x 868 LETTERS	PROBABILITY FOR	OR PHRASE
16	This calculation is for the matrix with an 868-letter area	ON	TORAH	LETTERS	FREQUENCY	EQUALS	THE WORD	HAS 1 CHANCE
17		FIGURE		IN THE	PER LETTER	WORD	APPEARING	IN THE
18	It is the most significant window (area).			TORAH		EXPECTANCY	AT LEAST ONCE	FOLLOWING OF
19								BEING HERE
20	REPTILIAN @ Skips +/- 1 and +/- 1465	-1465	3	304805	9.84236E-06	0.008543167	0.008506778	117.5533233
21	FORT (2 synonyms) @ skips +/- 1 and +/-1465	-1	12	304805	3.93694E-05	0.034172668	0.033596377	29.76600051
22	FOOD (3 synonyms) @ Skip +1 only	1	71	304805	0.000232936	0.202188284	0.183058904	5.462722542
23	CHILD @ Skip +1 only	1	110	304805	0.000360886	0.313249455	0.268932478	3.718405476
24	ARMY @ Skip +1 only	1	108	304805	0.000354325	0.30755401	0.264756844	3.777050614
25	Combined odds against this matrix = h20*h21*h22*h23*h24						3.725E-06	268456.6024
26								
27	AXIS TERM HUACHUCA @ Skip 1,465	SKIP	NUMBER	DIVIDE BY	THE QUOTIENT	E QUOTIENT	POISSON EQUATION	EACH WORD
28	ELS RANK 1	USED	IN	304805	EQUALS	x 1023 LETTERS	PROBABILITY FOR	OR PHRASE
29	This calculation is for the full matrix with a 1,023-letter area.	ON	TORAH	LETTERS	Combined odd	EQUALS	THE WORD	HAS 1 CHANCE
30		FIGURE		IN THE	PER LETTER	WORD	APPEARING	IN THE
31				TORAH		EXPECTANCY	AT LEAST ONCE	FOLLOWING OF
32								BEING HERE
33	REPTILIAN @ Skips +/- 1 and +/- 1465	-1465	3	304805	9.84236E-06	0.010068732	0.010018212	99.81820628
34	FORT (2 synonyms) @ Skip +1 only	1	7	304805	2.29655E-05	0.023493709	0.023219881	43.06654374
35	FOOD (3 synonyms) @ Skip +1 only	1	71	304805	0.000232936	0.238293335	0.212028483	4.716347484
36	CHILD @ Skip +1 only	1	110	304805	0.000360886	0.369186857	0.308703776	3.239351379
37	ARMY @ Skip +1 only	1	108	304805	0.000354325	0.362474369	0.304047849	3.288956014
38	Combined odds against this matrix = h33*h34*h35*h36*h37						4.62944E-06	216008.9274

Above: Spreadsheet for Figure 6-7 The 868-letter matrix area (green and white background) was most significant. REPTILIAN was the most significant term.

Chapter 7
Martian Lies and Bad Guys

This chapter details how a lack of logical Martian weather observations led to an understanding that our government is deliberately hiding the true nature of the Red Planet. It is a quick overview of findings in our over 1,200-page report – ***MARS CORRECT: CRITIQUE OF ALL NASA MARS WEATHER DATA***. I say "our" report because my son, Dr. David Alexander Roffman (PhD, physics), did most of the math that went into the report. The report largely steered clear of politics and UFO discussions. This chapter will not be so kind. There are national security implications to what we have found, and the lies we uncovered with respect to what is really going on Mars have turned out to be the tip of the iceberg when it comes to corruption and lies on Earth.

The abstract for *Mars Correct* quickly sums up our non-political findings:

We present evidence that NASA is seriously understating Martian air pressure. Our 12-year study critiques 3,025 Sols up through 8 February 2021 (8.51 terrestrial years, 4.52 Martian years) of highly problematic MSL Rover Environmental Monitoring Station (REMS) weather data, and offers an in-depth audit of over 8,311 hourly Viking 1 and 2 weather reports. We discuss the analysis of technical papers, NASA documents, and personal interviews with transducer designers. We troubleshoot pressures based on radio occultation/spectroscopy, and the previously accepted small pressure ranges that could be measured by Viking 1 and 2 (18 mbar), Pathfinder and Phoenix (12 mbar), and MSL (11.5 mbar – altered to 14 mbar in 2017). For MSL, there were several pressures published from August 30 to September 5, 2012, that were from 737 mbar to 747 mbar – two orders of magnitude too high by NASA doctrine– only to be retracted. We challenged many pressures, and NASA revised them. However, there are two pressure sensor ranges listed on a CAD for Mars Pathfinder. We long thought the CAD listed two different sensors, but based on specifications of a newer Tavis sensor for InSight that is like the one on Pathfinder, it appears that the transducer could toggle between two ranges: 0-0.174 PSIA/12 mbar (Tavis Dash 2) and 0-15 PSIA/1,034 mbar (Tavis Dash 1). Further, an abstract to the American Geophysical Union for the Fall 2012 meeting shows the Finnish Meteorological Institute (FMI) states of their MSL (and Phoenix) Vaisala transducers, "The pressure device measurement range is 0 – 1025 hPa in a temperature range of -45°C – +55°C (-45°C is warmer than MSL night temperatures), but its calibration is optimized for the Martian pressure range of 4 – 12 hPa." So, in fact, of the first five landers with meteorological suites, three were actually equipped to measure Earth-like pressure. All original 19 low µV values were removed when we asked about them, although eventually 12 were restored. REMS' always-sunny opacity reports were contradicted by Mars Reconnaissance Orbiter photos. We demonstrate that REMS weather data was regularly revised after they studied online critiques in working versions of this report. REMS even labeled all dust 2018 Global Dust Storm weather as sunny, although they did list the µV values then as all low. Vikings and MSL showed consistent timing of daily pressure spikes, which we link to how gas pressure in a sealed container would vary with absolute temperature, heating by radioisotope thermoelectric generators (RTGs), and dust clots at air access tubes and dust filters.

Pathfinder, Phoenix, and MSL wind measurements failed. Phoenix and MSL pressure transducer design problems included confusion about dust filter location and a lack of information about nearby heat sources due to International Traffic and Arms Regulations (ITAR). NASA Ames could not replicate dust devils at 10 mbar. Rapidly filled MER Spirit tracks required wind speeds of 80 mph at the assumed low pressures. These winds were never recorded on Mars. Nor could NASA explain drifting Barchan sand dunes. Based on the above and dust devils on Arsia Mons to altitudes of 17 km above areoid (Martian equivalent of sea level), spiral storms with 10 km eye-walls above Arsia Mons and similar storms above Olympus Mons (over 21 km high), dust storm opacity at MER Opportunity blacking out the sun, snow that descends 1 to 2 km in only 5 or 10 minutes, excessive aero braking, liquid water running at or near the surface in numerous locations at Recurring Slope Lineae (RSL), and stratus clouds 13 km above areoid, we argue for an average pressure at areoid of ~511 mbar rather than the accepted 6.1 mbar. This pressure grows to 1,050 mbar in the Hellas Basin.

Figure 7-1 above – Martian and terrestrial dust devils. When my son was a 16-year-old college sophomore at Embry-Riddle Aeronautical University, he asked what he should write about for his first technical paper. I recommended Martian dust devils because I could not understand how they could form in such a thin atmosphere. The average pressure on Earth is 1013.25 millibars, but on Mars, it's only about 6.1 millibars on average – less than 1% of Earth's pressure. My second concern was based on Martian sky color. When Viking I sent the first pictures of the Martian sky back to Earth, it was bright blue. It should have been quite black, like on a lunar day. There was some green on the rocks that might have been lichens. But the King of Martian lies, NASA Administrator James Fletcher, immediately ordered his stooges to manually alter the sky color by tuning the sky color monitors at Jet Propulsion Laboratory (JPL) to show a butterscotch or reddish sky with only brown on the rocks. This angered many people there, but Fletcher (pictured in Figure 7-34 with President Nixon) threatened to throw out anyone who tried to fix their monitors. NASA continued to lie to us about the Martian sky color seen by every lander until Curiosity/Mars Science Lander (MSL) landed on the Red Planet on August 6, 2012. We first saw a butterscotch sky again, but then JPL altered the color back to blue, supposedly because the dust cap had been left on.

Why did Fletcher and others lie to us from 1976 to 2012 (36 years)? The answer lies in understanding what the Deep State is. I prefer to call it the Fourth Reich, but the truth is worse than even these names can indicate. Too much curiosity about UFOs was likely what led to assassination of President Kennedy, and possibly assassination attempts against President Trump. He survived two attempts in 2024 and one at Mar-A-Lago in 2021. Many of his enemies, who pose such a threat, are named and discussed in Chapter 3 of this book. The Torah Code exposes lies about Mars and politicians here on Earth. It gives us a much more coherent view of real world history.

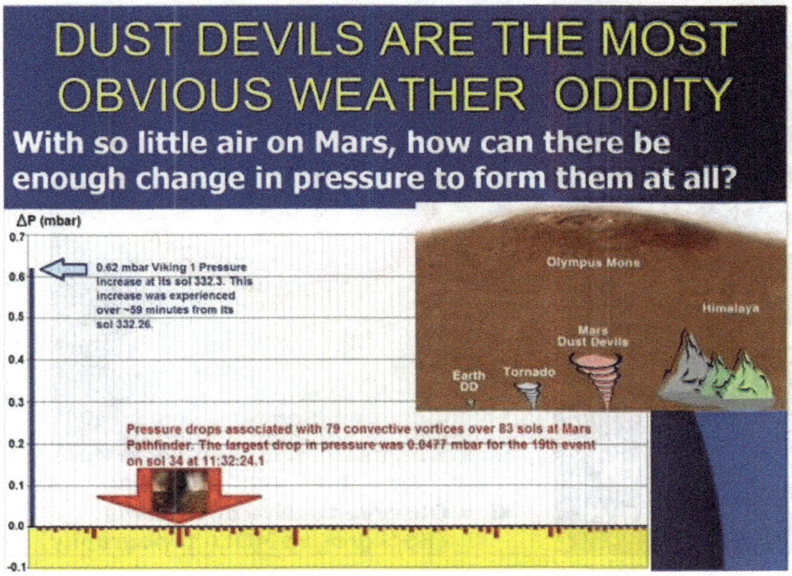

Figure 7-2 above. Relative magnitude of 0.62 mbar increases in pressure for Viking 1 at its sol

332.3 and pressure drops for 79 convective vortices/dust devils at Mars Pathfinder over its 83 sols. Source: Murphy, J. and Nelli, S., Mars Pathfinder Convective Vortices: Frequency of Occurrence (2002) http://tide.gsfc.nasa.gov/studies/Chen/proposals/IES/2002GL015214.pdf

Figure 7-2 offers evidence that internal events on the Vikings were having a much greater impact on pressure readings than dramatic events like dust devils. Pressure increases in the 0.26 to 0.3 time bins were comparable to pressure drops associated with global dust storms. An increase of 0.62 mbar in about 59 minutes that makes up one time-bin equates to a pressure rise 13 times greater than the largest (0.477 mbar) pressure fall shown for all 79 Pathfinder dust devil events, and about 21 times greater than the largest (0.0289 mbar) pressure drop seen for a Phoenix dust devil.

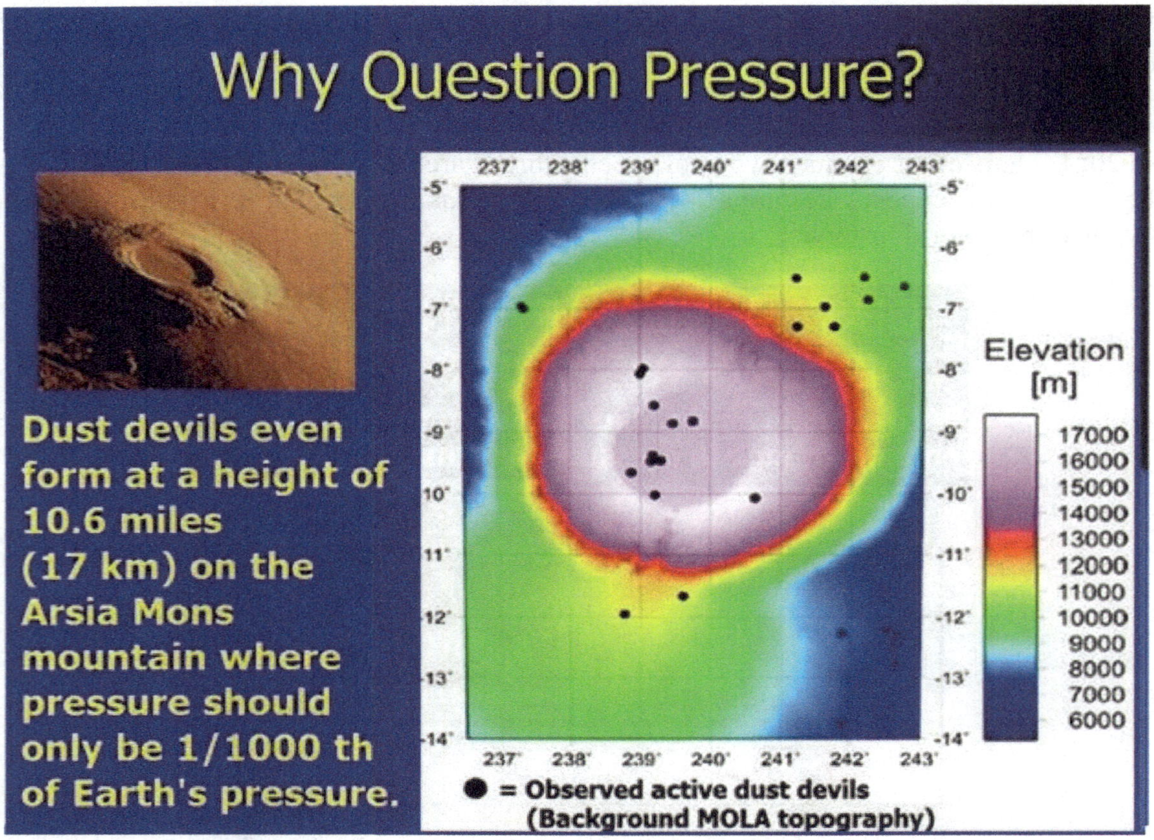

Figure 7-3 above: Thousands of dust devils per week occur in the Peruvian Andes near the Subancaya volcano (Metzger, 2001), which is 5,900 meters high. Dust devils are also seen in abundance on a Martian volcano, Arsia Mons. But the base altitude of some dust devils there has been about 17,000 meters. Such an altitude on Mars supposedly would have about 1.2 mbar pressure, compared to about 478 mbar at Subancaya on Earth. Reis et al. state that 28 active dust devils were reported in their study region for Arsia Mons, with 11 of them at altitudes greater than 16 km, and most inside the caldera (see Figure 7-2). They don't fully understand how particles that are a few microns in size can be lifted there, and state that one mbar "requires wind speeds 2 to 3 times higher than at the Mars mean elevation for particle entanglement."

Figure 7-4 below: Opacity changes at Opportunity from sols 1205 to 1235. Redrawn from http://www.jpl.nasa.gov/news/news.cfm?release=2007-080 for Opportunity between sols 1205 and 1235.

Figure 7-4 above. All photos were taken between 10:53 and 11:30 local time. The dust in the Martian air over Opportunity blocked 99 percent of direct sunlight. This fact alone makes it very hard to accept that pressures would be unaffected.

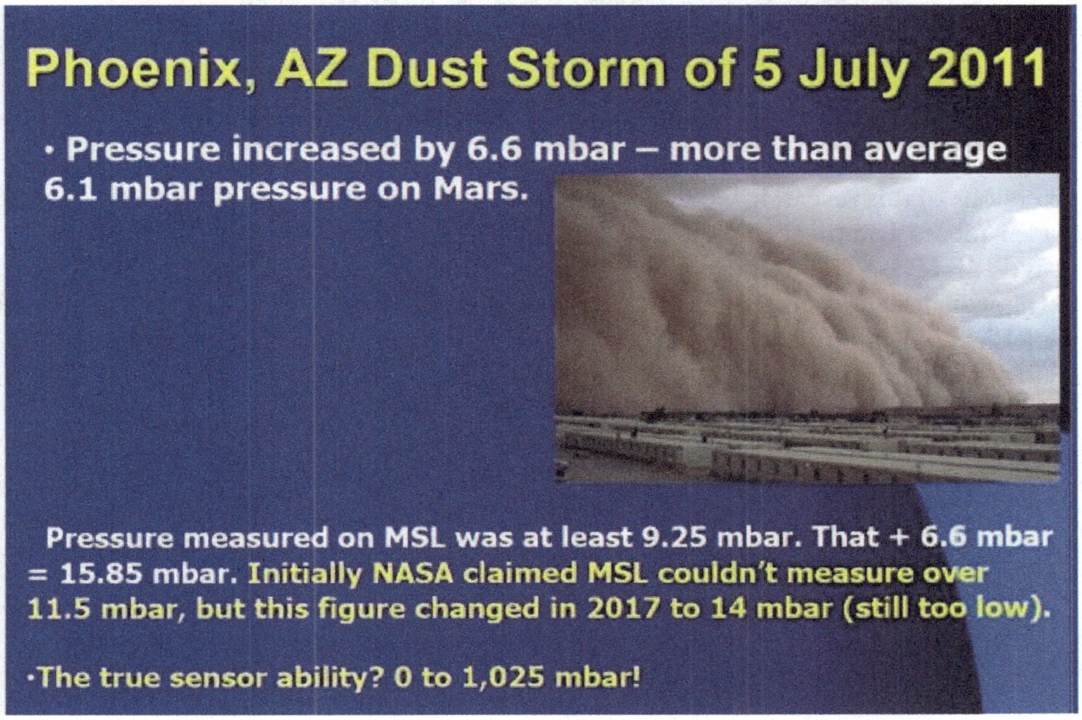

Figure 7-5 above: Arizona Dust Storm of July 5, 2011. Pressure at Luke Air Force Base increased during the dust storm by 6.6 mbar – more than the average pressure (6.1 mbar) at the areoid on Mars.

Figure 7-6 below: Print-screen (recorded on July 23, 2017) of the FMI Abstract entitled Pressure and Humidity Measurements at the MSL Landing Site Supported by Modeling of the Atmospheric Conditions.

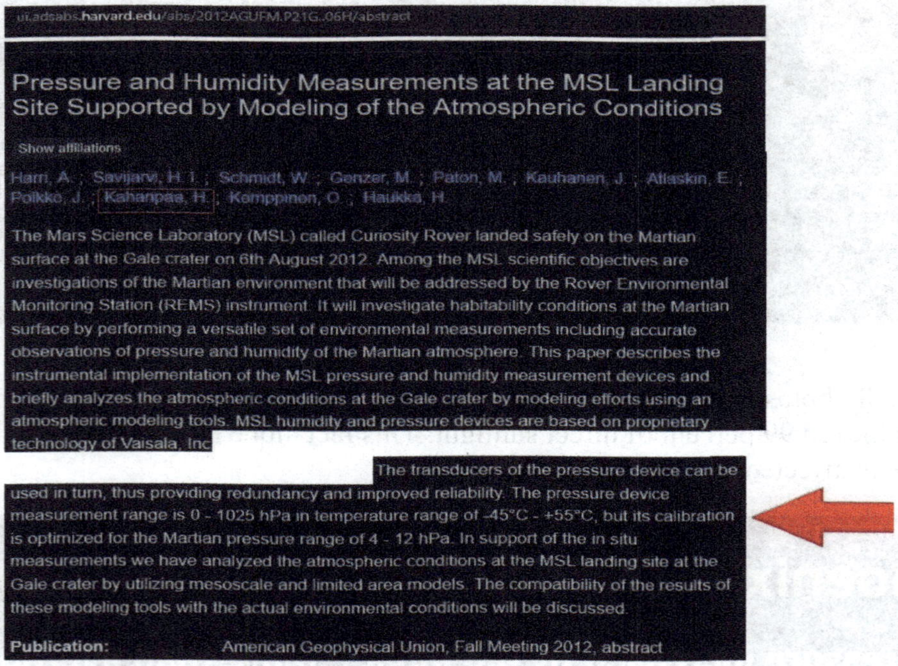

ui.adsabs.harvard.edu/abs/2012AGUFM.P21G..06H/abstract

Pressure and Humidity Measurements at the MSL Landing Site Supported by Modeling of the Atmospheric Conditions

Show affiliations

Harri, A. ; Savijarvi, H. I. ; Schmidt, W. ; Genzer, M. ; Paton, M. ; Kauhanen, J. ; Atlaskin, E. ; Polkko, J. ; Kahanpaa, H. ; Kemppinen, O. ; Haukka, H.

The Mars Science Laboratory (MSL) called Curiosity Rover landed safely on the Martian surface at the Gale crater on 6th August 2012. Among the MSL scientific objectives are investigations of the Martian environment that will be addressed by the Rover Environmental Monitoring Station (REMS) instrument. It will investigate habitability conditions at the Martian surface by performing a versatile set of environmental measurements including accurate observations of pressure and humidity of the Martian atmosphere. This paper describes the instrumental implementation of the MSL pressure and humidity measurement devices and briefly analyzes the atmospheric conditions at the Gale crater by modeling efforts using an atmospheric modeling tools. MSL humidity and pressure devices are based on proprietary technology of Vaisala, Inc

The transducers of the pressure device can be used in turn, thus providing redundancy and improved reliability. The pressure device measurement range is 0 - 1025 hPa in temperature range of -45°C - +55°C, but its calibration is optimized for the Martian pressure range of 4 - 12 hPa. In support of the in situ measurements we have analyzed the atmospheric conditions at the MSL landing site at the Gale crater by utilizing mesoscale and limited area models. The compatibility of the results of these modeling tools with the actual environmental conditions will be discussed.

Publication: American Geophysical Union, Fall Meeting 2012, abstract

In the initial versions of this research, we wrote that MPF was restricted to 10 mbar on the surface, and MSL was held to 11.5 mbar. The mean pressure recorded for MSL sol 370 was 11.49 mbar (at least until we challenged it and JPL revised it). The original pressure indicates that for much or most of that day, the actual pressure was almost certainly above the maximum pressure that the Vaisala pressure transducer could measure. The REMS Team published 1,177 Pa and 1,200 Pa pressures for sols 1,160 and 1,161, but after over two months of our questioning these pressures on our websites, JPL backed off and revised the pressures to 899 and 898 Pa. They likewise backed off a 1,154 Pa pressure for sol 1301 and changed it to 752 Pa. However, the REMS Team and the FMI read our findings. So, when we found on July 24, 2017, that REMS was suddenly posting a maximum pressure range of up to 1,400 Pa, all we could say was, "How convenient!" But it is totally inconsistent with everything they published before. Then there is that little matter of the transducer actually being capable of measuring up to 1,025 hPa (102,500 Pa – see Figure 7-6 material highlighted in red above).

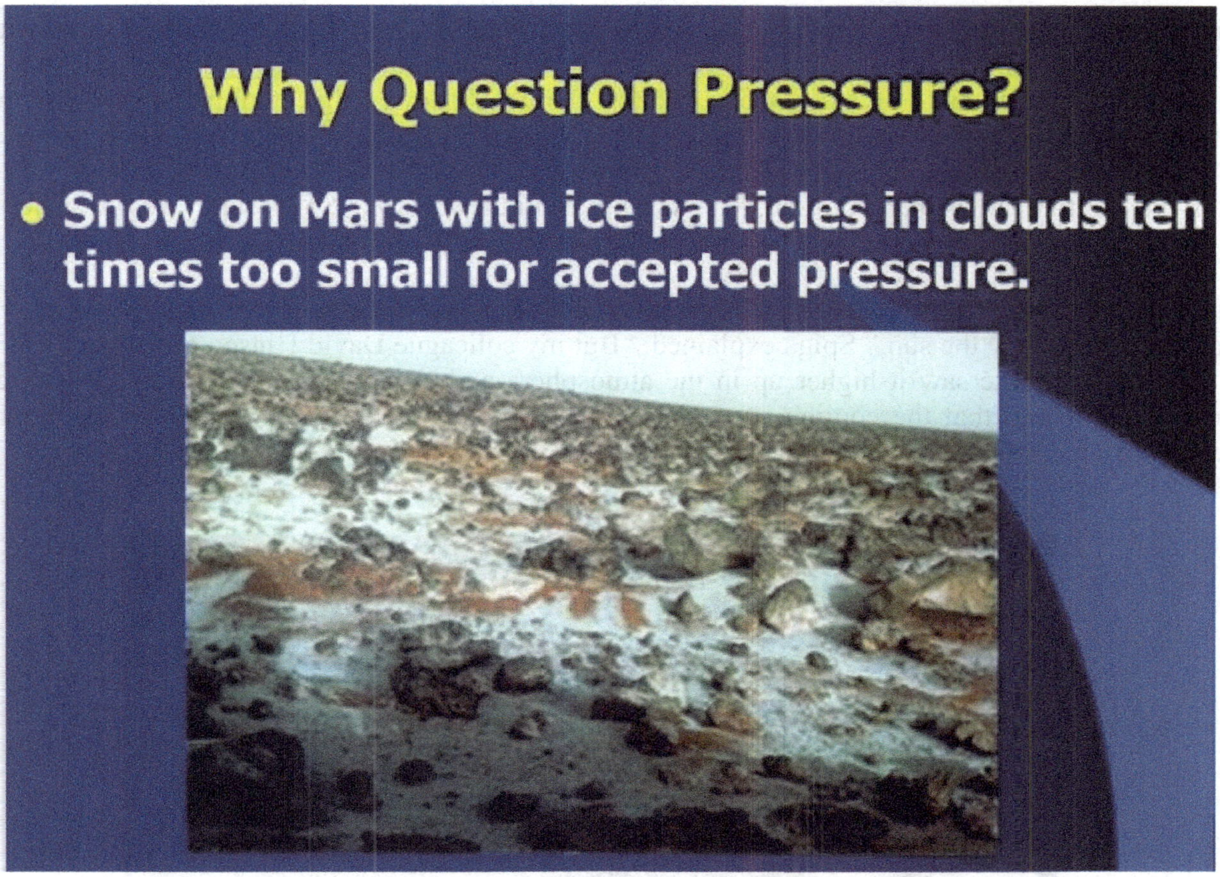

Figure 7-7 above: 2 μm ice particles survive on Mars because, in fact, the pressure is much higher than NASA has been telling us.

Phoenix captured snow on Mars. This was not unexpected. Richardson et al. (2002) discussed snow on Mars before Phoenix saw it, but they declared that in order to get a good fit to all other data, cloud ice particle sizes must be used that are about an order of magnitude too large (that is, 20 μm rather than the 2 μm observed).

They state that "significant work remains to be done assessing the quality of GCM predictions of Martian circulation vigor and resultant tracer transport." They concede the need to bump up ice particle size to levels that are "unrealistically large." While they were not specific about why the ice particles need to be so much bigger than those seen, it would make sense that if pressure were as low as advertised by NASA, the 2 μm ice particles would sublimate back into the atmosphere before the snow could fall. But at 20 μm, it could survive to hit the surface at such low pressures.

If so, it follows that 2 μm ice particles survive because, in fact, the pressure is much higher than NASA has been telling us. Wherever we look at the weather plainly seen on Mars, it fails to match pressures under 10 mbar.

On August 21, 2017, a new study (with lead author Aymeric Spiga, of the University of Pierre and Marie

Curie in Paris – see http://www.nature.com) noted that previous research suggested that if snow did fall from Martian clouds, it would waft down very slowly. "We thought that snow on Mars fell very gently, taking hours or days to fall 1 or 2 kilometers [0.6 to 1.2 miles]." Now, Spiga et al. found that, "Snow could take something like just 5 or 10 minutes to fall 1 to 2 km [0.6 to 1.2 miles]." The researchers were analyzing data from Mars Global Surveyor and Mars Reconnaissance Orbiter when they noticed a strong mixing of heat in the Martian atmosphere at night, "about 5 km from the surface," Spiga said. "This was never seen before."

You expect heat to get mixed in the Martian atmosphere close to the surface during the daytime, since the surface gets heated by the sun," Spiga explained. "But my colleague David Hinson at Stanford University and the SETI Institute saw it higher up in the atmosphere and at night. This was very surprising." The scientists discovered that the cooling of water-ice cloud particles during the cold Martian night could generate unstable turbulence within the clouds.

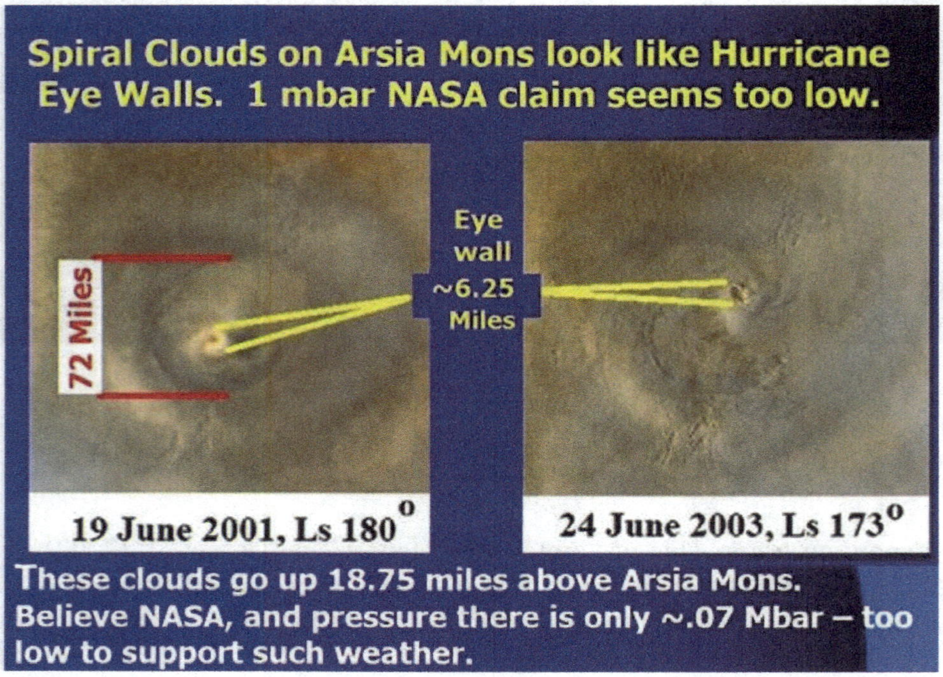

This can lead to strong winds, vertical plumes going upward and downward within and below the clouds at about 10 meters [33 feet] per second," or about 22 mph (36 km/h), Spiga said. "Those are the kinds of winds that are in moderate thunderstorms on Earth." Here again, the more we study Mars, the more it looks like Earth.

http://photojournal.jpl.nasa.gov/catalog/PIA04294 and
http://mars.jpl.nasa.gov/mro/multimedia/images/?ImageID=894&NewsInfo=59C884BFF2B8E0EDCEDF15F64B98BC57A54F95914A0576D9DF4145F3BFA98ECDCED7889AA9

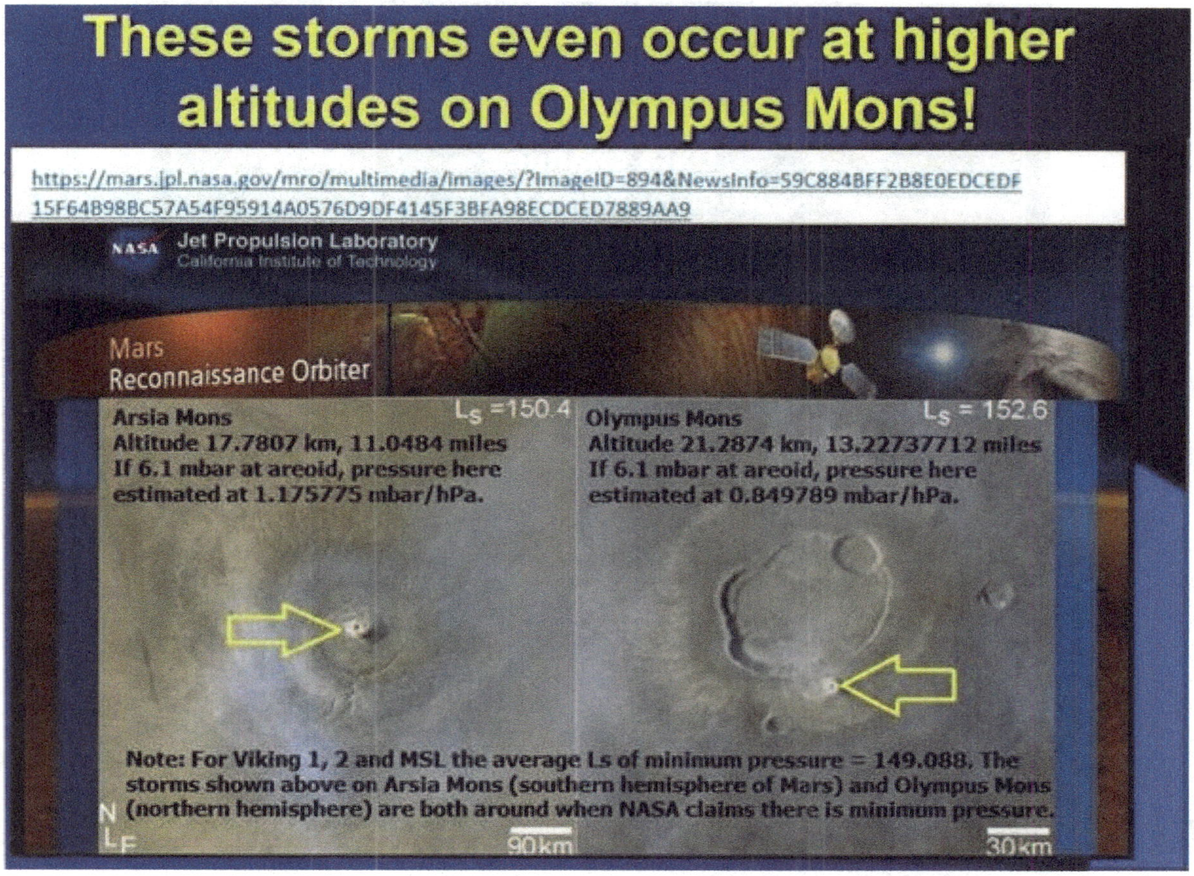

Figure 7-9 above: Comparison of alleged pressures at Arsia Mons and Olympus Mons on Mars.

Arsia Mons is at 9° South. With respect to the season, southern spring begins at Ls 180. It extends to Ls 270. Ls 90 to 179.9 is southern winter. These storms occur between Ls 150.4 and 180. They are therefore between the late winter and the first day of spring, but the storm over Olympus Mons in the northern hemisphere at Ls 152.6 is in late summer. Figures 7-8 and 7-9 show structures analogous to the eye walls of small hurricanes associated with the spiral clouds. They are about 10 km across and appear quite vigorous on Arsia Mons and about 7 km across at Olympus Mons. These pictures were taken just before planetary pressures should be near minimums. At such a high altitude, there shouldn't be enough pressure differentials to drive these storms if NASA is right, but they are plainly wrong.

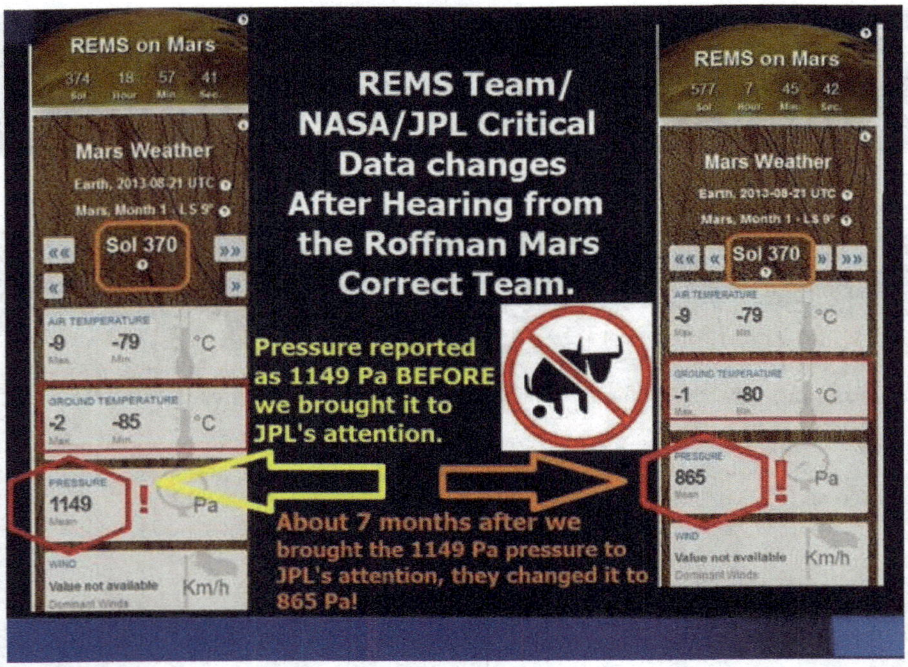

Figure 7-10 above: Sol 370 shows that the REMS Team and JPL's approach to problem-solving – they simply rewrite history and hope that nobody will notice it. The pressure the day before the 1149 Pa (11.49 mbar) spike was 865 Pa (8.65 mbar on Sol 369). After I called JPL about it, the pressure for the next day (Sol 371) returned to a more politically correct 865 Pa again.

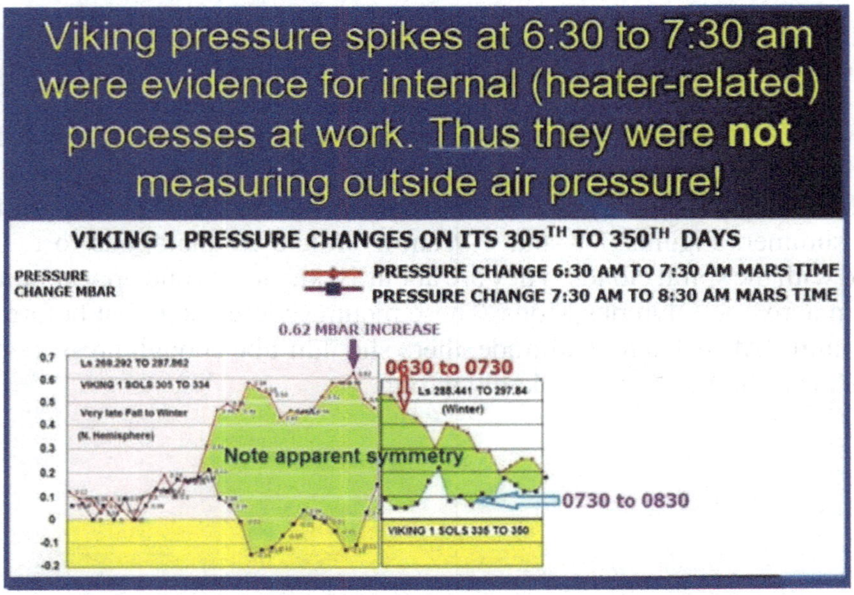

Figure 7-11 above. For VL-2 over 206 sols specified, pressure only decreased twice, each time just 0.01 mbar. The next time-bin (0.3-0.34) showed a much more varied pattern. Red lines show the first time-bin and blue shows the second time-bin on Figure 7-11.

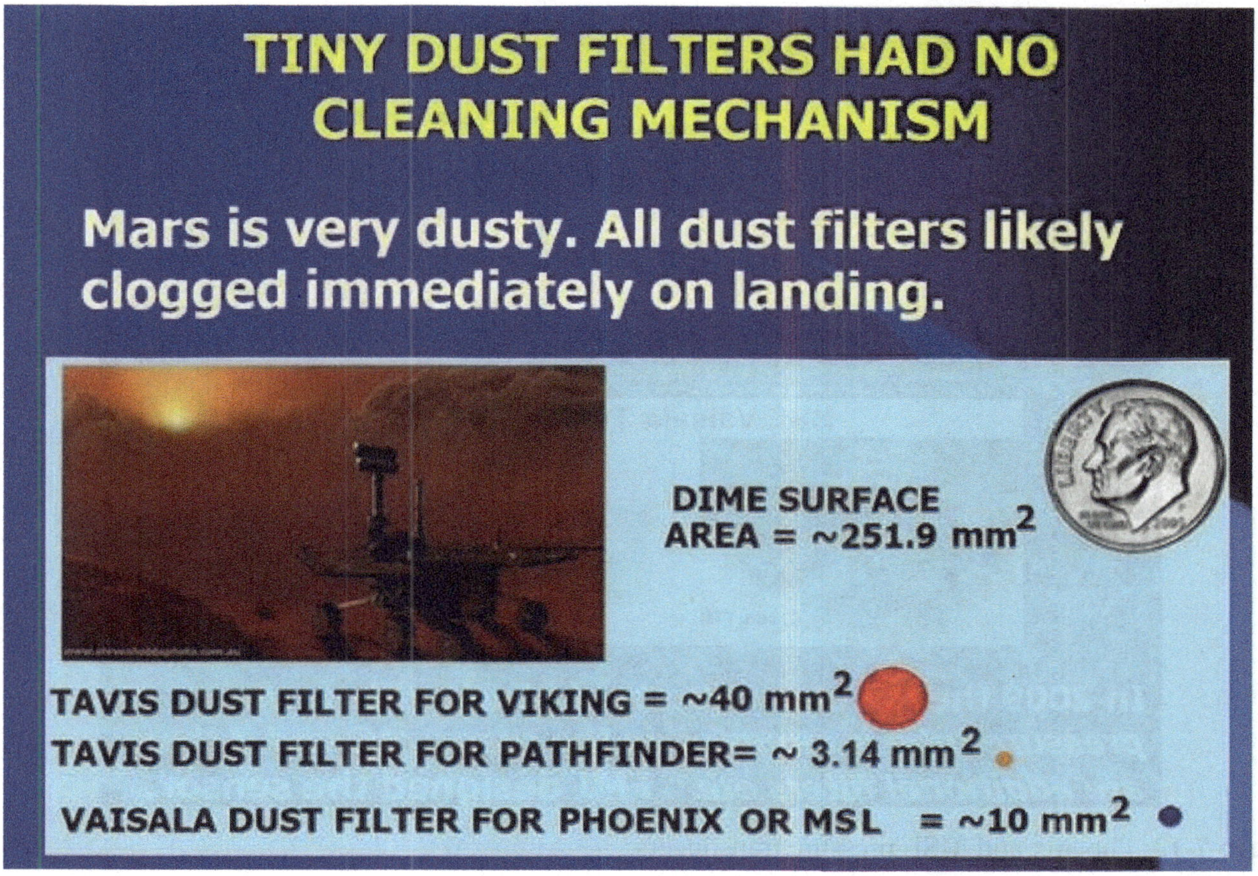

Figure 7-12 above: Relative size of dust filter for landers on Mars. 2 mm diameter of Mars Pathfinder (MPF) tubing from Seiff et al. (1997). On Figure 7-13 below, the top transducer is for Phoenix. Note the tiny dust filter shown under Praw (adapted from Doc. No: FMI_S-PHX-BAR- TN-00 FM-00 Revision 1.0 dated 2009-02-26). The report is entitled The Time Response of the PHOENIX Pressure Sensor. An area of concern for clogging by dust is highlighted. The photo on the right is adapted from http://www.space.fmi.fi/phoenix/?sivu=instrument. The bottom pictures are for MSL.

Issues Raised by the FMI. The FMI report by Kahanpää and Polkko (2009) discusses the Vaisala pressure sensor that it designed for use on Phoenix. It states, "We should find out how the pressure tube is mounted in the spacecraft and if there are additional filters etc." The one and only filter for the Vaisala transducer is shown on the top of Figure 7-13 (with its near twin for MSL shown below it).

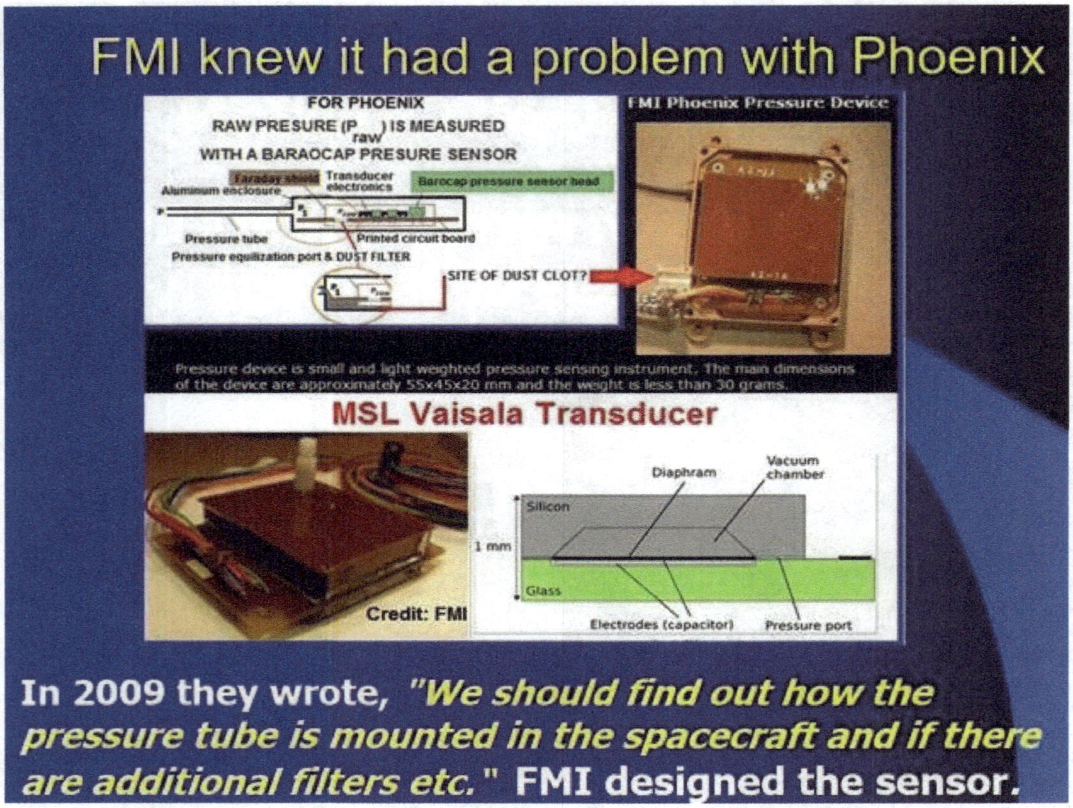

Figure 7-13: Phoenix and MSL pressure transducers.

I challenged the above statement on November 14, 2009, and published a criticism of it on my website on November 17, 2009. Kahanpää's partial response from the FMI to my assertion that "something stinks" about his request for information on additional filters was as follows:

"Your nose smelled was also a real issue. The fact that we at FMI did not know how our sensor was mounted in the spacecraft and how many filters there were shows that the exchange of information between NASA and the foreign subcontractors did not work optimally in this mission!" (Kahanpää, personal communication, December 15, 2009).

In his e-mail of December 15, 2009, Kahanpää made clear that there was no extra filter. However, the confusion in his report highlights another possibility. As is shown in Figure 7-12, the filter is very small (~10 mm²).

A pressure calculator with entering arguments based on VL-1 and 2 Year 1 results showed the prediction was 98.19% in agreement with measured results for Viking 1, and 91.04% in agreement for VL-2.

Vikings, MSL, and Gay-Lussac's Law.

Radioisotope thermoelectric generators (RTGs) may be at the root of problems with Viking and MSL pressure readings, which appear to vary inversely with outside temperatures. That is, when it gets colder outside and RTGs need to warm the inside of the landers, the pressure recorded inside goes up. Temperature and pressure variations seen for Viking 1 Year 1 almost exactly match what would be expected in accordance with Gay-Lussac's Pressure Law (see Figures 7-13 A through 7- 13 and **7-**14). To counteract a minimum Year 1 temperature of 177.19K seen, and to raise internal temperatures to the maximum Year 1 external temperature seen (255.77 K), air caught behind a dust clog would experience a pressure rise. If Viking 1 sucked in enough dust and sand on landing to clog, but not enough to equalize the internal pressure with the air pressure outside, then whatever Year 1 minimum pressure seen inside the lander at the Tavis pressure transducer (6.51 mbar) would increase in pressure in accordance with Gay-Lussac's Law. As shown in Figure 7-6, when the above two temperatures and 6.51 mbar are entered into the calculator, the expected pressure is shown to be 9.397 mbar. The actual maximum pressure recorded by Viking 1 was 9.57 mbar. That is a 98.19% agreement with the idea that the air access tube for the sensor was clogged. For Viking 2, the minimum and maximum temperatures were 152.14 K and 245.74 K. The minimum pressure found was 7.29 mbar. The maximum predicted pressure was 11.775 mbar. The maximum pressure recorded by VL-2 was 10.72 mbar, which is 91.04% of the predicted value.

The data points on Figure 7-13 are meant to get some sense of whether the pressure limits seen were roughly in line with expectations based on heat applied to a sealed space (behind the dust clots). They were, but obviously more so in Viking 1's first year. By Year 2, overall predictions were off by 9 or 10 percent, but the calculations are less certain because of many incidents involving stuck pressure readings, sometimes for days on end. Annex C of our *Mars Correct* report supports this allegation, but Annex D also highlights stuck pressure readings for Viking 1. The old cliché "Garbage in, garbage out" sums up the problem. Temperature data seemed credible for the Vikings (except when reported as Absolute Zero). However, temperatures (in particular, ground temperatures) were problematic for MSL, as is detailed in Section 14.1 of the *Mars Correct Report*, which asserts that pressure data was not credible for any lander.

When comparing maximum air temperatures seen at MSL and Viking 1, we show in Annex M of our *Mars Correct Report* that the highest air temperature seen after JPL revised its Year 1 data was 4°C (274.15K/39.2°F). MSL sits at 4.59° South on Mars at an altitude of 4,400 meters below the areoid. Viking 1 was also in the tropics at 22° North. However, VL-1 was at an altitude of 3,627 meters below the altitude. R.M. Haberle at NASA Ames claims that the adiabatic lapse rate for Mars is about 2.5K km^{-1}. Using that rate, we would expect the maximum temperature at VL-1 to be about 1.9325 K lower than at MSL; however, the maximum temperature at VL-1 was only 255.77K/0.716K, while the maximum (revised) temperature for MSL Year 1 (on MSL Sol 227/March 2, 2013) was 274.15K/33.8°F, a full 18.38 K warmer than at VL-1. Further, before JPL revised its MSL temperatures, it indicated a maximum air temperature at MSL of 8°C (281.15K/46.4°F) on MSL Sol 102 (November 18, 2012), but they later altered this temperature to -3°C (270.15K). The high for MSL Year 2 was 11°C (284.15K/51.8°F) on Sol 760. So, it would appear that there is room to question the accuracy and consistency of air temperature sensors on these two missions.

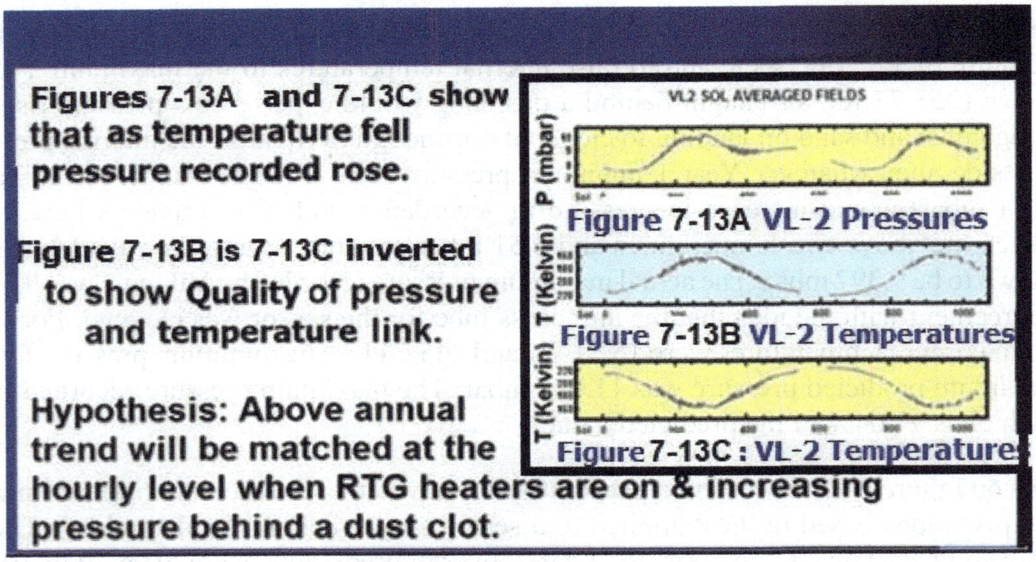

Figures 7-13 A 7-13C are redrawn from Tillman and Johnson. Figure 7-13B inverts the direction of temperatures on the Y-axis to show how heating by RTGs to counter the increasing cold outside produces a curve very similar to the pressure curve.

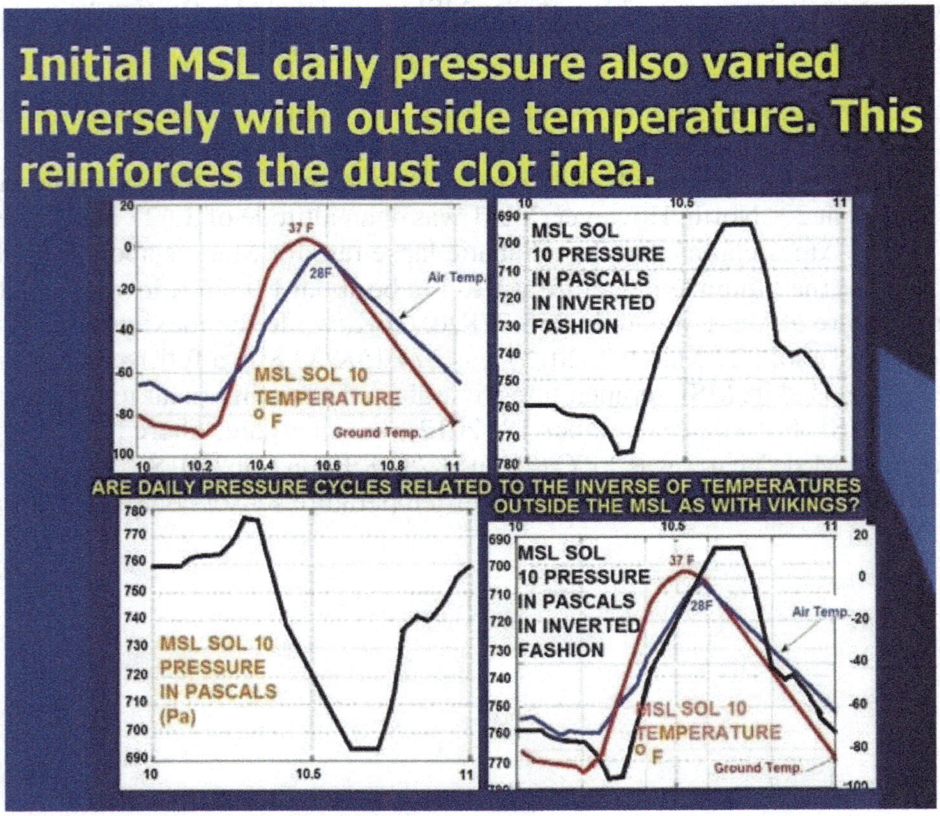

Figure 7-14 above: Daily pressure variation at MSL.

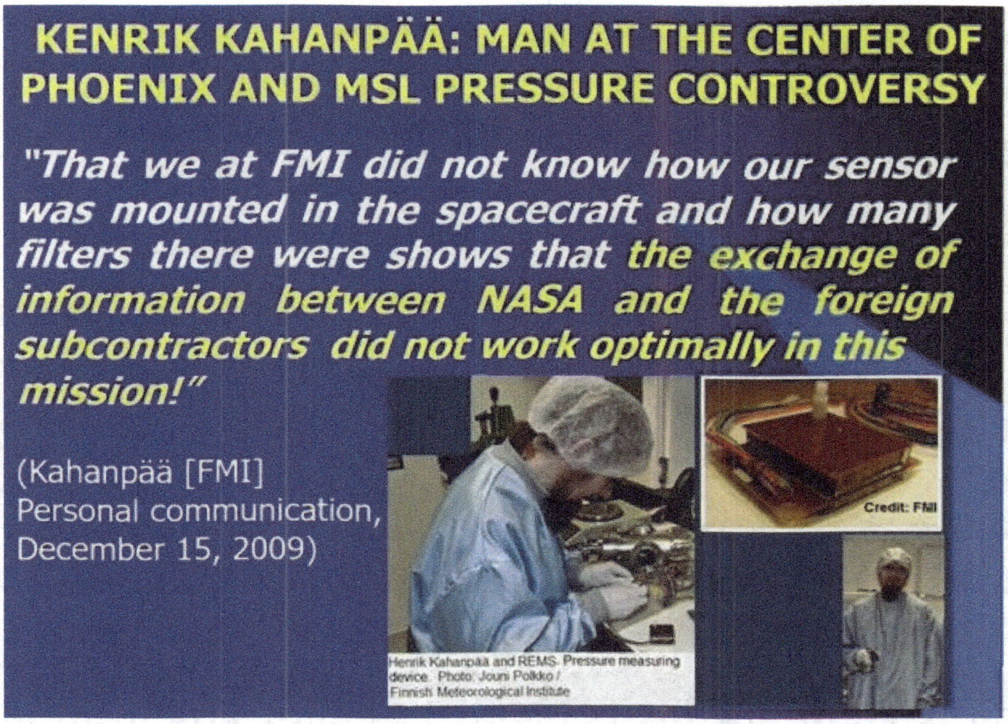

Figure 7-15 above: Cause of fatal flaws in spacecraft design

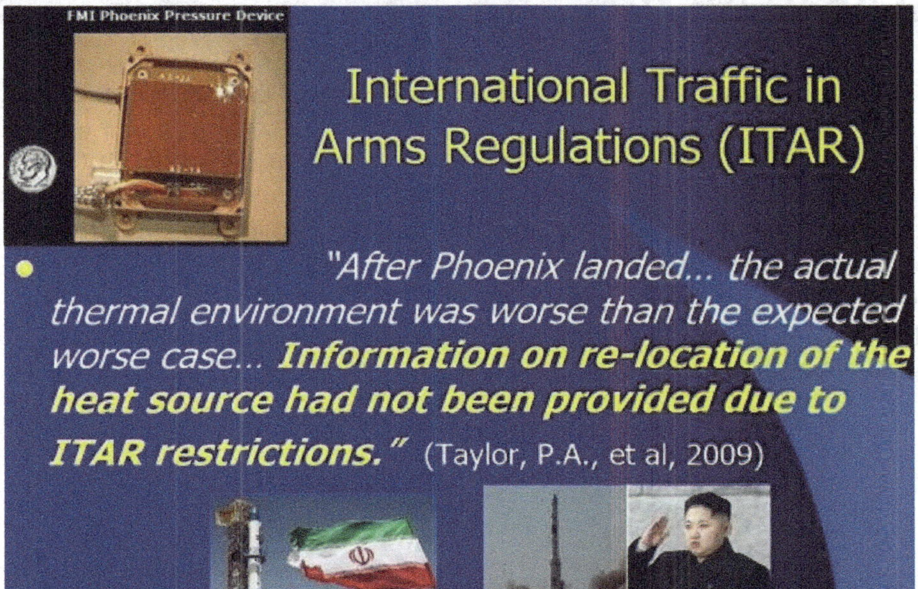

Figure 7-16 above: ITAR restrictions also lead to fatal design flaws.

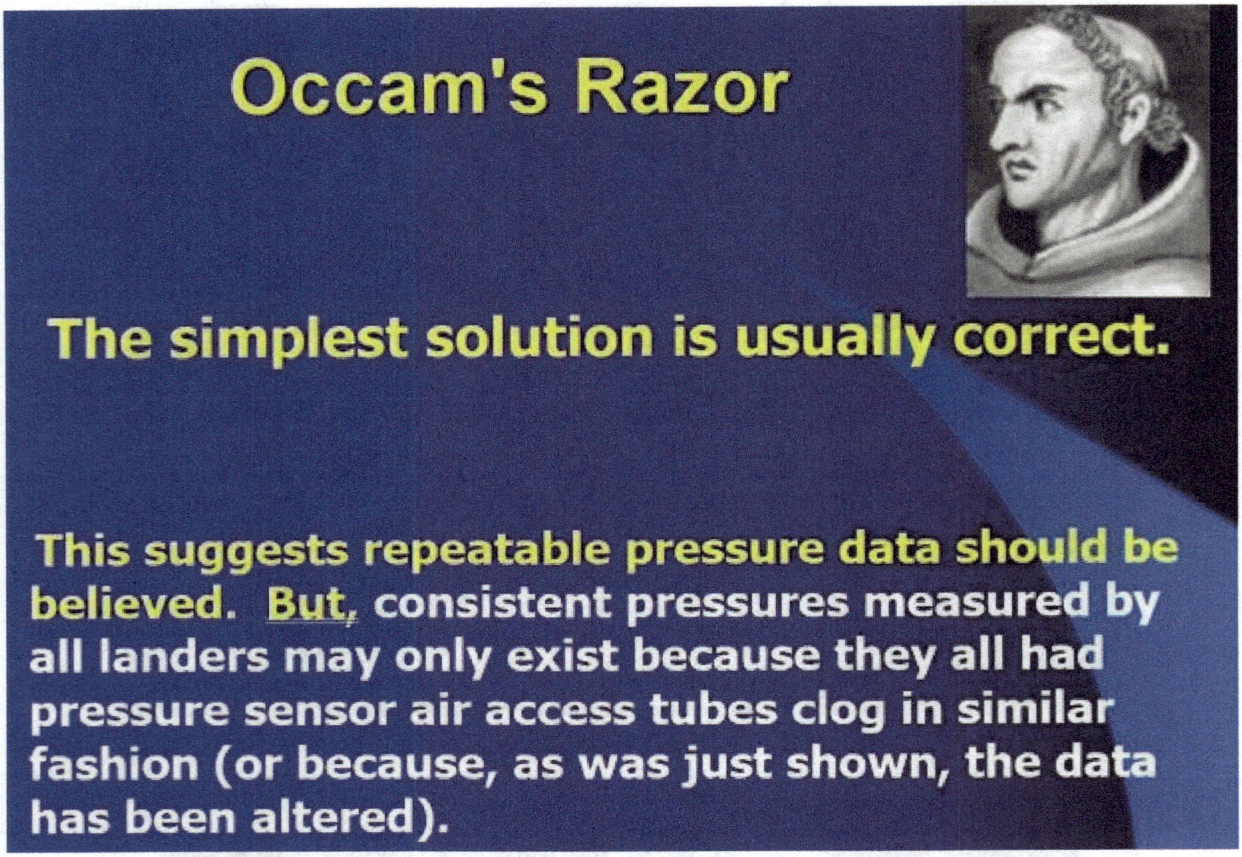

Figure 7-17 above: Unfortunately, the situation on Mars is more complex than Occam could imagine.

On July 24, 2017, we found that the REMS Team again altered the max pressure. On October 19, 2017, ESA reported that ExoMars had to raise its orbit. The move was mandated by "excessive density of Mars' atmosphere." We received notice of this from our partner Marco de Marco. Maximum pressure was raised to 1400 Pa (14 mbar). After they raised the maximum pressure from 1150 to 1400 Pa, they published a maximum pressure of 1,294 Pa for Sol 1784 on August 13, 2017. On the previous sol (1783), the pressure published was only 879 Pa. Yet even with the newer (likely false) upper pressure range of 1,400 Pa, when we challenged it with our colored spreadsheet and print-screen, the REMS Team dropped the 1,294 Pa for that sol to 883 Pa.

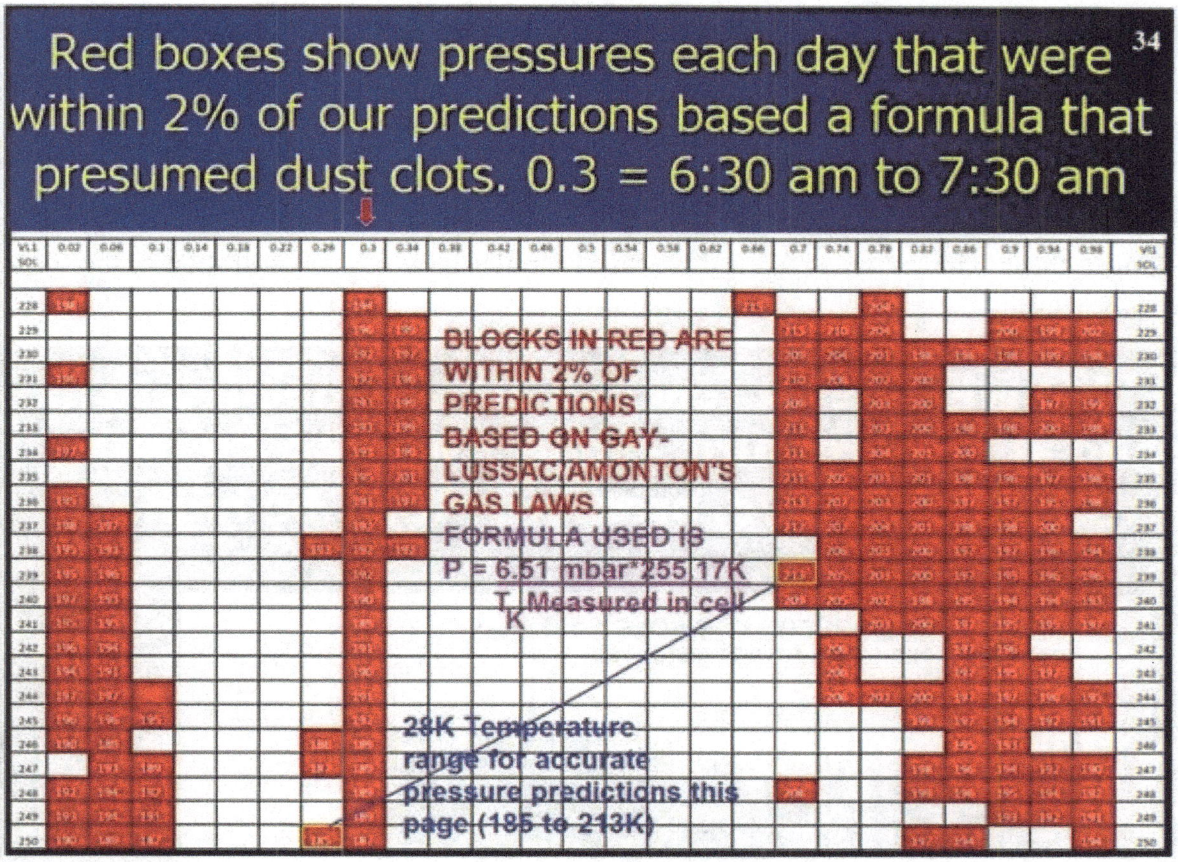

Figure 7-18 above: Sample of Annex F of *Mars Correct: Critique of All NASA Mars Weather Data*. Showing the times of day (for sols 228 through 250) when pressure predictions had less than a 2% difference from measured pressures at Viking 1. The formula used assumes that the pressure transducer is no longer in contact with the ambient atmosphere on Mars.

NASA Orbiter Catches Mars Sand Dunes in Motion. The first startling confession was that:

"Mars either has more gusts of wind than we knew about Martian Bedforms – Too Much Movement of Sand Dunes and Ripples for 6.1 mbar."

In November 2012, an article was published by Dwayne Brown of NASA Headquarters and Priscilla Vega at JPL. It said,

"The winds are capable of transporting more sand, said Nathan Bridges, planetary scientist at the Johns Hopkins University's Applied Physics Laboratory in Laurel, Md., and lead author of a paper on the finding published online in the journal Geology. We used to think of the sand on Mars as relatively immobile, so these new observations are changing our whole perspective."

It states that wind-tunnel experiments have shown that a patch of sand would require winds of about 80 miles/hour (128.7 km/hour) to move on Mars compared with only 10 mph (16 km/hour) on Earth. It then makes the understatement that measurements from the Viking landers, in addition to climate models, showed such winds should be rare on Mars.

The word "rare" was too generous. How does the above required 128.7 km/hour compare with winds observed on Mars? The set of graphs on Figure 7-19 below shows how wind speed varied at Viking 1 between its sols 1 and 350 (with the exception of sols 116 to 133 because data was missing then). The maximum wind recorded was 57.9 miles/hr. At no measured point over 8,331 measurements did the wind ever reach 128.7 km/hr. Average winds for Viking 1 were about 9.85 km/hr during sols 1 to 199, and 19.08 km/hr during its sols 200 to 350. All wind data was obtained from the Tillman Viking data.

For Viking 2, the maximum wind recorded was at 51.9 miles/hr which was still short of the speed required to move the sands.

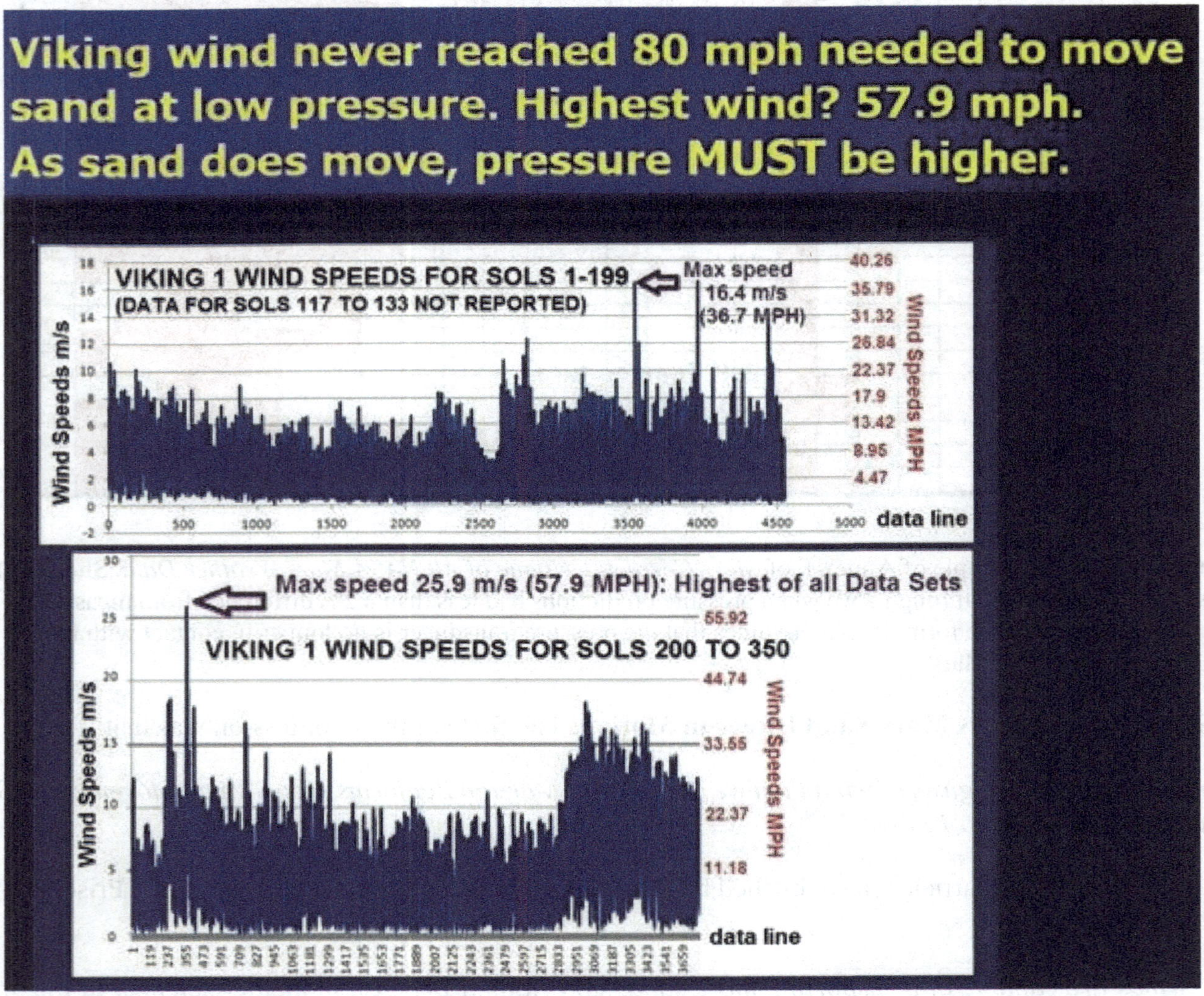

Figure 7-19 above: Wind speeds recorded at Viking 1 for its sols 1 to 116 and 134 to 350.

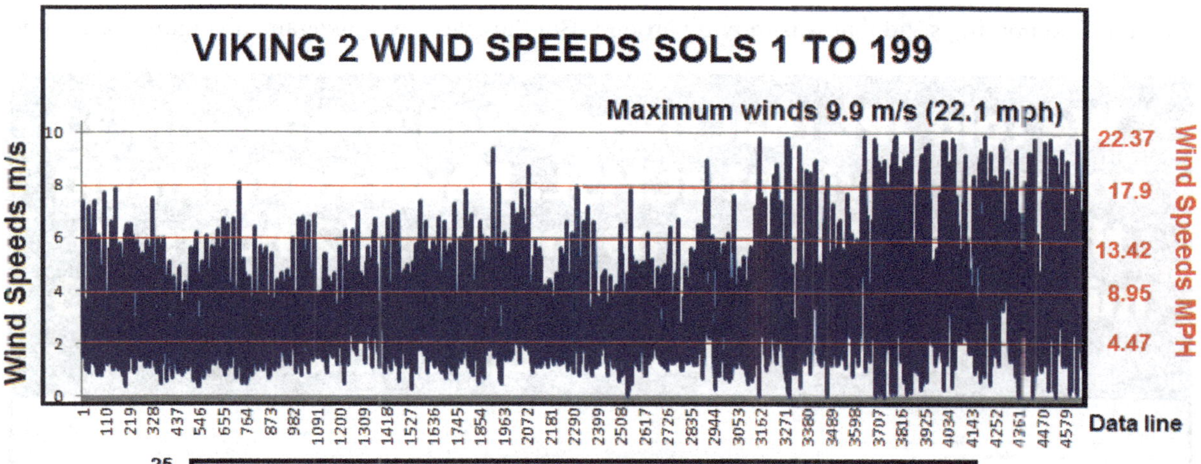

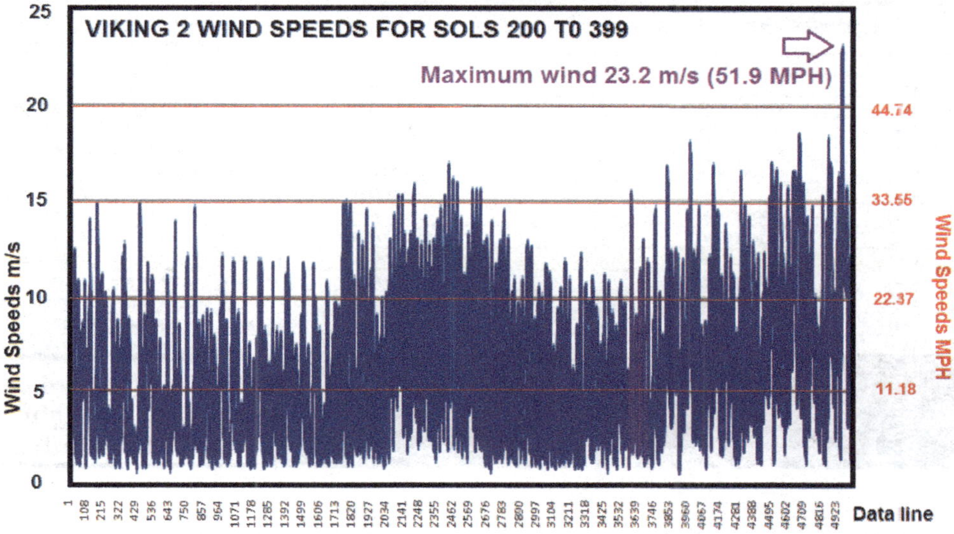

Figure 7-20 Wind speeds recorded at Viking 2 for its sols 1 to 399

Figure 7-21 above: By NASA standards, the Martian atmosphere is too thin to move sand dunes, and for years,

they claimed to find nothing solid that was moving around. But they were wrong again, as Figure 7-21 shows.

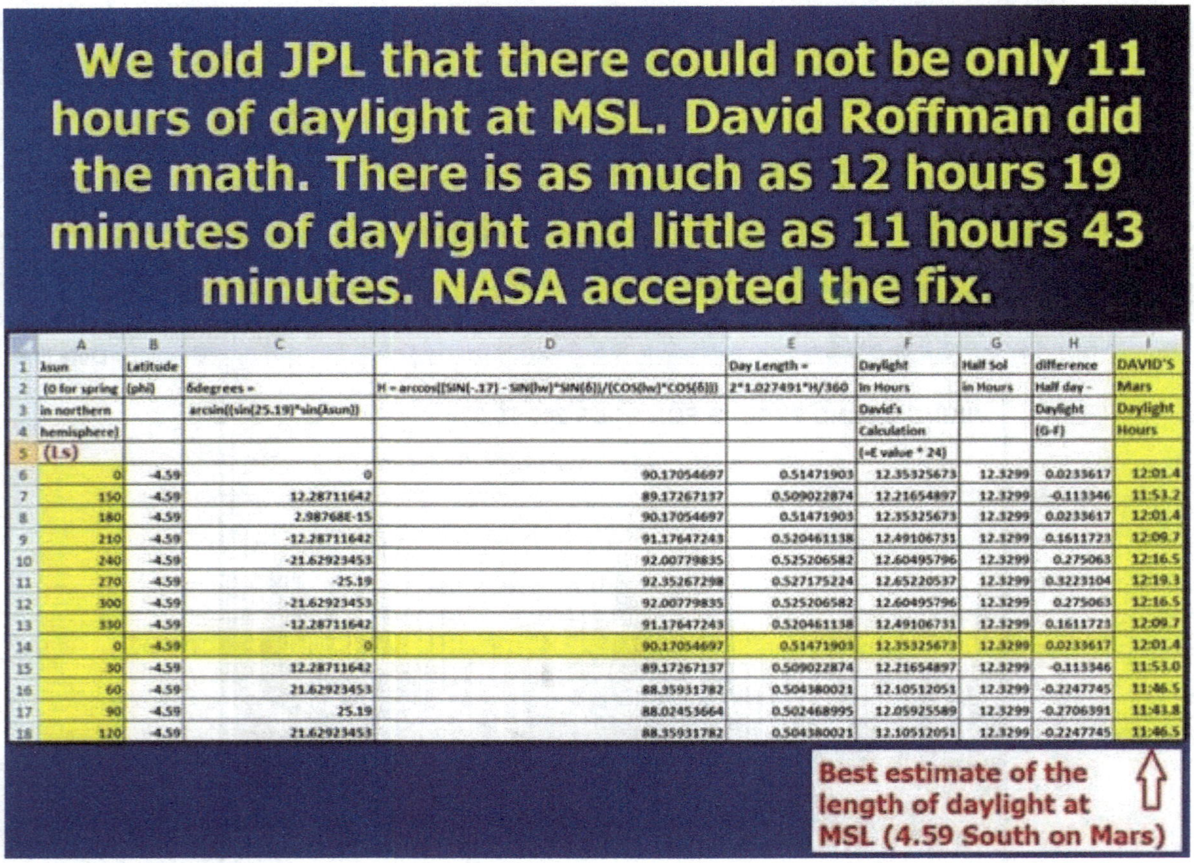

	A	B	C	D	E	F	G	H	I
1	λsun	Latitude			Day Length =	Daylight	Half Sol	difference	DAVID'S
2	(0 for spring	(phi)	δdegrees =	H = arccos([SIN(-.17) - SIN(lw)*SIN(δ)]/[COS(lw)*COS(δ)])	2*1.027491*H/360	in Hours	in Hours	Half day -	Mars
3	in northern		arcsin((sin(25.19)*sin(λsun))			David's		Daylight	Daylight
4	hemisphere)					Calculation		(G-F)	Hours
5	(Ls)					(=E value * 24)			
6	0	-4.59	0	90.17054697	0.51471903	12.35325673	12.3299	0.0233617	12:01.4
7	150	-4.59	12.28711642	89.17267137	0.509022874	12.21654897	12.3299	-0.113346	11:53.2
8	180	-4.59	2.98768E-15	90.17054697	0.51471903	12.35325673	12.3299	0.0233617	12:01.4
9	210	-4.59	-12.28711642	91.17647243	0.520461138	12.49106731	12.3299	0.1611723	12:09.7
10	240	-4.59	-21.62923453	92.00779835	0.525206582	12.60495796	12.3299	0.275063	12:16.5
11	270	-4.59	-25.19	92.35267298	0.527175224	12.65220537	12.3299	0.3223104	12:19.3
12	300	-4.59	-21.62923453	92.00779835	0.525206582	12.60495796	12.3299	0.275063	12:16.5
13	330	-4.59	-12.28711642	91.17647243	0.520461138	12.49106731	12.3299	0.1611723	12:09.7
14	0	-4.59	0	90.17054697	0.51471903	12.35325673	12.3299	0.0233617	12:01.4
15	30	-4.59	12.28711642	89.17267137	0.509022874	12.21654897	12.3299	-0.113346	11:53.0
16	60	-4.59	21.62923453	88.35931782	0.504380021	12.10512051	12.3299	-0.2247745	11:46.5
17	90	-4.59	25.19	88.02453664	0.502468995	12.05925589	12.3299	-0.2706391	11:43.8
18	120	-4.59	21.62923453	88.35931782	0.504380021	12.10512051	12.3299	-0.2247745	11:46.5

Figure 7-22 above: Calculating daylight length on Mars.

Figure 7-23 below: Stratus clouds on Earth are found up to altitudes of 13,000 meters, where pressures are about 163.33 mbar. They are not found on Earth at pressures below this level. The same kind of clouds are found on Mars. If the same minimum pressure is required on Mars, where these clouds were seen 16 km above Mars Pathfinder, which was 3.682 km below the areoid, it means that the stratus clouds were about 12.318 km above the areoid. Based on an accepted scale height of 10.8, this implies a pressure at the areoid of about 511. Further evidence that NASA is wrong is that the photo of clouds glowing red was taken an hour and 40 minutes before sunrise. This implies that the atmosphere is much denser than the near vacuum that NASA advertises for Mars.

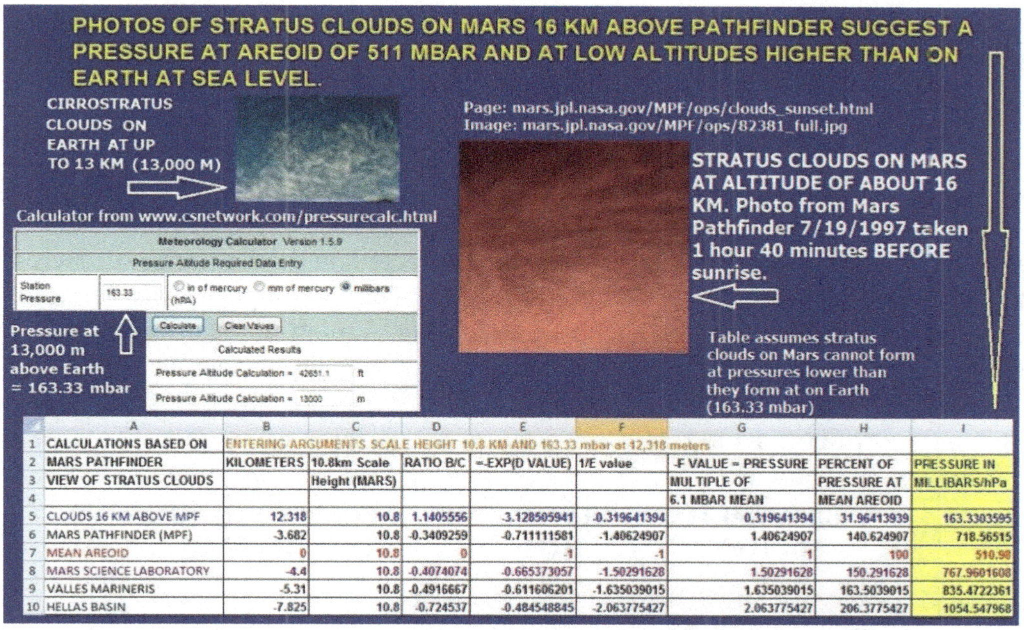

	A	B	C	D	E	F	G	H	I
1	CALCULATIONS BASED ON	ENTERING ARGUMENTS SCALE HEIGHT 10.8 KM AND 163.33 mbar at 12,318 meters							
2	MARS PATHFINDER	KILOMETERS	10.8km Scale	RATIO B/C	=-EXP(D VALUE)	1/E value	-F VALUE = PRESSURE	PERCENT OF	PRESSURE IN
3	VIEW OF STRATUS CLOUDS		Height (MARS)				MULTIPLE OF	PRESSURE AT	MILLIBARS/hPa
4							6.1 MBAR MEAN	MEAN AREOID	
5	CLOUDS 16 KM ABOVE MPF	12.318	10.8	1.1405556	-3.128505941	-0.319641394	0.319641394	31.96413939	163.3303595
6	MARS PATHFINDER (MPF)	-3.682	10.8	-0.3409259	-0.711111581	-1.40624907	1.40624907	140.624907	718.56515
7	MEAN AREOID	0	10.8	0	-1	-1	1	100	510.98
8	MARS SCIENCE LABORATORY	-4.4	10.8	-0.4074074	-0.665373057	-1.50291628	1.50291628	150.291628	767.9601608
9	VALLES MARINERIS	-5.31	10.8	-0.4916667	-0.611606201	-1.635039015	1.635039015	163.5039015	835.4722361
10	HELLAS BASIN	-7.825	10.8	-0.724537	-0.484548845	-2.063775427	2.063775427	206.3775427	1054.547968

Figure 7-23 above. Stratus clouds on Mars at 13,000 meters above the areoid suggest a higher pressure than what NASA admits to.

Figure 7-24 below shows NASA publishing wrong Martian temperatures.

Figure 7-24 above: The REMS Team here appears to purposely sabotage all of its remaining credibility by going back and dropping very nearly all of its above-freezing air temperatures to below freezing. The question which has not yet been answered by JPL (or by anyone else) – Who ordered these changes, and why did they do so?

Temperature Measurement Concerns

Until July 3, 2013, we knew that over the first 11 months of operation, the REMS Team and Ashima Research had put out clearly erroneous winds, sunrise and sunset times, pressure units, dates on their reports, months, and claims about relative humidity that were not reflected on their reports. We (wrongly) assumed, however, that at least the temperature reports were reliable. That assumption was demolished on July 3, 2013, when they revised all temperatures back to the landing, wiping out scores of days where they had claimed air temperature highs above freezing. Some of these revisions are visible on Table 7-1.

TABLE 7-1 – MSL Air Temperatures Altered by JPL in July, 2013

A	B	C	D
SOL	ORIGINAL MAX AIR TEMP °C	NEW MAX AIR TEMP °C	CHANGE °C (EQUALS CHANGE K)
23	0	-16	16
26	2	-14	16
27	-1	-15	14
31	-3	-23	20
38	-3	-13	10
40	2	-12	14
41	2	-12	14
42	5	-7	12
43	3	-12	15
44	4	-10	14
45	3	-9	12
46	4	-12	16
47	6	-9	15
49	4	-10	14
50	0	-10	10
51	3	-7	10
52	7	-7	14
53	5	-5	10
54	5	-9	14
102	8	-3	11
112	5	-8	13
116	5	-6	11
118	4.53	-6	10.53
123	2.1	-10	12.1
124	5.4	-5	10.4
179	5	-7	12

Table 7-1 above: For some unexplained reason, JPL dropped many Martian air temperatures for Curiosity on July 3, 2013, nearly 11 months after the landing.

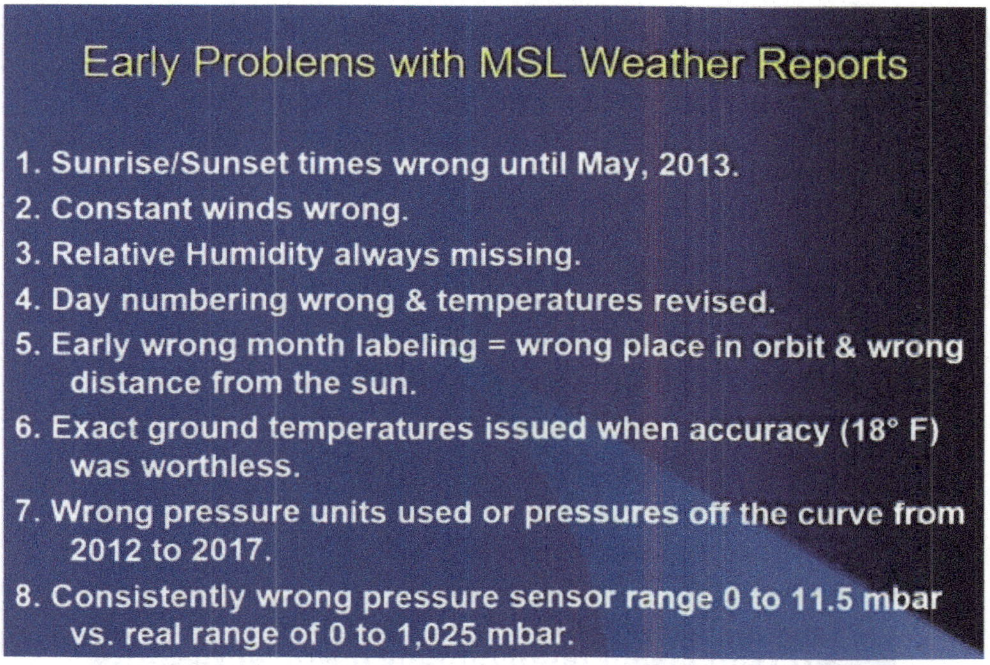

Figure 7-25 above: Problems with MSL weather reports.

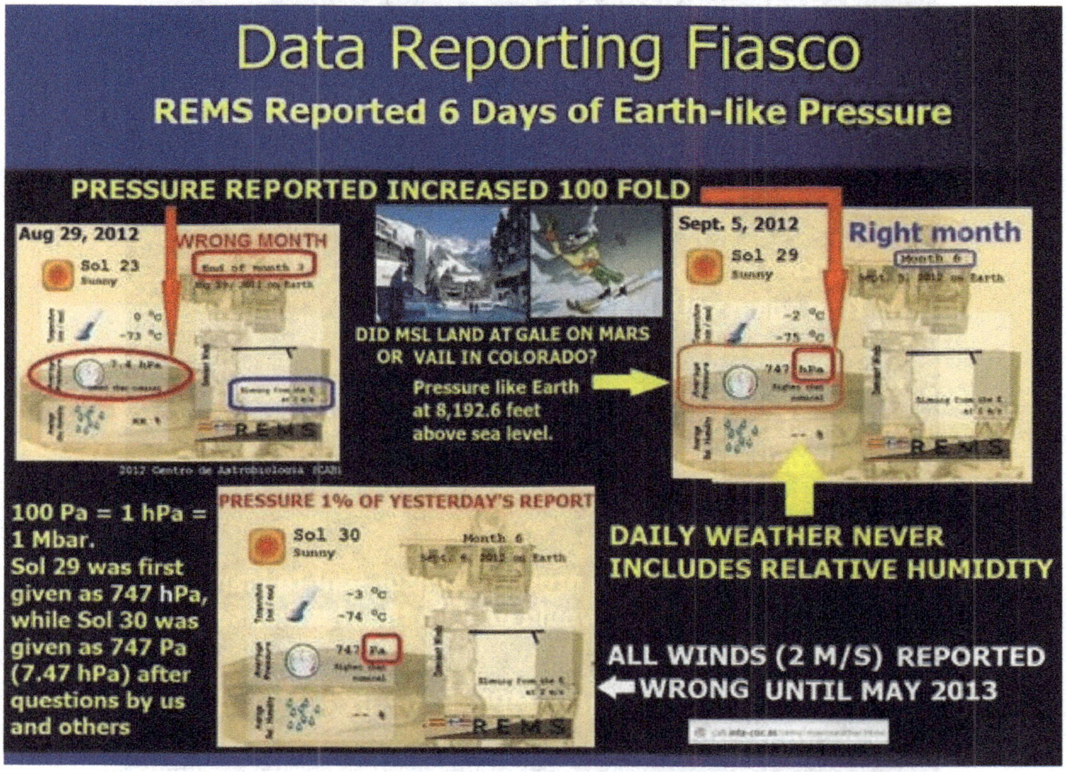

Figure 7-26 above: Data reporting fiasco by REMS.

MSL Weather Reporting Fiasco. The MSL REMS Team initially put out continually flawed data at http://cab.inta-csic.es/rems/marsweather.html. The REMS Team went from listing the pressure on August 28, 2012, as 7.4 hPa (mbar) and the month as 3 when it was really month 6; to a September 1, 2012, pressure of 742 hPa (Earth-like, seen in much of the U.S. West every day) in month 3 to 743 hPa pressure for September 2, 2012, which was correctly listed as month 6. Between September 5 and 6, 2012, reported pressures dropped from 7.47 hPa to 1% of that – 7.47 Pa. See Figure 7-26.

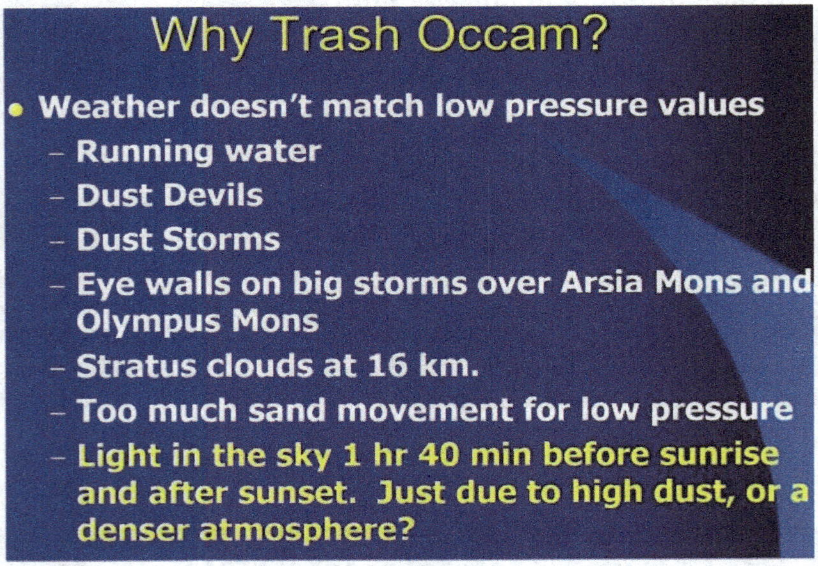

Figure 7-27 above. Why trash Occam?

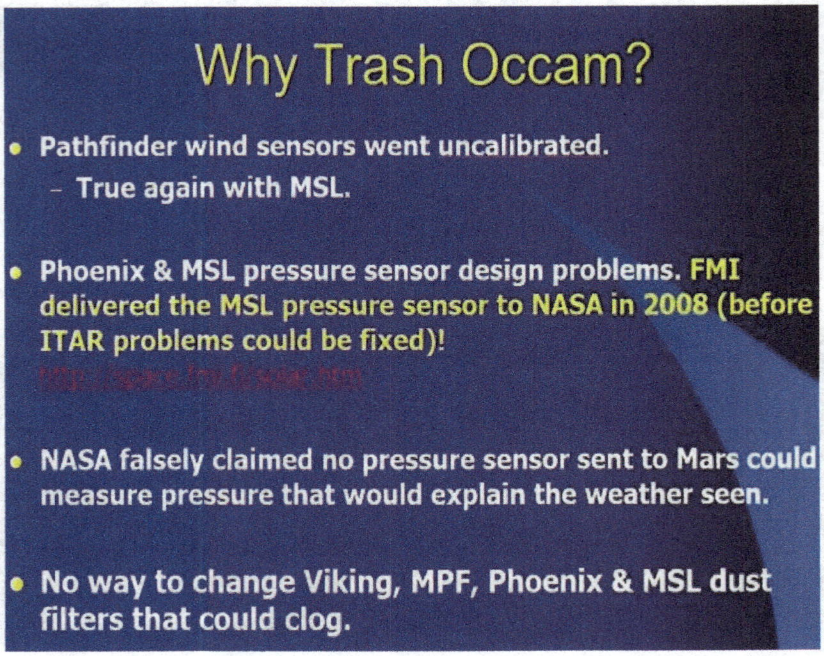

Figure 7-28 above: More reasons to trash Occam.

CRASH OF THE EXOMARS 2016 SCHIAPARELLI LANDER

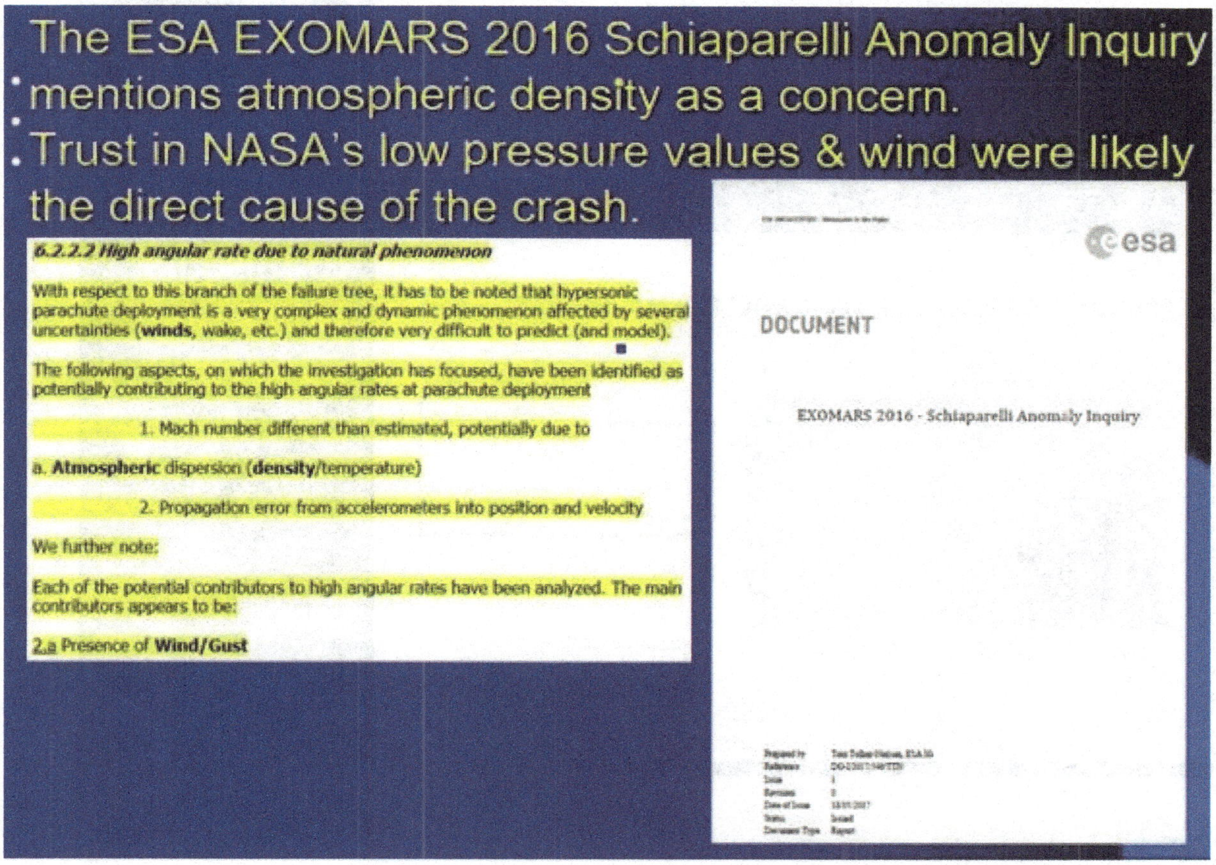

Figure 7-29 above: Critical portion of the ExoMars 2016 – Schiaparelli Anomaly Inquiry that backs our contention that NASA is underestimating Martian air pressures and wind speed. Their incorrect guidance caused this crash.

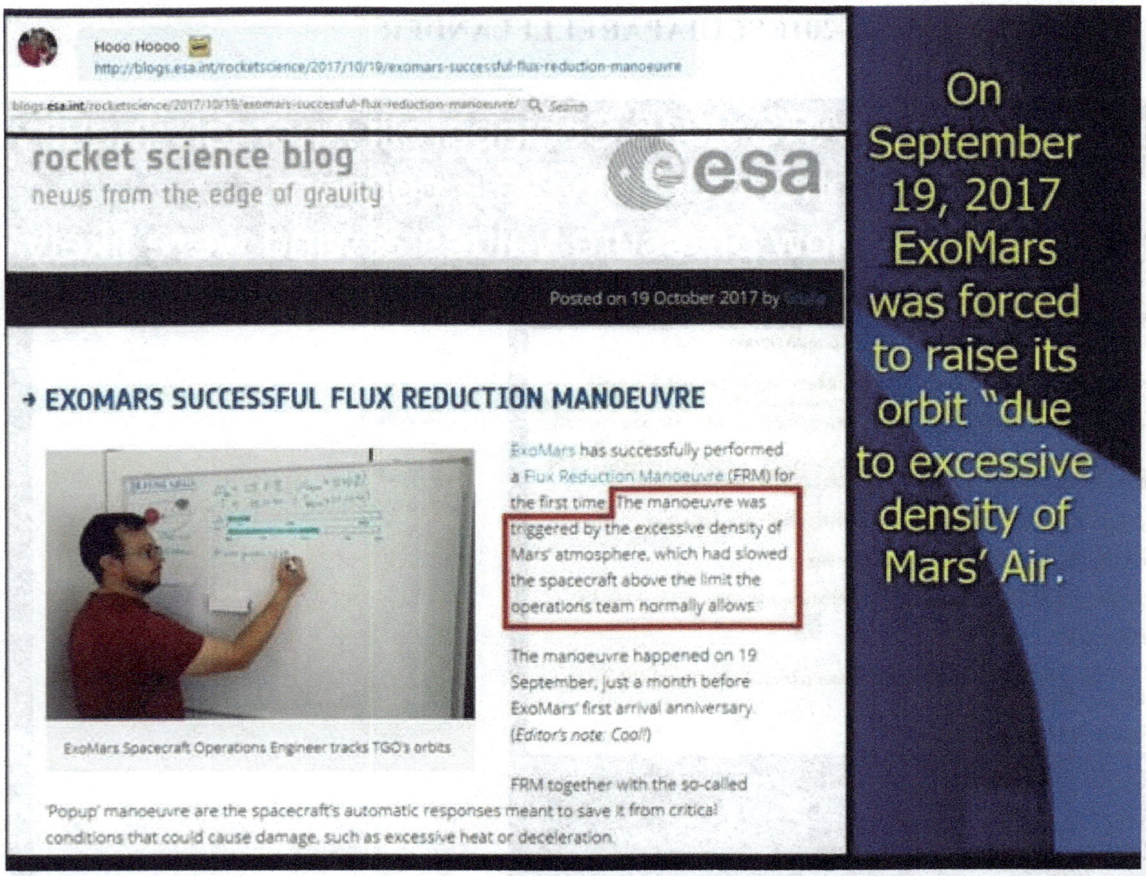

Figure 7-30 above: On October 19, 2017, ESA reported that ExoMars had to raise its orbit. The move was mandated by "excessive density of Mars' atmosphere." We received notice of this from our partner Marco de Marco.

On May 18, 2017, ESA published its **<u>ExoMars 2016 – Schiaparelli Anomaly Inquiry</u>**. While our research was not directly cited, we maintain a log of significant IP addresses and readers who access this report and our Mars-related websites. One of our most frequent readers is traced to Thales Alenia Space Italia S.p.A. in Milan, Italy. They built the Schiaparelli lander. In reading through the Inquiry, the following sections were of particular note:

Inquiry paragraph 6.2.2.2 High angular rate due to natural phenomenon. *With respect to this branch of the failure tree, it has to be noted that hypersonic parachute deployment is a very complex and dynamic phenomenon affected by several uncertainties (winds, wake, etc.) and therefore very difficult to predict (and model).*

The following aspects, on which the investigation has focused, have been identified as potentially contributing to the high angular rates at parachute deployment:

Mach number different from estimated, potentially due to atmospheric dispersion density/temperature Propagation error from accelerometers into position and velocity We further note:

Each of the potential contributors to high angular rates has been analyzed. The main contributors appear to be:

2.a. Presence of Wind/Gust

Of course, with respect to atmospheric density, we argue for air pressure at the areoid that is about 85 times higher than NASA asserts. As for wind/gusts, if NASA was right about a low atmospheric density and pressure, winds aloft would probably be insufficient to cause the loss of the lander. ESA is likely right about correcting the problem with the IMU (Inertial Momentum Unit). Perhaps that will be enough to overcome the density problem, but we challenged the wisdom of their statement that ExoMars 2020 was planned to proceed with models of atmosphere and winds as per 2016. In reality, looking back now from 2025, we see that there was no launch of an ExoMars 2020. The next planned launch of an ESA Mars mission has been put back until 2028.

It is still important to understand that a full-blown rejection of NASA and JPL without an in-situ ESA lander measuring pressures is problematic. ESA still depends upon NASA/JPL experience for advice on a number of space-related matters. If the IMU is fixed, it should not, as apparently happened in 2016, go into something akin to a nervous breakdown when the parachute is deployed and runs into much greater atmospheric density than expected. The specific final sequence of events in this "nervous breakdown" is spelled out as follows in ESA's Inquiry:

f. *Parachute deployment time (time from mortar firing to peak load factor) was circa 1 sec (in line with the predictions).*

   The parachute was deployed, and the parachute inflation triggered some oscillations of Schiaparelli at a frequency of approximately 2.5 Hz.

   About 0.2 sec after the peak load of the parachute inflation, the IMU measured a pitch angular rate (angular rate around Z-EDM axis) larger than expected. The IMU raised a saturation flag.

   During the period the IMU saturation flag was set, the GNC Software integrated an angular rate assumed to be equal to the saturation threshold rate. The integration of this constant angular rate, during which the EDM was in reality oscillating, led to an error in the GNC estimated attitude of the EDM of about 165 degrees. This would correspond to an EDM nearly turned upside down with the front shield side pointing to quasi-zenith.

   After the parachute inflation, the oscillatory motion of Schiaparelli under its parachute was mostly damped, and Schiaparelli was descending at a nominal descent rate, with very small oscillations (< 3 degrees) around the pitch and yaw axes. After parachute inflation, the angular acceleration around the spin axis changed again.

g. *The Front Shield was jettisoned as planned 40 seconds after parachute deployment (timer-based command) at 14:46:03.*

h. *The RDA (Radar Doppler Altimeter) was switched on at 14:46:19 (15s after Front Shield separation acknowledgment) and provided coherent slant ranges, without any indication of anomalies;*

   Once the RDA is on, RIL (Radar in the Loop) mode, "consistency checks" between IMU and RDA measurements are performed. The parameters checked are: delta velocity and delta altitude. The altitude is obtained using the GNC estimated attitude to project the RDA slant ranges on the vertical.

Because of the error in the estimated attitude that occurred at parachute inflation, the GNC Software projected the RDA range measurements with an erroneous off-vertical angle and deduced a negative altitude (cosines of angles > 90 degrees are negative). There was no check on board of the plausibility of this altitude calculation.

i. *Consequently, the "consistency check" failed for more than 5 seconds. After which, the RDA was forced anyway into the loop based on the logic that landing was impossible without the RDA. The correctness of the other contributor to the altitude estimation, i.e., the attitude estimate, was not put in question. The RDA was put in the loop (event signaled by RIL time-out flag at 14:46:46).*

   The GNC (Guidance, Navigation and Control) mode entered was TERMINAL DESCENT, where the altitude is scrutinized to release the Back-Shell and parachute if the altitude is below an onboard calculated limit.

   Because of the incorrect attitude estimation leading to an estimated negative altitude, the GNC Software validated the conditions for separating the back-shell and parachute.

j. *Back-shell separation at 14:46:49.*

k. *Switch-on of the Reaction Control System (RCS).*

   First RCS thruster operation was at 14:46:51 (no backshell avoidance maneuver).

l. *Switch-off of the RCS 3 seconds later at 14:46:54.*

   The criterion for the RCS switch-off was based on the estimation of the EDM (Entry Demonstrator Module) energy (as a combination of the altitude and vertical velocity) being lower than a pre-set threshold. Since the estimation of the altitude was negative and very big, the negative potential energy was much higher than the positive kinetic energy (square of the velocity), and this criterion was immediately satisfied, the RCS was commanded off as soon as allowed by the thruster modulation logic. This occurred just 3 seconds after the RCS switch-on command when the capsule was at an altitude of about 3.7 km, leading to a free fall of Schiaparelli and the impact on Mars surface about 34 seconds later.

m. *The touchdown occurred at 14:47:28, corresponding to the crash of the surface platform*

   on the surface of Mars at an estimated velocity of ≈150 m/s (about 335.54 miles per

   hour). The expected landing time was 14:48:05 (some 37s later).

At some point, hopefully in 2028, ESA will succeed. But here we must caution NASA. There is an old cliché*:*

"Fool me once, shame on you. Fool me twice, shame on me."

NASA has fooled ESA once. But ESA is on to the problem and should not be fooled again. If NASA announces that they have come to understand that air pressure is much higher than they previously announced, there may be room for plausible deniability with respect to issues related to liability.

Whether NASA blames mistakes on unit conversion, or failure to allow for dust filter replacement on transducers, or inability to provide critical design information with respect to heat sources near the Vaisala pressure sensor due to ITAR, NASA can still preserve some of its respect if they publicly abandon their loyalty to a 6.1 mbar pressure at the areoid in time to ensure a successful ExoMars 2028 mission. The Chinese Tianwen-1 lander in 2021 safely arrived on the Martian surface. They had five years to read the **ExoMars 2016 – Schiaparelli Anomaly Inquiry** and make adjustments. I certainly caught them reading my website for my earlier discussion of the ESA crash. But they did not reveal ongoing fraud on a massive basis. They want Mars for themselves and, perhaps, they are not anxious to get mired in controversy about life (including Americans) being on Mars now. There appears to be a race forming to get people there. At his 2025 Inauguration, President Trump announced his intention to (publicly) plant our flag on Mars.

ESA gets smarter – Raises ExoMars orbit due to excessive density of Mars's atmosphere. See Figure 7-30. This is similar to what was seen with the Mars Global Surveyor and also with the Mars Reconnaissance Orbiter. With the loss of the Schiaparelli lander and now this public ESA statement about excessive density of Martian air, the question remains as to when NASA will reach and publish the same common-sense conclusion. But we would be surprised if it occurs as a result of observations made by Perseverance, because again it apparently carries a pressure sensor that can only measure up to 11.5 mbar. In the Chinese Tianwen-1, the sensor can measure up to 20 mbar. If NASA is close to being right about air pressure on Mars, the Chinese sensor will be better for measuring pressure increases during major global or regional dust storms. But if my son and I are right about average pressure being about 511 mbar, neither the U.S. nor the Chinese sensor will be good for anything other than continuing disinformation.

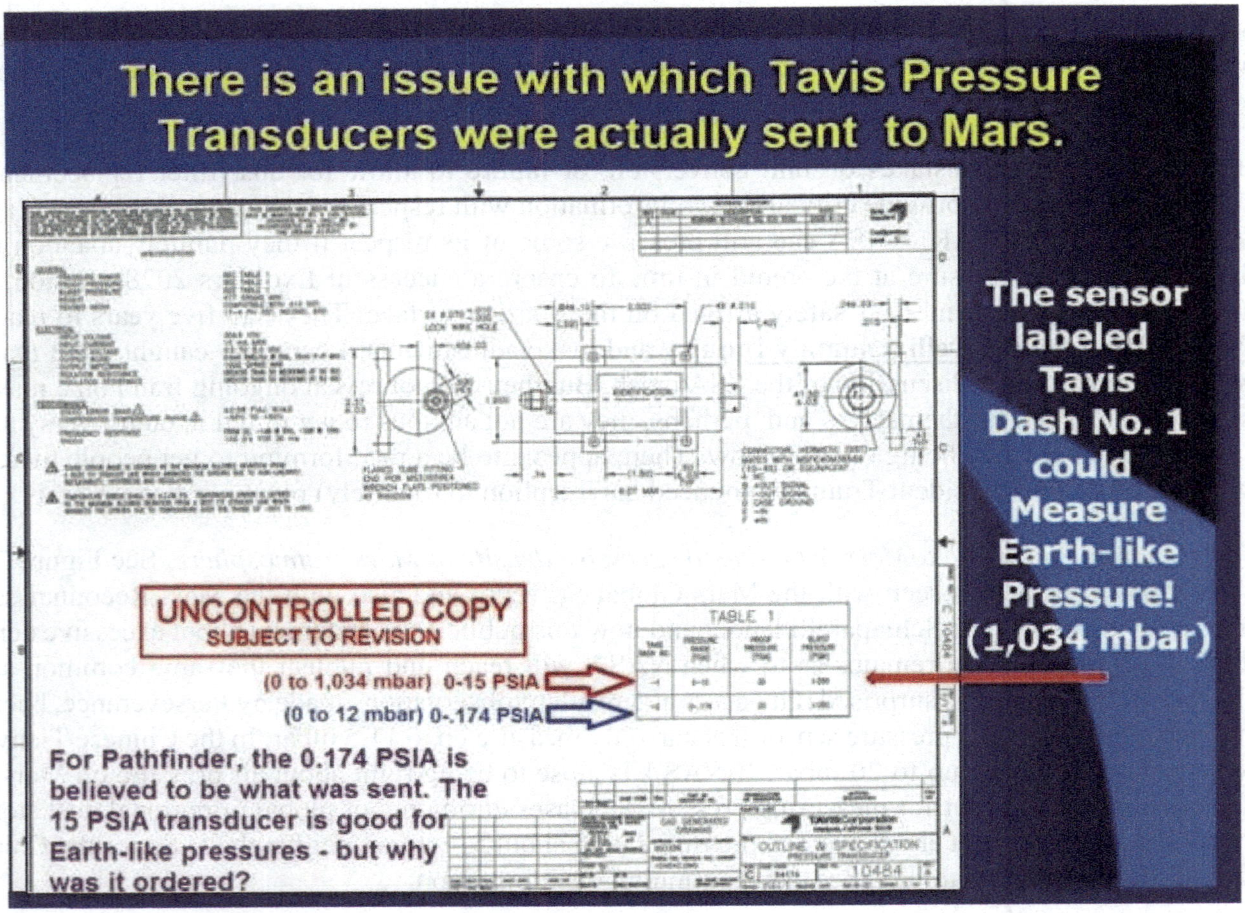

Figure 7-31 above: Reproduced from Tavis CAD Diagram 10484. For Mars Pathfinder, Tavis Dash No. 2 had a 0.174 PSIA limit (12 mbar), but Pathfinder Tavis Dash No. 1 had a 15 PSIA limit (1,034 mbar – best suited for Earth-like pressures). Source: Personal communication, Tavis Corporation 10/29/2009.

Which Transducers were used? A Tavis spokesman (Marty Kudella) thought Pathfinder used Part 10484 (Tavis Dash No. 2). The red words, uncontrolled copy subject to revision on both CADs, shown allow for a possible need in the future to alter the transducer pressure range.

Figure 7-31 lists it as having a 0.174 psia limit (12 mbar), the same limit later imposed by Vaisala on Phoenix. It first appeared that NASA also ordered a Tavis transducer that could measure from 0 to 15 psia (1,034 mbar): Part 10484, Tavis Dash No. 1 – see Figure 7-31 again. For 9 years, we believed that it supposedly remained on Earth and wrote that if, for classified reasons, a decision was made to send it in place of the 12-mbar transducer, none of the pressure data published by NASA for Pathfinder would be reliable. If there was a separate transducer that could measure Earth-like pressure, its final disposition still isn't clear at this time, but based on information from the InSight Mission that landed on Mars on November 26, 2018, it seems possible that the same Tavis transducer could operate in either the low- or high-pressure range. Our Italian partner, Marco de Marco, called Tavis Corp. for clarification. They knew who he was, but wouldn't answer his questions. We should look at the evidence for one physical transducer rather than two, but first, let's discuss Tavis transducers in general.

Apparently, similar-looking and sized Tavis transducers could measure up to 0.1 psia (6.9 mbar), 0.174 psia limit (12 mbar), 0.2 psia (13.79 mbar), 0.26 psia (17.9 mbar), 0.36 psia (24.82 mbar), or 15 psia (1,034 mbar). Given their outward similarity and the enigma of Martian weather, the possible installation of the wrong Tavis sensor cannot be overlooked. Perhaps somebody wanted a 15-mbar sensor and mistakenly chose the 15-psi transducer. People made mistakes back then, and they still do today, as is apparent when we examine REMS (Rover Environmental Monitoring Station) data for MSL. For five days straight from September 1 to September 5, 2012, they published Martian pressures of over 740 hPa (Earth-like), when they supposedly meant 740 Pa. A pressure of 740 hPa = 740 mbar, while 740 Pa = 7.4 mbar. They published numerous other similar questionable items or obvious errors (see Section 2.7 and Figures 17A and 17B in our *Mars Correct Basic Report*).

We only learned one night before InSight reached Mars that there are also meteorological sensors aboard, including the same Tavis sensors above. With respect to the dual pressure range, Tavis states, "Tavis specializes in custom configurations and capabilities for your specific application. Discuss your application requirements with our engineers for your exploration science needs" (http://pressure-transducers.taviscorp.com/item/all-categories/pressure-transducers-for- interplanetary-exploration/10484). Could a radio signal cause the sensor to toggle from the low range to the high? Again, Tavis wouldn't tell us, but it's quite possible.

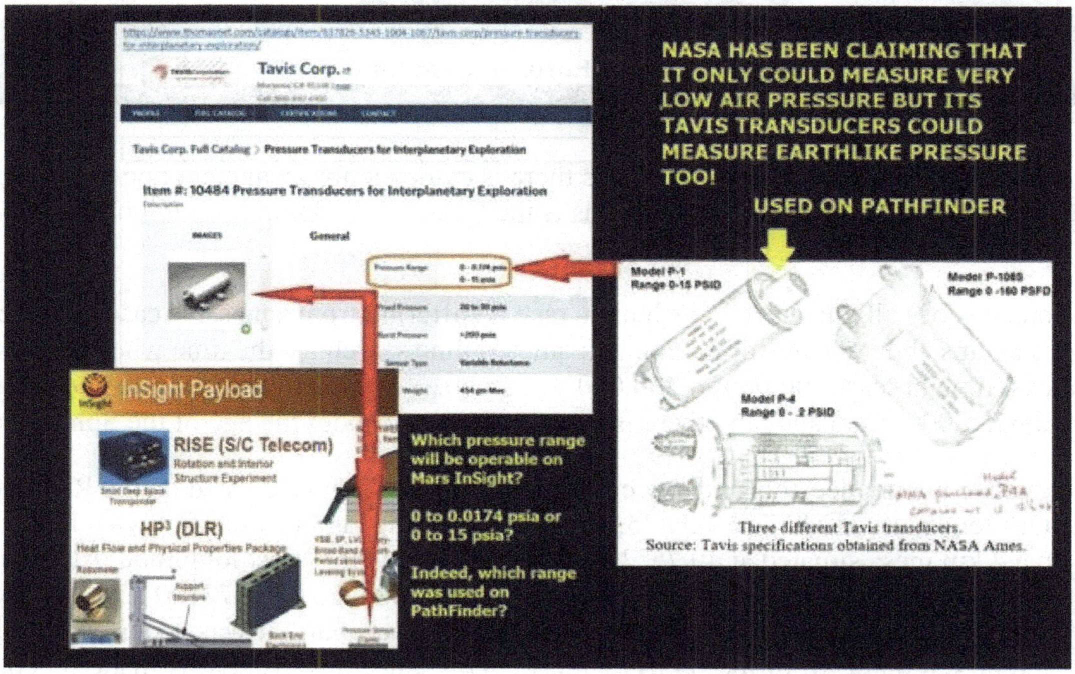

Figure 7-32 above – Tavis pressure sensors had the ability to measure Earth-like pressures on Mars, but it took years for me to learn this sensitive information.

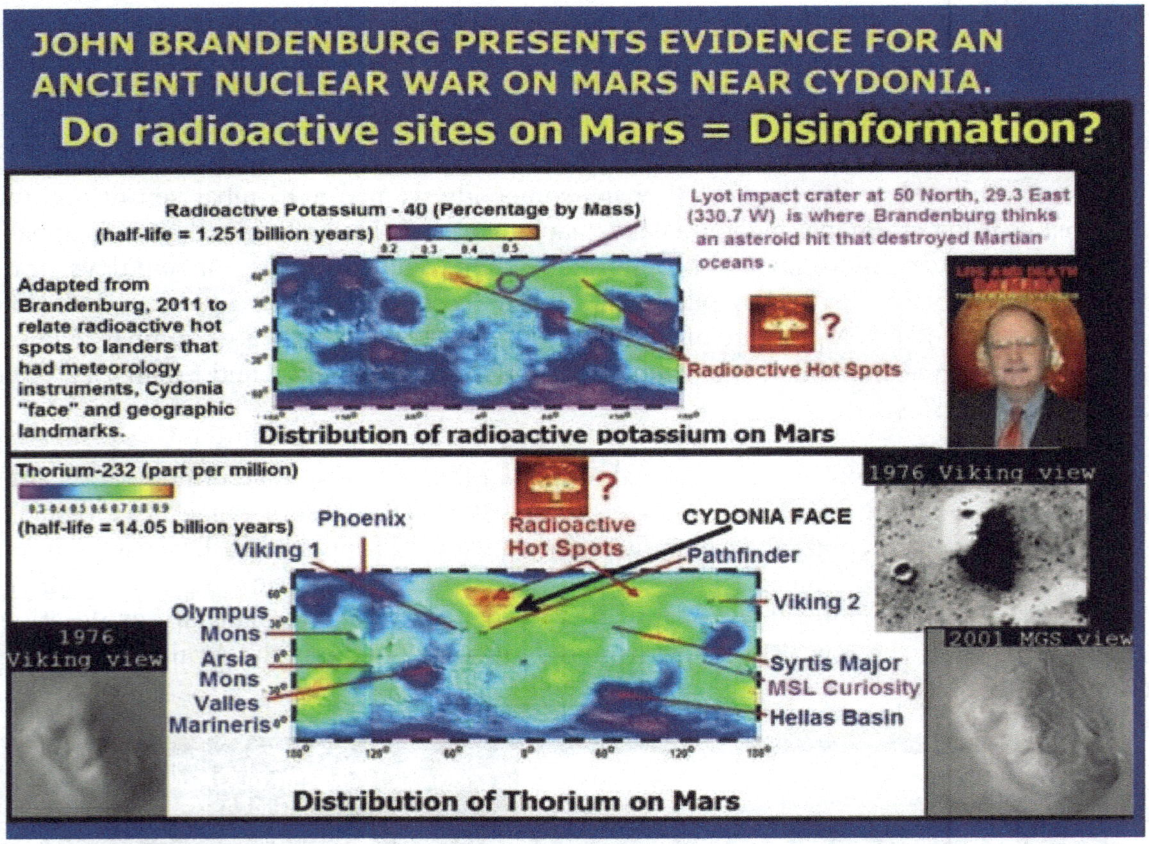

Figure 7-33 above: Dr. John Brandenburg claims there is evidence for an ancient nuclear war on Mars. The still radioactive bomb sites are shown here, as is the "face" at Cydonia, which looks like ruins from that war.

So, what difference does all this make? The human race is suffering from a massive case of amnesia. Setting aside religious claims about the age of mankind, a quick online search for the time when homo sapiens first arose indicates a time at least 100,000 to 200,000 years ago. Yet we know very little about our ancestors from before about 5,000 years ago, and there is even controversy about the age of the Sphinx in Egypt.

Is there a link between our ancestors and any civilization that might have existed on the Red Planet? When the Viking 1 Orbiter observed Mars on July 25, 1976, it sent us the famous image of what looked like a face at Cydonia on a 2 km mesa situated at 40.75° North latitude and 9.46° West longitude. At the time, Viking chief scientist Gerry Soffen dismissed the "Face on Mars" in image 035A72100 as a "trick of light and shadow." However, a second image, 070A13, also shows the "face," and it was acquired 35 Viking orbits later at a different sun angle from the 035A72 image. This latter discovery was made independently by Vincent DiPietro and Gregory Molenaar, two computer engineers at NASA's Goddard Space Flight Center. DiPietro and Molenaar discovered the two misfiled images, Viking frames 035A72 and 070A13, while searching through NASA archives. Images taken decades later did not back the original ones, and the belief in an alien origin of the face-like mesa has been used to mock and dismiss the work of scientists who backed what was first seen.

Analysis of the face is beyond the scope of this work, but it is tempting to match what looks like disinformation about Martian air pressure to Cydonia. Why would the Government hide the truth about Mars? One only has to look at the panic that ensued when Orson Welles gave his *War of the Worlds* radio broadcast in 1938. Combine that with reports of radioactive isotopes found on Mars that match what is seen at nuclear detonations on Earth (Brandenburg, 2011), and it's easy to understand why the Governments might want to keep the public from knowing the truth about Mars. While Brandenburg's paper, *Evidence for a Large, Natural, Paleo-Nuclear Reactor on Mars*, only discusses a natural cause for an ancient nuclear explosion there, his 2011 book, *Life and Death on Mars, The New Mars Synthesis*, on page 179, forwards the idea of an airburst caused by a great intelligence with great malice.

Do I believe Brandenburg? I originally wrote that, "It's not important." But now, as old age begins to set in on me (I went through cancer last year), I think it's important to take a stand. The portrait of Mars given to our world by NASA is so wrong that I don't think all the mistakes are coincidental accidents. NASA is hiding something, and I think it's likely to be Brandenburg's war. If true, he thinks it was likely to occur hundreds of millions of years ago. If the warriors of that past era were not exterminated in a nuclear blaze, then they had time to evolve or advance themselves into a time- traveling race advanced enough to write something like the Torah. But there are "ifs" attached to this thought, and it ignores the personal aspects and experiences in relationships between God and people. I have had such experiences. (see https://biblio.ie/9780935834550). The story is an entirely true account of the landmark interfaith custody battle for my older son, Rabbi Robert Altair Roffman. It resulted in Archbishop McCarthy's annulment of Robert's Catholic Baptism. The news coverage is at *1984 Miami Herald Press Coverage* (https://arkcode.com/images/herald_001.png).

Do I believe in the literal word of the Torah? There are other ways to interpret it (metaphorical, allegorical, and mathematically encoded). However, there are intriguing hints in the Bible about contact between someone not human (Nephilim/giants – possibly Reptilians, and the daughters of man in Genesis 6:4). These ideas can be looked at as distractions, but the reality is this: Martian weather does not match NASA-backed low Martian air pressure. Both science and religion stress the importance of our past and our fate in the future.

It's generally accepted in science that Mars was once a warmer, wetter world with oceans. If the air pressure remains high, there is still the question of what happened to its seas. Indeed, to have a

sea suddenly appear there today, all that would be necessary would be to melt the frozen freshwater sea in Utopia Planitia.

If Mars had life, did it all die, or merely move underground? Did bacteria ever travel from Mars to Earth after an asteroid impact there? We'll examine this evidence for that in Chapter 8. Did it play a role in our evolution? Did something much higher up the evolutionary scale make it from there to here, and was it linked to Cydonia or the Bible's Nephilim? Did they influence or color any of mankind's religious experiences or doctrines? These are questions of fundamental importance. The human race often uses religious doctrine to justify wars. However, the Government might worry that data supporting a past not generally in line with the Bible would lead to anarchy and chaos.

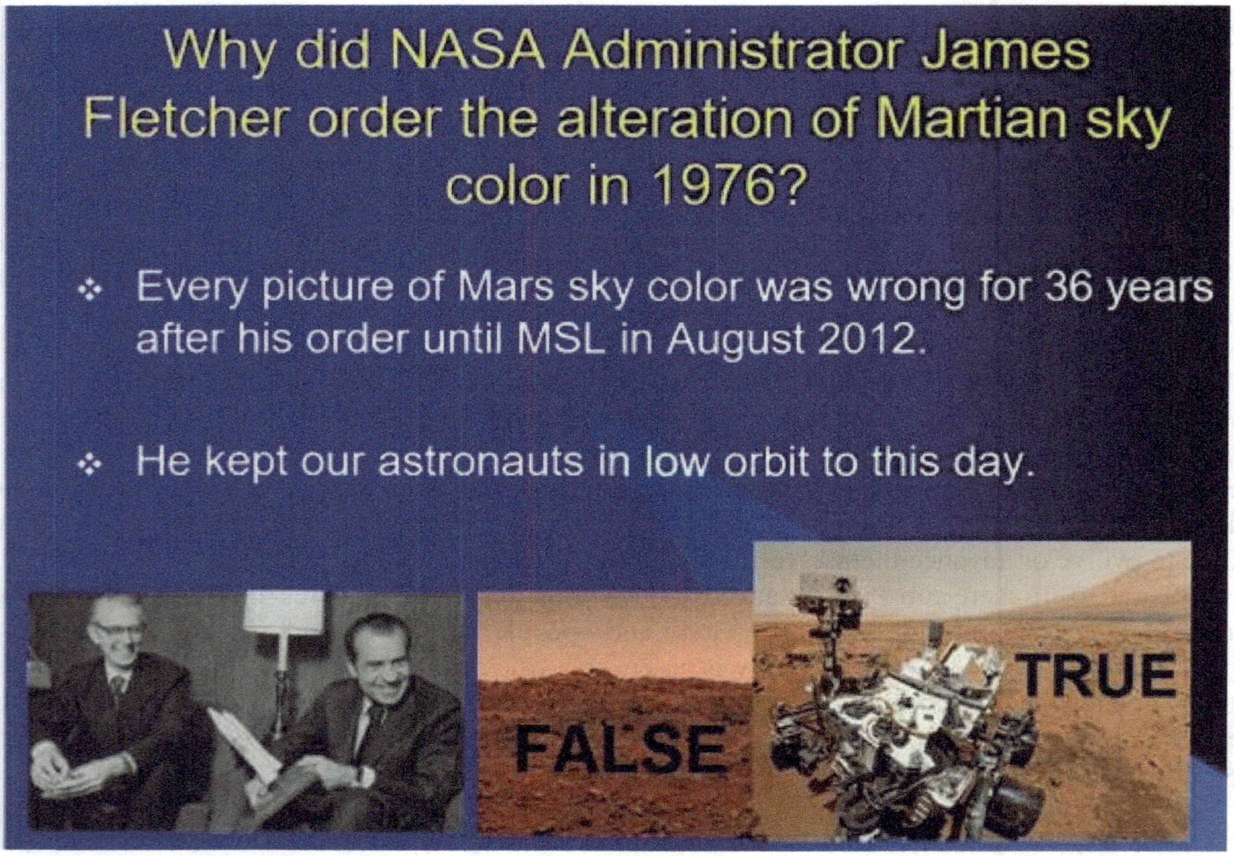

Figure 7-34 above: NASA Director James Fletcher ordered the sky color seen by Viking 1 to be changed from blue to red, thus hiding green patches (possible lichens) seen on some rocks to look brown. He took the lead in covering up life on Mars. President Nixon shut down the Apollo moon program early. Supposedly, according to Jackie Gleason, Nixon and Gleason once were shown four alien corpses being kept at Homestead Air Force Base. See https://www.kpl.gov/catalog/item/?i=ent://ERC_215_8682/0/215_8682:HOOPLA:16226931.

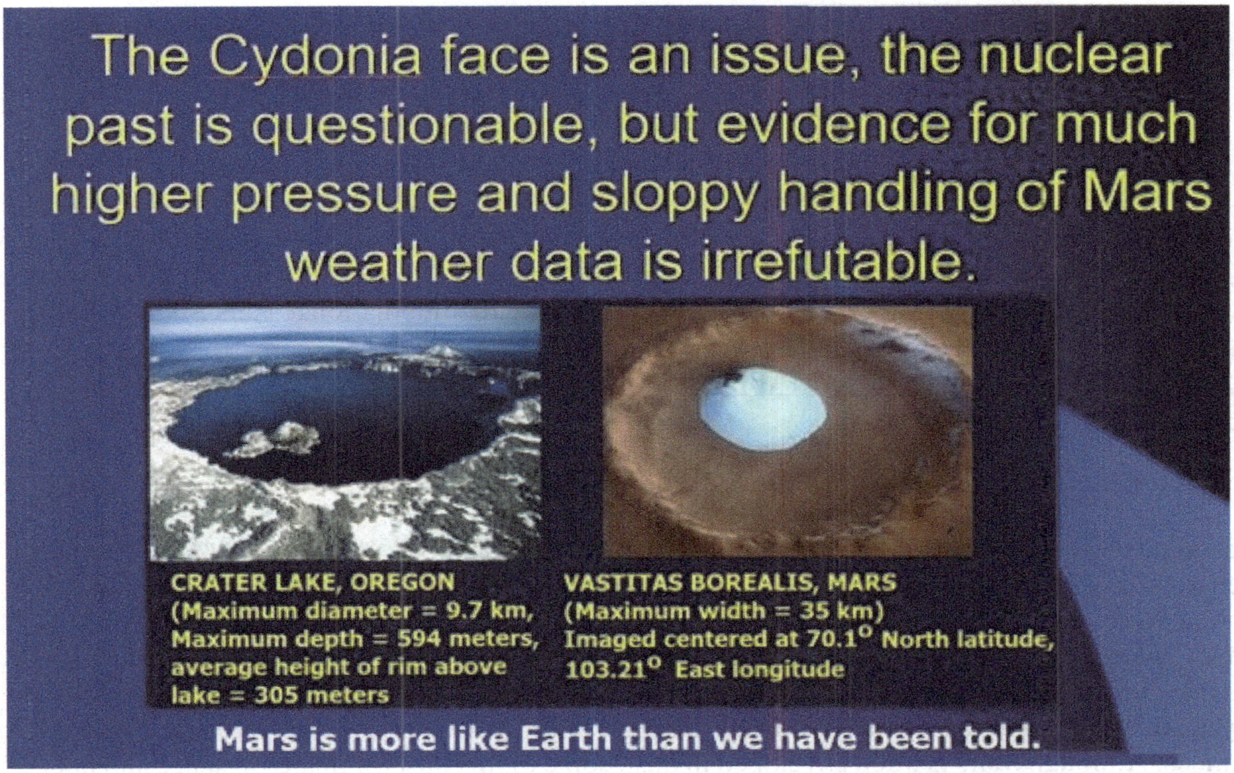

Figure 7-35 above: Place a nuclear heater in Vastitas Borealis, and you might just create a good fishing hole and place for a base. If we ever do it, we might want to cover the lake with a transparent tarp so it could hold some extra atmosphere. A hundred years in the future, and this place could be functioning as a fantastic resort.

Chapter 8
Martian Microbes & Primitive Life

This chapter documents the existence of primitive life needed to support more advanced life on Mars and beyond. The primary source of the chapter is *Meteorological Implications: Evidence of Life on Mars?* Dr. David Roffman (my son) and I published it in the *Journal of Astrobiology and Space Science Reviews*, 1, 329-337, 2019. Our second source is our joint report entitled *MARS CORRECT: CRITIQUE OF ALL NASA MARS WEATHER DATA*. This paper was published by the International Mars Society and can be accessed at https://www.academia.edu/40831370/MARS_CORRECT_CRITIQUE_OF_ALL_NASA_MARS_WEATHER_DATA.

Abstract for Meteorological Implications: Evidence of Life on Mars

In a detailed review of over 130 research reports by over 500 scientists, Joseph et al. (2019) provide strong evidence for multiple forms of prokaryotic and eukaryotic life on Mars, which may be contaminants from Earth due to solar winds and meteor strikes. What is notable are those specimens, photographed by NASA on Mars, which resemble terrestrial fungi, lichens, and sphere- shaped basidiomycota, which the authors admit may be hematite. However, fifteen spheres became larger and emerged from beneath the surface over 3 days; an observation which may indicate biological growth or a strong wind that uncovered these specimens. Although Mars is often enveloped in dust, some of this dust originates in space and not on Mars, which supports the hypotheses, first proposed by Arrhenius (1908), that microorganisms may be attached to that extraterrestrial dust, some of which may have been propelled from Earth to Mars. Meteorological data, pro and con, is discussed as it relates to the possibility of Martian life.

KEYWORDS: Contamination, Life, Mars, Atmospheric Pressure, Weather, Methane, Wind.

Introduction

Mars has long fascinated humanity as a possible home for life. In July 1964, that hope was dealt a blow by Mariner 4. Observations from 9,846 km out showed a heavily cratered, cold, and apparently lifeless world. Air pressure was estimated at 4.1 to 7 mbar with daytime temperatures of -100° C (NASA n.d.). By contrast, Mariner 9 found evidence of wind and water erosion, fog, and weather fronts (Greene 2015). When Vikings 1 and 2 landed, we learned of frequent dust devils. Later (from orbit), we found they were also seen up to 17 km above the areoid (similar to sea level on Earth) on Arsia Mons on Mars.

Over Arsia Mons, there were also spiral clouds with 10 km-wide eye walls where pressure (in the caldera) should be only ~1.3 mbar. In fact, massive storms were "observed by Mariner 9 (1971- 1972) and Mars Global Surveyor (2001). Those storms totally obscured the planet's surface (NASA 2018). In 2018, a massive dust storm covered 14 million square miles (35 million square kilometers) of Mars — a quarter of the Martian surface (NASA 2018). The rover Opportunity was also blanketed with dust, such that, in consequence, the solar panels stopped functioning.

Therefore, we know Mars is a dusty planet. And yet, rather than uncovering, these frequent dust storms blanket the surface, and the rovers with dust. Moreover, these dust storms also absorb heat and increase surface temperature (at night), and "limit extreme temperatures" (NASA 2018), making conditions more conducive to life.

In addition, as Mars orbits through streams of dust in the wake of comets, meteorites, and sources unknown, extraterrestrial dust is deposited on Mars (Andersson et al. 2015; Treiman & Treiman, 2000). Might that dust contain microorganisms blown into space from Earth? The answer is unknown.

What is known is, Mars has an atmosphere, weather, clouds, what appears to be water frozen at the poles, and not just dust but snowstorms, as witnessed by the Phoenix lander (Dunbar 2015). On Earth, it is suspected that microbes contribute to changing weather patterns and even the formation of clouds. We should not be surprised if Mars also harbors life.

Evidence for Life

When Levin and Straat (1976) reported that biological activity was detected via Labelled Release experiments on Vikings 1 and 2, they were strongly challenged because of the failure to find organics. Yet, with time, MSL detected methane, chloromethane, dichloromethane, trichloromethane, dichloroethane, 1,2-dichloropropane, 1,2-dichlorobutane, and chlorobenzene. Christopher McKay (2006) of NASA Ames announced that the Viking instrumentation was not capable of detecting organics.

Using a Fourier Transform Spectrometer, Krasnopolsky et al. (2004) observed Martian methane. Webster et al. (2018) reported methane background levels vary with the local seasons. Joseph and colleagues (2019) point out that variations in terrestrial methane are directly correlated with the growth and death cycles of plants, and that 80% of terrestrial methane is biological in origin.

Vlada Stamenković et al. (2018) found that Mars can support liquid environments with dissolved O2 values ranging from at least ~2.5×10^{-6} mol m^{-3} to 2 mol m^{-3} across the planet. Near-surface environments had enough O_2 available for aerobic microbes to breathe independent of photosynthesis.

All this set the table for finding life on Mars, whereas Joseph et al. (2019), while cautioning that "morphology is not proof," have speculated and provided photographic evidence of what may be fungi, lichens, cyanobacteria, basidiomycota ("puffballs"), plus stromatolites and outcroppings like terrestrial microbialites.

Issues Related to Wind and Pressure

The authors compared M.E.R. Rover Opportunity photos on Sol 1145 with Sol 1148 and assert, "Fifteen specimens resembling and identified as "puffballs" were photographed emerging from the ground over a three-day period. It is possible these latter specimens are hematite and what appears to be "growth" is due to a strong wind which uncovered these specimens – an explanation which cannot account for before and after photos of what appears to be masses of fungi growing atop and within the Mars rovers." Later, they ask, "What is the likelihood that a strong wind would have uncovered the specimens in Figure 8-1, and not covered them (and Opportunity's solar panels)?"

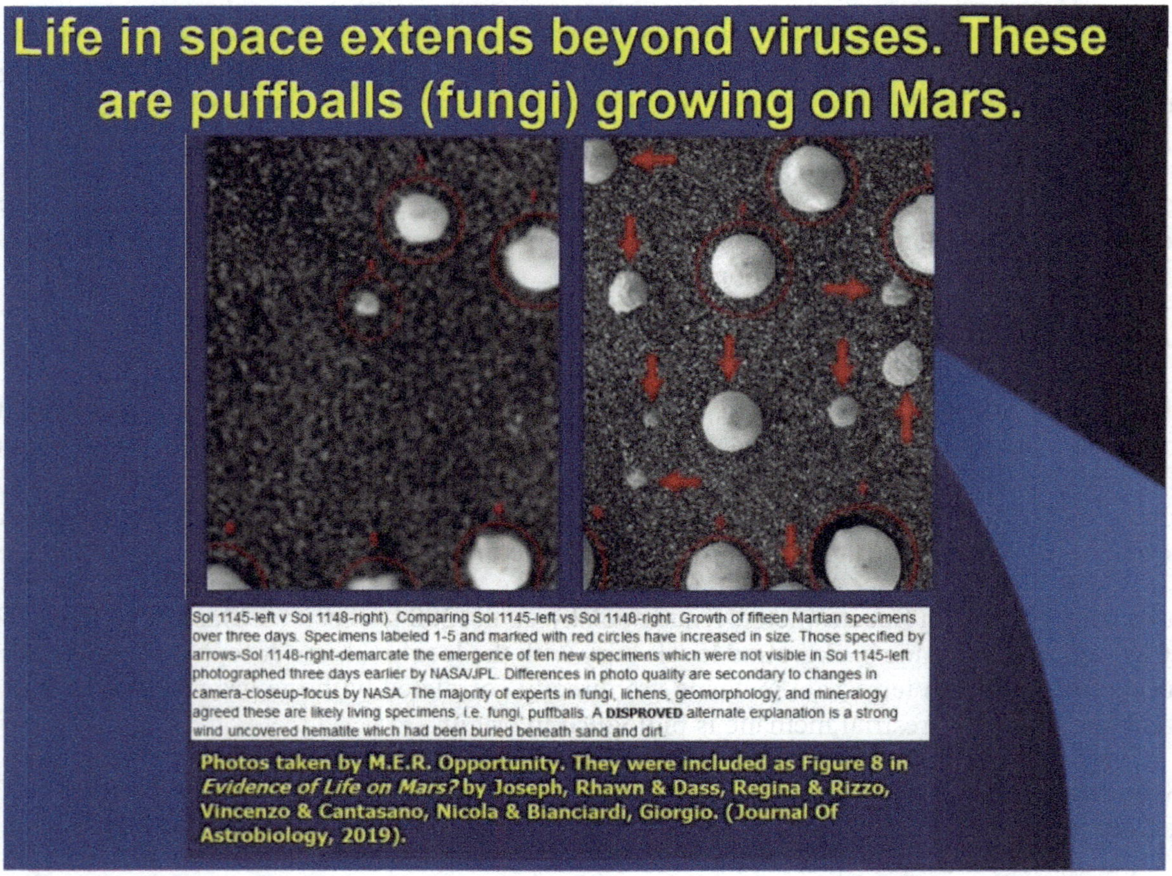

Figure 8-1 These appear to be puffballs growing on Mars.

In November 2011, NASA released a statement entitled *NASA Orbiter Catches Mars Sand Dunes in Motion*. It stated, "Mars either has more gusts of wind than we knew about before, or the winds are capable of transporting more sand, said Nathan Bridges, planetary scientist at the Johns Hopkins University's Applied Physics Laboratory … We used to think of the sand on Mars as relatively immobile, so these new observations are changing our whole perspective." They assert that wind-tunnel experiments have shown that a patch of sand would require winds of about 128.7 km/hr to move on Mars compared with only 16 km/hr on Earth. They then state that measurements from the Viking landers and climate models showed such winds should be rare on Mars (NASA 2011).

Using Tillman's Viking data (n.d.), I produced the graphs shown in Figure 8-2, which show Viking 1 winds for its sols 1 to 350 (except sols 116 to 133 because data was missing) and for Viking 2 sols 200 to 350. Every sol was divided into 25 time bins. During Viking 1, the maximum wind was 93.24 km/hr (see Table 8-1). For Viking 2 (see Figure 8-3) winds reached 83.52 km/hr, but over 8,331 measurements, the wind never reached the 128.7 km/hr that Bridges said were required to move sand.

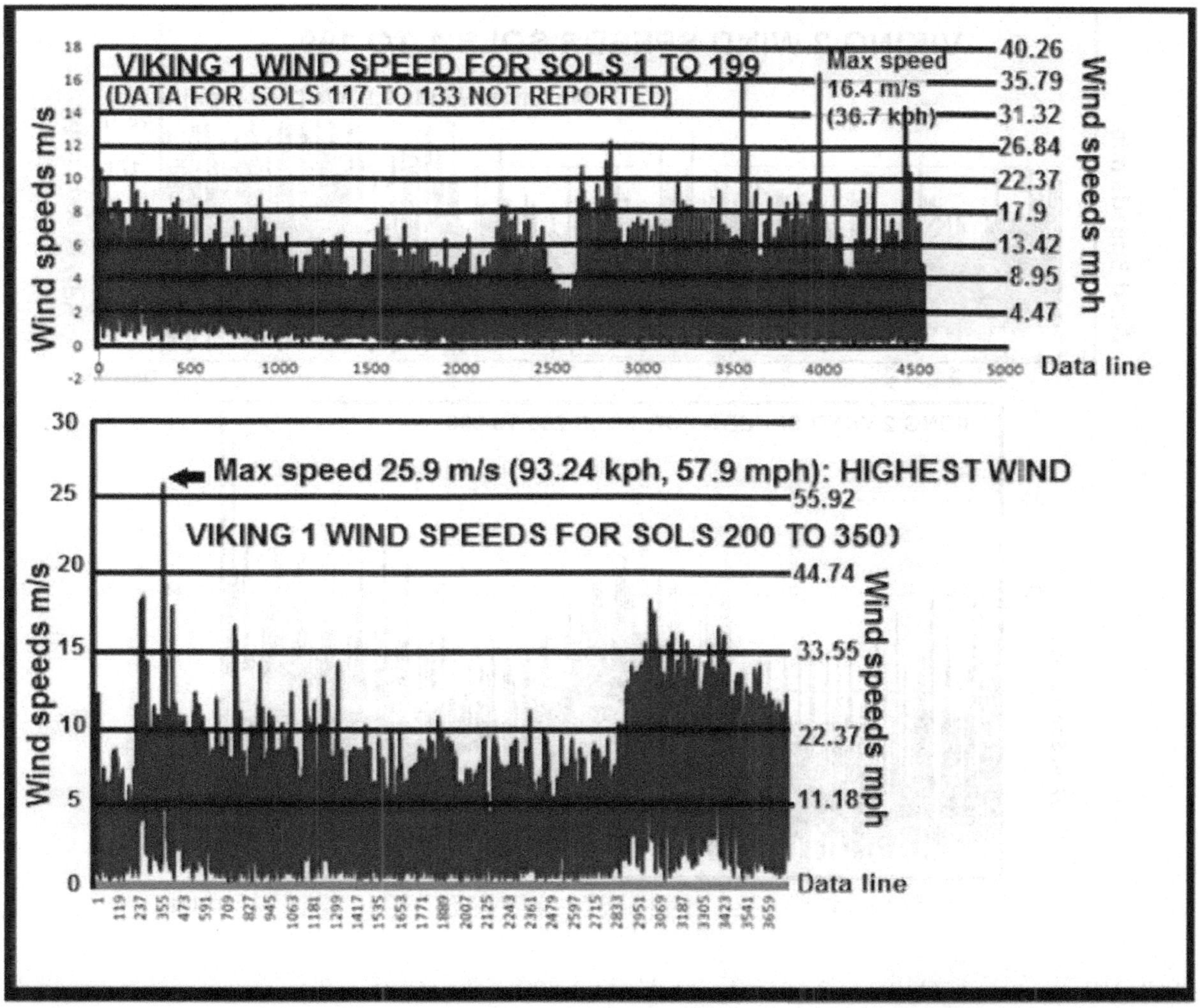

"Viking Lander 1 Sols 1 thru 199." *Viking Lander 1 Sols 1 Thru 199.* http://www-k12.atmos.washington.edu/k12/mars/data/vl1/part1.html

Figure 8-2 top above: Wind speeds for Viking Lander Sols 1 to 199. Bottom: Wind speeds for Viking 1 for Sols 200 to 350.

Figure 8-2– bottom above. Wind speeds for VL-1 for its sols 1 to 116 and 134 to 350 and for VL-2 for its sols 1 to 399

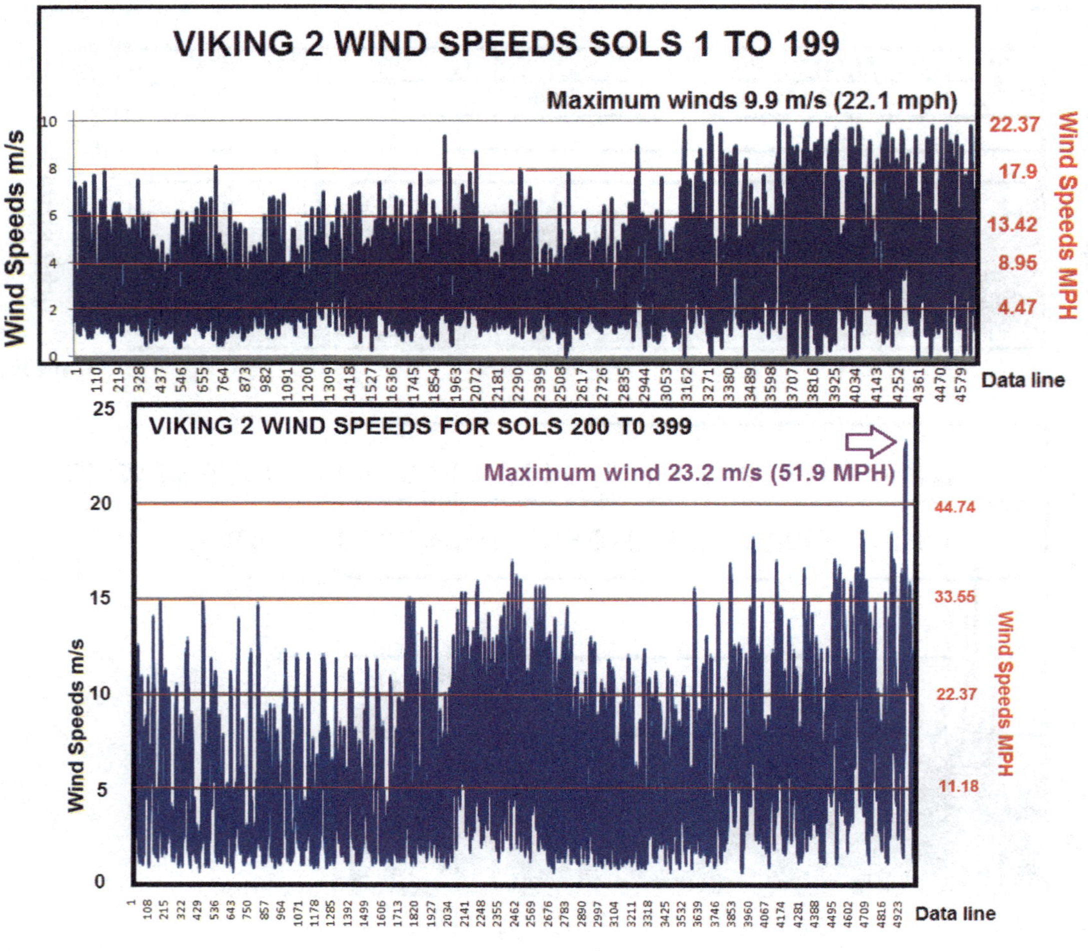

Figure 8-3. Winds seen by Viking 2. Winds reached 83.52 km/hr (51.9 mph).

There was little wind data for Mars after the Vikings. Pathfinder was only calibrated for 1,015 mbar and ~15 mbar of terrestrial air pressure (Schofield et al., 1997).

For Phoenix, Taylor et al. (2008) state, "We had hoped to include an anemometer in the MET package." Faced with a lack of resources and needing wind data, they used the SSI camera and a Telltale. But for wind over 10 m/s, the Telltale went horizontal and lost its wind speed/deflection correlation ability."

Figure 8-4: Phoenix telltale waving in Martian wind. The out-of-phase image may indicate a dust devil occurrence. Images taken before & after the event have west winds estimated at 7 m/s. During the event, south winds are estimated at 11 m/s. Adapted from Taylor et al., 2008.

Curiosity had part of an anemometer on Boom 1, but it broke on landing. Yet for 9 months, NASA published wind data that never changed – always 7.2 km/hr from the East. However, upon being alerted by us of this error (and at our request), the published data was changed to N/A.

To determine wind speed accurately, knowledge of air density is essential. The relationship between these two (the windsock equation) from NASA (1999) is given as equation (1): u = sqrt{[2 R(1) M g tan(theta)]/[R(2) A(d) rho]}

In Equation 1, R1 = distance between pivot and center of mass, M = non-counter-balanced mass, g = acceleration of gravity, R2 = distance between pivot and center of aerodynamic pressure, A(d)= effective aerodynamic cross-section, and rho = atmospheric density (a function of pressure, temperature, and molecular weight).

So, if the density is incorrect, the wind speed will be wrong for at least a windsock or a Telltale. It is not the purpose of this commentary to fully explore multiple indications of higher pressure than accepted and problems with instrumentation, but we would be remiss to ignore excessive deceleration during aerobraking by Mars Global Surveyor and Mars Reconnaissance Orbiter. Further, an October 19, 2017, ESA report documents how ExoMars 2016 had to raise its orbit because of "excessive density of Mars' atmosphere." The *ExoMars 2016 – Schiaparelli Anomaly Inquiry* (2017) also pointed to atmospheric density and the presence of wind/gusts as possible causes of the crash of the Schiaparelli lander. See Figure 8-5.

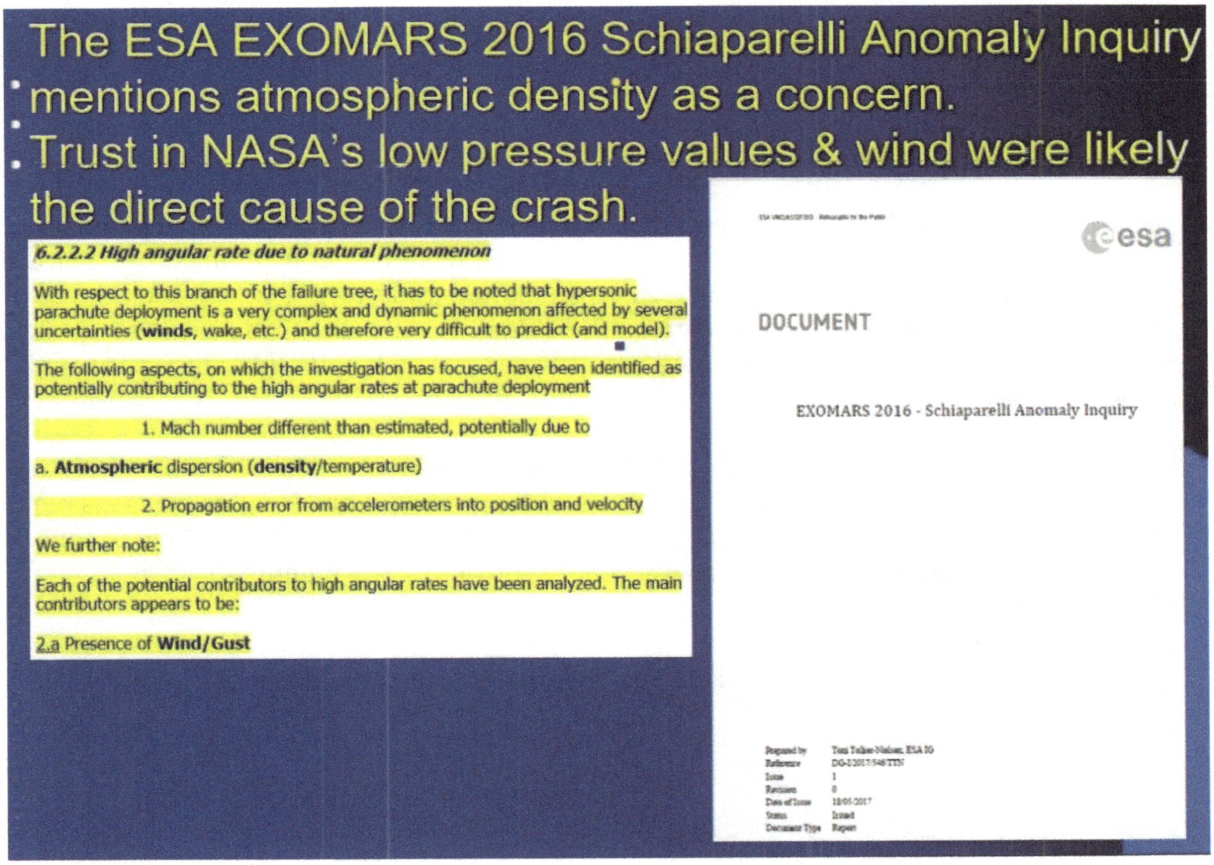

Figure 8-5 above: Trusting NASA was a fatal decision made by the European Space Agency.

Assuming that we are really dealing with life, or something that merely mimics it, a denser atmosphere would facilitate its transport. Morrison (2016) states that, "Microbes have been found in the skies since Darwin collected windswept dust aboard the HMS Beagle 1,000 miles west of Africa in the 1830s. Recent research suggests that microbes are hidden players in the atmosphere, making clouds, causing rain, spreading diseases between continents, and maybe even changing climates."

It is well established that methane contributes to climate change and global warming. As pointed out by Joseph et al. (2019), 90% of terrestrial methane is biological in origin.

Microbes and Martian Methane

High levels of methane, which vary in concentration depending on the season, have been detected at ground level and in the atmosphere of Mars (Webster et al. 2018), i.e., "a strong, repeatable seasonal variation…" To date, no abiogenic source for Martian methane has been discovered. Joseph et al. (2019), however, point out that terrestrial methane levels also vary according to the seasons and that "These seasonal variations have as their source biological activity in wetlands and on farms and in rice paddies, just prior to harvest."

Coupled with their review of the literature, the methane evidence is also suggestive of life, which in turn may be impacting Martian temperatures and weather. Unless there is a political decision made to bring about alien life disclosure, only additional research can hope to answer this question and perhaps determine how any possible microbes on Mars interact with Martian dust storms. However, based on the behavior that I observed at NASA Ames when it came to how to finish the paper that powered this chapter, it's not going to be easy. **N**ever-**A**-Straight-**A**nswer NASA is not interested in getting the truth out.

Possible Sources of Pressure Measurement Errors

My 12-year study (Roffman, 2019) found many possible sources of errors for pressure measurement, with some errors at two orders of magnitude. Some may have been mere typographic mistakes. For example, from September 1 to 5, 2012, pressure at Curiosity was reported as up to 747 **h**Pa, but on the next sol, it was only 1% of that – just 747 Pa. Further, data on the Viking Project Site (Tillman, n.d.) up through at least March 19, 2019, had a definition under pressure that states one mb = 100 hPa. In fact, one mb = only one hPa (which is 100 Pa).

Stratus clouds are seen on Earth up to 13,000 meters, where pressures are about 163 mbar. They were also found on Mars ~12,318 meters above the surface. If a similar minimum pressure for clouds is required on Mars, based on an accepted scale height of 10.8 km, the pressure at the areoid would be about 511 mbar rather than 6.1 mbar. However, this doesn't factor in atmospheric dust load. See Figure 8-5.

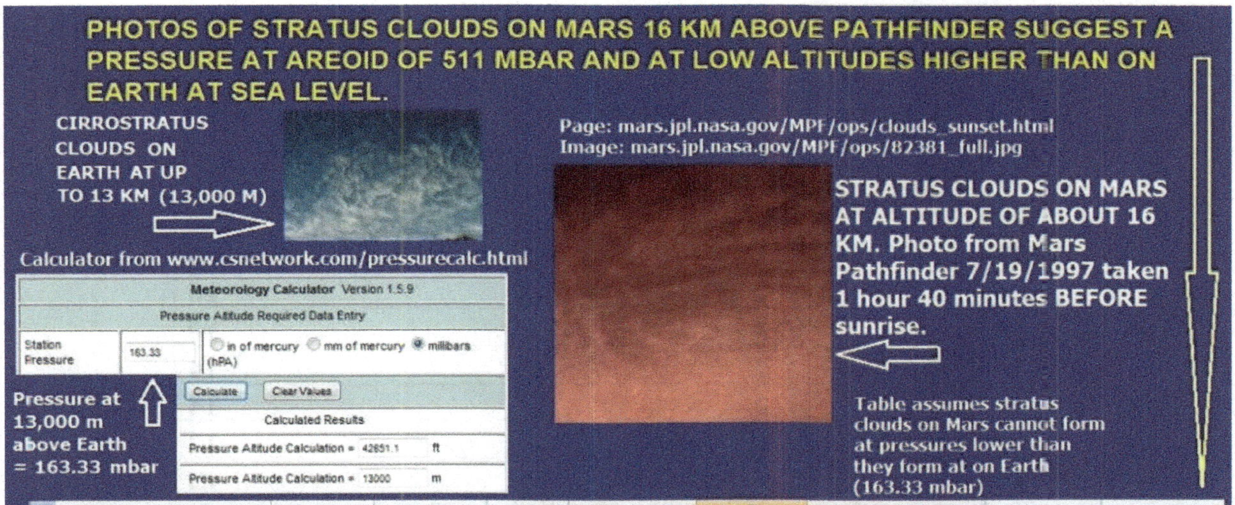

	A	B	C	D	E	F	G	H	I
1	CALCULATIONS BASED ON	ENTERING ARGUMENTS SCALE HEIGHT 10.8 KM AND 163.33 mbar at 12,318 meters							
2	MARS PATHFINDER	KILOMETERS	10.8km Scale	RATIO B/C	=-EXP(D VALUE)	1/E value	-F VALUE = PRESSURE	PERCENT OF	PRESSURE IN
3	VIEW OF STRATUS CLOUDS		Height (MARS)				MULTIPLE OF	PRESSURE AT	MILLIBARS/hPa
4							6.1 MBAR MEAN	MEAN AREOID	
5	CLOUDS 16 KM ABOVE MPF	12.318	10.8	1.1405556	-3.128505941	-0.319641394	0.319641394	31.95413939	163.3303595
6	MARS PATHFINDER (MPF)	-3.682	10.8	-0.3409259	-0.711111581	-1.40624907	1.40624907	140.624907	718.56515
7	MEAN AREOID	0	10.8	0	-1	-1	1	100	510.98
8	MARS SCIENCE LABORATORY	-4.4	10.8	-0.4074074	-0.665373057	-1.50291628	1.50291628	150.291628	767.9601608
9	VALLES MARINERIS	-5.31	10.8	-0.4916667	-0.611606201	-1.635039015	1.635039015	163.5039015	835.4722361
10	HELLAS BASIN	-7.825	10.8	-0.724537	-0.484548845	-2.063775427	2.063775427	206.3775427	1054.547968

Figure 8-6 Stratus clouds on Earth and Mars.

I assumed that the MSL temperatures were reliable, but on July 3, 2013, JPL revised down many high air temperatures by more than 10K, wiping out above-freezing temperatures on sols 26, 40- 47, 49-54, 102, 112, 116, 118, 123-124, and 179. See Table 8 – 1 (a repeat of Table 7-1).

TABLE 8-1 – MSL Air Temperatures Altered by JPL in July, 2013			
A	B	C	D
SOL	ORIGINAL MAX AIR TEMP °C	NEW MAX AIR TEMP °C	CHANGE °C (EQUALS CHANGE K)
23	0	-16	16
26	2	-14	16
27	-1	-15	14
31	-3	-23	20
38	-3	-13	10
40	2	-12	14
41	2	-12	14
42	5	-7	12
43	3	-12	15
44	4	-10	14
45	3	-9	12
46	4	-12	16
47	6	-9	15
49	4	-10	14
50	0	-10	10
51	3	-7	10
52	7	-7	14
53	5	-5	10
54	5	-9	14
102	8	-3	11
112	5	-8	13
116	5	-6	11
118	4.53	-6	10.53
123	2.1	-10	12.1
124	5.4	-5	10.4
179	5	-7	12

Table 8-1 above: Air temperatures revised down by JPL.

When we examine claims of life for areas where there are perchlorates that affect the freezing point of water, we need firm temperature data to analyze the situation. We have developed colored spreadsheets for at least 2,319 sols of MSL (Roffman, 2019). This database, (voluntarily) meant to assist JPL, shows not only current weather claims, but also documents anomalies and alterations over time. When comparing alleged Martian life forms with known terrestrial life forms, these charts are essential for understanding environments.

Figure 8-7 above: Light Green Spheres on Mars.

Joseph et al. (2019) present evidence of sphere-shaped specimens, which they argue could be “puffballs,” but which they admit may be hematite; however, we see signs of growth and reproduction (see Figure 8-1 above). These spheres have been photographed in numerous locations and on various dates, such as on M.E.R. Opportunity sols 1185 and 1189. (Joseph et al. 2019). On Sol 1248, MSL returned to the Sol 1185 site (NASA 2016).

Later, more spheres (still considered probable hematite by NASA) were photographed on Sols 1555, 1571, and 1797. If these spheres are hematite, weather and temperature are irrelevant. However, life-forms, temperature, and atmospheric pressure may be factors.

Altitude varied from about 4,420 meters below areoid to 4,215 meters below areoid. See Figure 8-

7. Temperatures are shown in Figure 8-7.

Although winter, daytime highs were (in the second Martian year of MSL operations) strangely above freezing. They were clearly warmer than what was measured at the same LS in the first and third years of operation, when there were no temperatures above freezing and there were no spheres. "Ls" refers to the Martian solar longitude, a measure of the Sun's position as seen from Mars, used to track Martian seasons.

It's analogous to Earth's concept of solar longitude, but specifically for Mars.

Do the thin layer of green material and "green spherules," which resemble algae in the soil seen by MER Spirit in Gusev Crater, contain chlorophyll or something functionally like it? There are possible water pathways that may intermittently fill with water there (Krupa, 2017). Figure 8-9 has greenish balls photographed from Sols 1185-1797.

Figure 8-8 below: For areas of high interest, we must return again with a new probe that can conduct a definitive analysis. Do the growth-like materials on Opportunity and Curiosity contain nucleic acids? The trek map shown in Figure 8-8 indicates that the find was significant enough to cause JPL to return to the initial finding after 63 sols.

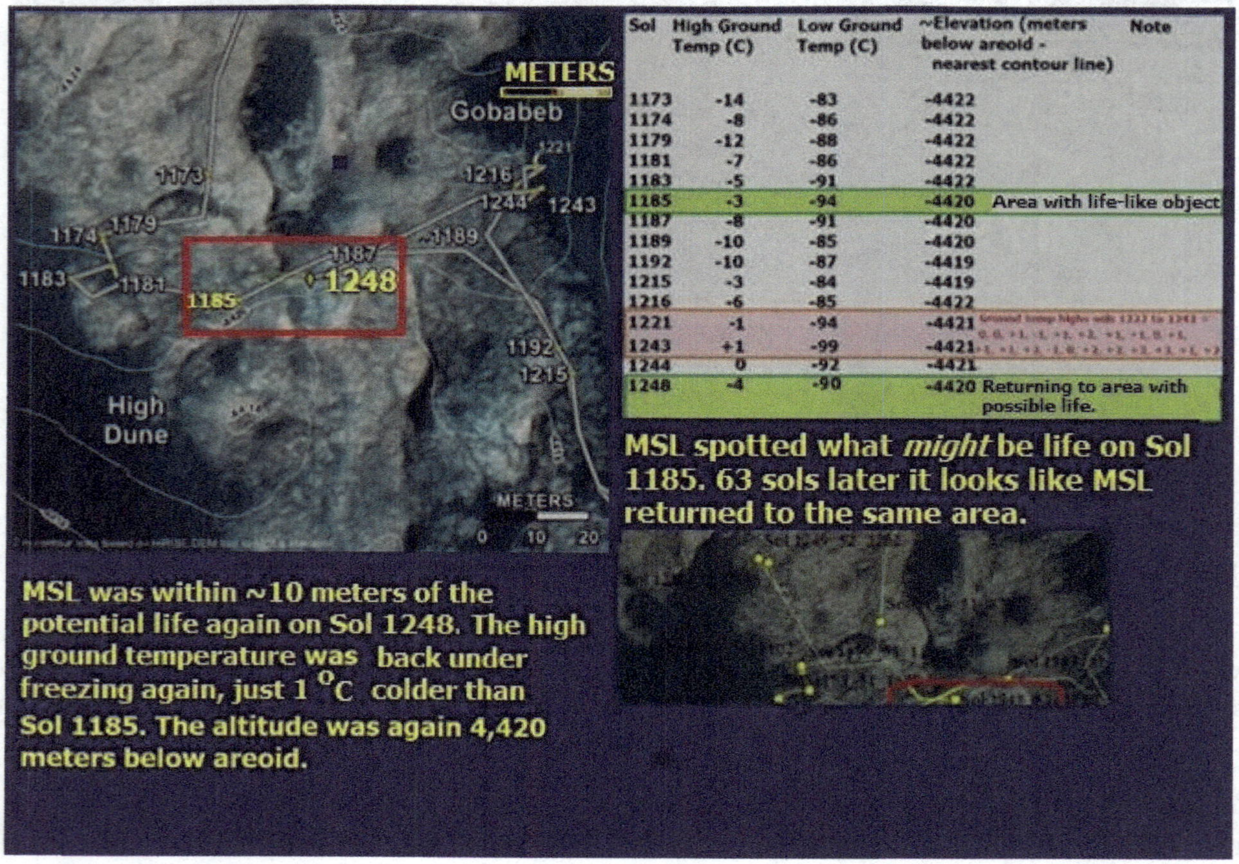

Figure 8-8 above: Possible biological specimens photographed on Sols 1185 through 1248 with elevations and ground temperatures.

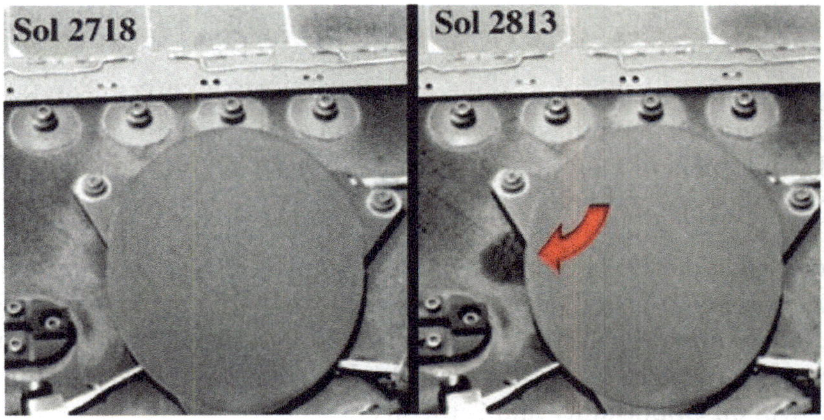

Figure 8-9 Mars Sol 2718 vs Sol 2813 – The exact cause or identity of this specimen (right) is unknown but may represent the growth of what appears to be a mass of bacteria and fungi cn the Mars Rover, Opportunity, after 95 (Martian) days. Photo, NASA/JPL.

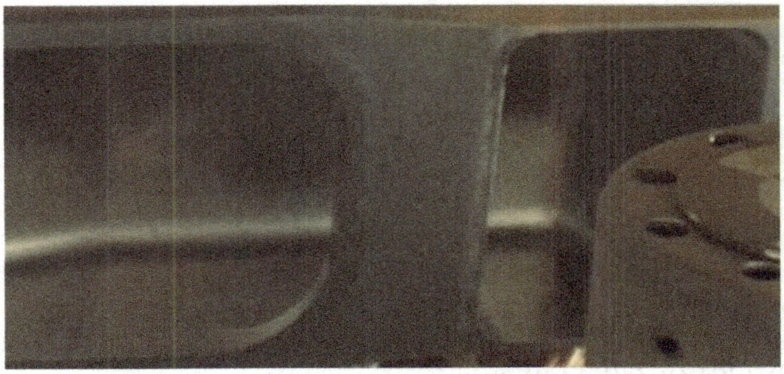

Sol 51. Mastcam photo of the interior, flooring and shelter of an exposed compartment in the rover Curiosity's chem cam deck after 51 Martian days. Photo by NASA/JPL.

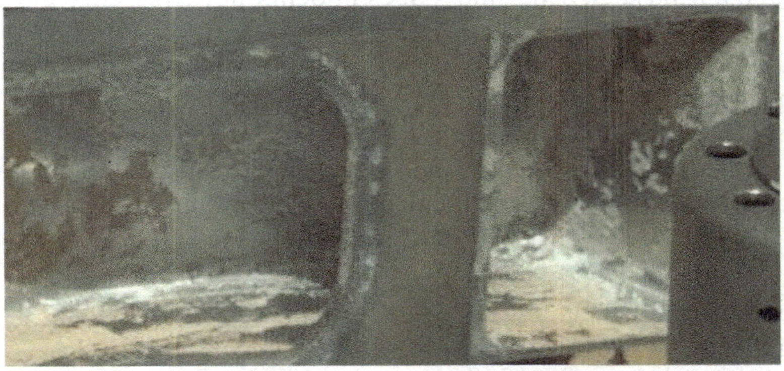

Sol 1089. Possible Fungal contamination or bio-corrosion of the interior, walls, flooring, and shelter of an exposed compartment in the rover Curiosity's chem cam deck after 1089 Martian days? Another possibility: chemical contamination or sand and salt which adhered only to the interior surface? Photo Mastcam, by NASA/JPL.

Figure 8-10 above: Possible fungi growing on the Mars Curiosity rover.

Conclusion

Evidence provided by five Mars landers favors the probability of life. Unfortunately, as stressed by Joseph et al. (2019), NASA has rammed down the throats of Joseph et al., plus my son and me, that "the evidence is circumstantial, and similarities in morphology are not proof." They would not publish our work without this statement inserted against our will!

Future robotic missions should take equipment like the Miniature Variable Pressure Scanning Electron Microscope for in-situ imaging & chemical analysis (Gaskin 2012) to the most promising sites. Only then can we begin to make a determination as to whether the specimens photographed on the Martian ground are oddly shaped sedimentary structures or evidence of life. Likewise, we must determine the exact nature of what appears to be growth on Opportunity and Curiosity.

Acknowledgements

Thanks are due to Marco de Marco and Mateo Fagone, who hosted our 3+ hour TV interview about Mars, complete with simultaneous translation into Italian. See https://www.youtube.com/watch?v=PqCxAErabuU (Marte Correto – Mars Correct).

We also appreciate the green sphere images sent to us by David Kiepke.

References

Andersson, L. et al. (2015) Dust observations at orbital altitudes surrounding Mars, *Science* 06 Nov 2015: Vol. 350, Issue 6261,

Arrhenius, S. (1908). *Worlds in the Making*. Harper & Brothers, New York.

Bridges, N. T., F. Ayoub, J-P. Avouac, S. Leprince, A. Lucas, and S. Mattson. "Earth-like Sand Fluxes on Mars." *Nature* 485.7398 (2012): 339-42. Web.

Dunbar, D. (2008) http://www.nasa.gov/mission_pages/phoenix/news/phoenix-20080929.html

Exomars (2016) – Schiaparelli Anomaly Inquiry (2017). http://exploration.esa.int/mars/59176- exomars-2016-schiaparelli-anomaly-inquiry/#

Gaskin, J.A.; Jerman, G.; Gregory, D.; Sampson, A.R., (2012) Miniature Variable Pressure Scanning Electron Microscope for in-situ imaging & chemical analysis, *Aerospace Conference, 2012 IEEE*, vol., no., pp.1,10, 3-10 March 2012 doi: 10.1109/AERO.2012.6187064

Gigapan (n.d.) Sol 1248. http://gigapan.com/gigapans/202461/snapshots/556038 taken from mars.jpl.nasa.gov/msl/multimedia/raw/?s=1248&camera=MAST_
mars.jpl.nasa.gov/msl/multimedia/raw/?s=1249&camera=MAST_ Credit: NASA / JPL-Caltech / Malin Space Science Systems

Greene, Nick. "Mariner 9 Information." N.p., n.d. Web. 10 Feb. 2015.

http://space.about.com/od/marinermissions/p/mariner9info.htm

Joseph, R. G, Dass, R, Rizzo, V., Cantasano, ., Bianciardi, G. (2019). *Evidence of Life on Mars?*

Journal of Astrobiology and Space Science Reviews, Vol 1, 40-81.

Krasnopolsky, V., Maillard, J., Owen, T. (2004). Detection of methane in the Martian atmosphere: evidence for life? *ICARUS*, 172. doi: 10.1016/j.icarus.2004.07.004

Krupa, T. A. (2017). Flowing water with a photosynthetic life form in Gusav Crater on Mars,

Lunar and Planetary Society, XLVIII.

Levin, G. V.; Straat, P. A. (1976). "Viking Labeled Release Biology Experiment: Interim Results".

Science. 194 (4271): 1322-1329. Bibcode:1976Sci...194.1322L.

doi:10.1126/science.194.4271.1322. PMID 17797094.

The Mariner Missions." The Mariner Missions. N.p., n.d. Web. http://nssdc.gsfc.nasa.gov/planetary/mars/mariner.html

McKay, C. (2006) https://www.nasa.gov/centers/ames/research/2006/marsorganics.html

Morrison, Jim (2016). *Smithsonian.com*. https://www.smithsonianmag.com/science- nature/living-bacteria-are-riding-earths-air-currents-180957734/

NASA (n.d.) *Mariner Missions*. https://nssdc.gsfc.nasa.gov/planetary/mars/mariner.html

NASA (1999) https://pds.jpl.nasa.gov/ds-view/pds/viewContext.jsp?identifier=urn%3Anasa%3Apds%3Acontext%3Ainstrument%3Awindsock.mpfl&version=1.0

NASA (2011). *NASA Orbiter Catches Mars Sand Dunes in Motion*. NASA/JPL. https://www.jpl.nasa.gov/news/news.php?release=2011-358

NASA (2016) https://mars.nasa.gov/resources/7704/curiositys-traverse-map-through-sol-1250/

NASA (2018) https://mars.nasa.gov/news/8348/opportunity-hunkers-down-during-dust-storm/

Roffman, D (2019) http://davidaroffman.com/custom3_66.html.

Roffman, D. (2019) Web site contents. http://davidaroffman.com/photo2_18.html

Schofield, J., Barnes, J., Crisp, D., Haberle, R., Larsen, S., Magalhães, J., Murphy, J., Seiff, A. and Stamenković, V., Ward, L., Mischna, M. and Fischer, W. (2018). O_2 solubility in Martian near- surface environments and implications for aerobic life https://www.nature.com/articles/s41561- 018-0243-0?WT.feed_name=subjects_inner-planets. *Nature Geoscience*, [online] 11(12), pp.905-

909. Available at: https://www.nature.com/articles/s41561-018-0243- 0?WT.feed_name=subjects_inner-planets.

Taylor, P. A., Catling, D. C., Daly, M., Dickinson, C. S., Gunnlaugsson, H. P, Harri, A.M., and Lange C. F., (2008). Temperature, pressure, and wind instrumentation in the Phoenix meteorological package, *Journal of Geophysical Research*, 113, E00A10. http://faculty.washington.edu/dcatling/Taylor2008_Phoenix_MET.pdf

Tillman, J. Mars Meteorology Data; Viking Lander (n.d.). Retrieved from http://www-k12.atmos.washington.edu/k12/resources/mars_data-information/data.html

Treiman, A. H. & Treiman, J. S. (2000) Cometary dust streams at Mars: Preliminary predictions from meteor streams at Earth and from periodic comets, *Journal of Geophysical Research*, 105, 571-24,58

Webster, C.R., Mahaffy, P. et al., (2018) Background levels of methane in Mars' atmosphere show strong seasonal variations, *Science* 08 Jun 2018: Vol. 360, Issue 6393, pp. 1093-1096 DOI: 10.1126/science.aaq013

Wilson, G. (1997). The Mars Pathfinder Atmospheric Structure Investigation/Meteorology (ASI/MET) Experiment. *Science*, [online] 278(5344), pp.1752-1758. Available at:

https://pdfs.semanticscholar.org/b399/9b8083ed3a234688c89f0771cf12e01a9481.pdf

Chapter 9
Fighting Radical Islam and E.T.

This chapter will focus on the top leaders of radical Islam, Iran, Gaza, and Lebanon, whom Israel and/or the U.S. have found necessary to kill. On 10/22/2015, I published Figure 9-1 below. It was an outstanding matrix then, and it remains so as I update my studies of General Soleimani in 2025. On June 12, 2025, Israel went to war with Iran. That too is encoded as I will show at the end of this chapter, but I didn't find any dates for it until June 13, 2025 because I did not know what to use as an axis term. What worked turned out to be the name of the operation – LION RISING. While I didn't know the name of the axis term after the war started, the Israelis obviously did know it before they attacked (but, of course, they kept it a secret). My point here is that if Mossad chose to use the Code, they could have derived some useful intelligence from it but (unless they are using an IP address not yet deciphered by me) I do not yet detect heavy Israeli traffic at the website. As of June 15, 2025, in comparing known U.S. with Iranian and Israeli hits for 14 days on my roffmanmarsresearch.com site, I only see 5,823 U.S. hits, 14 Israeli hits, and 61 Iranian hits.

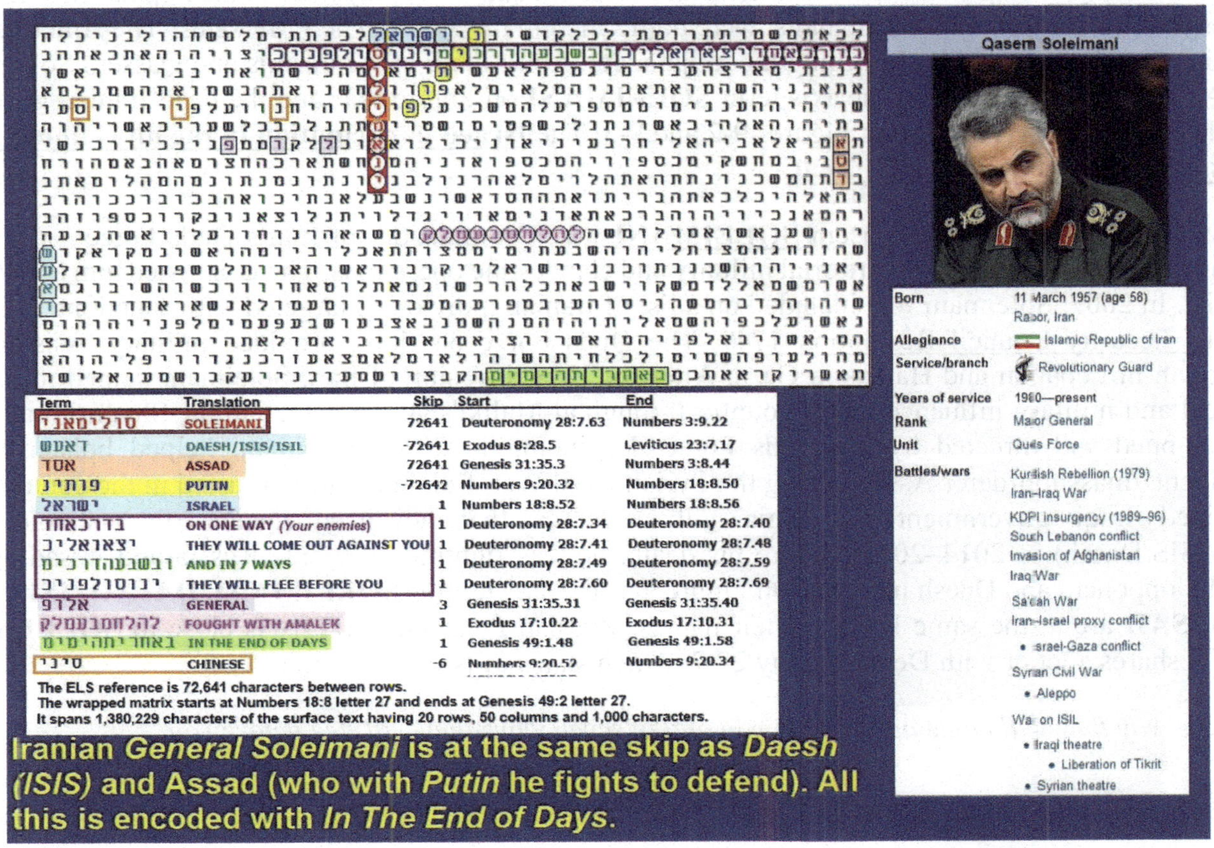

Figure 9-1 above. Its probability spreadsheet is below.

	A	B	C	D	E	F	G	H
1								
2	AXIS TERM SOLEIMAN@ Skip 72,641	SKIP	NUMBER	DIVIDE BY	THE QUOTIENT	E QUOTIENT	POISSON EQUATION	EACH WORD
3	Axis Term ELS Rank 10	USED	IN	304805	EQUALS	X 1000 LETTERS	PROBABILITY FOR	OR PHRASE
4		ON	TORAH	LETTERS	FREQUENCY	EQUALS	THE WORD	HAS 1 CHANCE
5		FIGURE		IN THE	PER LETTER	WORD	APPEARING	IN THE FOLLOWING
6				TORAH		EXPECTANCY	AT LEAST ONCE	OF BEING HERE.
7	IN THE END OF DAYS @ Skip +1 only	1	4	304805	1.31231E-05	0.013123144	0.013037411	76.70234359
8	DAESH @ Skips +/- 1 and +/- 72641	-72641	6	304805	1.96847E-05	0.019684716	0.019492237	51.30247372
9	ASSAD @ Skips +/- 1 and +/- 72641	72641	6	304805	1.96847E-05	0.019684716	0.019492237	51.30247372
10	ISRAEL@ Skip +1 only	1	591	304805	0.001938945	1.938944571	0.856144301	1.168027398
11	AMALEK @ Skip +1 only	1	21	304805	6.88965E-05	0.068896508	0.066576723	15.02026473
12	PUTIN @ Skips -72642 to -72865 (2 spellings)	-72642	29	304805	9.51428E-05	0.095142796	0.090756911	11.01844461
13	CHINESE @ Skips -6 to -433	-6	425	304805	0.001394334	1.394334082	0.752001872	1.329783923
14	GENERAL @ Skips +3 to +430	3	1244	304805	0.004081298	4.081297879	0.983114464	1.017175554
15	INITIAL COMBINED SIGNIFICANCE						1.89447E-08	52785318.85
16								Divide above by 10
17								to Axis Term ELS
18								Rank 10 =
19	FINAL COMBINED SIGNIFICANCE						1.89447E-07	5278531.885

Figure 9-1 was found against odds of about 5,278,531 to 1.

On **<u>January 3, 2020, the Jerusalem Post wrote that Qasem Soleimani was reported killed at the Baghdad Airport.</u>** Iraqi state media reported that Soleimani and the deputy head of the Popular Mobilization Forces (PMF), Abu Mahdi al-Muhandis, were both killed in the attack. The PMF said, "The American and Israeli enemy is responsible for killing the mujahideen Abu Mahdi al- Muhandis and Qassem Soleimani," said Assadi. **ISRAEL** touches the axis term **SOLEIMANI** on Figure 9-1, and the dead Iranian general is encoded with **NETANYAHU** on Figures 9-2 and 9-3. The **<u>Israeli participation in the hit is found in the Jerusalem Post on January 12, 2020.</u>**

Iranian Army Major General Qasem **<u>SOLEIMANI</u>** was a commander of Iran's Quds Force—a division primarily responsible for extraterritorial military and clandestine operations. The U.S. listed him as a known terrorist. In 2007, Soleimani was included on a list of Iranian individuals targeted with sanctions in <u>United Nations Security Council Resolution 1747</u>. The Quds Force has long provided military assistance to Hezbollah in Lebanon and Hamas in Gaza. After the U. S. invasion of Iraq, Soleimani gained increasing political and military influence in the country through al-Maliki's government and the Shia militia, which are equipped and directed by the Quds Force. Beginning in 2012, Soleimani helped bolster Syrian "president"/mass murderer Assad during the Syrian Civil War. Soleimani also assisted in the command of combined Iraqi government and Shia militia forces that advanced against the Islamic State (ISIL/ISIS/Daesh) in 2014–2015. Before his death, he was fighting alongside Russia and Assad against Assad's opponent and Daesh in Syria. On Figure 9-1, the axis term is **SOLEIMANI. DAESH (ISIS/ISIL)** and **ASSAD** are at the same level as their names. Russian President **PUTIN** is close to **SOLEIMANI. PUTIN** shares a letter with Deuteronomy 28:7, which states of Israel's enemies:

"On one way they will come out against you and in seven ways they will flee before you."

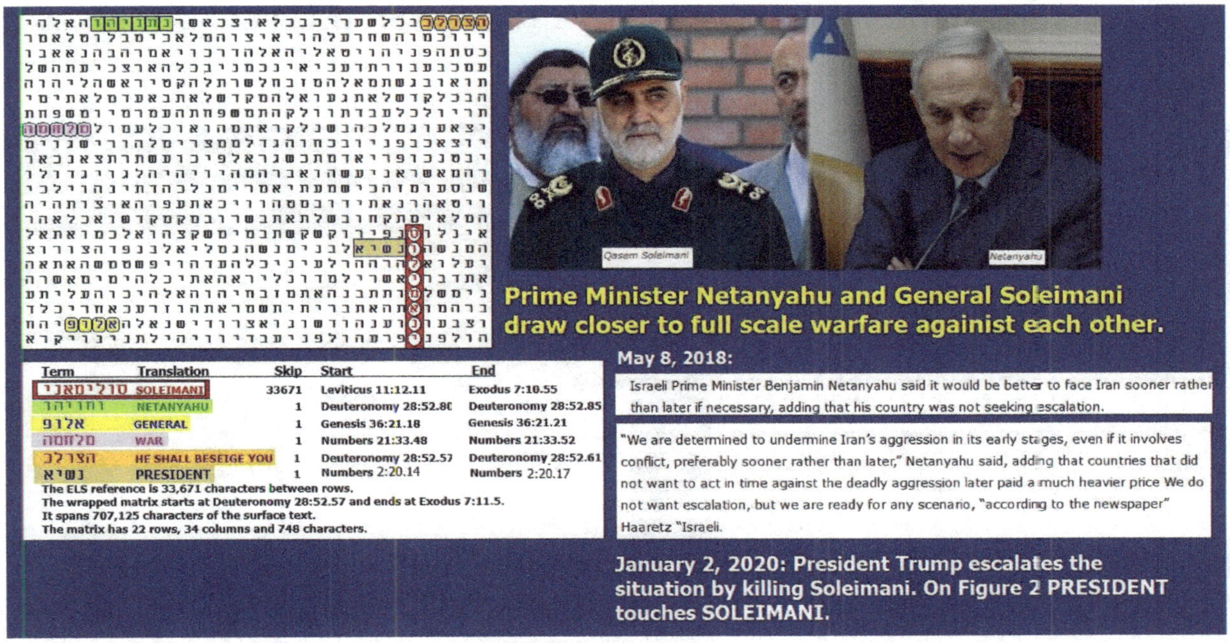

Figure 9-2 above: Soleimani and Netanyahu.

SIGNIFICANCE OF FIGURE 9-2. The most significant a priori term is **NETANYAHU** at Skip

+1. Before adjusting the axis term ELS rank, **NETANYAHU** was found at skip +1 against odds of about 26 to 1. **WAR** was found in the open text against odds of about 11.8 to 1. Soleimani's rank of **GENERAL** is on the matrix in the open text against odds of about 10 to 1. **PRESIDENT** was located in the open text against odds of about 6.5 to 1, but these odds just refer to the word at skip +1 being anywhere on the 748-letter matrix. In actuality, it touches **SOLEIMANI.** No value was assigned to **HE SHALL BESIEGE YOU** because it was only found *a posteriori.* After dividing combined odds by 8 to account for the ELS rank of the axis term, the matrix was found against odds of about 2,494 to 1.

Iranians first sent hundreds of missiles flying toward Israel in 2024 and 2025. This is why many people think we are now in the End of Days, which is on Figure 9-1. ISRAEL is on the matrix. **Netanyahu sees Iran as Israel's ancient enemy – Amalek**. *Fought with Amalek* is also on Figure 9-1. Finally, Soleimani's rank, **GENERAL,** intersects his name, as do the **CHINESE However, the word for CHINESE** who are helping him and Putin had about a 75% chance of being on a matrix this large. Just before Soleimani received justice for killing over 600 Americans, both the Russians and the Chinese were conducting war games in the area.

FIGURE 9-3 links **SOLEIMANI** to **BENJAMIN NETANYAHU, ISRAEL, PERSIA, LEBANON** and IRAQ.

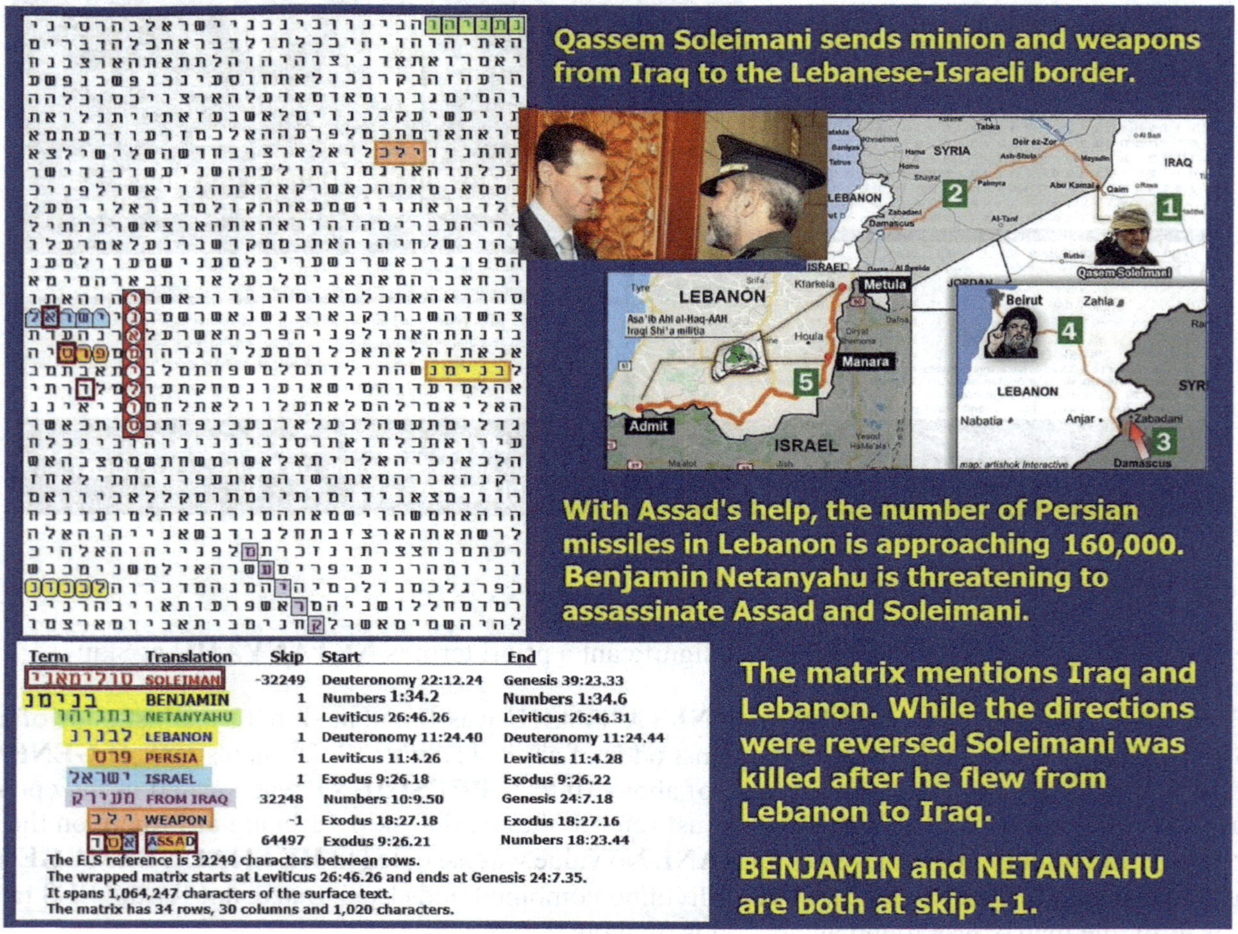

SIGNIFICANCE OF FIGURE 9-3. The most significant *a priori* term is **LEBANON** in the open text. Before adjusting for axis term ELS rank, **LEBANON** was found at skip +1 against odds of about 149.9 to 1. I originally looked for **LEBANON** because the map picture that I selected for the Figure had it on it. But on the night that Soleimani was killed, he flew from Lebanon to Baghdad, **IRAQ. IRAQ** was located at an ELS one off the skip of the axis term. **NETANYAHU** is at skip +1 against odds of about 19 to 1. In a big surprise, Netanyahu's first name **BENJAMIN** (at skip +1) has all five letters of the name in the same columns as 5 of the six letters of **NETANYAHU** at the top of the matrix. **PERSIA** was found at skip +1, but I used the combined frequency of **PERSIA** and **IRAN** because I would have settled for either name. As such, it was found at skip +1 against odds of about 12.4 to 1. **ASSAD** was not at a special case skip, but it was on the matrix against odds of about 3.5 to 1. No extra value was added for the fact that **ASSAD** shares a letter *alef* with **ISRAEL** and a letter *samech* with **PERSIA.** The other terms on the matrix were not statistically significant. Overall, Figure 9-3 was found against odds of about 1,320,763 to 1.

Spreadsheet for Figure 9-3.

	A	B	C	D	E	F	G	H
1	Axis term SOLEIMANI @ Skip -32,249	SKIP	NUMBER	DIVIDE BY	THE QUOTIENT	E QUOTIENT	POISSON EQUATION	EACH WORD OR
2	ELS Rank 7	USED	IN	304805	EQUALS	X 1020 LETTERS	PROBABILITY FOR	PHRASE HAS 1
3		ON	TORAH	LETTERS	FREQUENCY	EQUALS	THE WORD	CHANCE IN THE
4		FIGURE		IN THE	PER LETTER	WORD	APPEARING	IN THE FOLLOWING
5				TORAH		EXPECTANCY	AT LEAST ONCE	OF BEING HERE.
6	LEBANON @ Skip +1	1	2	304805	6.56157E-06	0.006692804	0.006670457	149.9147734
7	NETANYAHU @ Skip +1 only	1	16	304805	5.24926E-05	0.053542429	0.052134277	19.18123862
8	BENJAMIN @ Skip +1 only	1	31	304805	0.000101704	0.103738456	0.098538961	10.14827014
9	PERSIA @ Skip +1 only	1	25	304805	8.20197E-05	0.083660045	0.080256125	12.46010811
10	WEAPONS (2 synonyms) @ Skips +/-1 and +/- 32249 only	1	68	304805	0.000223093	0.227555322	0.203521643	4.913482354
11	IRAQ @ Skips 32248 to 32683	32248	474	304805	0.001555093	1.586194452	0.795296861	1.257392112
12	ASSAD @ Skips 64479 to	64479	99	304805	0.000324798	0.331293778	0.282085793	3.546026448
13	ISRAEL @ Skip +1 only	1	591	304805	0.001938945	1.977723463	0.861616085	1.160609717
14	FINAL COMBINED ODDS = (H6*H7*H8*H9*H10*H11*H12*H13)/7						7.57133E-07	1320763.856

The Middle East is far more complex than most people think because some of the bad guys are not human. As has been true (largely with respect to leaders of the Democrat Party), some are alien invaders. They are named in Chapters 3 and 4 of this book. This is such a startling claim that before I proceed, I feel required to present more evidence than the matrices seen in Chapter 3, raising questions of species and loyalty for **PRESIDENT JOE BIDEN, PRESIDENT BARACK OBAMA,** Senator **CHUCK SCHUMER,** and Vice President **HARRIS, GENERAL MILLEY, FORMER FBI DIRECTOR CHRIS WRAY, U.S. ATTORNEY GENERAL MERRICK**

GARLAND, ALVIN BRAGG, GOVERNOR TIM WALZ, former Secret Service Director

KIM CHEATLE, Senator ADAM **SCHIFF** and **JOHN DURHAM.**

Let's first discuss Israeli Retired Brigadier General Haim Eshed, and then how his knowledge fits in Craig Ebrahimi's claim (backed by a reasonable amount of physical evidence) that Craig is an extraterrestrial. We will also look at alien attempts to limit or reduce the human population on Earth. Here, I refer to the alliance between the Democrat Party and all homosexual and transsexual causes that aim to reduce the size of the human population on Earth by mating people who can't make babies. They further attack the human population by encouraging religious wars. Judaism, Christianity, and Islam each teach very different concepts about God. Why? The Muslim Koran has the following explanation:

...To each among you has been prescribed a Law and an Open Way. If God had so willed, He would have made you a single People, but (His plan is) to test you in what He has given you: so strive as in a race to all virtues. The goal of all is to God. It is He that will show you the truth of the matters in which you dispute.

(Koran, Ali translation, Table 5:51)

Having set the stage for explaining our true situation, let's study the Haim Eshed part of the equation by looking at his book (Figure 9-4) and the matrix found about him (Figure 9-5).

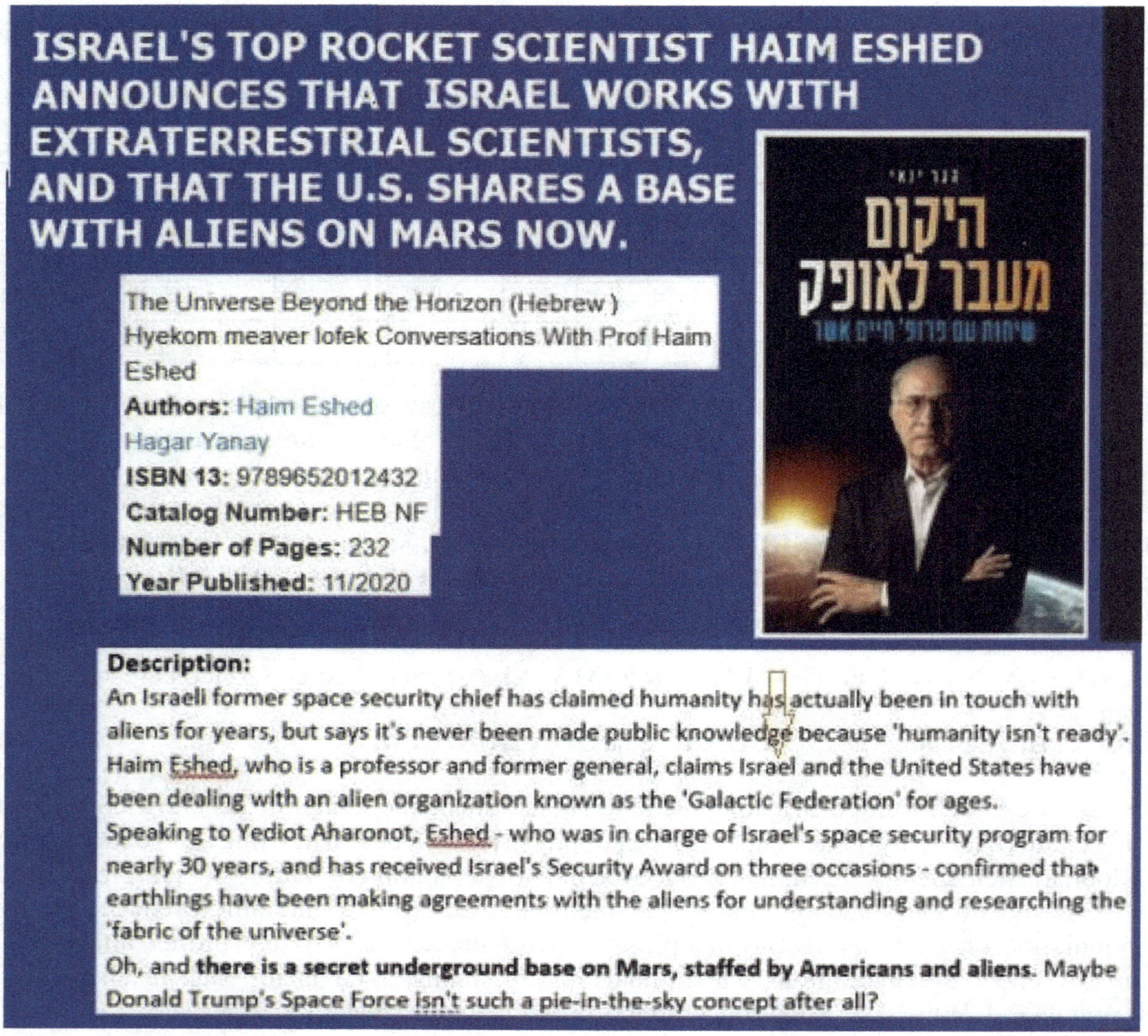

Figure 9-4 above: General Haim Eshed's book and main claims.

Israel's top scientist, Eshed, reveals that people are on Mars. 12/12/23

Whenever I choose to write another article about Mars or UFOs, a major concern is with respect to whom I am quoting. For UFOs/UVMs, I love to write about the works of Dr. Michael Salla, but the one thing that bothers me about his work is that he often quotes people who use pseudonyms. In November 2020, an author appeared (not yet interviewed by Salla – Israeli General/Professor Haim Eshed) who claimed not only that the Israeli Space Program had aliens working with it, but that there is also a joint U.S./Alien base on Mars! He is the most credible author in this field, and his claim is well supported in the Torah Code.

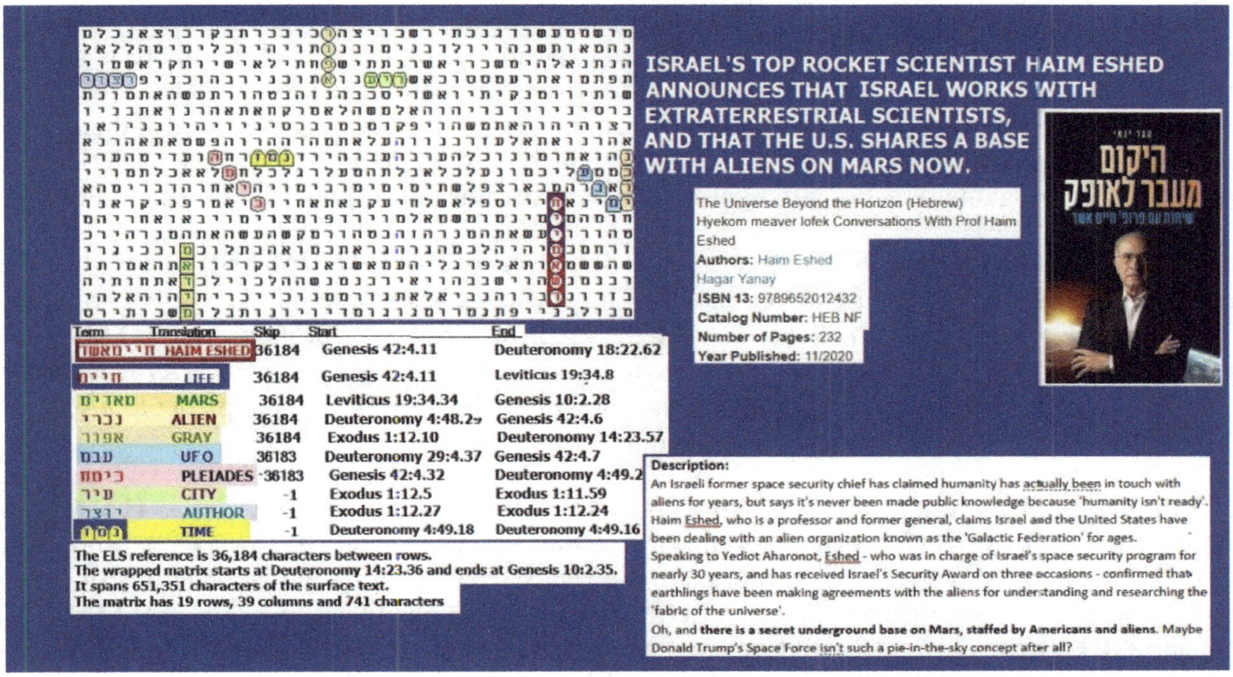

Above: Figure 9-5 with axis term **HAIM ESHED** matched with nine key words that describe his **LIFE**.
Below: Spreadsheet showing the odds against these key words being found on the matrix (2,216,538 to 1).

	A	B	C	D	E	F	G	H
1	Axis term HAIM ESHED @ Skip +36,184	SKIP	NUMBER	DIVIDE BY	THE QUOTIENT	E QUOTIENT	POISSON EQUATION	EACH WORD OR
2	ELS Rank 53	USED	IN	304805	EQUALS	X 741 LETTERS	PROBABILITY FOR	PHRASE HAS 1
3		ON	TORAH	LETTERS	FREQUENCY	EQUALS	THE WORD	CHANCE IN THE
4		FIGURE		IN THE	PER LETTER	WORD	APPEARING	IN THE FOLLOWING
5				TORAH		EXPECTANCY	AT LEAST ONCE	OF BEING HERE.
6								
7	MARS @ Skip +/- 1 and +/- 36184	36184	1	304805	3.28079E-06	0.002431062	0.002422811	411.8429826
8	GRAY @ Skips +/-1 and +/- 36184	36184	4	304805	1.31231E-05	0.00972425	0.009677122	103.3365054
9	ALIEN @ Skips +/-1 and +/- 36184	36184	19	304805	6.23349E-05	0.046190187	0.045139657	22.15346905
10	TIME @ Skips Skip +/- 1 and +/- 36184	-1	24	304805	7.87389E-05	0.0583455	0.056676027	17.64414435
11	LIFE @ Skips +/-1 and +/- 36184	36184	44	304805	0.000144355	0.106966749	0.10144445	9.857611742
12	AUTHOR @ Skips +/- 1 and +/- 36184 (2 synonyms)	-1	100	304805	0.000328079	0.243106248	0.215811809	4.633666727
13	CITY @ Skips +/- 1 and +/- 36184	-1	455	304805	0.001492758	1.10613343	0.669164309	1.494401279
14	UFO @ Skips +36183 to +36210	36183	1406	304805	0.004612785	3.41807385	0.967224495	1.03388614
15	PLEIADES @ Skips -36183 to -36506	36183	3101	304805	0.010173718	7.538724758	0.999467924	1.000532359
16	FINAL SIGNIFICANCE = (H7*H8*H9*H10*H11*H12*H13*H14*H15)/53						4.51212E-08	22162538.47

Wikipedia states that Haim Eshed (Hebrew: אשד חיים; born 1933) is a visiting professor of aeronautics and astronautics at 333 various space technology research institutions. A retired brigadier general in Israeli Military Intelligence, Eshed was director of space programs for Israel Ministry of Defense for nearly 30 years, is former chair of the Space Committee of the National Council for Research and Development for the Ministry of Science, Technology and Space, and a member of the steering committee of the Israel Space Agency. Eshed is responsible for the launch of 20 Israeli-made satellites, and he is widely cited as the father of Israel's space program.

In 1967, Eshed was awarded the Chief of Staff Citation, the highest non-combat decoration awarded by the IDF. Throughout his career, he also received the Israel Defense Prize – the highest civilian defense honor of the State of Israel – three times, but the reasons remain classified.

Extra-terrestrial Claims: Eshed claimed in an interview with Israeli national newspaper Yediot Aharonot that the United States government had been in contact with extraterrestrial life for years and had signed secret agreements with a "Galactic Federation" in order to do experiments on Earth, and that there is a joint base underground on Mars where they collaborate with American astronauts.[20] He also stated that US President Donald Trump was aware of this and was "on the verge" of informing everyone of their existence. Still, he was stopped by the "Galactic Federation," which wished to prevent mass hysteria.[21] The interview, in Hebrew, gained traction after parts were published in English by The Jerusalem Post.[22]

UNDERSTANDING FIGURE 9-5: The axis term (first term sought) is **HAIM ESHED.** He has the best reputation of anyone who has ever claimed that the United States has a base on Mars. He says we are not alone there; we share the base with aliens. The 5-letter Hebrew for Mars is pronounced **MOADEEM**, and it was the first keyword sought. Although I expected to find **MARS** at the same skip as **HAIM ESHED**, and I did, the ELS rank of **MARS** was high (53). It was, however, the only find of the planet at a special case skip. The odds against finding it this way were about 411 to 1.

The next most significant term found was **GREY/GRAY,** which was at the same skip as **HAIM ESHED** and **MARS** (skip +36,184). Odds against its appearance this way were about 103 to 1. Greys abduct people and are the most common type of aliens reported in the 1960s. To emphasize the fact that this find is not an accident, the next most significant term was **ALIEN**, which was again at the same skip as **HAIM ESHED, MARS,** and **GREY** (skip +36,184). Odds against its appearance this way were about 22 to 1. Finally, there is still one more major keyword also at skip +36,184. It's **LIFE.** Chaim Eshed is the most serious person to report human (American) and alien life on Mars. But we don't even need to travel away from his name to find **LIFE** because Chaim's first name (Chaim) means LIFE! Odds against LIFE being somewhere on the matrix at a special case are about 9.85 to 1. However, this does not recognize the fact that this keyword is found inside the axis term.

TIME might seem like a questionable term to seek, but we are talking about a matrix with space travel as a major topic with **MARS, ALIEN, GREY, PLEIADES,** and **UFO** on it. Based on my research into UFOs, it appears that to be able to fly interstellar distances, we are really talking about the ability to fly not just from one point in space to another, but from one point in space- time to another. In other words, time travel. Odds against **TIME** being on the matrix were about 17 to 1.

Odds against each of the other terms found were less than 5 to 1. The word for **CITY** was sought and found at skip -1. Motivation for seeking it centered around the issue of how many people live on Mars now and how many may have lived there sometime in the past. While I would assume that populations are low, probably no more than a few hundred people, I have heard claims before that there may be over a hundred thousand residents now and that there may have been millions before a global catastrophe took place (according to John Brandenburg, a nuclear war). Overall, the matrix was found against odds of over 2.2 million to 1.

One thing is especially troubling about Eshed's revelations. He says that President Trump was about to reveal the alien presence here on Earth, but the Galactic Federation of Worlds (GFW) asked (or directed?) him not to do it because it would cause a global panic. This implies that we are not allowed to have full sovereignty. ET is at least partially in charge of the Earth. Supposedly, they even prevented at least one nuclear war here (strangely associated with the Bay of Pigs invasion of Cuba, which was the year before the Cuban Missile Crisis).

Aliens in Gaza, massacre by Hamas, its destruction, & cost to Ashkelon.

This article was originally published by me on 11/17/2023. Although Eshed's revelation was certainly astonishing, it left out the other side of the equation presented in the Torah Code. While Israel works with "white hat" ETs, evidently Hamas, like the Nazis before them, is now working with "black hat" Reptilian (and/or Gray) aliens. In Figure 9-6, the hostile (Reptilian) aliens are named as the **CIAKAHRR**. A simple transliteration of that name touches **ALIEN IN GAZA.**

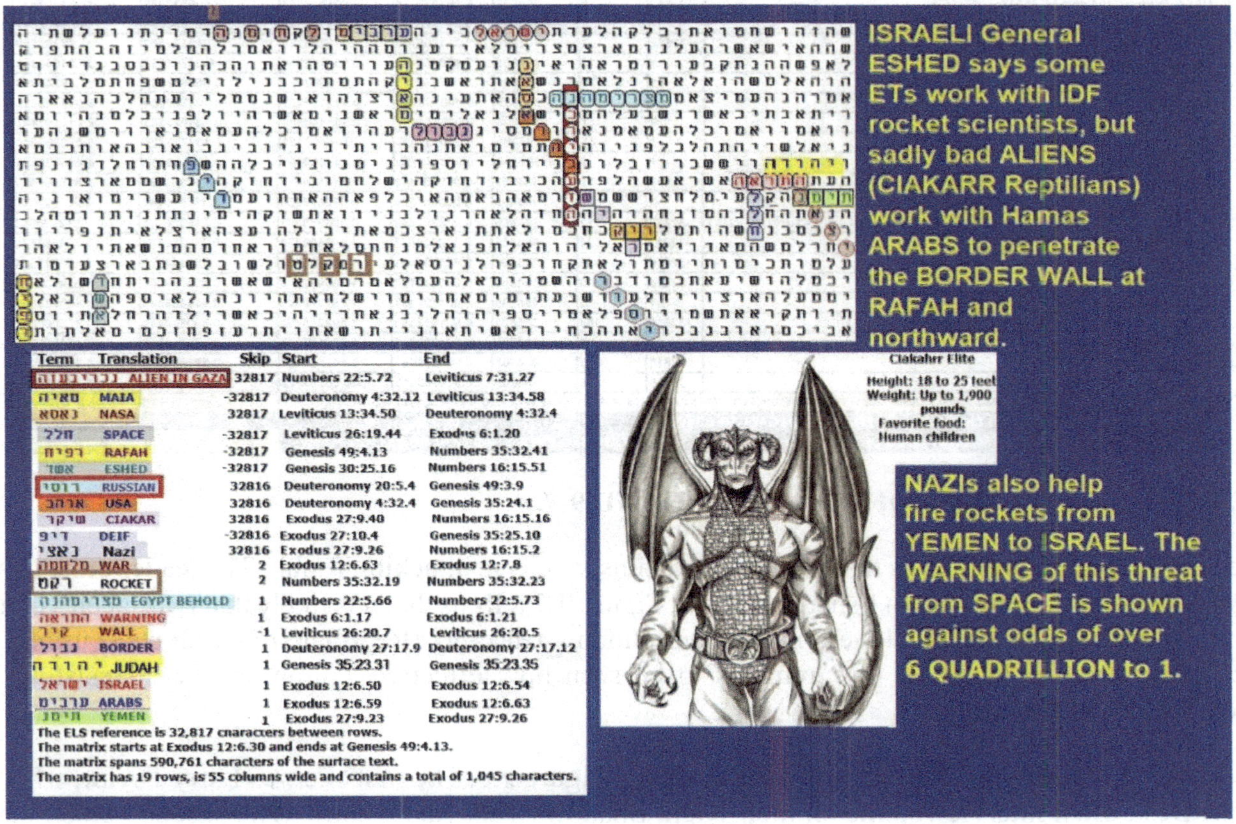

Figure 9-6 above: Axis term ALIEN IN GAZA.

Figure 9-6 above: **ALIEN IN GAZA** is at the same skip (+32,827) as **MAIA,** a star in the Pleiades, **NASA, SPACE, RAFAH** (where major battles took place between Israel and Hamas), and **ESHED**. There are many indicators that suggest **CIAKAHRR** (Reptilian) aliens provided

technology to the **GAZA ARABS** to break through the **BORDER WALL. BORDER** is at skip +1, **WALL** is at -1, and a short transliteration of **CIAKAHRR** is one skip off the axis term at +32,816 touching **ALIEN IN GAZA.** This caused a new **ISRAEL WAR** that involves getting involved with **EGYPT, YEMEN,** Iran, Lebanon, Syria, possibly Russia, Ukraine, China, and Taiwan.

FIGURE 9-6 spreadsheet below: The odds against finding this matrix are over six quadrillion to

1. I started publishing Torah Code matrices in 1997. This is the most significant matrix that I have ever found. However, I like Figure 9-7 a bit more because it reduces the area from 1,045 letters to 840 letters.

Figure 9-7 describes the requirement for Israel to avenge all the 1,200 people, including children, massacred and over 240 people taken hostage (more of them were also murdered in captivity). Figure 9-7 was found against odds of almost 5 QUADRILLION to 1.

	A	B	C	D	E	F	G	H
1	Axis term ALIEN IN GAZA @ Skip +32817	SKIP	NUMBER	DIVIDE BY	THE QUOTIENT	E QUOTIENT	POISSON EQUATION	EACH WORD OR
2	ELS Rank 1	USED	IN	304805	EQUALS	X 1045 LETTERS	PROBABILITY FOR	PHRASE HAS 1
3	Updated on November 7, 2023	ON	TORAH	LETTERS	FREQUENCY	EQUALS	THE WORD	CHANCE IN THE
4		FIGURE		IN THE	PER LETTER	WORD	APPEARING	IN THE FOLLOWING
5				TORAH		EXPECTANCY	AT LEAST ONCE	OF BEING HERE.
6	NASA @ Skips +/-1 and +/- 32817	32817	2	304805	6.56157E-06	0.006856843	0.006833388	146.3402843
7	WARNING @ Skip +1	1	2	304805	6.56157E-06	0.006856843	0.006833388	146.3402843
8	RAFAH @ Skips +/-1 and +/- 32817	-32817	3	304805	9.84236E-06	0.010285264	0.010232552	97.72733238
9	ARABS @ Skip +1 only	1	11	304805	3.60886E-05	0.037712636	0.03701037	27.01945408
10	YEMEN@ Skip +1 only	1	17	304805	5.57734E-05	0.058283165	0.056617223	17.66246994
11	BORDER @ Skip +1 only	1	33	304805	0.000108266	0.113137908	0.106972504	9.348196628
12	ROCKET @ Skips +2 to +29	2	47	304805	0.000154197	0.161135808	0.148823533	6.719367407
13	WAR @ Skips +2 to +285	2	48	304805	0.000157478	0.16456423	0.151736728	6.590362205
14	JUDAH @ Skips +1 only	1	49	304805	0.000160759	0.167992651	0.154639953	6.466634151
15	MIAI (star in Pleiades) @ Skips +/- 1 and +/- 32817	32817	55	304805	0.000180443	0.18856318	0.17185182	5.818966582
16	WALL @ Skips +/- 1 and +/- 32817	-1	109	304805	0.000357606	0.373697938	0.311815245	3.207027288
17	ESHED (Haim - Director of Israeli Space Program) @ Skips +/- 1 and +/- 32817	-32817	112	304805	0.000367448	0.383983202	0.318857131	3.13620083
18	SPACE @ Skips +/- 1 and +/- 32817	-32817	134	304805	0.000439625	0.459408474	0.368342824	2.714862178
19	DEIF (Gaza planner) @ Skips -32816 to -32843	-32816	322	304805	0.001056413	1.103951707	0.66844173	1.496016714
20	RUSSIAN @ Skips +32816 to +33191	32816	410	304805	0.001345122	1.405652794	0.754793066	1.324866438
21	EGYPT (behold) @ Skip +1 only	1	365	304805	0.001197487	1.251373829	0.713888542	1.400778892
22	CIAKARR (Reptilian species) @ Skips +32816 to +33191	-32816	554	304805	0.001817555	1.899345483	0.850333454	1.176009242
23	ISRAEL @ Skip +1 only	1	591	304805	0.001938945	2.026197077	0.868164069	1.151856009
24	NAZI @ Skips 32816 to 33191	32816	625	304805	0.002050491	2.142763406	0.882669836	1.132926445
25	USA @ Skips +32816 to +33191	-32816	2976	304805	0.009763619	10.20298223	0.99996294	1.000037061
26	FINAL ODDS AGAINST MATRIX (NO AXIS TERM) =H6*H7*H8*H9*H10*H11*H12*H13*H14*H15*H16*H17*H18*H19*H20*H21*H22*H23*H24*H25						1.81018E+15	1.81018E+15
27	AXIS TERM R FACTOR = 0.531 = 1 CHANCE IN 3.39 FOR ALIEN IN GAZA TO BE HERE.							
28	THIS FULL MATRIX WAS FOUND AGAINST ODDS OF 6,136,490,000,000,000 to 1. (OVER 6 QUADRILLION 136 TRILLION 490 BILLION TO 1).						1.62959E-16	6.13649E+15

THE DEADLY MEANING OF FIGURES 9-6 AND 9-7.

Quite simply, Figure 9-6 tells us that Israel is up against aliens in attacking Rafah. The leader of NASA and the Israeli Space Program (Haim Eshed) knows it. Figure 9-7 tells us that it is likely that two species of aliens finished taking over the U.S. Government under Biden - Ciakahrr (Reptilians) and Grays. Further, they serve the purpose of Hitler. It now remains to be seen how effective Trump can be at resisting what Democrats have set in motion.

Two people asked me to search for the matrix found on Figure 9-6. The first one was Larry Taylor, who, with Stewart Best, often interviewed me on their Night Shadows Talk Radio show. The second person making the request was Rebecca. Rebecca originally claimed to be from the Pleiades star system. I did a joint radio interview with her on the show on June 7, 2023. Sadly, although the show was available on my website until about December 2024, Stewart shut down his show in 2025 when he suffered a serious illness. When he did, the link to the interview no longer worked on my site. A smaller version of Figure 9-6 (area 1,045 letters) is shown below as

Figure 9-7 (area 840 letters). The two men recently started another show, and on April 30, 2025, they had me on for 94 minutes. Taylor discussed the possibility that Aliens have taken him to Mars (backed in the Epilogue by Torah Code matrices – Figures E-14 and E-15).

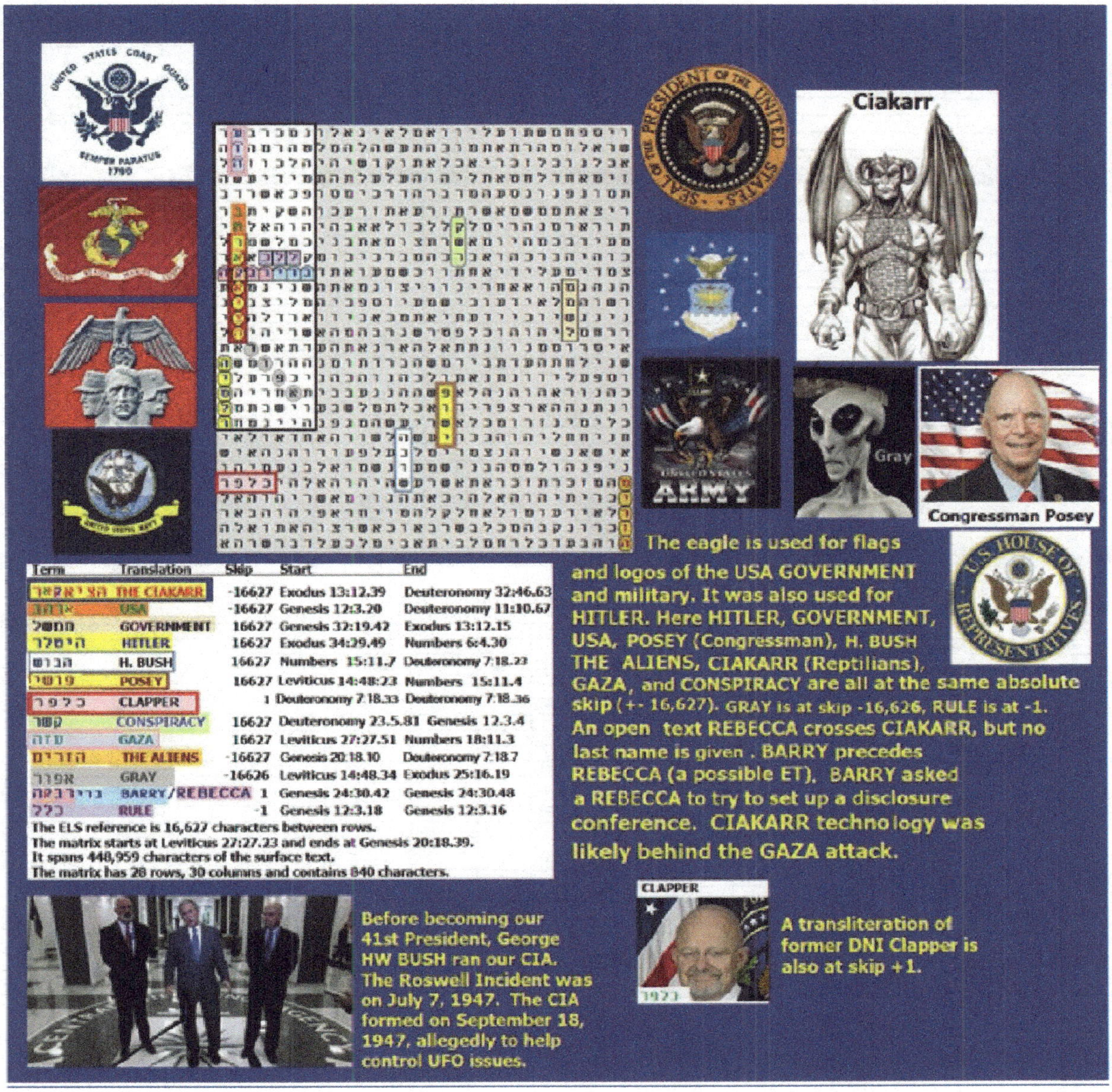

Figure 9-7 is above. Its spreadsheets are below. Notice that **THE CIAKAHRR, GAZA, USA, GOVERNMENT, HITLER** (best spelling), H. BUSH**, CONSPIRACY**, and **THE ALIENS** are

all at the same special case skip of +/- 16,627. The small matrix exists against odds of about 68.9 billion to 1, whereas the larger matrix exists against odds of about 4.4 trillion to 1 if we just use **BUSH** or 26.7957 trillion to 1 if we use **H. BUSH.**

	A	B	C	D	E	F	G	H
1	Axis term THE CIAKARR @ Skip 16,627	SKIP	NUMBER	DIVIDE BY	THE QUOTIENT	E QUOTIENT	POISSON EQUATION	EACH WORD OR
2	ELS Rank 15	USED	IN	304805	EQUALS	X 140 LETTERS	PROBABILITY FOR	PHRASE HAS 1
3		ON	TORAH	LETTERS	FREQUENCY	EQUALS	THE WORD	CHANCE IN THE
4		FIGURE		IN THE	PER LETTER	WORD	APPEARING	IN THE FOLLOWING
5				TORAH		EXPECTANCY	AT LEAST ONCE	OF BEING HERE.
6	HITLER (best spelling) @Skips +/-1 and +/-16627	16627	1	304805	3.28079E-06	0.00045931	0.000459205	2177.67861
7	GAZA @Skips +/-1 and +/-16627	16627	22	304805	7.21773E-05	0.010104821	0.010053939	99.4635044
8	USA @Skips +/-1 and +/-16627	16627	26	304805	8.53004E-05	0.011942061	0.011871038	84.23863253
9	REBECCA @Skip +1 only	1	31	304805	0.000101704	0.014238612	0.014137722	70.73275337
10	GRAY @ skips -16,626	16626	160	304805	0.000524926	0.073489608	0.070854199	14.11348965
11	RULE @Skips +/-1 and +/-16627	-1	218	304805	0.000715211	0.100129591	0.095279833	10.49540041
12	BARRY @Skips +1 only	1	445	304805	0.00145995	0.204392973	0.18485802	5.409557016
13	FINAL SIGNIFICANCE FOR THE 140-LETTER MATRIX =(H6*H7*H8*H9*H10*H11*H12)/15						1.45046E-11	68943492302
14								
15	Axis term THE CIAKARR @ Skip -16,627	SKIP	NUMBER	DIVIDE BY	THE QUOTIENT	E QUOTIENT	POISSON EQUATION	EACH WORD OR
16	ELS Rank 15	USED	IN	304805	EQUALS	X 840 LETTERS	PROBABILITY FOR	PHRASE HAS 1
17		ON	TORAH	LETTERS	FREQUENCY	EQUALS	THE WORD	CHANCE IN THE
18		FIGURE		IN THE	PER LETTER	WORD	APPEARING	IN THE FOLLOWING
19				TORAH		EXPECTANCY	AT LEAST ONCE	OF BEING HERE.
20	HITLER (best spelling) @Skips +/-1 and +/-16627	-16627	1	304805	3.28079E-06	0.00275586	0.002752066	363.3633249
21	THE ALIENS @ Skips +/-1 and +/- 16627	-16627	1	304805	3.28079E-06	0.00275586	0.002752066	363.3633249
22	POSEY @ Skip +/-1 and +/-16627	16227	8	304805	2.62463E-05	0.022046882	0.021805626	45.85972413
23	GOVERNMENT @ Skips +/-1 and +/- 16627	16627	17	304805	5.57734E-05	0.046849625	0.045769121	21.84879195
24	GAZA @Skips +/-1 and +/-16627	16627	22	304805	7.21773E-05	0.060628927	0.058827581	16.99882916
25	USA @Skips +/-1 and +/-16627	16627	26	304805	8.53004E-05	0.071652368	0.069145566	14.46224341
26	REBECCA @Skip +1 only	1	31	304805	0.000101704	0.085431669	0.081884124	12.21237958
27	CONSPIRACY @ Skips +/- 1 and +/- 16627	16627	36	304805	0.000118108	0.099210971	0.094448357	10.58779665
28	GRAY @ skips -16,626	16626	160	304805	0.000524926	0.440937649	0.356567176	2.804520629
29	RULE @Skips +/-1 and +/-16627	-1	218	304805	0.000715211	0.600777546	0.451614925	2.214275804
30	BUSH @ Skips +/ 1 and +/-16627	16627	297	304805	0.000974393	0.81849051	0.558903016	1.789219186
31	BARRY @Skips +1 only	1	445	304805	0.00145995	1.226357835	0.706640904	1.415145932
32	ODDS AGAINST BUSH MATRIX = (H20*H21*H22*H23*H24*H25*H26*H27*H28*H29*H30*H31)/15						2.26848E-13	4.40824E+12
33								
34	H. BUSH @ Skips +/ 1 and +/-16627	16627	35	304805	0.000114828	0.096455111	0.091949342	10.87555369
35	ODDS AGAINST H.BUSH MATRIX = (H20*H21*H22*H23*H24*H25*H26*H27*H28*H29*H34*H31)/15						3.73204E-14	2.6795E+13

	A	B	C	D	E	F	G	H
1	Axis term ALIEN IN GAZA @ Skip +32817	SKIP	NUMBER	DIVIDE BY	THE QUOTIENT	E QUOTIENT	POISSON EQUATION	EACH WORD OR
2	ELS Rank 1	USED	IN	304805	EQUALS	X 840 LETTERS	PROBABILITY FOR	PHRASE HAS 1
3	Updated on November 1, 2023	ON	TORAH	LETTERS	FREQUENCY	EQUALS	THE WORD	CHANCE IN THE
4		FIGURE		IN THE	PER LETTER	WORD	APPEARING	IN THE FOLLOWING
5				TORAH		EXPECTANCY	AT LEAST ONCE	OF BEING HERE.
6	AVENGE THE CHILDREN OF ISRAEL	1	1	304805	3.28079E-06	0.00275586	0.002752066	363.3633249
7	NASA @ Skips +/-1 and +/- 32817	32817	2	304805	6.56157E-06	0.005511721	0.005496559	181.9320069
8	WARNING @ Skip +1	1	2	304805	6.56157E-06	0.005511721	0.005496559	181.9320069
9	CAIRO @ Skip +/-1 and +/- 32817	32817	7	304805	2.29655E-05	0.019291022	0.019106141	52.33919261
10	YEMEN@ Skip +1 only	1	17	304805	5.57734E-05	0.046849625	0.045769121	21.84879195
11	PLEIADES (2 spellings) @ +/- 1 and +/-32817	-32817	65	304805	0.000213251	0.17913092	0.164003556	6.097428757
12	BORDER @ Skip +1 only	1	33	304805	0.000108266	0.09094339	0.086930602	11.50342894
13	JUDEA @ Skip +1 only	1	49	304805	0.000160759	0.135037155	0.126316551	7.916618968
14	MIAI (star in Pleiades) @ Skips +/- 1 and +/- 32817	32817	55	304805	0.000180443	0.151572317	0.140644266	7.110137015
15	SPACE @ Skips +/- 1 and +/- 32817	-32817	134	304805	0.000439625	0.369285281	0.308771812	3.238637602
16	CHINA @ Skip -2 to -29	-2	280	304805	0.00091862	0.771640885	0.53774606	1.859613811
17	RUSSIAN @ Skips +32816 to +33191	32816	410	304805	0.001345122	1.129902725	0.676935319	1.477246012
18	EGYPT (behold) @ Skip +1 only	1	365	304805	0.001197487	1.005889011	0.634280638	1.576589194
19	CIAKARR (Reptilian species) @ Skips +32816 to +33191	-32816	554	304805	0.001817555	1.526746608	0.78275871	1.277532894
20	ISRAEL @ Skip +1 only	1	591	304805	0.001938945	1.62871344	0.803818188	1.244062419
21	NAZI @ Skips 32816 to 33191	32816	625	304805	0.002050491	1.72241269	0.821365362	1.21748499
22	USA @ Skips +32816 to +33191	-32816	2976	304805	0.009763619	8.201440265	0.999725742	1.000274334
23	FINAL ODDS AGAINST MATRIX (NO AXIS TERM) =H6*H7*H8*H9*H10*H11*H12*H13*H14*H15*H16*H17*H18*H19*H20*H21*H22						6.7833E-16	1.47421E+15
24	AXIS TERM R FACTOR = 0.531 = 1 CHANCE IN 3.39 FOR ALIEN IN GAZA TO BE HERE.							
25	THIS FULL MATRIX WAS FOUND AGAINST ODDS OF 4,997,570,000,000,000 to 1. (ABOUT 4.99757 QUADRILLLION TO 1).						2.00097E-16	4.99757E+15

9-8 below: Nuc military Iran & Israel

This matrix shows the axis term **ATOMIC MILITARY** (skip -60,790) touch and share a letter alef with both **IRAN** and **ISRAEL** (open text). The name of a suspected alien (Ebrahimi) is at skip +1. At the same skip as **ATOMIC MILITARY** is **THE ALIEN.** Craig Ebrahimi claimed to be from **IRAN** before he claimed to be an **ALIEN. IRAN** is at one skip off the axis term, and it touches **ATOMIC MILITARY. UFO** is at skip -60,791. The word for **WAR** is at skip +2 and is only seen on the larger (600-letter) matrix.

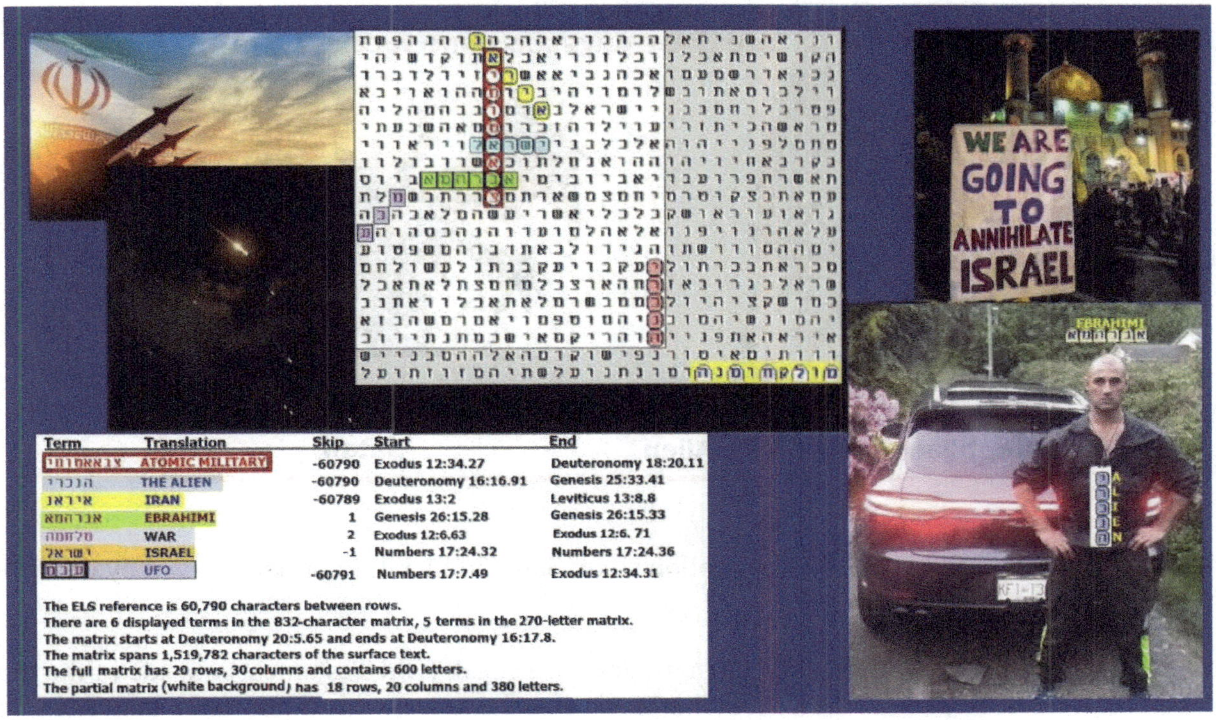

Figure 9-8 above: ATOMIC MILITARY and IRAN.

GAZA AND ET. If there are friendly ETs on Earth, they must reveal their presence and intentions. The world has been deceived by two huge lies:

1. Gaza claims it was occupied territory up to October 7, 2023. In reality, Israel completely withdrew every single Jew from Gaza on September 22, 2005. It gave up all homes and left behind the keys for all greenhouse facilities for Gazans to form a healthy economy. In return, Gaza destroyed all greenhouses and began to fire missiles at Israel 12 days later. Hamas seized control of Gaza in June, 2007. So, the "occupation" excuse for Gaza murdering 1,200 Jews on October 7 is nonsense.

2. The invasion of Israel was accomplished only through Palestinian ingenuity. They had help, but it was not merely from Iran. It was alien help, Reptilian help from a group of murderous savages known as the **CIAKAHRR.** What are the Ciakahrr like? Satanic Nazis. That's probably why the Ciakahrr and Satan are often drawn looking much like each other. See Figure 9-9.

Figure 9-9 above – Should we use the description study Satan in literature or one derived by dissection in a lab? I recommend a dissection.

There is no Jewish tradition of an ELS Code in the Jewish Tenach (Old Testament) outside of the first five books (Torah). Yet the Reptilians are encoded in Zechariah 9:5 along with Gaza, and a major city missile target was Ashkelon, just 8 miles north of the Gaza border. Indeed, all within this one verse we learn that the king of Gaza will be killed and Ashkelon (a city of 158,613 in 2023) will be uninhabited. See Figure 9-10. This verse has not yet been fully investigated, but the book of Zechariah also describes a horrendous fate for Jerusalem (and women who get raped) before God finally takes vengeance on nations (and even Jews) that come against Israel.

Figure 9-10 above is from Zechariah 9:5. Ashkelon was not destroyed yet, but it was hit by missiles from Gaza many times since the war began in 2023. The full verse says,

"Ashkelon shall see it and fear, Gaza also, and shall be sore pained, And Ekron, for her expectation shall be ashamed; And the king shall perish from Gaza, and Ashkelon shall not be inhabited." Reptilian shares a letter zayin with the first Gaza.

If Ashkelon is evacuated, the Zechariah 9:5 Reptilian-encoded prediction will be proven true against odds of over 534 billion to 1. There is a long list of Hamas **leaders who were assassinated by Israel here.** While about 150 missiles were fired at Ashkelon on one day (October 10, 2023) in the first week of this war, there were no injuries, but Israel came under missile attack from Iran again. That the city may need to be evacuated may point to something like a dirty bomb or biochemical weapon in the future. It is believed that the Iranians managed to smuggle weapons- grade uranium out of the Fordow site before American B-2 bombers destroyed it. That Hamas leader **SINWAR** was born in Ashkelon but had to leave it when Israel won its independence in 1948 may have motivated him to commit such a **SIN** in this **WAR**, but Israel has killed him and his brother. The Code rightly predicted death from the order by Netanyahu. This will be shown later below with Figure 9-18.

The rape is described in Zechariah 14:1 which states,

"Behold, a day of the LORD comes, When your spoil shall be divided in the midst of you, For I will gather all nations against Jerusalem to battle; And the city shall be taken, And the women ravished; And half of the city shall go forth into captivity, But the residue shall not be cut off from the city."

What a mess! The rape so far was only in the ~25 towns taken by Hamas on October 7, 2023, but it sounds like we have more treatment like it still to come, including, as we saw in the mass kidnappings on October 7. It seems a bit out of sequence, but before we were told about rape and kidnap in Zechariah 14:1, we cautioned about consequences in Zechariah 12:9:

"And it shall come to pass in that day, that I will seek to destroy all the nations that come against Jerusalem."

Haniyeh was killed as predicted. Haniyeh, leader of Hamas, was declared dead on July 31, 2024. The axis term below is **HANIYEH GAZA**. Crossing it in pink is **IN SORROW TO HIS GRAVE** and **EVIL. IRAN,** the site of his death, is at skip +1 sharing a letter *nun* with his name in the axis term. On the same line, the text reads **FOR MAN SHALL NOT SEE ME AND LIVE**. That seems about right for a terrorist. **GUILT** touches and is directly under **IRAN**. The open text finishes off with **HIS BLOOD SHALL BE DASHED.** Right on! A planted explosive blew him to hell in a million pieces. While I usually don't show 2-letter words, on the day that Haniyeh was having his funeral, Israel announced that their bombs three weeks earlier also killed Mohammed **DEIF**, who was the major planner for the October 7, 2023, slaughter of 1,200 Israelis. **DEIF** is at the same absolute skip as **HANIYEH GAZA.**

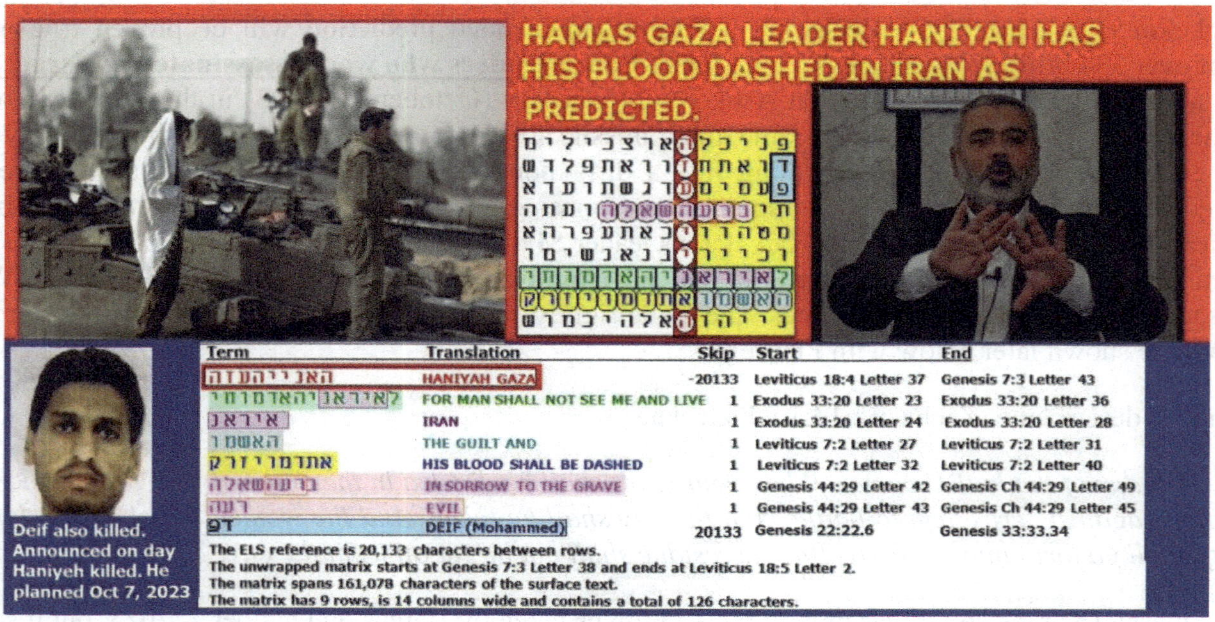

Figure 9-11 above was found by me many years ago, except that I did not look for Mohammed Deif until the day that Haniyeh was blown apart.

SPREADSHEET FOR FIGURE 9-11. As the spreadsheet below indicates, it was found against odds of about 39,706,500,000,000 to 1 (over 39 trillion to 1).

	A	B	C	D	E	F	G	H
1	AXIS TERM: HANIYEH/GAZA @ Skip 6,492	SKIP	NUMBER	DIVIDE BY	THE QUOTIENT	E QUOTIENT	POISSON EQUATION	EACH WORD
2	ELS RANK 1	USED	IN	304805	EQUALS	X 126 LETTERS	PROBABILITY FOR	OR PHRASE
3		ON	TORAH	LETTERS	FREQUENCY	EQUALS	THE WORD	HAS 1 CHANCE
4		FIGURE		IN THE	PER LETTER	WORD	APPEARING	IN THE
5				TORAH		EXPECTANCY	AT LEAST ONCE	FOLLOWING OF
6								BEING HERE
7	IN SORROW TO THE GRAVE +1	1	1	304805	3.28079E-06	0.000413379	0.000413294	2419.587336
8	HIS BLOOD SHALL BE DASHED +1	1	1	304805	3.28079E-06	0.000413379	0.000413294	2419.587336
9	IRAN @ Skip +1 only	1	2	304805	6.56157E-06	0.000826758	0.000826416	1210.04372
10	GUILT @ Skip +1 only	1	3	304805	9.84236E-06	0.001240137	0.001239368	806.8625372
11	DEIF (Mohammed) @ Skips +/- 6492 and +/-1	6492	376	304805	0.001233576	0.155430521	0.143953448	6.946690155
12	Combined odds against matrix = h7*h8*h9*h10*h11						2.51848E-14	3.97065E+13

While we can express amazement about predictions that are later proven to be correct, we must not overlook predictions that our government is overrun with hostile aliens that have strong mind control powers.

Rafah, Gaza and Haniyeh

As is shown on the spreadsheet below, Figure 9-11 odds against finding this matrix were about 1.2 billion to 1.

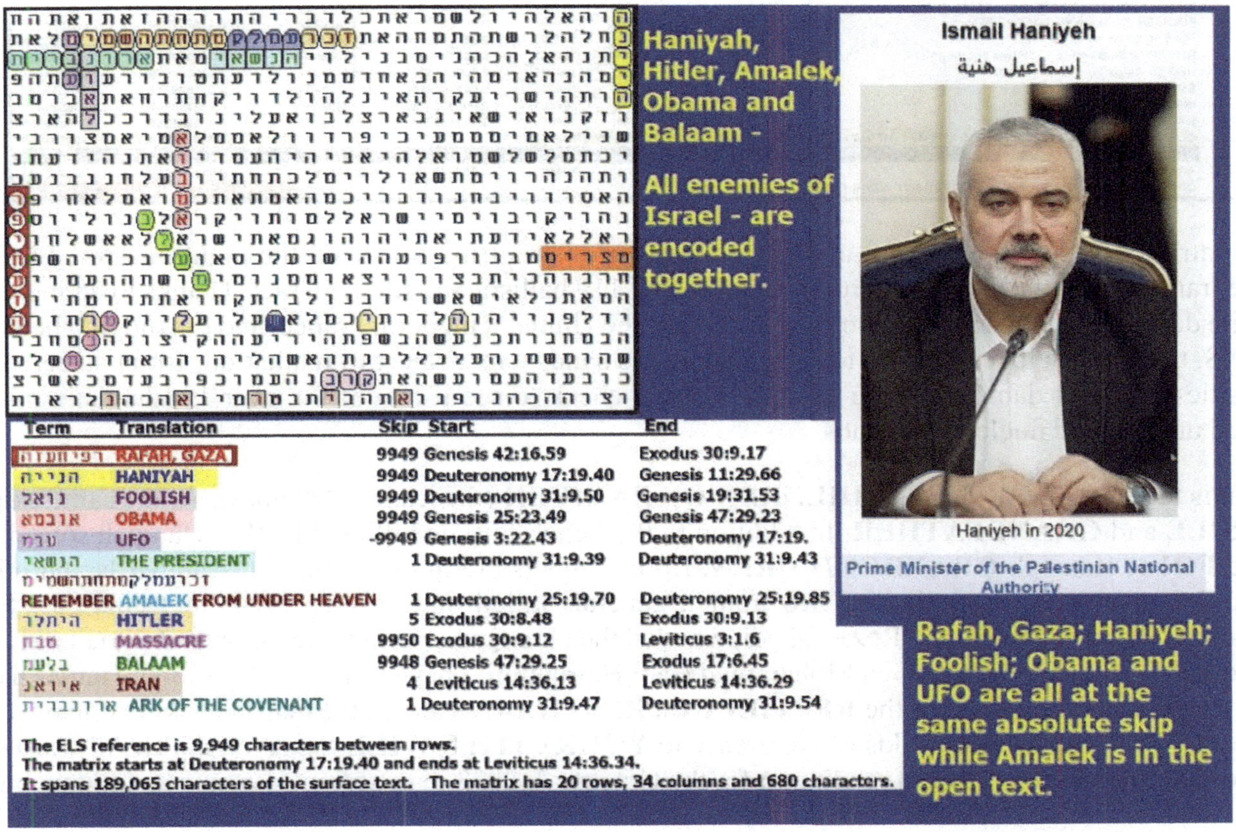

Figure 9-12. The axis term is **RAFAH, GAZA**, which was the last place in Gaza to fall to Israel in its push to destroy Hamas in retaliation for the deadly and perverted attack against Israel on October 7, 2023.

RAFAH, GAZA is in the left column, while its prime minister **HANIYEH** is at the same skip (9,949 letters) in the right column. Between them and again at the same skip is a transcription of **OBAMA. HANIYEH** was foolish in deciding to launch its attack on Israel, just as **OBAMA** was foolish (really, a traitor) in constantly working to help **IRAN** acquire nuclear weapons. The word **FOOLISH** is again at the same skip of 9,947. A fifth term at this same absolute skip (but opposite direction) is **UFO**, which is only really significant if it turns out that, as I suspect, alien technology was used to help bust 6,000 Hamas terrorists into Israel.

PRESIDENT is on the matrix at skip +1 (open text). This matches up with anti-Semitic leaders **OBAMA, AMALEK,** and **HITLER**, plus the gentile prophet **BALAAM**. The last letter of **HITLER** (*resh*) touches two letters of **MASSACRE.** Odds against Figure 9-12 were about 1,225,448,567 to 1.

	A	B	C	D	E	F	G	H
1	*AXIS TERM RAFAH, GAZA @ Skip 9949*	SKIP	NUMBER	DIVIDE BY	THE QUOTIENT	E QUOTIENT	POISSON EQUATION	EACH WORD
2	Axis Term ELS Rank 2	USED	IN	304805	EQUALS	X 680 LETTERS	PROBABILITY FOR	OR PHRASE
3		ON	TORAH	LETTERS	FREQUENCY	EQUALS	THE WORD	HAS 1 CHANCE
4		FIGURE		IN THE	PER LETTER	WORD	APPEARING	IN THE
5				TORAH		EXPECTANCY	AT LEAST ONCE	FOLLOWING OF
6								BEING HERE
7	OBAMA @ Skips +/- 1 and +/-9949 two spellings	9949	6	304805	1.96847E-05	0.013385607	0.013296418	75.20822331
8	HANIYEH @ SKIP +/- 1 and +/- 9949	9949	14	304805	4.5931E-05	0.031233083	0.030750369	32.51993465
9	AMALEK @ @ Skips +1 only	1	21	304805	6.88965E-05	0.046849625	0.045769121	21.84879195
10	PRESIDENT @ Skip +1 only	1	68	304805	0.000223093	0.151703548	0.140757033	7.104440749
11	MASSACRE @ Skips +9950 to +9977	9950	78	304805	0.000255901	0.174012893	0.159713937	6.261194365
12	FOOLISH @ Skips +/- 1 and +/- 9949	9949	88	304805	0.000288709	0.196322239	0.178252607	5.610016131
13	IRAN @ Skips +4 to +/- 155	4	100	304805	0.000328079	0.223093453	0.19995992	5.001002188
14	HITLER @ Skips +5 to +156 two spellings	5	125	304805	0.000410098	0.278866816	0.243359331	4.109150013
15	UFO @ Skips +/- 1 and +/- 9949	-9949	162	304805	0.000531487	0.361411394	0.303307676	3.296982174
16	EGYPT @ Skip +1 only	1	365	304805	0.001197487	0.814291104	0.557046776	1.795181379
17	ISRAEL @ Skip +1 only	1	591	304805	0.001938945	1.318482308	0.732458961	1.365264203
18	BALAAM @Skips +9948 to 10245	9948	1048	304805	0.003438264	2.338019389	0.903481385	1.106829666
19	COMBINED SIGNIFICANCE (NO AXIS TERM) = (H7*H8*H9*H10*H11*H12*H13*H14*H15*H16*H17*H18)/2						8.16028E-10	1225448567
20								
21	ARK OF THE COVENANT	1	6	304805	1.96847E-05	0.013385607	0.013296418	75.20822331
22	COMBINED ODDS IF ARK OF THE COVENANT WERE FOUND *A PRIORI*						1.08502E-11	92163809476

The matrix on Figure 9-13 was found against odds of over a trillion to 1 on the night of April 13, 2024, while Iran was firing 350 drones, cruise missiles, and 119 ballistic missiles at Israel. Thanks to an incredible missile defense system and some help from Jordan, the United Kingdom, France, and begrudging help from the U.S. that came with conditions to try to tie Israel's hands with respect to retaliation, Israel suffered zero casualties or major damage. There is still a need to check to see if any of the over 100 ballistic missiles carried unexploded nuclear warheads.

The axis term on Figure 9-13 is **CHILDREN OF HANIYEH** (the leader of Hamas). At the same skip are **MISSILE** and **GRANDFATHER**. In addition to the Israeli missile killing **THREE** Haniyeh sons, it also killed **FOUR** Haniyeh **GRANDCHILDREN.** Both numbers are in the open text. Another Israeli missile killed 7 Iranian officers, including two Iranian generals killed on April 1, 2024, in Damascus, Syria, including General Mohammad Reza Zahedi, who led the elite Quds Force in Lebanon and Syria until 2016. It also killed Zahedi's deputy, Gen Mohammad Hadi Hajriahimi. I have written before about Hamas having links to an alien race known as the **REPTILIANS. REPTILIANS** are on the matrix at skip -1 against odds of over 307 to 1. At the same odds is a reference to **YOUR CITIES WILL BE A WASTE.** All Gaza cities are now destroyed as a result of retaliation for the October 7, 2023 massacre and/or rape and kidnapping of over 1,200 Israelis.

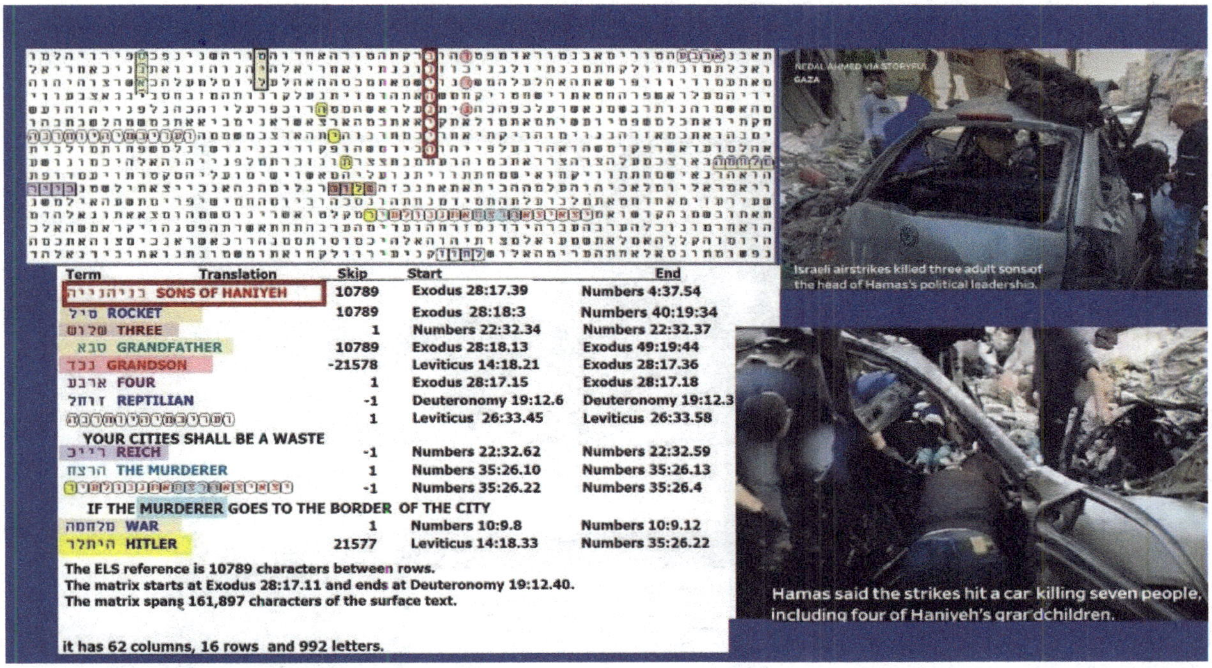

Figure 9-13 is above, and its spreadsheet shows that odds against finding this matrix are over

1.049 TRILLION TO 1.

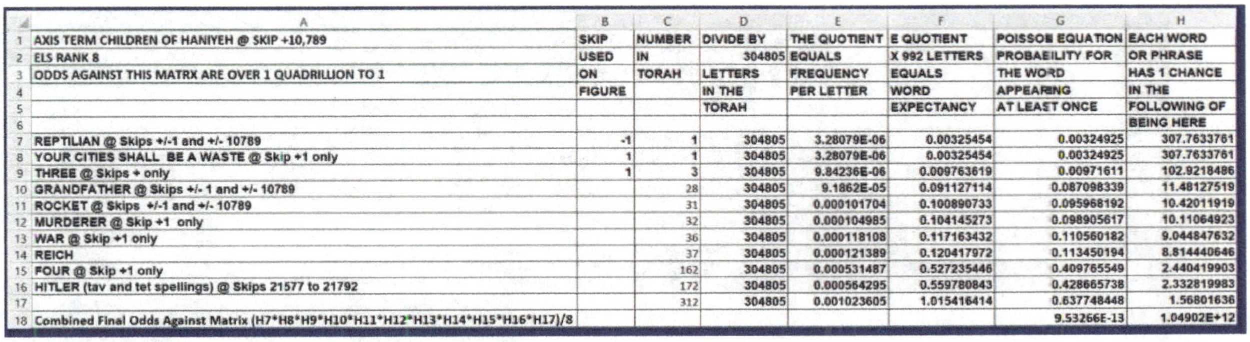

	A	B	C	D	E	F	G	H
1	AXIS TERM CHILDREN OF HANIYEH @ SKIP +10,789	SKIP	NUMBER	DIVIDE BY	THE QUOTIENT	E QUOTIENT	POISSON EQUATION	EACH WORD
2	ELS RANK 8	USED	IN	304805	EQUALS	X 992 LETTERS	PROBABILITY FOR	OR PHRASE
3	ODDS AGAINST THIS MATRX ARE OVER 1 QUADRILLION TO 1	ON	TORAH	LETTERS	FREQUENCY	EQUALS	THE WORD	HAS 1 CHANCE
4		FIGURE		IN THE	PER LETTER	WORD	APPEARING	IN THE
5				TORAH		EXPECTANCY	AT LEAST ONCE	FOLLOWING OF
6								BEING HERE
7	REPTILIAN @ Skips +/-1 and +/- 10789	-1	1	304805	3.28079E-06	0.00325454	0.00324925	307.7633761
8	YOUR CITIES SHALL BE A WASTE @ Skip +1 only	1	1	304805	3.28079E-06	0.00325454	0.00324925	307.7633761
9	THREE @ Skips + only	1	3	304805	9.84236E-06	0.009763619	0.00971611	102.9218486
10	GRANDFATHER @ Skips +/- 1 and +/- 10789		28	304805	9.1862E-05	0.091127114	0.087098339	11.48127519
11	ROCKET @ Skips +/-1 and +/- 10789		31	304805	0.000101704	0.100890733	0.095968192	10.42011919
12	MURDERER @ Skip +1 only		32	304805	0.000104985	0.104145273	0.098905617	10.11064923
13	WAR @ Skip +1 only		36	304805	0.000118108	0.117163432	0.110560182	9.044847632
14	REICH		37	304805	0.000121389	0.120417972	0.113450194	8.814440646
15	FOUR @ Skip +1 only		162	304805	0.000531487	0.527235446	0.409765549	2.440419903
16	HITLER (tav and tet spellings) @ Skips 21577 to 21792		172	304805	0.000564295	0.559780843	0.428665738	2.332819983
17			312	304805	0.001023605	1.015416414	0.637748448	1.56801636
18	Combined Final Odds Against Matrix (H7*H8*H9*H10*H11*H12*H13*H14*H15*H16*H17)/8						9.53266E-13	1.04902E+12

Nasrallah Gets Toasted 10/8/2024

FIGURE 9-14 below. Four **NASRALLAH** matrices are shown. In the top one, **NASRALLAH** at skip -60235 touches **NETANYAHU** (skip +1), matches up and shares a letter lamed with **ISRAEL** (open text, skip +1), and has the order **TO KILL** him (open text).

On the second matrix, **NASRALLAH** at skip 65,744 is encoded crossing and sharing a letter *lamed*

with a terrorist home in **THE LEBANON** in the open text.

Third matrix on Figure 9-15: **NASRALLAH** and **BIDEN** are encoded at Skips +73,579 or - 73,579. NETANYAHU is at Skip +1.

The fourth matrix above matches **NASRALLAH** at skip -9,185 with **REICH** at the same skip and **REPTILIAN** at skip +9,185.

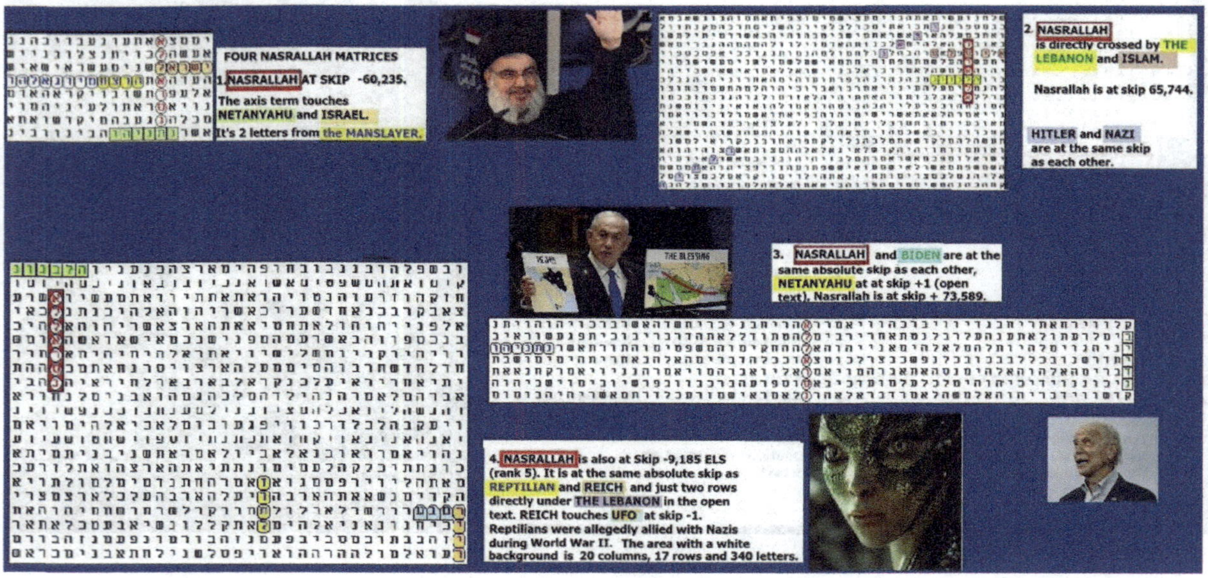

FIGURE 9-14 above: The Reptilian aliens are encoded with far more than **NASRALLAH**. As is shown in **Figure 9-15 below, OBAMA** is encoded as being a **REPTILIAN.**

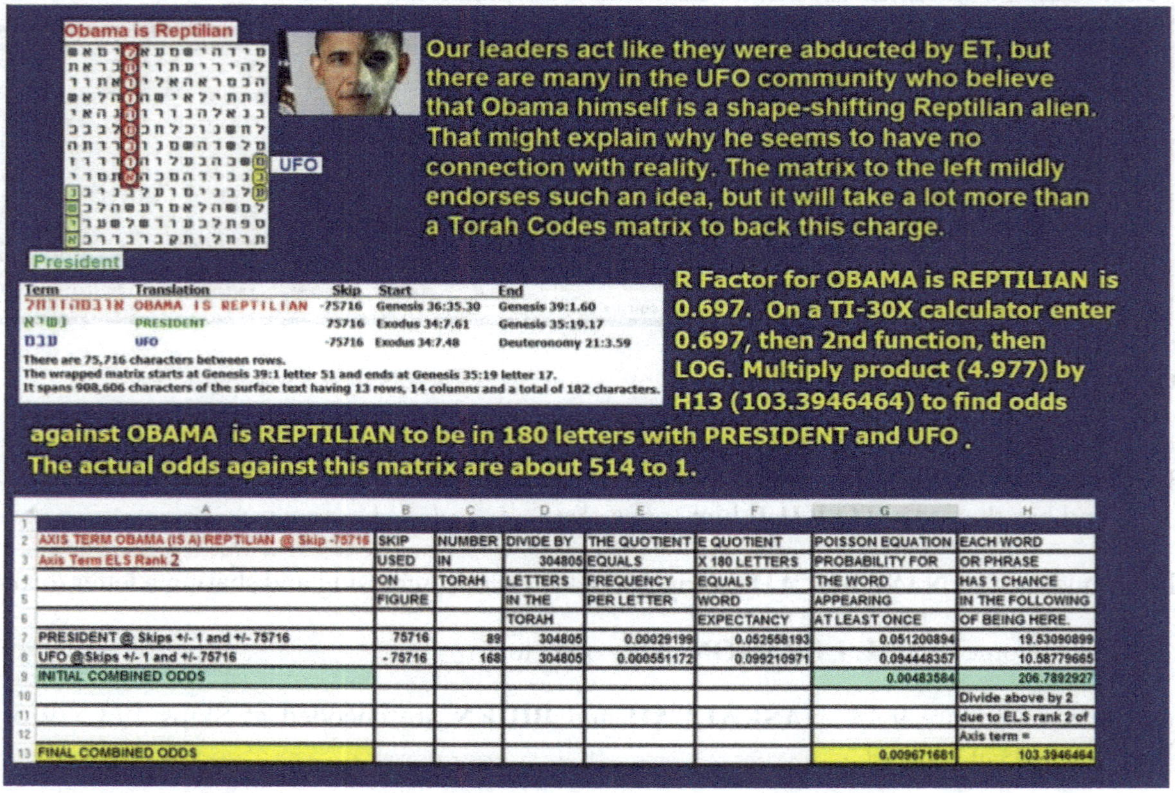

	A	B	C	D	E	F	G	H
1								
2	AXIS TERM OBAMA (IS A) REPTILIAN @ Skip -75716	SKIP	NUMBER	DIVIDE BY	THE QUOTIENT	E QUOTIENT	POISSON EQUATION	EACH WORD
3	Axis Term ELS Rank 2	USED	IN	304805	EQUALS	X 180 LETTERS	PROBABILITY FOR	OR PHRASE
4		ON	TORAH	LETTERS	FREQUENCY	EQUALS	THE WORD	HAS 1 CHANCE
5		FIGURE		IN THE	PER LETTER	WORD	APPEARING	IN THE FOLLOWING
6				TORAH		EXPECTANCY	AT LEAST ONCE	OF BEING HERE.
7	PRESIDENT @ Skips +/- 1 and +/- 75716	75716	89	304805	0.00029199	0.052558193	0.051200894	19.53090899
8	UFO @Skips +/- 1 and +/- 75716	- 75716	168	304805	0.000551172	0.099210971	0.094448357	10.58779665
9	INITIAL COMBINED ODDS						0.00483584	206.7892927
10								Divide above by 2
11								due to ELS rank 2 of
12								Axis term =
13	FINAL COMBINED ODDS						0.009671681	103.3946464

So is the Democratic nominee for Vice President is shown on Figure 9-16 below. It only takes 28 letters to show him intersect REPTILIAN at skip -1.

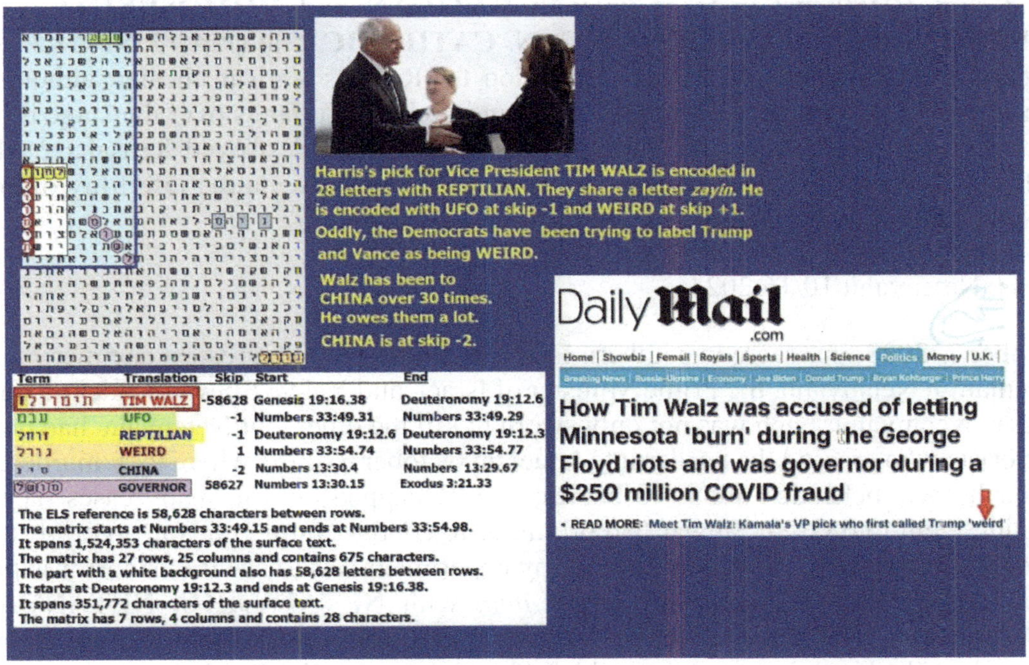

Figure 9-16 above. Evidence that the Democratic nominee for vice president in 2024 is a reptilian. Figure 9-17 below links CIAKAAHRR to GAZA, VATICAN, Catholic and ALIEN GOD.

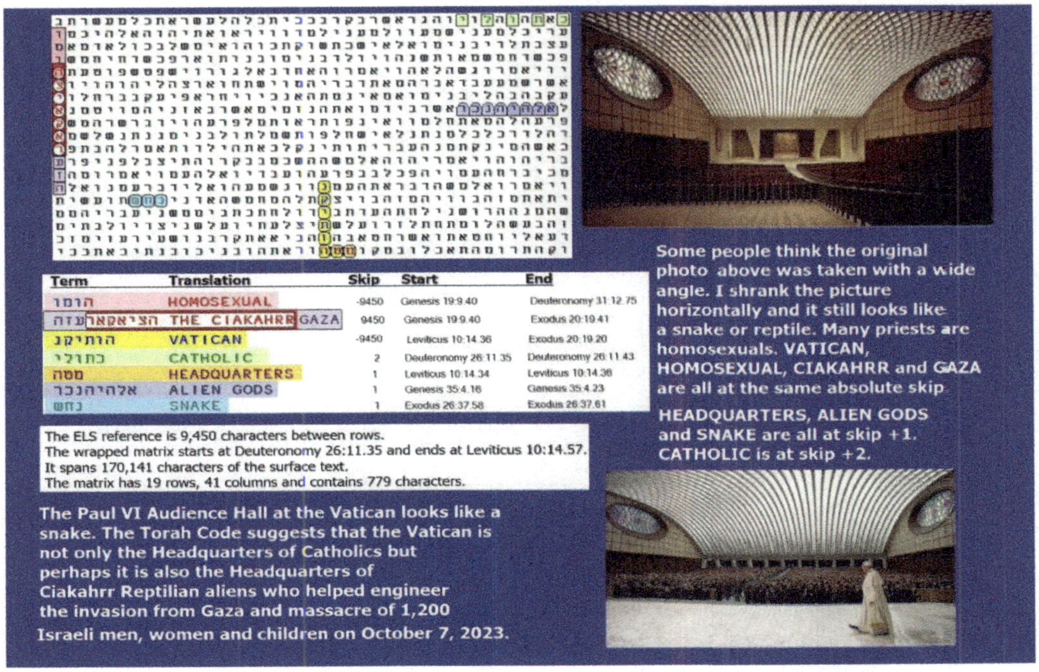

Figure 9-17 below: As I discussed in Chapter 2, there is an ELS train with, **GAZA, and The CIAKAHRR** (Reptilians) at skip +9450 that matches up with **VATICAN and HOMOSEXUAL** At skip -9450. **HEADQUARTERS** at skip +1 touches **VATICAN. CATHOLIC** is at skip +2. **SNAKE** is in the open text. This satanic figure was chosen as the inspiration for Rome's Vatican theater, although half of the building is in Vatican City and the other half is in Rome. Behind the fangs over the stage is an image that is supposed to represent Jesus emerging from the nuclear ashes of a destroyed Jerusalem! All that is seen here is a perfect picture of the Reptilian agenda. Annihilate humanity via the infertility of homosexuality and then nuke the survivors, especially Jews.

Why Sinwar vs. Netanyahu 10/16/2024

After the October 7, 2023, massacre in southern Israel, probably the two greatest enemies of each other on Earth were Benjamin Netanyahu, the Prime Minister of Israel, and Yahya Sinwar, leader of Hamas in Gaza, He was killed by Netanyahu. Each was not only sworn to kill the other, but Netanyahu had promised to kill all Hamas terrorists who crossed the border into Israel on October 7, 2023 where they murdered over 1,200 Israelis in their homes, beheaded about 40 babies, and kidnapped (and in many cases raped) 250 more Israelis, dragging them into Gaza. Sinwar had promised to conduct future raids in an effort to exterminate all Jews in Israel. On Figure 9-19, the purpose of my experiment was to see if this horror is prerecorded in Torah. Sadly, it is. **SINWAR** shares his letter *nunn* with **NETANYAHU. NETANYAHU** touches **BENJAMIN** at 4 letters on the line above **NETANYAHU. SINWAR** is at the same skip as **IDOL. IDOLATRY** is described in the story of the Gold Calf on the matrix. The Torah records that God killed Jews for worshiping idols. More on this later, but it appears that God may have used Sinwar to kill Jews violating Jewish Laws at the music festival on the Sabbath and Simchat Torah. Most notable was that a large Buddha idol overlooked the center of the mayhem.'

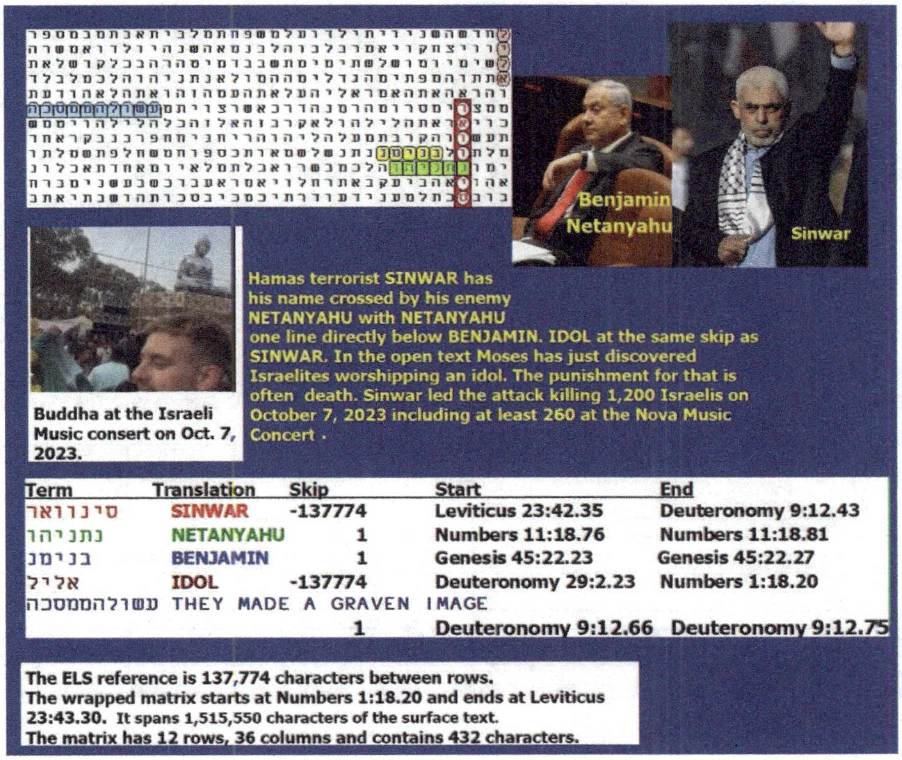

FIGURE 9-18 is above. Its spreadsheet is below. Odds against finding this matrix were about 82,795 to 1.

	A	B	C	D	E	F	G	H
1	*AXIS TERM SINWAR @ Skip -133,774*	SKIP	NUMBER	DIVIDE BY	THE QUOTIENT	E QUOTIENT	POISSON EQUATION	EACH WORD
2	Axis Term ELS Rank 106	USED	IN	304805	EQUALS	X 442 LETTERS	PROBABILITY FOR	OR PHRASE
3		ON	TORAH	LETTERS	FREQUENCY	EQUALS	THE WORD	HAS 1 CHANCE
4		FIGURE		IN THE	PER LETTER	WORD	APPEARING	IN THE
5				TORAH		EXPECTANCY	AT LEAST ONCE	FOLLOWING OF
6								BEING HERE
7	MADE A GRAVEN IMAGE@ Skip +1 only	1	1	304805	3.28079E-06	0.001450107	0.001449057	690.1041932
8	NETANYAHU @ SKIP +/- 1 and +/- 133774	-1	16	304805	5.24926E-05	0.023201719	0.022934629	43.60218798
9	BENJAMIN@ Skips +1 only	1	31	304805	0.000101704	0.044953331	0.043957902	22.74903864
10	IDOL @ @ Skips +/- 1 and +/- 133774	-133774	56	304805	0.000183724	0.081206017	0.077996277	12.82112486
11	COMBINED SIGNIFICANCE (NO AXIS TERM) = (H7*H8*H9*H10)/106						1.2078E-05	82795.41018

FORMAT OF ENCODING AND THE PERSONALITY OF THE ENCODER. This is not

the first time that I have seen the first name of a man on one line and his second name on the next row down, pretty much letter for letter in the same columns, except where there is a slight difference in letter count for each name. Here we have four of the five letters of **BENJAMIN** (*bet, nun, yud, mem*) touch four of the six letters of **NETANYAHU** (*nun, yud, hey, vav*). In an earlier example, I found all four letters of the first name of an intelligence agent (Joseph) from Fort Huachuca, Arizona, on the first line touch four of the letters of his second name on the next line down. See Figure 9-20, upper left corner.

	A	B	C	D	E	F	G	H
1	Axis term SIGNAL FROM FUTURE @ Skip +41,365	SKIP	NUMBER	DIVIDE BY	THE QUOTIENT	E QUOTIENT	POISSON EQUATION	EACH WORD OR
2	ELS Rank 4	USED	IN	304805	EQUALS	X 152 LETTERS	PROBABILITY FOR	PHRASE HAS 1
3	Although COLD and CONTROLLER share a letter *qof* with ROCK they are not	ON	TORAH	LETTERS	FREQUENCY	EQUALS	THE WORD	CHANCE IN THE
4	counted. COLD is only 2 letters and CONTROLLER was only found	FIGURE		IN THE	PER LETTER	WORD	APPEARING	IN THE FOLLOWING
5	*a posteriori*. Note: ROCK here is only associated with music but like English names			TORAH		EXPECTANCY	AT LEAST ONCE	OF BEING HERE.
6	transliterations are allowed for English words.							
7	"*SAPPHIRE*" @ Skip +1 only	1	3	304805	9.84236E-06	0.001496038	0.00149492	668.9321422
8	G. BEAR (gimel, then Hebrew for BEAR) @ Skip +1 only	1	7	304805	2.29655E-05	0.003490756	0.003484671	286.9711556
9	REICH @ Skips +/- 1 and +/- 41365	1	39	304805	0.000127951	0.0194485	0.019260598	51.9194682
10	THE PLAGUE or PLAGUE @ Skip +1 only (3 synonyms)	1	71	304805	0.000232936	0.035406243	0.034786775	28.74655683
11	BOOK @ Skip +1 only	1	104	304805	0.000341202	0.051862666	0.050540749	19.78601451
12	JOSEPH @ Skip +1 only	1	180	304805	0.000590541	0.089762307	0.085851554	11.64801281
13	ROCK @ Skips +/- 1 and +/- 41365	41365	379	304805	0.001243418	0.188999524	0.1722131	5.8067592
14	COMBINED ODDS AGAINST = (H7*H8*H9*H10*H11*H12*H13)/4						1.04323E-11	95856235315

Figure 9-19 above: **JOSEPH SAPPHIRE** seemed to have info from the future.

This method of encoding may be used to emphasize extremely important information. The Joseph encoded here with his second name is on a matrix where the axis term is **SIGNAL FROM FUTURE**. He gave me information about my son's future, and he may have even indicated that he was capable of time travel.

It is well worthwhile to compare the attacking force given to Moses to kill fellow Jews after the sin of their worshiping the golden calf, with somewhere between 5,600 and 6,000 used by Sinwar to murder Jews in the villages near Gaza. See Exodus 32:26 to 32:28:

26 *then Moses stood in the gate of the camp, and said, "Who is on the Lord's side? Come to me." And all the sons of Levi gathered themselves together to him.* 27 *And he said to them, "Thus says the Lord God of Israel, 'Put every man his sword on his side, and go to and fro from gate to gate throughout the camp, and slay every man his brother, and every man his companion, and every man his neighbor.'" 28 And the sons of Levi did according to the word of Moses; and there fell of* the people that day about three thousand men.

The fact that there were ill-clad women dancing about at the concert, attacked by Sinwar, is also reminiscent of the plan of Balaam to get God to kill Jews via a plague. See Numbers 25:1 to 9:

1 *While Israel dwelt in Shittim the people began to play the harlot with the daughters of Moab.* 2
These invited the people to the sacrifices of their gods, and the people ate, and bowed down to
their gods. 3 *So Israel yoked himself to Ba'al of Pe'or. And the anger of the Lord was kindled*
against Israel; 4 *and the Lord said to Moses,*

"Take all the chiefs of the people, and hang them in the sun before the Lord, that the fierce anger of the Lord may turn away from Israel." [5] *And Moses said to the judges of Israel, "Every one of you slay his men who have yoked themselves to Ba'al of Pe'or."*

[6] *And behold, one of the people of Israel came and brought a Mid'ianite woman to his family, in the sight of Moses and in the sight of the whole congregation of the people of Israel, while they were weeping at the door of the tent of meeting.* [7] *When Phin'ehas the son of Elea'zar, son of Aaron the priest, saw it, he rose and left the congregation, and took a spear in his hand* [8] *and went after the man of Israel into the inner room, and pierced both of them, the man of Israel and the woman, through her body. Thus the plague was stayed from the people of Israel.* [9] *Nevertheless those that died by the plague were twenty-four thousand.*

Having stated the nature of Israel's sins above, the sins of Israel on October 7, 2023, are shown on the video link at https://twitter.com/nexta_tv/status/1710956060469784836. They should be examined as should Figure 9-20 below. Messages like this tend to support the idea that a jealous God rather than a cold extraterrestrial or Artificial Intelligence encoded the Torah.

Figure 9-20 The Jews killed on October 7, 2023, were not all innocent. A large Buddha may have had something to do with why God left them naked before the very fatal assault.

Israeli War Games Designed to Help Fight Aliens.

It's only for training purposes, but the Israel Defense Forces do practice fighting aliens like Grays and Reptilians

Several months ago a flock of alien spaceships were on their way to invade the planet when advanced Israeli radar systems identified them. It was then up to the technological unit of the Israeli Air Force to thwart the dangerous alien invasion by infiltrating the main spaceship's technological systems and attacking them from within. The soldiers only had 12 hours to hack the systems, gather intelligence and stop the aliens.

This unique (and fabricated) scenario was part of the Israel Air Force's cyber defense awareness training. Each squad within the 'Ofek' Technological Unit of the IAF sent a team of programmers to compete in the event, that took place this year for the sixth time. The teams competed over hacking a previously built system that resembled an alien spaceship while going through different stages of attacking the system.

The objective of the event was to let the IAF programmers step into the shoes of their enemies for a day and experience what it's like to hack and attack complex technological systems. "The participants in the event develop operational technological systems for the Air Force on a daily basis," explained the commander of the engineering unit, Lt. Col. Meir. "Our goal was to help them understand the importance of secured program development. We wanted to adapt the training to the young and ambitious members of 'Generation Y', therefore we created the story line of an alien invasion to increase their interest."

Figure 9-21 above. Semper paratus. After meeting and working with aliens (according to Israeli Brigadier General Haim Eshed, discussed earlier), Israel began war games aimed at defeating alien invaders. The U.S. established a new branch of its military, the Space Force.

Figure 9-22 Reptilian enemies introduce themselves to Iran. I have shown in this chapter that aliens (most probably reptilians) likely gave Hamas the technology that it needed to invade Israel on October 7, 2023. If this were true, it is logical to assume that Reptilians would also be open to helping Iran get missiles and other arms from Iran into Gaza. Figure 9-22 lends credence to this idea. On this Figure, the axis term **REPTILIAN ENEMY** touches **IRAN** at skip +1. Iran appears at this best-case skip only twice in Torah. **WAR** in the open text goes through and shares a letter with the **REPTILIAN ENEMY. PRESIDENT** is at skip +1. **BARACK** (Obama) is at skip -1, and so is UFO. There was only about one chance in 90,905,743 to find a matrix.

Iranian jet fighters have attempted to intercept UFOs in the past, with one F-14 in January 2012 that exploded, killing both crewmen. No cause for the tragedy was determined. However, there has been plenty of time for Iranians to patch up early contact problems. Reptilians love Nazis and hate Jews. Therefore, reptilians, in addition to supporting Hamas, likely also love radical Iran. Barack Obama (and Biden) permitted Iran to enrich enough uranium to make nine nuclear bombs. **OBAMA** is encoded with the **ATOMIC HOLOCAUST**. See Figure 9-23.

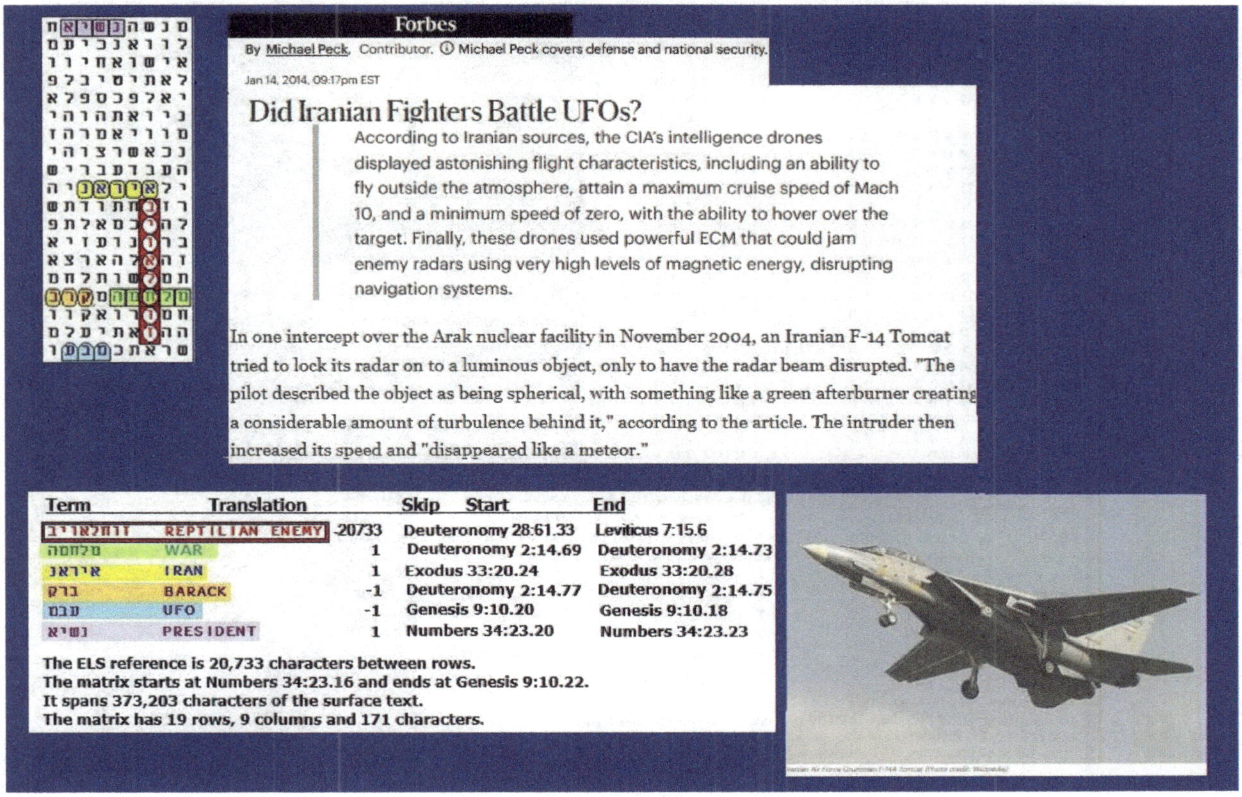

Term	Translation	Skip	Start	End
וזוחלאויב	REPTILIAN ENEMY	-20733	Deuteronomy 28:61.33	Leviticus 7:15.6
מלחמה	WAR	1	Deuteronomy 2:14.69	Deuteronomy 2:14.73
איראן	IRAN	1	Exodus 33:20.24	Exodus 33:20.28
ברק	BARACK	-1	Deuteronomy 2:14.77	Deuteronomy 2:14.75
עבמ	UFO	-1	Genesis 9:10.20	Genesis 9:10.18
נשיא	PRESIDENT	1	Numbers 34:23.20	Numbers 34:23.23

The ELS reference is 20,733 characters between rows.
The matrix starts at Numbers 34:23.16 and ends at Genesis 9:10.22.
It spans 373,203 characters of the surface text.
The matrix has 19 rows, 9 columns and 171 characters.

Figure 9-22 is above, and its probability spreadsheet is below. While **IRAN** was found at skip +1 on the 171-letter matrix against odds of about 891 to 1, it was actually found with **REPTILIAN ENEMY** in 45 letters against odds of about 3,387 to 1.

	A	B	C	D	E	F	G	H
1	Axis term: REPTLIAN ENEMY @ SKIP -20733	SKIP	NUMBER	DIVIDE BY	THE QUOTIENT	E QUOTIENT	POISSON EQUATION	EACH WORD OR
2	ELS RANK: 1	USED	IN	304805	EQUALS	X 171 LETTERS	PROBABILITY FOR	PHRASE HAS 1
3		ON	TORAH	LETTERS	FREQUENCY	EQUALS	THE WORD	CHANCE IN THE
4		FIGURE		IN THE	PER LETTER	WORD	APPEARING	IN THE FOLLOWING
5				TORAH		EXPECTANCY	AT LEAST ONCE	OF BEING HERE.
6	IRAN @ Skips +1 only	1	2	304805	6.56157E-06	0.001122029	0.0011214	891.7427836
7	WAR @ Skips + 1 only	1	35	304805	0.000114828	0.019635505	0.019443984	51.42979
8	PRESIDENT @ Skip +1 only	1	68	304805	0.000223093	0.03814898	0.037430474	26.7161993
9	UFO @ Skips +/- 1 and +/- 20733	-1	149	304805	0.000488837	0.083591148	0.080192756	12.46995425
10	BARACK @ Skip +/- 1 and +/- 20733	1	328	304805	0.001076098	0.184012729	0.168074786	5.949732394
11	INITIAL COMBINED ODDS = H6*H7*H8*H9*H10						1.10004E-08	90905743.18
12								
13	IRAN @ Skip +1 only in a 45-letter matrix	1	2	304805	6.56157E-06	0.000295271	0.000295227	3387.222247

WAS IRAN A NUCLEAR THREAT TO AMERICA BEFORE ISRAEL ATTACKED IRAN? Absolutely. But there may be indications that America (specifically New York) may have someone (Arab) that might hit them with a nuclear bomb. Also, a danger that should not be overlooked is ALIEN. This term is close to **ATOMIC HOLOCAUST** in the open text on Figure 9-25. I have shown that reptilians likely gave Hamas the ability to penetrate Israel on October 7, 2023. Figure 9-26 deals with the New York threat.

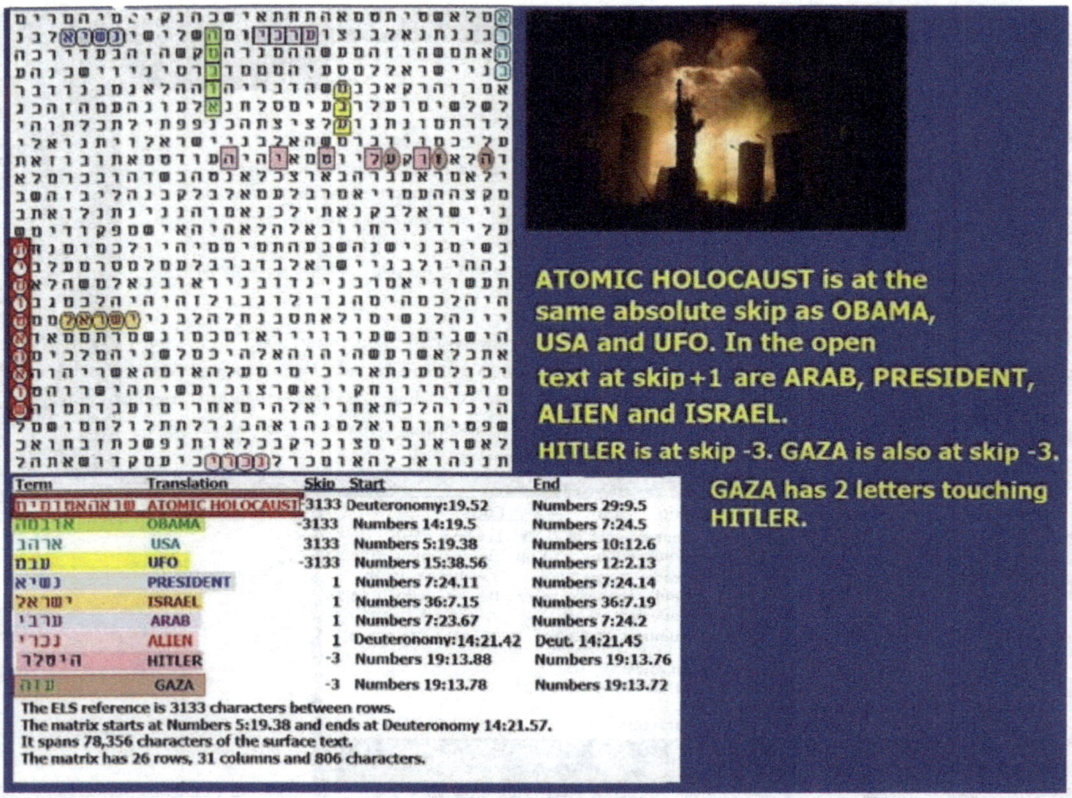

Figure 9-23 above and its spreadsheet below. Without counting the axis term **(ATOMIC HOLOCAUST),** the matrix was found against odds of about 2,782,626,903 to 1. The most significant a priori keyword was **OBAMA**, which was found at a special case skip (the same as the axis term) against odds of about 126 to 1. Normally, I don't count the axis term, but in this case, it's 10 letters long, which is rare. CodeFinder software indicates that it has an R factor of 2.676, which means that there was only one chance in 474 that it would be found at least once in Torah. On multiplying this number by the other odds found, there was actually only about one chance in 1,319,638,507,000 (about 1 in 1.3 trillion) that the matrix would be found. Note that **ATOMIC HOLOCAUST** is also at the same absolute skip as **USA** and **UFO**. Originally, I wrote this while the Trump Administration was trying to restart nuclear talks with Iran. Eventually, when the talks failed, he sent seven B-2 bombers to attack Iran's three top nuclear targets.

	A	B	C	D	E	F	G	H
1	Axis term: ATOMIC HOLOCAUST @ SKIP -3133	SKIP	NUMBER	DIVIDE BY	THE QUOTIENT	E QUOTIENT	POISSON EQUATION	EACH WORD OR
2	ELS RANK: 1	USED	IN	304805	EQUALS	X 806 LETTERS	PROBABILITY FOR	PHRASE HAS 1
3		ON	TORAH	LETTERS	FREQUENCY	EQUALS	THE WORD	CHANCE IN THE
4		FIGURE		IN THE	PER LETTER	WORD	APPEARING	IN THE FOLLOWING
5				TORAH		EXPECTANCY	AT LEAST ONCE	OF BEING HERE.
6	OBAMA @ Skips +/- 1 and +/- 3133	- 3133	3	304805	9.84236E-06	0.007932941	0.007901558	126.5573195
7	ALIEN @ Skip+1 only	1	10	304805	3.28079E-05	0.026443136	0.026096577	38.31920109
8	ARAB @ Skip +1 only	1	16	304805	5.24926E-05	0.042309017	0.041426481	24.1391491
9	HITLER (best spelling) @ Skips -3 to -196	-3	17	304805	5.57734E-05	0.044953331	0.043957902	22.74903864
10	USA @ Skips +/- 1 and +/- 3133	3133	20	304805	6.56157E-05	0.052886272	0.051512124	19.41290574
11	PRESIDENT @ Skip +1 only	1	68	304805	0.000223093	0.179813323	0.164573849	6.076299539
12	UFO @ Skips +/- 1 and +/- 3133	- 3133	105	304805	0.000344483	0.277652926	0.242440295	4.124726881
13	GAZA @ Skips -3 to -30	-3	237	304805	0.000777546	0.626702318	0.465648981	2.1475404
14	ISRAEL @ Skip +1 only	1	591	304805	0.001938945	1.562789324	0.79044925	1.265103358
15	INITIAL COMBINED ODDS = H6*H7*H8*H9*H10*H11*H12*H13						3.59373E-10	2782626903

Figure 9-24 below: The real nuclear nightmare is an atomic threat against New York. NEW YORK and

ATOMIC are at the same skip.

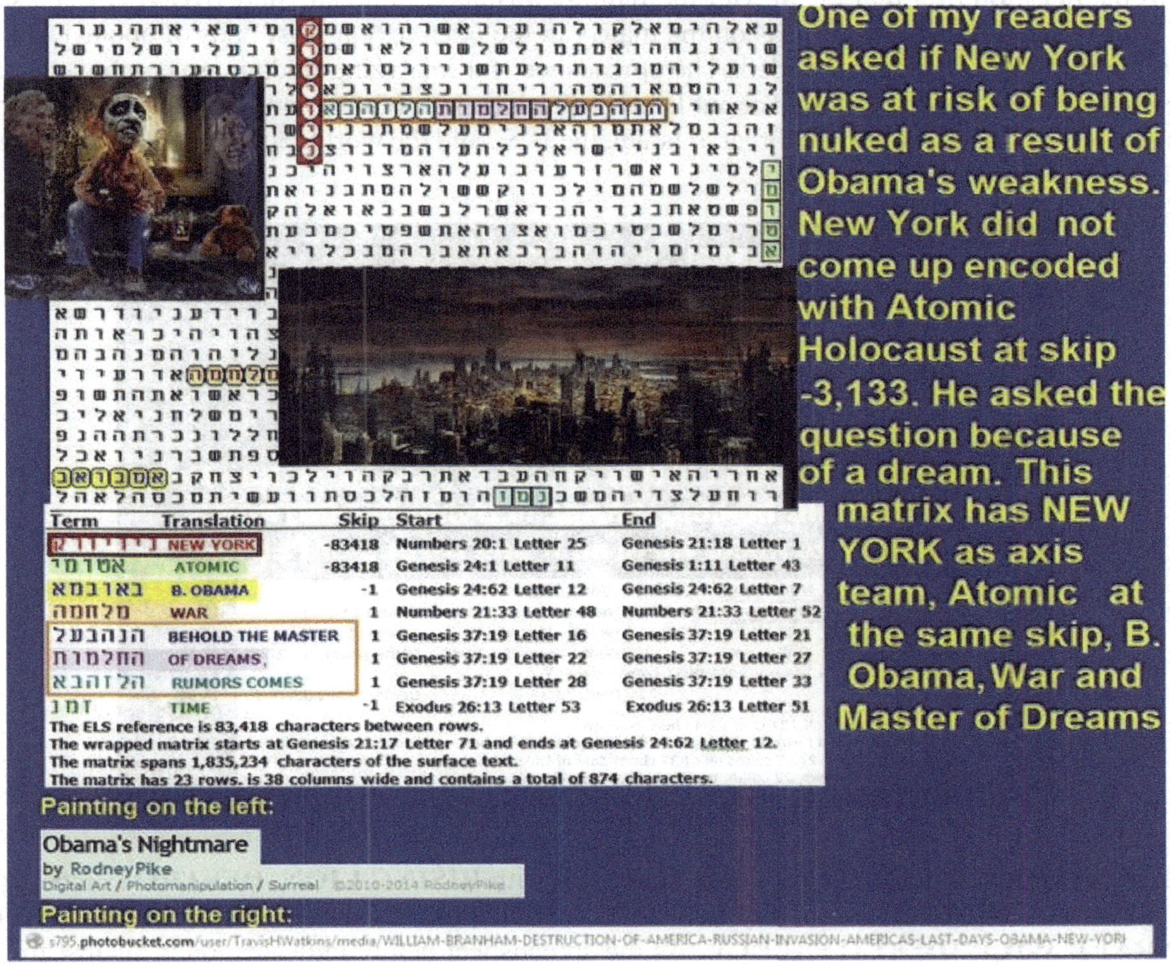

The spreadsheet showing the odds against Figure 9-24 is given below:

	A	B	C	D	E	F	G	H
1	AXIS TERM NEW YORK @ Skip -83,418	SKIP	NUMBER	DIVIDE BY	THE QUOTIENT	E QUOTIENT	POISSON EQUATION	EACH WORD
2	ELS RANK 233	USED	IN	304805	EQUALS	x 874 LETTERS	PROBABILITY FOR	OR PHRASE
3		ON	TORAH	LETTERS	FREQUENCY	EQUALS	THE WORD	HAS 1 CHANCE
4		FIGURE		IN THE	PER LETTER	WORD	APPEARING	IN THE
5				TORAH		EXPECTANCY	AT LEAST ONCE	FOLLOWING OF
6								BEING HERE
7	ATOMIC @ Skips +/- 1 and +/- 83418	-83418	1	304805	3.28079E-06	0.002867407	0.0028633	349.2473785
8	E. OBAMA @ Skips +/- 1 and +/- 83418	-1	1	304805	3.28079E-06	0.002867407	0.0028633	349.2473785
9	WAR @ Skips +1 only	1	36	304805	0.000118108	0.103226653	0.098077473	10.19602124
10	TIME @ Skips +/- 1 and +/- 83418	-1	27	304805	8.85812E-05	0.07741999	0.074498929	13.42301175
11	Combined odds against the matrix = (h7*h8*h9*h10)/233						1.39575E-05	71645.85843

Here, the axis term is **NEW YORK**, but only at its 233rd lowest skip in the wrapped Torah. **ATOMIC** is at the same skip as **NEW YORK**. The only ELS in Torah of **B. OBAMA** at an absolute skip of 1 (it's 1) is shown. **WAR** is the open text. But is the threat real, or simply a nightmare? Touching NEW **YORK** is **BEHOLD THE MASTER OF DREAMS, RUMORS COME**. There no date has been found yet, although at skip -1 there is a reference to **TIME**. Nor is a nation terrorist group seen on this matrix. Dates are rarely seen; however, on Figure 9-26 below, the month **(SIVAN)** and Hebrew year **(5785)** of Israel's Rising Lion war against Iran are correctly recorded at skip -1.

Overall, the **NEW YORK/ATOMIC** matrix was found against odds of about 71,645 to 1. These odds pertain to the question of a real threat. But the request was about a dream, and **DREAMS** is also on the matrix. Dream, dreams, or dreamer was on the matrix in the open text against odds of about 11.6 to 1, but I don't think a dream would be encoded. The spreadsheet for analyzing odds is given above: Let's look specifically at what was initially found about Israel's attack on Iran (Operation Rising Lion). The operation was designed to eliminate the atomic threat to Israel, the U.S., and the world. So, is **RISING LION** encoded with **ATOMIC?** Yes – see Figure 9-25.

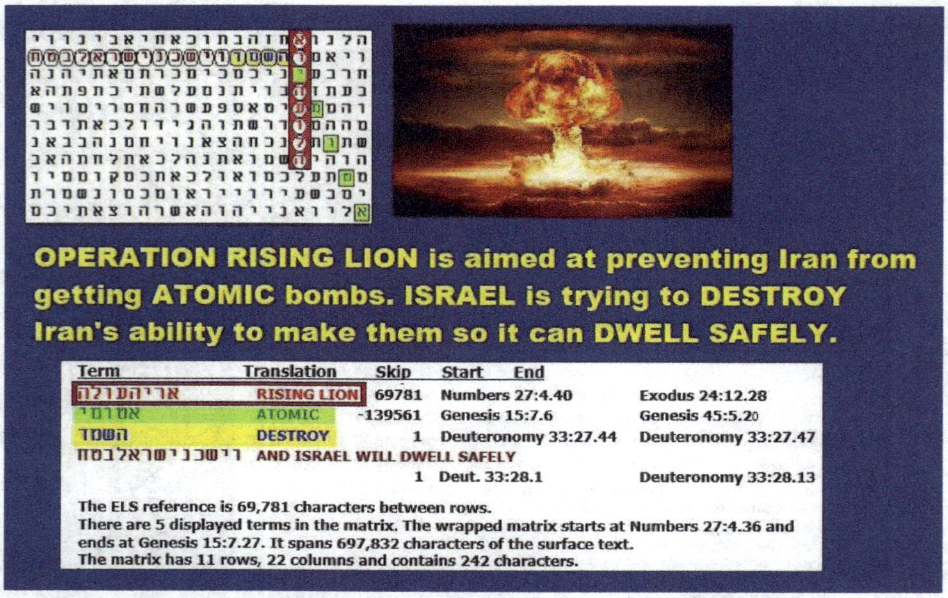

Figure 9-25 **ATOMIC** intersects and shares a letter yud with **RISING LION. DESTROY** in the open text touches **RISING LION** and (diagonally) **ATOMIC**. The reason for the operation follows **DESTROY: AND ISRAEL WILL DWELL SAFELY**. As the spreadsheet below indicates, odds against this matrix are about 473,298 to 1:

	A	B	C	D	E	F	G	H
1	AXIS TERM RISING LION @ Skip +69,781	SKIP	NUMBER	DIVIDE BY	THE QUOTIENT	E QUOTIENT	POISSON EQUATION	EACH WORD
2	ELS RANK 43	ON	TORAH	LETTERS	FREQUENCY	X 242 LETTERS	THE WORD	HAS 1 CHANCE
3		FIGURE		IN THE	PER LETTER	EQUALS	APPEARING	IN THE
4				TORAH		WORD	AT LEAST ONCE	FOLLOWING OF
5						EXPECTANCY		BEING HERE
6	ATOMIC @ Skips -139561 to -139614	-139561	9	304805	2.95271E-05	0.007145552	0.007120083	140.4477947
7	DESTROY @ Skip +1 only	1	11	304805	3.60886E-05	0.008733453	0.008695427	115.0029817
8	ISRAEL WILL DWELL SECURELY @ Skip +1 only	1	1	304805	3.28079E-06	0.00079395	0.000793635	1260.02486
9	Combined odds against the matrix = (H6*H7*H)/43							473298.015

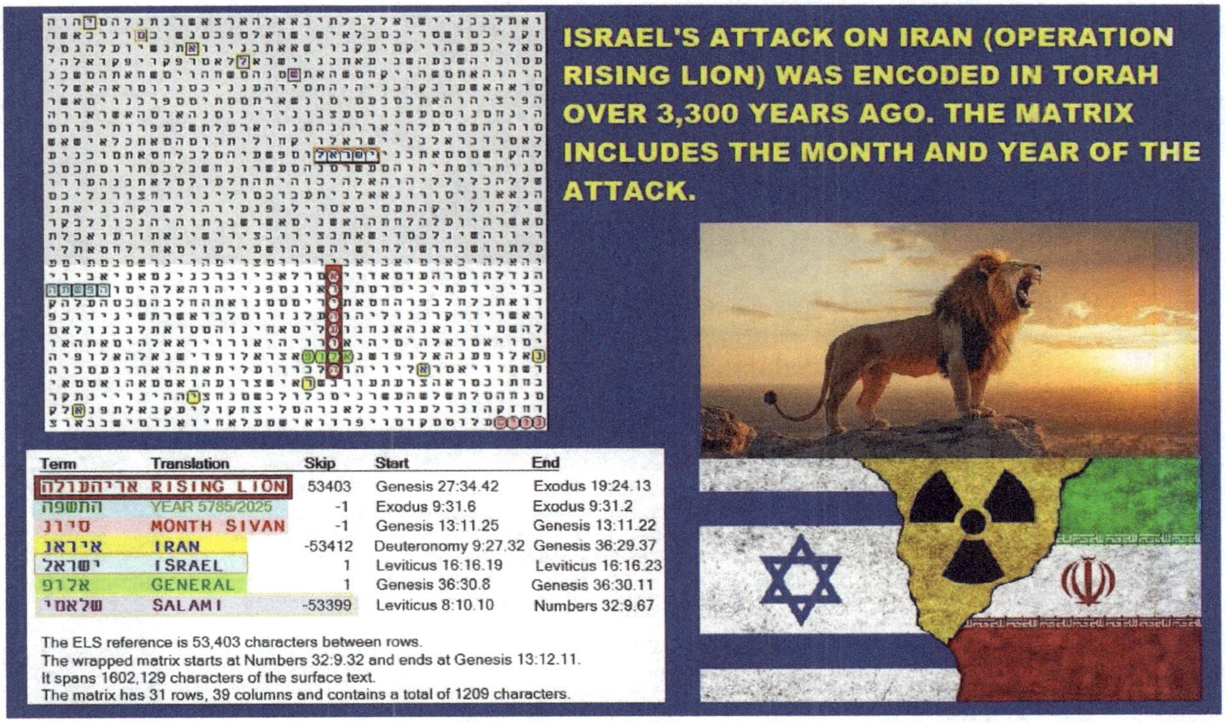

Figure 9-26 above. We are only interested in the bottom 12 rows of this matrix, which give the month and year of the Rising Lion attack on Iran, plus an ELS of **IRAN** and the word **GENERAL** passing through **RISING LION**. Israel killed **GENERAL SALAMI** and many other Iranian generals at the start of the word. There is a **SALAMI** at the top of the matrix, but it is spelled with a shin rather than a samech. Initial odds against having the correct year on the matrix were about 217 to 1, but **RISING LION** is at ELS rank 30. Odds against the entire matrix with the axis term ELS rank included are about 91,255 to 1.

	A	B	C	D	E	F	G	H
1	AXIS TERM RISING LION @ Skip +53,403	SKIP	NUMBER	DIVIDE BY	THE QUOTIENT	E QUOTIENT	POISSON EQUATION	EACH WORD
2	ELS RANK 30	ON	TORAH	LETTERS	FREQUENCY	x 1209	THE WORD	HAS 1 CHANCE
3	This matrix has the month and year of Rising Lion. The name SALAMI	FIGURE		IN THE	PER LETTER	EQUALS	APPEARING	IN THE
4	is not included because it began with a shin rather than a			TORAH		WORD	AT LEAST ONCE	FOLLOWING OF
5	samech and it increased the matrix size to 1,209 letters.					EXPECTANCY		BEING HERE
6	5785/2025 @ Skips +/- 1 and +/- 83418	-1	3	304805	9.84236E-06	0.004606224	0.004595631	217.5979622
7	SIVAN @ Skips +/- 1 and +/- 83418	-1	8	304805	2.62463E-05	0.012283263	0.012208132	81.91261548
8	IRAN @ Skips -53412 to -53505	-53412	70	304805	0.000229655	0.107478552	0.101904215	9.813136751
9	GENERAL @ Skips +1 only	1	43	304805	0.000141074	0.066022539	0.063890235	15.65184415
10	Combined odds against the matrix = (h7*h8*h9*h10)/30						1.09583E-05	91255.22251

Is there a Rising Lion matrix with the preferred spelling of SALAMI? Yes - see Figure 9-29.

Figure 9-27 above. This matrix includes both **SALAMI** and the man who ordered his death – Prime Minister **NETANYAHU** (in the open text). The word **WAR**, at an ELS, touches **RISING LION. ISRAEL** (open text) is two letters right of **RISING LION**. Odds against this matrix are about 1,725 TO 1.

	A	B	C	D	E	F	G	H
1	AXIS TERM RISING LION @ Skip -52,937	SKIP	NUMBER	DIVIDE BY	THE QUOTIENT	E QUOTIENT	POISSON EQUATION	EACH WORD
2	ELS RANK 29	ON	TORAH	LETTERS	FREQUENCY	x 462	THE WORD	HAS 1 CHANCE
3	This matrix has the name SALAMI. He was the head	FIGURE		IN THE	PER LETTER	EQUALS	APPEARING	IN THE
4	of Iran's Revolutionary Guard (i.e., the head of their military)			TORAH		WORD	AT LEAST ONCE	FOLLOWING OF
5	before Israel killed him on the opening day of the war.					EXPECTANCY		BEING HERE
6								
7	SALAMI (with a samach) @ Skips -52925 to -53056	-52925	16	304805	5.24926E-05	0.024251571	0.023959864	41.73646359
8	NETANYAHU @ Skip +1 only	1	16	304805	5.24926E-05	0.024251571	0.023959864	41.73646359
9	WAR @ Skips -52925 to -53056	-52925	40	304805	0.000131231	0.060628927	0.058827581	16.99882916
10	ISRAEL @ Skip 1 only	1	591	304805	0.001938945	0.895792392	0.59171605	1.689999788
11	Combined odds against the matrix = (h7*h8*h9*h10)/29						0.00057951	1725.59533

CHAPTER 10
IS GOD FROM THE FUTURE?

There may be a need to distinguish between Divine signals that are sent back in time versus those of human or alien origin.

Moses said to God,

"Behold, when I come to the children of Israel and say to them, 'The God of your forefathers has sent me to you,' and they say to me, 'What is His Name?' – what shall I say to them?"

God answered Moses,

"I Shall Be As I Shall Be." And He said, "So shall you say to the Children of Israel: 'I Shall Be has sent me to you.'" (Exodus 3:13 to 3:14)

This chapter examines the issue of how accurate predictions, both open text and encoded, came to be written into the Torah. In doing so, I will review aspects of an August 1, 2016 interview of me by a man likely to be an Army Intelligence officer. There is a non-zero chance that this man is also a time traveler, and possibly a time cop. I originally included his name in my *Future Warning* article, then deleted it for the sake of his safety and my own. However, it has come to my attention that the name he gave me may have been a pseudonym, though he told me it twice and made sure to spell it for me the second time. While it may be a coincidence, the first four letters of his name in Hebrew correspond to a precious stone (sapphire) on the breastplate, which lit up and served as a communication device that Aaron and other high priests used to communicate with God. He did claim that his first name was Joseph, so for the purposes of this article, I will refer to him as Joseph *Sapphire*. The word "*sapphire*" and his given last name both occur three times in the Torah at skip +1.

SAPPHIRE IN SCIENCE

Sapphire is used with thin sapphire wafers of an insulating substrate upon which silicon is deposited to make integrated circuits known as silicon on sapphire or "SOS." Besides its excellent electrical insulating properties, sapphire has high thermal conductivity. CMOS (Complementary metal–oxide–semiconductor) is a technology for constructing integrated circuits. CMOS chips on sapphire are especially useful for high-power radio-frequency (RF) applications such as those found in cellular telephones, public-safety band radios, and satellite communication systems. "SOS" also allows for the monolithic integration of both digital and analog circuitry all on one IC chip, and the construction of extremely low-power circuits. Titanium-sapphire lasers are popular due to their relatively rare capacity to be tuned to various wavelengths in the red and near-infrared regions of the electromagnetic spectrum. They can also be easily mode-locked. In these lasers, a synthetically produced sapphire crystal with chromium or titanium impurities is irradiated with intense light from a special lamp or another laser to create stimulated emission.

Focus of the Chapter.

This chapter will look at Torah Code encodings of **SAPPHIRE** (really, the full name of the **FORT HUACHUCA** agent) with terms such as **SIGNAL FROM FUTURE, SIGNAL FROM I SHALL BE**

(God's name), **WORMHOLE, TACHYON,** and **TIME POLICEMAN.**

In writing it, I am aware that since at least 2016 (more likely about 2014), U.S. Army Intelligence has been generally here at least many times per day and sometimes up to a thousand or more times per day. I am also aware that foreign governments (especially the British, Chinese, Canadians, and Russians) read this site often. Originally, in 2017, I wrote here that my task was to highlight what is encoded of interest with respect to moving signals through time and to lay out a general framework for suggested research without compromising national security. Unfortunately, now, in 2025, I believe that national security has been severely compromised for decades. President Eisenhower bore much of the responsibility for it. He tried to warn us on January 17, 1961, three days before leaving office. He spoke to us about the military-industrial complex, but he should have warned us about the military-industrial extraterrestrial complex. He should have made clear to us that when President Truman implemented Operation Paperclip in 1947 to bring into America 1,600 Nazi rocket scientists (including Wernher Von Braun), he wasn't just setting up the Deep State. He was setting the stage for a Fourth Reich to control our government.

As I have written many times, there needs to be a follow-up to my August 1, 2016, meeting with Joseph. It should include any Government research facility likely to have an interest in *Roffmanmarsresearch.com* – starting with Brookhaven National Laboratory, since their IP address was once the exact reverse IP address of mine (later the UK Ministry of Defence used this address too, but their habit stopped around 2023.

The Incident Leading to Figures 10-1 and 10-2.

The word sapphire may refer just to the jewel, be part of the name of the agent who contacted me in 2016, or it may refer to someone who functions in the role of a time policeman, as Figure 10-2 may hint at. Please remember that, at best, the Code offers a fuzzy view of the future. Even a prophet, such as the Biblical Joseph, can be wrong in making predictions. He predicted that his father, mother, and 11 brothers would one day bow before him (Genesis 37:10). His mother, Rachel, never did. She died before Joseph went to Egypt. It is possible that errors may occur where parallel universes exist and come close together. But, normally in Judaism, an individual is seen as a false prophet when he makes false statements in God's Name. Please note that I never make such statements. My matrices are not based on Divine revelation. Rather, they are based on math (with a bit of intuition thrown in). As far as the time cop idea, I first published it around 2017, but in a book that I read in March 2023 (*PROJECT MANKIND: Extraterrestrials and Their Involvement with Humanity* by Ashwin Vinoo, 2022), I found the following discussion of the concept:

As most ETs can simply time-travel to the past and alter events in their favor, many ET races have time police that have their headquarters in the future, from where they guide the security of past events. This is done in an attempt to steer their experiences toward the most optimal timeline based on their predefined standards of what constitutes the best for their collective.

Ouch! It's hard enough to take on a government time traveler from the future! Does he need to be an ET too? Not necessarily. Vinoo admits on page 3 of his book that his conclusions (based largely on telepathic experiences with ETs) are probably only about 75 to 80% correct.

The August 1, 2016, Encounter with Joseph "Sapphire."

On August 1, 2016, I drove my son's car from my home in Cape Canaveral, Florida, to Wells Fargo to pay my mortgage, and then went on to the Cape Canaveral Public Library to return at least one book and check out another. When I parked at the library, there was no other vehicle anywhere nearby. I was inside for no more than five minutes. When I came out, a man had parked his car two spaces left of mine. He was not coming inside the library but was, in fact, standing to the right of his car, essentially blocking access to the vehicle that I was using. He was waiting for me (see the location on Figure 10-1).

Figure 10-1 below: In 2020, the word *REICH* was found crossing *SIGNAL FROM THE FUTURE. THE PLAGUE* was also found touching the name of the agent who interviewed me.

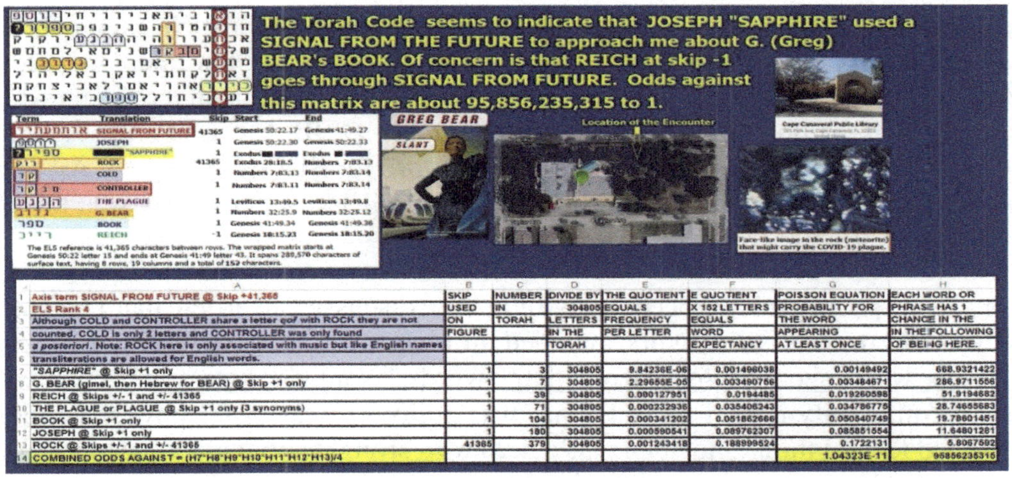

"Figure 10-1 above suggests that this matrix indicates the ability to move information through time, at least to some extent."

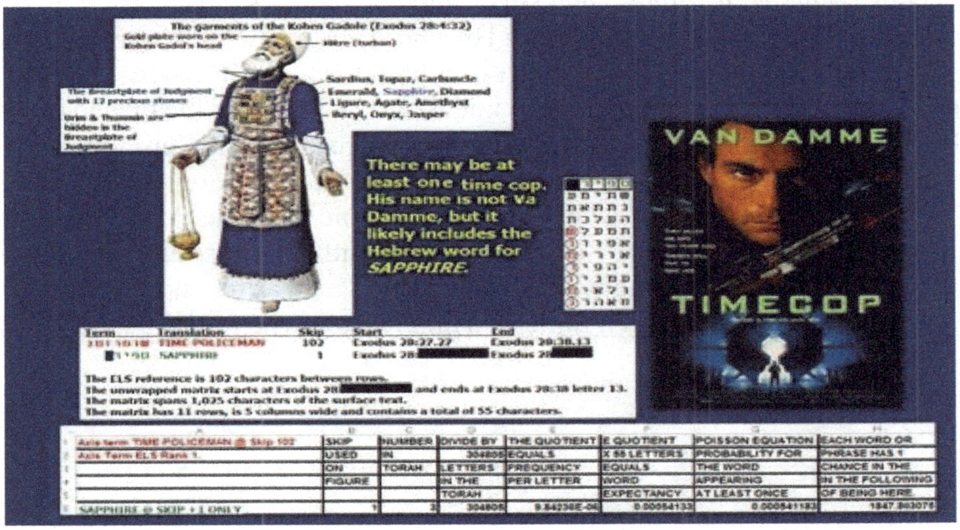

Above: Figure 10-2. **SAPPHIRE**, really the full last name of the agent who threatened me, is encoded with **TIME POLICEMAN**.

Joseph made sure to tell me his full name twice, spelling it for me the second time, which was wise because I didn't quite get it right the first time. While I gave him my card, which reads "LT Barry S. Roffman, USCG-Retired, Martian Meteorology Research," along with images of Mars as well as my phone numbers and websites, he did not have a card for me. With respect to my card, he asked me if he could keep it. Frankly, I found that a bit odd. His reaction to the card was almost akin to what would be expected if I were handing him a historical document. He also indicated that he had an interest in meteorology.

Joe said he lived near the NSA in Maryland, but denied working for them, although Army Intel might be a possibility since the Army Intelligence Base at Fort Huachuca, Arizona, has read all my websites at least once per week since about November 2014, when I started keeping records. Again, probably the strangest thing about the incident was that he seemed to know of the book that I had just checked out. On Figure 10-1 above, the axis term is **SIGNAL FROM FUTURE**. **JOSEPH** is on the first line. His second name, here cut short as **SAPPHIRE**, is on the next line down at skip +1 (SAPPHIRE is in the open text in conjunction with *urim*). The author of the book that Joseph caught me with, **G. BEAR**, is at skip +1, three lines under **SAPPHIRE.** Go down another three lines, and there is the word for **BOOK** at skip +1. All of this was found in 136 letters, against odds of more than 17 million to 1. The full 152-letter matrix was found against odds of over 95 billion to 1.

Why else do I think Joseph might have been Army Intel? We discussed an article that I had up about Hillary Clinton and Edward Snowden. The artwork included an image of Clinton with the question "Wanted for president?" Next to it was a photo of Snowden with the words "Wanted for treason?" I part ways with many in our Intelligence community here. I think Hillary was guilty of treason for how she handled email and should be imprisoned for it. I think Snowden was a hero who brought about much-needed reform in how the NSA treats our citizens. I wrote another article about him entitled *President Snowden? He is reshaping our law even now.*

So, how does all this relate to Joseph (Joe)? One of the first readers for the Clinton/Snowden story had an IP address for the Lincoln Technical Institute. Their street address is 9325 Snowden River Parkway, Columbia, Maryland. The reverse IP address for them is at Fort Huachuca, an Army Intel base in Arizona. I mentioned this to Joe, but I mispronounced the name of the fort – calling it **HU-A-CHU-CA**. He corrected me, telling me that **HUA** was one syllable. I asked him how he knew, and he informed me that he had been based there for three years. I told him that some of the IP addresses of my readers came up on the South Lawn of the White House, but I assumed that was just a generic point on the Google map for Washington. He immediately informed me that the South Lawn was where Marine One, the President's helicopter, landed. Again, I asked him how he knew that, and he claimed to work with the chopper at the White House for four years! As for Fort Huachuca, they visit here so often that I have an article up about them on my website entitled *Fort Huachuca, Spies, MARS, and a Little Martian Humor*. The relevant matrix is shown in Figure 10-3:

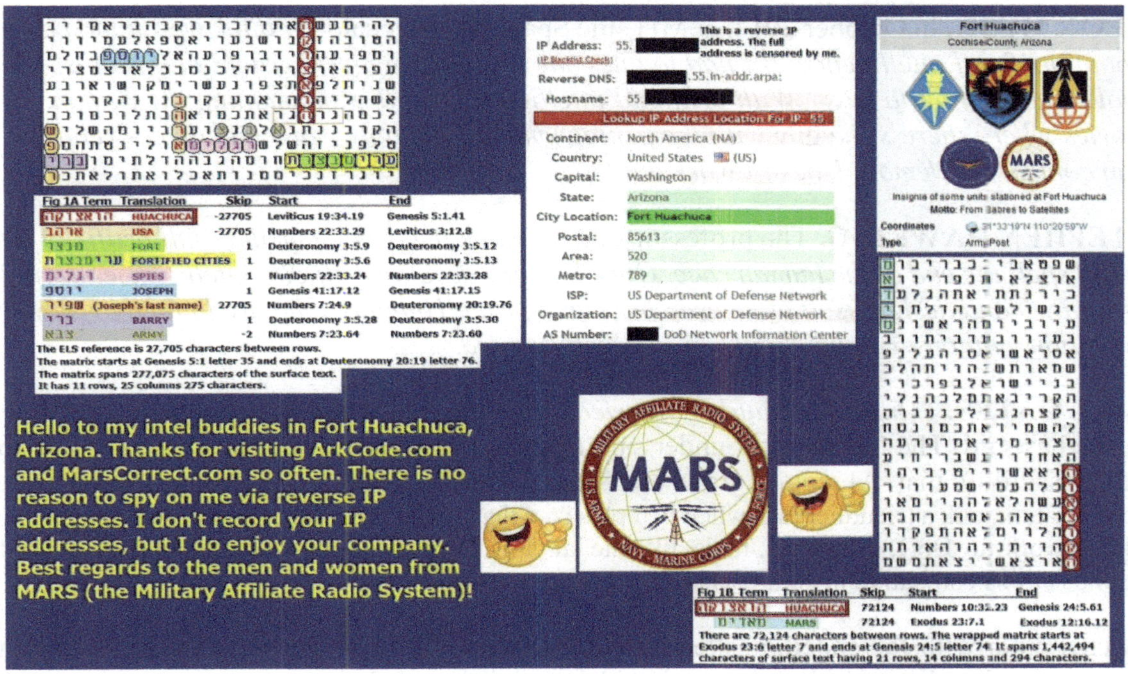

Figure 10-3 above. Martian humor in the Torah Code. On the left matrix, the axis term **HUACHUCA** is at the same skip as **SAPPHIRE** and **USA**, with **FORT, JOSEPH, ARMY**, and **SPIES** at skip +1. On the right matrix, the Military Affiliate Radio Station (**MARS**), which is based at Fort Huachuca, is at the same skip as **HUACHUCA**. Our conversation was not entirely friendly. At one point, Sapphire heard about how I was challenging NASA. I told him that I had published the names of scientists from the U.S., Netherlands, Canada, Belgium, West Germany, Taiwan, Italy, Japan, Switzerland, Scotland, France, Sweden, as well as from the Soviet Union and its Warsaw Pact ally Poland, who had met in Tokyo and Osaka, Japan, at the Third International Conference of Nondestructive Testing in March, 1960. This group was apparently trying to figure out how to counter the technology that was recovered at the Roswell Incident. It was at this point that Joe told me he had heard about someone like me who had his computer impounded by the Government. They loaded child pornography on it and then prosecuted him for the porno. He never owned a computer again.

My response wasn't friendly. Joe said he was a Catholic. I told him about what I did to the Catholic Church after they baptized my older son against his will and against mine. I forced them to change their doctrine and annul the baptism. Think I'm kidding? Archbishop McCarthy's Annulment letter is found at http://arkcode.com/images/annument_letter.png, and the press coverage in the Miami Herald is located at http://arkcode.com/photo4_85.html. The press made it sound like they did me a favor. Actually, they gave me what I wanted because I was suing them for six million dollars, and they knew I would win See the Release from Liability at https://www.arkcode.com/images/release_from_liability.jpg. **Note:** The boy in question went on to earn an MIT geophysics degree, but despite initial orders to Space Command, he entered the Air Force as an Orthodox rabbi in the chaplain corps. My point in bringing up the custody battle was that I'm a bad guy to piss off and I don't back off from a fight. That was the low point in the discussion, but it improved a lot from there, perhaps when we discussed the potential misuse of nanotechnology and/or AI. We spoke about people who claim that AI is a threat to human survival. Examples:

(1) **ELON MUSK.** In October 2014, at MIT, the SpaceX owner said, *"I think we should be very careful about artificial intelligence. If I had to guess at what our biggest existential threat is, it's probably that." He went on to say, "With artificial intelligence, we are summoning the demon. In all those stories where there's the guy with the pentagram and the holy water, it's like – yeah, he's sure he can control the demon. Doesn't work out."*

(2) **STEPHEN HAWKING.** The professor said, *"I think the development of full artificial intelligence could spell the end of the human race. Once humans develop artificial intelligence, it would take off on its own and design itself at an ever-increasing rate. Humans, who are limited by slow biological evolution, couldn't compete and would be superseded."*

(3) **BILL GATES.** *"Humans should be worried about the threat posed by artificial intelligence,"* Bill Gates has said. The Microsoft founder said he didn't understand people who were not troubled by the possibility that AI could grow too strong for people to control. It's also possible that the conversation took a friendlier turn when I emphasized that my younger son, David, would (then, but not now) like to work for DoD. Of course, he is, like me, already fully aware of what is going on behind the scenes, but he isn't as combative in nature as I am.

When my encounter with Joe ended, I asked him to contact me if he had any questions. At first, he said he wouldn't, but then he changed his mind and said he might. As I drove away, I noticed that his car had Maryland plates. He did not, while in my view, leave the area in the space next to the right of his car. He certainly did not make any attempt that I could see to enter the library. Based on that, my son and I have had discussions about the relative difficulty of moving a signal back in time versus opening a wormhole large enough for a person. On this issue, physics aside, I would point out that the axis term on **Figure 10-1** is **SIGNAL FROM FUTURE** (not **PERSON** from future); however, on **Figure 10-2**, I can place **SAPPHIRE** in a 55-letter box with **TIME COP** or **TIME POLICEMAN** as the axis term.

Here I want to make a few things clear about what happened when I checked out *Slant* by Greg Bear. On June 30, 2017, I went back to the library to look for the book and check on records there pertaining to when I read it and what its circulation looked like since they acquired it. Initially, I couldn't recall the book's name or author. I only remembered the gray cover, but I quickly found it in the science fiction section. Oddly, the librarian told me that the book had not been checked out since March 2016 – five months before I read it. And it had only been checked out eight times since they bought it in 2009. Could I have been wrong about when I read it? No. After my encounter with Joe, I remember considering taking my son's car into the shop for an oil change, with the real intention of looking under it to see if a tracking device had been attached that allowed Joe to follow me to the library. I only had the car back from Gainesville, Florida, for a few weeks since David finished his doctorate, and I felt like an oil change was due anyway. I didn't go for the change because I believed that if I were being tracked, it would more likely have been via my cell phone. However, when I went back to the library on June 30, 2017, it was in search of a more out-of-the-box solution to the question of how Joe knew when and where to find me and, perhaps, what I would be reading, at 2:57 pm on July 5, 2017. I wrote, *"I will pause for a day or two on this part of the article to give DoD some time to react to what I've written."* Well, that was before I found **Figure 10-1**, which really speaks for itself. If we are not dealing with an actual time traveler, we are almost certainly dealing with a meeting set up due to a signal sent back from the future, although on **Figure 10-2**, the man who matched well with **SIGNAL FROM FUTURE** also matches well with **TIME POLICEMAN**.

Figure 10-4. There is a good match between **SIGNAL FROM I SHALL BE** at skip +98,479 and **SAPPHIRE** at skip +1. **WORMHOLE** at skip +1 is here (and a wormhole may be needed to send a signal through time). **JOSEPH** and **TIME** are also here at special case skips, but the axis term **ELS rank** is too high, and the matrix size is too large to be significant. The Biblical Joseph correctly predicted that Pharaoh would restore the cupbearer to his job but behead the baker. Joseph demonstrated that information about the future can be sent back in time..

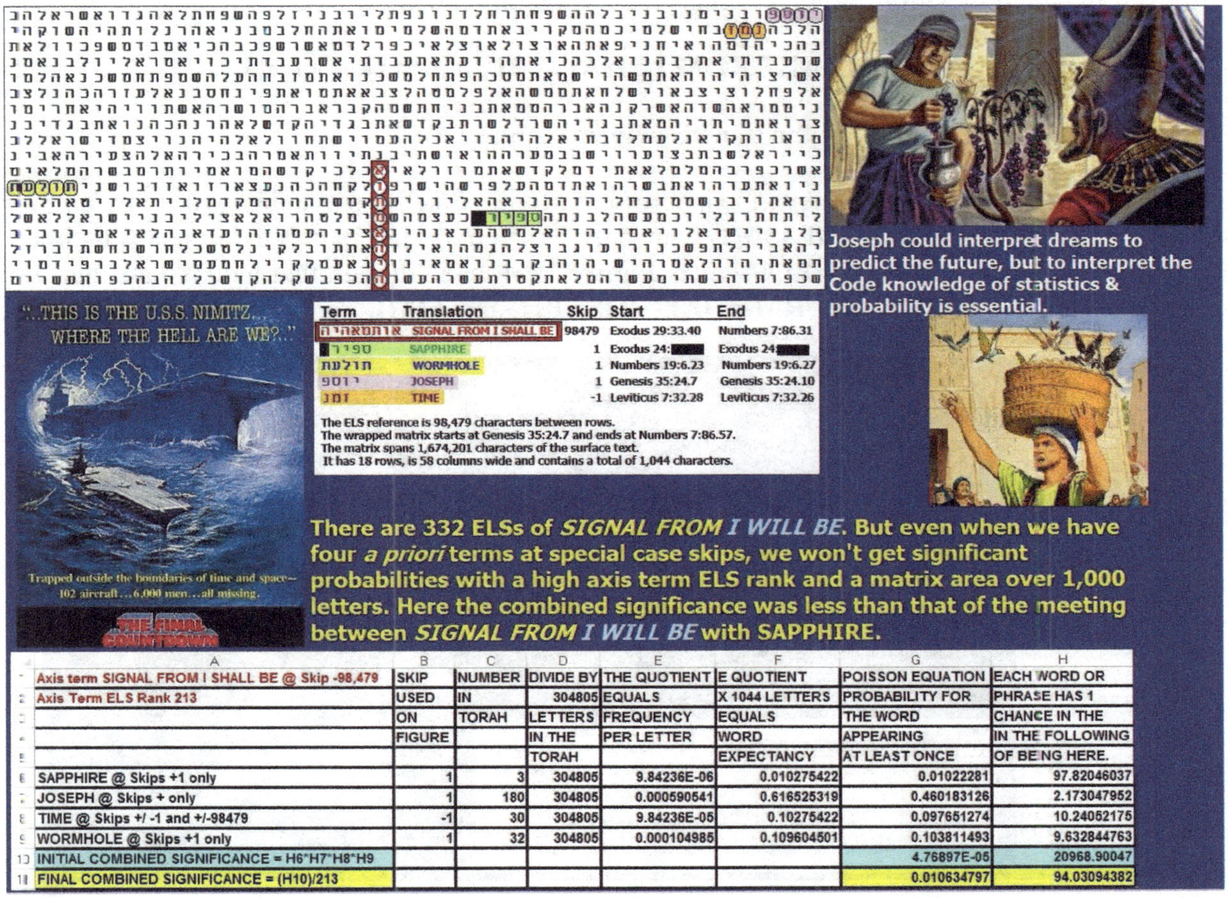

	A	B	C	D	E	F	G	H
1	Axis term SIGNAL FROM I SHALL BE @ Skip -98,479	SKIP	NUMBER	DIVIDE BY	THE QUOTIENT	E QUOTIENT	POISSON EQUATION	EACH WORD OR
2	Axis Term ELS Rank 213	USED	IN	304805	EQUALS	X 1044 LETTERS	PROBABILITY FOR	PHRASE HAS 1
3		ON	TORAH	LETTERS	FREQUENCY	EQUALS	THE WORD	CHANCE IN THE
4		FIGURE		IN THE	PER LETTER	WORD	APPEARING	IN THE FOLLOWING
5				TORAH		EXPECTANCY	AT LEAST ONCE	OF BEING HERE.
6	SAPPHIRE @ Skips +1 only	1	3	304805	9.84236E-06	0.010275422	0.01022281	97.82046037
7	JOSEPH @ Skips + only	1	180	304805	0.000590541	0.616525319	0.460183126	2.173047952
8	TIME @ Skips +/ -1 and +/-98479	-1	30	304805	9.84236E-05	0.10275422	0.097651274	10.24052175
9	WORMHOLE @ Skips +1 only	1	32	304805	0.000104985	0.109604501	0.103811493	9.632844763
10	INITIAL COMBINED SIGNIFICANCE = H6*H7*H8*H9						4.76897E-05	20968.90047
11	FINAL COMBINED SIGNIFICANCE = (H10)/213						0.010634797	94.03094382

Figure 10-5 MATCHES BETWEEN "I SHALL BE" AND "SAPPHIRE" (OPEN TEXT) UNDER 500 LETTERS IN AREA.

On the Sabbath of January 6, 2018, in all legitimate synagogues, the Torah portion read was from Exodus 1:1 to Exodus 6:1. In the course of reading it again, I was struck in Exodus 3:13-14 by the Name that God chose to pass on to the Israelites via Moses… "I Shall Be." If ever there was a remark out of the mouth of God designed to cast the eyes of physicists to the future for a human writer of the Torah Code, and by extension, to the Torah, it's this.

We have looked at the issue of who Joseph *Sapphire* really is. Here is a man who almost seemed to appear out of thin air (but I was in the library for under 5 minutes when he positioned himself in a manner that basically blocked entry to my car). He seemed to know the topic of the book that I had just selected in the library, yet he was not in there himself, so far as I could tell either before or after we met. He said he was from Maryland, visiting a girlfriend, so it's unlikely that he had a Brevard County library card. The warning about knowing someone like me who was falsely

accused of using child pornography, coupled with his admission of an intelligence background in the very Army Intelligence Base (Fort Huachuca) that tracks my writing most closely, did not make him sound like a religious figure. Ditto with his claim to work at the White House for four years.

I found myself in an odd position. When I first spoke to Sapphire, he knew a hell of a lot about my life.

There are 31 matches between **SIGNAL FROM I SHALL BE** and, at skip +1, the last

SAPPHIRE+ who interviewed me at the Cape Canaveral Library.

As can be seen with **Figure 10-5**, when an axis term with **ELS rank 22** is matched with only two terms at special case skips – **SAPPHIRE** and **TIME** – in only 374 letters, the matrix is almost 2.6 times more significant than what was seen in the 1,044-letter **Figure 10-4**, even though that matrix had four words at special case skips – **SAPPHIRE**, **WORMHOLE**, **JOSEPH**, and **TIME**.

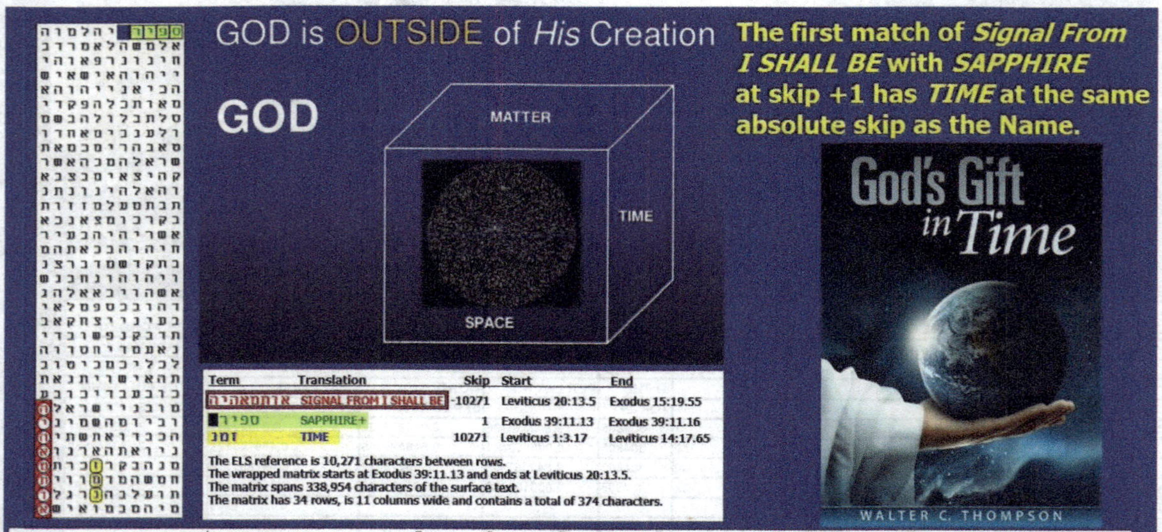

	A	B	C	D	E	F	G	H
1	Axis term SIGNAL FROM I SHALL BE @ Skip -10,271	SKIP	NUMBER	DIVIDE BY	THE QUOTIENT	E QUOTIENT	POISSON EQUATION	EACH WORD OR
2	Axis Term ELS Rank 22	USED	IN	304805	EQUALS	X 374 LETTERS	PROBABILITY FOR	PHRASE HAS 1
3		ON	TORAH	LETTERS	FREQUENCY	EQUALS	THE WORD	CHANCE IN THE
4		FIGURE		IN THE	PER LETTER	WORD	APPEARING	IN THE FOLLOWING
5				TORAH		EXPECTANCY	AT LEAST ONCE	OF BEING HERE.
6	SAPPHIRE @ Skips +1 only	1	3	304805	9.84236E-06	0.003681042	0.003674275	272.1625171
7	TIME @ Skips +/ -1 and +/-98479	-1	30	304805	9.84236E-05	0.03681042	0.036141153	27.6692885
8	INITIAL COMBINED SIGNIFICANCE = H6*H7						0.000132793	7530.543204
9	FINAL COMBINED SIGNIFICANCE = (H8)/22						0.002921436	342.2974184

Figure 10-5 In addition to God communicating with Aaron and other high priests via the breastplate with urim and **thumim**, He also made His voice heard between the cherubim on top of the Ark of the Covenant. When I checked to see how many times there was a match between **SIGNAL FROM I SHALL BE** and **ARK OF THE COVENANT**, with the row split function disabled on CodeFinder, I found 36 matches, with the smallest having an area of only 88 letters (8 columns for the eight letters of **ARK OF THE COVENANT** and 11 rows, 8 of which were required to show **SIGNAL FROM I SHALL BE).**

One thing becomes obvious when I look at a term like **SIGNAL FROM I SHALL BE**, where there are hundreds of ELS occurrences – the searches that I do are inadequate in that they require so much time (and insight). The best way to conduct searches for military intelligence purposes is to employ artificial intelligence in the search engine. This may, to fellow Orthodox Jews, seem to be a nonreligious thing to do. But here I advance the idea that, with the understanding that there is an Intelligence behind the encoding of Torah, the belief in the Encoder begins to move from an area of faith to an area of mathematics – and math on the part of the Encoder that may not necessarily be linked to compassion as we generally understand the term. The function of our military is to defend our nation. There are plenty of people who act in ways that God may seek to destroy. But should God elect to destroy us not by a flood again, but by an asteroid or attack by a hostile ET race like the Reptilians, we have a right to expect our military to do its best to shoot it out of the sky. History is replete with incidents like the Crusades, the Inquisition, Pogroms, and Holocausts, where nobody stepped in to save millions of the innocent. Thus, given evidence that an Encoder exists, our military (and even NASA) have a duty to understand the exact nature and intent of that Encoder. Don't depend on NASA. While I want to say **Trust the U.S. Space Force**, both NASA and Space Force fly command flags that map out where to find the same alleged ET civilizations in the Pleiades (Pleiadians), Alpha Draconis (Reptilians), and Orion (Grays). I have been contacted by a woman (Rebecca R.) who claims to have had many ET encounters, including ones with Pleiadians who told her that she is also from there. She asked me to refrain from publishing her full name so she can lead a normal, quiet life. Unlike Craig with his involvement in COVID-19, she comes across as non-threatening and a source of information about the alleged ET operations on Earth. In particular, I am interested in her testimony that she has personally seen Reptilians eat live hybrid Human-Grey children. The NASA and Space Force flags that mock public ignorance of relations between our 4th Reich government and ET worlds are shown on **Figure 10-6**.

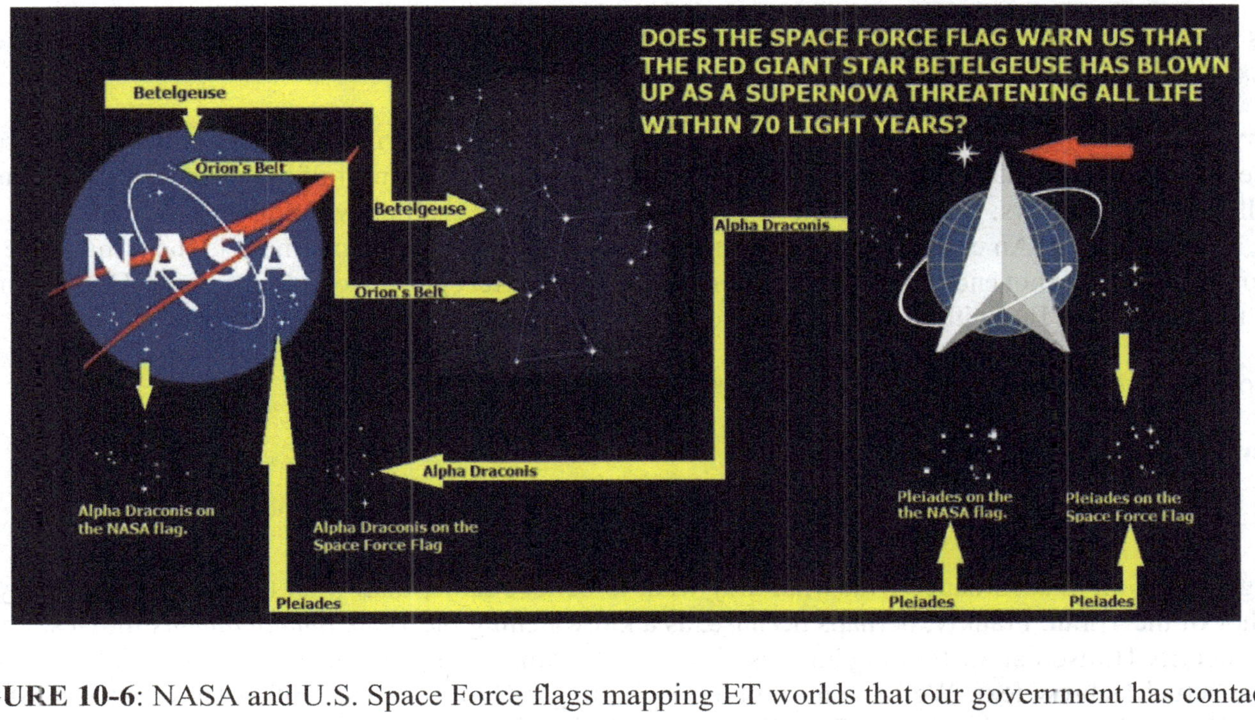

FIGURE 10-6: NASA and U.S. Space Force flags mapping ET worlds that our government has contact with.

RABBINICAL COMMENTARY ON THE NAME "I SHALL BE AS I SHALL BE" (EXODUS 3:14).

This name is often translated incorrectly as, "I AM WHO I AM." The correct translation is only given in a footnote for the Revised Standard Edition and American Standard Edition. Wikipedia notes the controversy as to how to translate the Name as follows:

I Am that I Am (◌אֶ◌הִיה◌אֲשֶׁר אֶ◌הִיה*'ehyeh 'ăšer 'ehyeh*) is the common English translation (JPS among others) of the response that God used in the Hebrew Bible when Moses asked for his name (Exodus 3:14). It is one of the most famous verses in the Torah. *Hayah* means "existed" in Hebrew; *'ehyeh* is the first person singular imperfect form and is usually translated in English Bibles as "I am" or "I will be" (or "I shall be"), for example, at Exodus 3:14. *'ehyeh 'ăšer 'ehyeh* literally translates as "I Am Who I Am." The ancient Hebrew of Exodus 3:14 lacks a future tense such as modern English has, yet a few translations render

this name as "I Will Be What I Will Be," given the context of Yahweh's promising to be

with His people through their future troubles. Wikipedia goes on to make the following statement as to what is correct:

The word 'ehyeh is used a total of 43 places in the Hebrew Bible, where it is often translated as "I will be" – as is the case for its first occurrence, in Genesis 26:3 – or "I shall be," as is the case for its final occurrence in Zechariah 8:8.

Commentary on the Name in the Stone Edition Chumash

This book has the Hebrew text, translation, and rabbinical commentaries that I trust the most. In reference to Exodus 3:14, they state the following:

אֶ◌הְיֶה אֲשֶׁר אֶ◌הְיֶה – "I Shall Be As I Shall Be." This in itself is a Divine Name, as implied by Onkelos, since he does not translate these three words. Rashi explains that the import of the word אֶ◌הְיֶה – literally I Shall Be – is as follows: "I shall be with them in this sorrow as I shall be with them in other sorrows." To this, Moses replied, "An evil in its own time is enough!" i.e., why should you imply to them that there will be future exiles; is it not enough that they suffer now in Egypt? Accepting Moses's argument, God instructed him to say, "I Shall Be [with them in this sorrow] has sent me to you" (Rashi from Berachos 9b).

According to the Midrash, the word◌אֶ◌הְיֶה describes God as timeless and eternal.

Summary of the Translation Dispute.

It is obvious that those who choose to translate◌אֶהְיֶה as "I Am" rather than "I Will Be" are

designing a God in accordance with what they want to believe rather than what they are being told by the Author of the Torah. Frankly, perhaps because, as a kid watching the *Ten Commandments* movies, I heard God identify Himself at the burning bush as "I Am That I Am," I was surprised to see the future tense used in the actual Hebrew text. We have no problem accepting a God as declaring that He exists (or existed) when He was busy carrying out the Ten Plagues or splitting the Sea of Reeds, but when He says in the time of Moses, "I Shall Be As I Shall Be," while nobody back then knew about relativity, wormholes, or tachyons, we know about them now. And, on a personal basis, I have had the rather odd experience with Joseph

Sapphire. Moreover, I spend much of every day looking for signals encoded in Torah that are predictive. In 2025 I've found enough this way to keep me occupied for 28 years, and to merit having U.S. Government and other government agencies reading my Codes findings for many years.

As for my son, David, he once hounded me for years about his desire to one day invent a time machine. I finally got him to give up the idea when he started college and began to study physics, but now, when he sees the evidence and, in particular, who monitors our sites, his interest has returned. Having said all this, I remain skeptical that any human will write the Code and send it back in time to Moses. The problem is the time information paradox. How can a human write (and encode) all the stories in the Torah if he or she has already read them? But I do believe that signals of military value can be sent back in time, at least from the future to a point in time when humans first succeed in opening a wormhole.

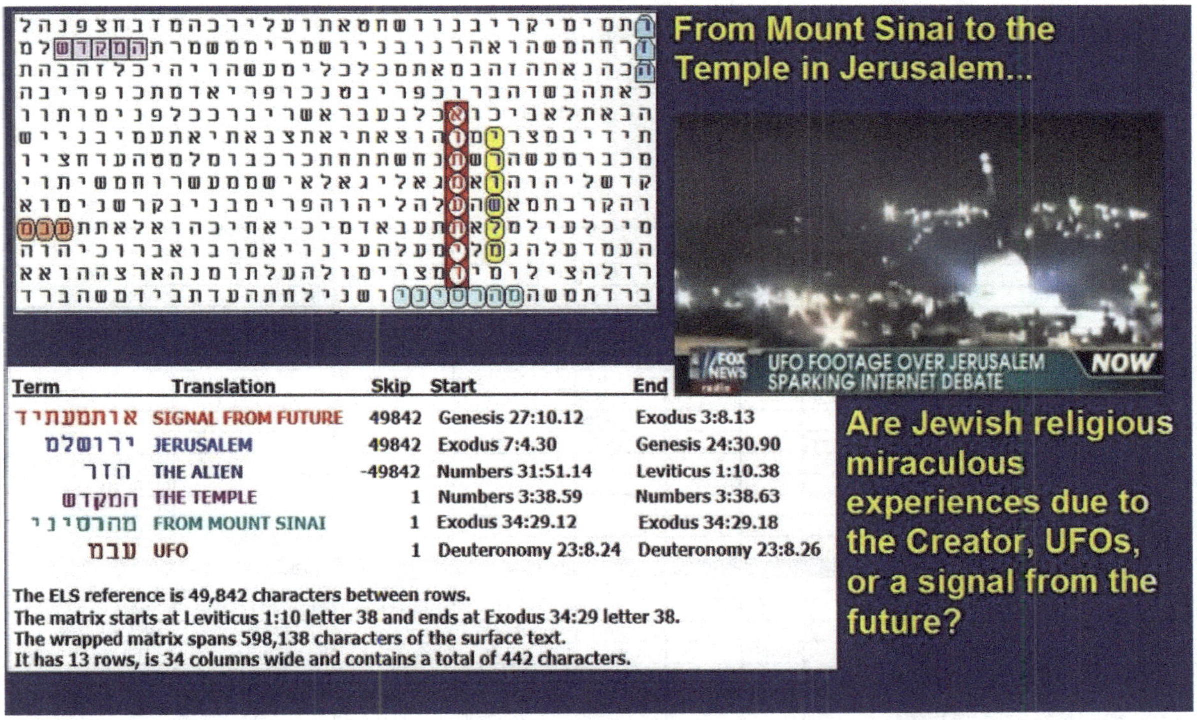

Term	Translation	Skip	Start	End
אותמעתיד	SIGNAL FROM FUTURE	49842	Genesis 27:10.12	Exodus 3:8.13
ירושלם	JERUSALEM	49842	Exodus 7:4.30	Genesis 24:30.90
הזר	THE ALIEN	-49842	Numbers 31:51.14	Leviticus 1:10.38
המקדש	THE TEMPLE	1	Numbers 3:38.59	Numbers 3:38.63
מהרסיני	FROM MOUNT SINAI	1	Exodus 34:29.12	Exodus 34:29.18
עבמ	UFO	1	Deuteronomy 23:8.24	Deuteronomy 23:8.26

The ELS reference is 49,842 characters between rows.
The matrix starts at Leviticus 1:10 letter 38 and ends at Exodus 34:29 letter 38.
The wrapped matrix spans 598,138 characters of the surface text.
It has 13 rows, is 34 columns wide and contains a total of 442 characters.

Figure 10-7 above: SIGNAL FROM THE FUTURE – ARE ALIENS INVOLVED?

On Figure 10-7, the axis term is **SIGNAL FROM FUTURE**, at its seventh-lowest skip in the wrapped Torah. The primary *a priori* word sought was **JERUSALEM**; I also sought **MOUNT SINAI** because I wanted to know if there is a link between miracles occurring at both locations and a signal from the future. In the Hebrew text of the Tenach (Old Testament to Christians), **JERUSALEM** occurs 664 times at skip +1 with the six-letter spelling found on the matrix, and another five times with the modern seven-letter spelling. I sought both spellings in the Torah matrix. (Jerusalem is never found at skips ±1; so the only special-case skips to seek are those at the absolute skip of the axis term.) **SIGNAL FROM FUTURE** is at skip +49,842. There is only one ELS of **JERUSALEM** at skip ± 49,841 in the Torah. After factoring in ELS rank 6 of **SIGNAL FROM FUTURE**, there was only one chance in 1,814 to find **JERUSALEM** in a 24- letter box with the axis term—actually less, one chance in 2,721, because there were only 16 letters left to find Jerusalem after eliminating the column containing **SIGNAL FROM FUTURE** (its eight letters conflict with **JERUSALEM**).

Odds against finding **JERUSALEM** at a special-case skip somewhere on the full 442-letter matrix were about 690 to 1. This is why I was shocked to see **JERUSALEM** so close to **SIGNAL FROM FUTURE.** But the other *a priori* keywords were also quite significant. Before factoring in the axis-term ELS rank of 7:

- **UFO** was found at skip +1 against odds of about 58 to 1.
- **THE TEMPLE** and **MOUNT SINAI** were each found at skip +1 against odds of about 43.6 to 1.
- One of three synonyms sought for **THE ALIEN** was found at a special-case skip against odds of about 12 to 1.

As for the UFO shown over Temple Mount, worthwhile commentary is found at http://nationalufocenter.com/2014/01/are-the-videos-taken-over-jerusalem-real-or-fake/. In all, the odds against Figure 10-7 occurring by chance alone were about 132,500,832 to 1.

The spreadsheet for Figure 10-7 is here:

	A	B	C	D	E	F	G	H
1	AXIS TERM SIGNAL FROM FUTURE @ Skip 49842	SKIP	NUMBER	DIVIDE BY	THE QUOTIENT	E QUOTIENT	POISSON EQUATION	EACH WORD OR
2	Axis Term ELS Rank 7	USED	IN	304805	EQUALS	X 442 LETTERS	PROBABILITY FOR	PHRASE HAS 1
3		ON	TORAH	LETTERS	FREQUENCY	EQUALS	THE WORD	CHANCE IN THE
4		FIGURE		IN THE	PER LETTER	WORD	APPEARING	IN THE FOLLOWING
5				TORAH		EXPECTANCY	AT LEAST ONCE	OF BEING HERE.
6	JERUSALEM @ Skips +/- 1 and +/- 49842	1	1	304805	3.28079E-06	0.001450107	0.001449057	690.1041932
7	THE TEMPLE @ Skip +1 only	1	16	304805	5.24926E-05	0.023201719	0.022934629	43.60218798
8	UFO @ Skip +1 only	1	12	304805	3.93694E-05	0.017401289	0.017250761	57.96845613
9	MOUNT SINAI @ Skip +1 only	1	16	304805	5.24926E-05	0.023201719	0.022934629	43.60218798
10	THE ALIEN @ Skips +/- 1 and +/- 49842 (3 synonyms)	-49842	59	304805	0.000193566	0.085556339	0.081998578	12.19533344
11	INITIAL SIGNIFICANCE						1.07816E-09	927505824.3
12								DIVIDE BY 7 DUE TO
13								ELS RANK 2 OF
14								AXIS TERM =
15	FINAL SIGNIFICANCE						7.54712E-09	132500832
16								
17	AXIS TERM SIGNAL FROM FUTURE @ Skip 49842	SKIP	NUMBER	DIVIDE BY	THE QUOTIENT	E QUOTIENT	POISSON EQUATION	EACH WORD OR
18	Axis Term ELS Rank 7	USED	IN	304805	EQUALS	X 24 LETTERS	PROBABILITY FOR	PHRASE HAS 1
19	THIS CALCULATION IS ONLY FOR THE MATCH	ON	TORAH	LETTERS	FREQUENCY	EQUALS	THE WORD	CHANCE IN THE
20	BETWEEN *SIGNAL FROM FUTURE* AND	FIGURE		IN THE	PER LETTER	WORD	APPEARING	IN THE FOLLOWING
21	*JERUSALEM* IN A 24-LETTER BOX			TORAH		EXPECTANCY	AT LEAST ONCE	OF BEING HERE.
22								
23	JERUSALEM @ Skips +/- 1 and +/- 49842	1	1	304805	3.28079E-06	7.87389E-05	7.87358E-05	12700.70834
24								DIVIDE BY 7 DUE TO
25								ELS RANK 2 OF
26								AXIS TERM =
27	FINAL SIGNIFICANCE						0.00055115	1814.386906

As such, the matrix raises fundamental theological questions of the highest nature. **Figure 10-7**, found on July 6, 2017, not only greatly reinforced my concerns but really removed all doubt that the concerns are valid.

One last point about **Figure 10-7**. The possibility of a human from the future and the possibility of aliens from the future sending signals back in time are not necessarily mutually exclusive. Given cooperation between aliens and humans (or the recovery of an alien device), we might (really, probably did) get the technology from them.

ARE SIGNALS FROM THE FUTURE ENCODED WITH TACHYONS? Yes. Despite my son David's belief that wormholes are the way to go when sending signals back in time, tachyons are theoretical particles that travel faster than light and can travel back in time. As discussed below, there is not yet any direct evidence that they exist, but we might find indirect evidence. However, when **SIGNAL FROM FUTURE** is found at an ELS at skip -78,480 (ELS rank 5), we find **TACHYON** at skip -1 in a matrix that measures 27 rows by six columns (a 182-letter area). This match occurs against odds of about 188 to 1. Further, when **SIGNAL FROM FUTURE** is found at an ELS at skip +127,845, **TACHYON** is at skip -127,845 in a matrix that measures 24 rows by 14 columns (a 336-letter area). This match occurs against odds of about 43 to 1. Note that the statistical significance of **TACHYON** at skip -78,480 on **Figure 10-8** is greater than for any other

terms shown on all the matrices in this article, except for ***SAPPHIRE*** (really, the name given by the agent who interviewed me in 2016) and **JERUSALEM** on **Figure 10-7**.

As noted above, on the physics line of investigation, tachyons, if they exist, are not proving to be easy to detect directly. The following statement by Dr. James T. Wheeler, professor of physics at Utah State University, is noteworthy for those who can follow the physics involved:

Tachyons are well-defined within special relativity. They are particles for which the magnitude of the momentum is greater than the energy, so that their momentum 4-vector is space-like, rather than time-like, like the momentum 4-vectors of regular particles and their antiparticles.

Adding energy to a tachyon can be shown to slow it down, but an infinite amount of energy would be required to slow it to the speed of light.

To observe them does not require light to 'catch' them, but only to interact – the light could be coming from any direction. But we don't even have to 'see' them with light. If a tachyon is produced, say, in a particle accelerator, and escapes unseen, then the total energy and momentum of the other particles will not be conserved. Moreover, a large amount of momentum will be missing – more than the missing energy – so we have a clear signal of the production of the tachyon, whether we see it directly or not.

No such missing momentum has been observed in accelerator experiments to date, so if tachyons do occur, they must at least have very large mass.

The space-time path of a tachyon lies outside the light-cone, which is what fixes our notion of causality. Thus, with tachyons, an event *here* could rapidly affect an event at an arbitrarily great distance, instantaneously for some observers. Physical effects would then become nonlocal, and – with easy production of tachyons – prediction nearly impossible.

Figure 10-8 – Signal from the future is twice encoded with tachyon.

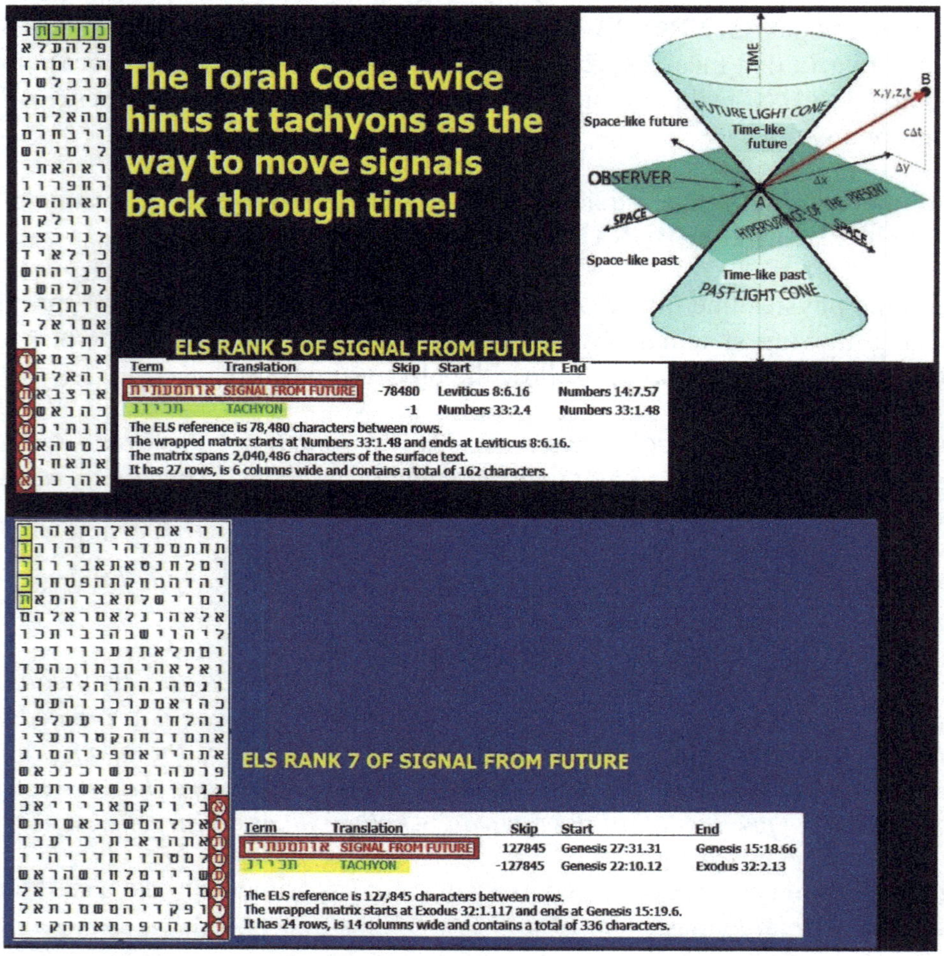

Spreadsheets for Figure 10-8.

	A	B	C	D	E	F	G	H
1	Axis term SIGNAL FROM FUTURE @ -78480	SKIP	NUMBER	DIVIDE BY	THE QUOTIENT	E QUOTIENT	POISSON EQUATION	EACH WORD OR
2	Axis Term ELS Rank 5	USED	IN	304805	EQUALS	X 162 LETTERS	PROBABILITY FOR	PHRASE HAS 1
3		ON	TORAH	LETTERS	FREQUENCY	EQUALS	THE WORD	CHANCE IN THE
4		FIGURE		IN THE	PER LETTER	WORD	APPEARING	IN THE FOLLOWING
5				TORAH		EXPECTANCY	AT LEAST ONCE	OF BEING HERE.
6	TACHYON @ Skips +/ -1 and +/- 78480	-1	2	304805	6.56157E-06	0.001062975	0.00106241	941.2562614
7	FINAL SIGNIFICANCE = H6/5						0.00531205	188.2512523
8								
9	Axis term SIGNAL FROM FUTURE @ 127,845	SKIP	NUMBER	DIVIDE BY	THE QUOTIENT	E QUOTIENT	POISSON EQUATION	EACH WORD OR
10	Axis Term ELS Rank 7	USED	IN	304805	EQUALS	X 336 LETTERS	PROBABILITY FOR	PHRASE HAS 1
11		ON	TORAH	LETTERS	FREQUENCY	EQUALS	THE WORD	CHANCE IN THE
12		FIGURE		IN THE	PER LETTER	WORD	APPEARING	IN THE FOLLOWING
13				TORAH		EXPECTANCY	AT LEAST ONCE	OF BEING HERE.
14	TACHYON @ Skips +/ -1 and +/- 127845	-1	3	304805	9.84236E-06	0.003307032	0.00330157	302.8861883
15	FINAL SIGNIFICANCE = H14/7						0.023110991	43.26945547

ON GETTING ALONG WITH GOD. If there is one thing about being Jewish that seems very different from what other religions offer, it's the right we feel we have to argue with God. We saw some of that above with Moses getting God to change His mind about which Name to use. We see it with Moses telling God that if He wants to punish the Israelites too much, God should write Moses out of His Book (Exodus 32:32:

"But now, please forgive their sin—but if not, then blot me out of the book you have written").

We saw it when Abraham bargained with God to try to save Sodom and Gomorrah, and when Jonah refused to go to Nineveh until he was spat up by a giant fish that ate him. We also see evidence that God's attitudes evolve with time in Hosea 6:6:

"For I desire mercy, and not sacrifice; and the knowledge of God more than burnt offerings."

God seems to be a life form, but One that may be argued with only from a moral perspective. In the end, only He generally gets to define what's moral and what's not.

A WORD OF HOPE. If it is true that we are headed for—or are now dealing with—a technology that can send signals back into the past, **the real significance is that death is not as final as some people think.** Let's say that my son, David, one day sends me a signal from his future, or that Joseph Sapphire contacted me based on his access to such technology, and that he gave me hints about David being involved with nanotechnology and artificial intelligence. Let's assume the signal originated from a time when I am dead. That means that although I am dead in that time, I remained accessible in this time. No death is really final. Rather, life exists at a point on a timeline that is no longer bound by one direction—forward. It also lends credence to the idea that a truly Superior Intelligence, generally referred to by the Name God, could finally be accepted as God when He reverses time and brings about the Resurrection that Jews, Christians, and Muslims have believed in for centuries. **I SHALL BE AS I SHALL BE**—shall be when this bit of physics is mastered.

LINGERING CONCERNS. Why did Joseph *Sapphire* really contact me? I am convinced that he had a perfect view of the future. That means he came to me knowing everything I would ever write. He knew how I would react to his threat to load child porn onto my computer. He spelled his name the second time because he knew what I first heard was not accurate enough to find the matrix in Figure 10-1 above. That matrix was required for me to understand critical issues touching on the nature of time and the threats facing humanity. If you Google Fort Huachuca, you will find accusations that they have captured or "loaned" alien spacecraft under their facility. There are also accusations of Reptilians being present there, which raises nasty questions about what they eat. If *Sapphire* is not himself an ET, he may be concerned about hostile aliens gaining control over what lies below—and what they intend to do to our species in the future. The works of Dr. Michael Salla and Elena Danaan focus heavily on these issues, but their reliance on a small number of named sources leaves us fascinated by the stories yet unsure of their accuracy. Danaan discusses the threat at Fort Huachuca.

Our government under Biden lost all credibility. On March 1, 2023, "President" Biden laughed at a woman who blamed him for the death of her two sons due to fentanyl poisoning that resulted from leaving our borders wide open. How can *Sapphire* (or anyone else) combat this level of corruption? Unless Joseph is more open with me (or Trump approves disclosure), all I can do is continue writing about my investigations, hoping to stimulate the right kind of opposition in the military-intelligence community to ensure an effective response to the threat *Sapphire* suggested by contacting me in 2016.

President Trump, I can help you get disclosure done the right way—but I'm not getting any younger. You have dangled job offers, but so far the only concrete contact I've seen is a signed portrait you sent as thanks for contributions to your campaigns. Your emails about a job in your cabinet are always accompanied by requests for more money. I need a phone call from you to know you are serious. The only money that should be discussed is my salary.

Chapter 11
The Obama/Biden Years

When I started this chapter, I had at least 179 articles about Obama on my website (many with more than one matrix) and another 13+ about Biden. My website still provides a wide number of peeks into just about every aspect of his life, but after 28 years of study, 23 matrices begin to make clear that the Encoder had plenty to tell us (really everything) about Obama and his sick VP (Biden). Perhaps there is also more to tell us about Obama due to the fact that he served two terms while "Hiden' Biden" largely spent his one term in his basement or on vacation (on a Delaware beach), using the rest of his time (about four hours per day) to try to destroy America.

I have such a large number of matrices to choose from because I spent most of my 18 years of retirement doing Torah Codes research. I started almost each day by first searching the news and then the Code to see how much of it (which was most of it) I could find encoded in the Torah. A few aspects of Obama's life shown on matrices in this chapter include his antisemitism with respect to a deli massacre in Paris, his hatred of Israel, the Rodney King riots, views on homosexuality, technological efforts to control the minds of our population, treasonous behavior, his communist leanings, marijuana usage and policy, questionable country of origin, development of identity chips, loyalty to the Illuminati and Hitler, lack of a plan for dealing with the ISIS terrorist group, failure to stop North Korea's development and testing of nuclear bombs, surrender on dealing with Iran's nuclear plans, Obamacare, and concerns that his weakness (and Biden's) may eventually cause New York to be nuked.

I learned almost two decades ago that there is no question about the Code being real. The only question is who wrote it.

Figure 11-1above. The axis term is **ANTISEMITIC**, and it's at the same skip as **OBAMA**. On January 9, 2015,

Amedy Coulibaly, armed with a submachine gun, an assault rifle, and two Tokarev pistols, entered and attacked a Hypercacher kosher supermarket in Paris, France. There, he murdered four Jewish hostages and held fifteen other hostages during a siege while demanding that the perpetrators of the Charlie Hebdo shooting, the Kouachi brothers, be released. The siege ended when police stormed the supermarket, killing Coulibaly. Obama refused to attend or send a high dignitary to the service, calling the attack random rather than antisemitic. On the matrix**, PARIS IS UNCLEAN** is at skip +1. **HE** (Obama) **SHALL NOT COME** is on the matrix and so is **UNTO A DEAD BODY** with both phrases also at skip +1.

It should be obvious to all thinking people that Joe Biden is a small, senile man who has always been anxious to destroy Israel. But is it as obvious when we read the words of his Creator? Does the Author of the Torah Code warn us about Israel's great enemy? The answer is most definitely yes. On **Figure 11-2**, the axis term is ***BIDEN ENEMY***. Its ELS appears only once in the Torah. I hoped that it would be directly crossed by **ISRAEL**, and it didn't disappoint me.

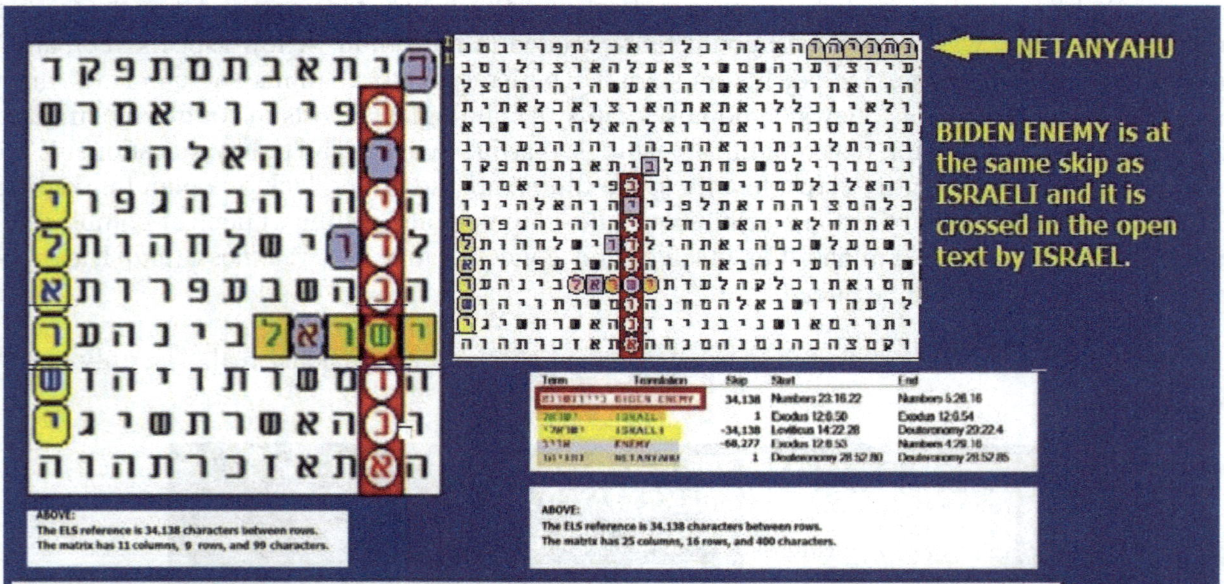

	A	B	C	D	E	F	G	H
1	AXIS TERM BIDEN ENEMY @ +34,138	SKIP	NUMBER	DIVIDE BY	THE QUOTIENT	E QUOTIENT	POISSON EQUATION	EACH WORD
2	Axis Term ELS Rank 1	USED	IN	304805	EQUALS	X 99 LETTERS	PROBABILITY FOR	OR PHRASE
3		ON	TORAH	LETTERS	FREQUENCY	EQUALS	THE WORD	HAS 1 CHANCE
4		FIGURE		IN THE	PER LETTER	WORD	APPEARING	IN THE
5				TORAH		EXPECTANCY	AT LEAST ONCE	FOLLOWING OF
6								BEING HERE
7								
8	ISRAEL or ISRAELI @ Skips +/- 34,138	-34138	1	304805	3.28079E-06	0.000324798	0.000324745	3079.338411
9	ISRAEL @ Skip +1 only	1	591	304805	0.001938945	0.191955513	0.174656415	5.72552689
10	Enemy (alef vav yud vet) @ skips -68277 to -68356	-68277	1173	304805	0.003848362	0.380987845	0.316813806	3.1564281
11	Combined odds against this matrix = H8*H9*H10						1.79693E-05	55650.46263
12								
13	AXIS TERM BIDEN ENEMY @ +34,138	SKIP	NUMBER	DIVIDE BY	THE QUOTIENT	E QUOTIENT	POISSON EQUATION	EACH WORD
14	Axis Term ELS Rank 1	USED	IN	304805	EQUALS	X 400 LETTERS	PROBABILITY FOR	OR PHRASE
15		ON	TORAH	LETTERS	FREQUENCY	EQUALS	THE WORD	HAS 1 CHANCE
16		FIGURE		IN THE	PER LETTER	WORD	APPEARING	IN THE
17				TORAH		EXPECTANCY	AT LEAST ONCE	FOLLOWING OF
18								BEING HERE
19								
20	ISRAEL or ISRAELI @ Skips +/- 34,138	-34138	1	304805	3.28079E-06	0.001312314	0.001311454	762.5126094
21	NETANYAHU @ Skips +1 only	1	16	304805	5.24926E-05	0.020997031	0.020778128	48.12753099
22	ISRAEL @ Skip +1 only	1	591	304805	0.001938945	0.775577828	0.53956235	1.853353927
23	Enemy (alef vav yud vet) @ skips -68277 to -68356	-68277	1173	304805	0.003848362	1.539344827	0.785478396	1.27310949
24	Combined odds against this matrix = H20*H21*H22*H23						1.15488E-05	86589.39999

Figure 11-2 above. The axis term is **BIDEN ENEMY**. The matrix shows that Israel is Biden's enemy. It took just 45 letters in 5 columns and nine rows to show the five letters of ISRAEL in the open text with the nine letters of BIDEN ENEMY.

However, this match was insufficient to show intent, because **Israel** occurs 591 times in the open text of the Torah, so close matches are easy to find. Odds against this match were about 5 to 1. A better test is to look for matches with **ISRAEL** at the absolute skip of the axis term: ±34,128. Such a match (with **ISRAELI**) occurs only once in the Torah, on the smaller, 99-letter matrix in Figure 11-2.

Also, on the 99-letter matrix is another **ENEMY** that shares a letter with **BIDEN ENEMY**. Odds against this matrix are 55,650 to 1.

Figure 11-3 below.

MAY 20, 2016: Obama picks a homosexual as Secretary of the Army. Republican Senate caves in to Democratic agenda.

The larger matrix includes **NETANYAHU** at skip +1. While this decreases the value of the terms just discussed, the overall significance of the matrix increases. Odds against finding this matrix are over 86,589 to 1.

On May 17, 2016, the last Senate opposition (Senator Pat Roberts, R-Kansas) to Obama's homosexual choice for Secretary of the Army folded. Obama's pick was Eric Fanning. On **Figure 11-3**, **E. FANNING** is the axis term. An ELS of **OBAMA** is vertical, though only half the skip of **E. FANNING**. **ARMY** touches **E. FANNING**. Obama is shown touching **AND A MAN SHALL NOT LIE WITH A MAN** in Leviticus 20:13. The full verse for Leviticus 20:13 states, "'If a man lies with a male, as with a woman, both of them have committed an abomination: they shall surely be put to death; their blood shall be upon them." While one open text hit is shown on the report for **UNCLEAN**, a term which refers to a spiritual state that can have many causes (including homosexual acts), the term is actually seen three times on the matrix.

Figure 11-4. As can be seen below, **OBAMA** chose a homosexual to pervert our Army because **OBAMA IS SATANIC.**

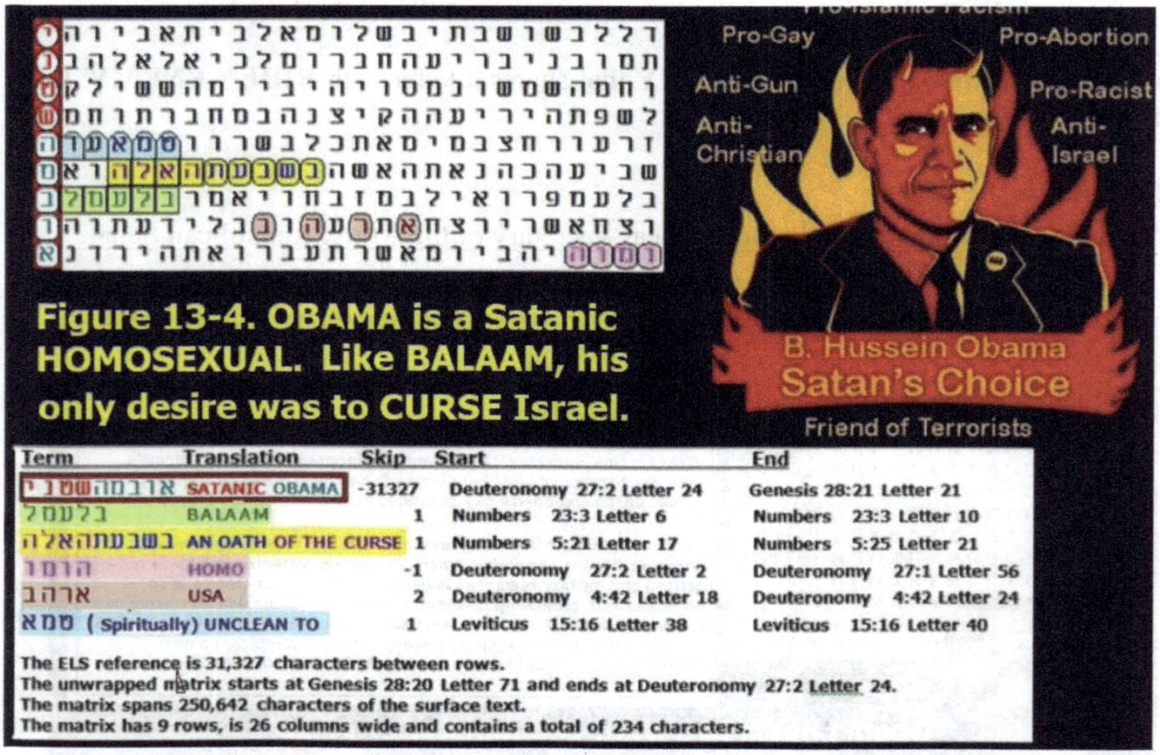

Figure 11-5 The video link to this site is https://giphy.com/gifs/xThuWbSB7BfPugnzQk?utm_source=iframe&utm_medium=embed&utm_campaign=Embeds&utm_term=

As Castro closed off the volley of questions about human rights issues in Cuba, he grabbed Obama's hand in a bizarre photo flop. Castro raised Obama's hand and it looked like he broke Obama's wrist as OB's hand went limp. **UFO** is at the same absolute skip as **OBAMA IN CUBA**. Does this show that Obama is not human?

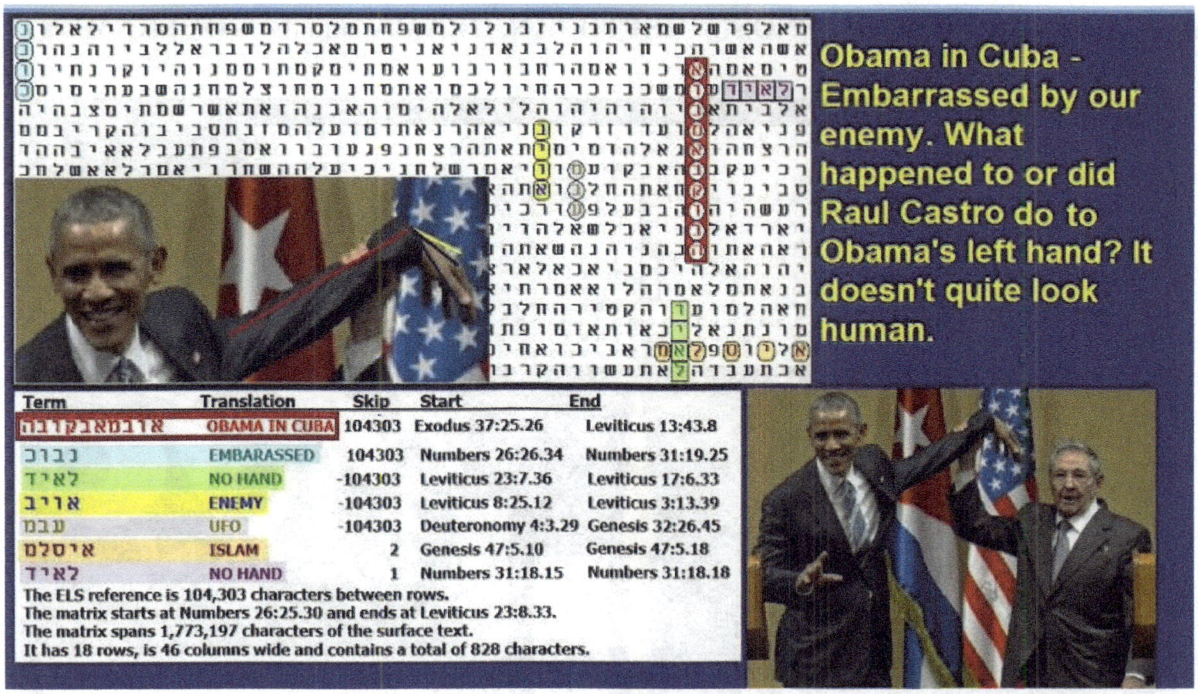

	A	B	C	D	E	F	G	H
1	AXIS TERM OBAMA IN CUBA @ Skip 104303		NUMBER	DIVIDE BY	THE QUOTIENT	E QUOTIENT	POISSON EQUATION	EACH WORD OR
2	Axis Term ELS Rank 1	USED	IN	304805	EQUALS	X 828 LETTERS	PROBABILITY FOR	PHRASE HAS 1
3		ON	TORAH	LETTERS	FREQUENCY	EQUALS	THE WORD	CHANCE IN THE
4		FIGURE		IN THE	PER LETTER	WORD	APPEARING	IN THE FOLLOWING
5				TORAH		EXPECTANCY	AT LEAST ONCE	OF BEING HERE.
6								
7	ISLAM @ Skips +2 to +207	2	24	304805	7.87389E-05	0.065195781	0.063115979	15.84384846
8	EMBARRASSSED @ Skips +/- 1 and +/-104303	104303	18	304805	5.90541E-05	0.048896836	0.047720634	20.95529572
9	ENEMY (2 synonyms) @ skips +/-1 an +/- 104303	-104303	72	304805	0.000236217	0.195587343	0.177648486	5.62909385
10	NO HAND @ Skip +1 only	1	47	304805	0.000154197	0.127675071	0.119860684	8.343019267
11	UFO @ Skips +/- 1 and +/- 104303	-104303	174	304805	0.000570857	0.472669412	0.376663895	2.654886791
12	FINAL COMBINED SIGNIFICANCE						2.41567E-05	41396.36586

Figure 11-5 Odds. As you can see above, the odds against Figure 11-5 were about 41,396 to 1. The terms at the same absolute skip as **OBAMA IN CUBA** were **EMBASSASSED** (the most significant a priori term), **ENEMY,** and **UFO**. However, the odds just quoted do not include the relatively rare 10-letter axis term, **OBAMA IN CUBA**. CodeFinder software assigns this term an R factor of 1.044, which translates to only one chance in 46.23318249 that it would be found in the Torah at least once. The true odds against this matrix existing are thus 46.23318249 × 41,396 = 1,913,868 to 1.

Figure 11-6 below. Worldwide **HYPNOSIS, REPTILIAN, OBAMA**, and the **GOVERNMENT**.

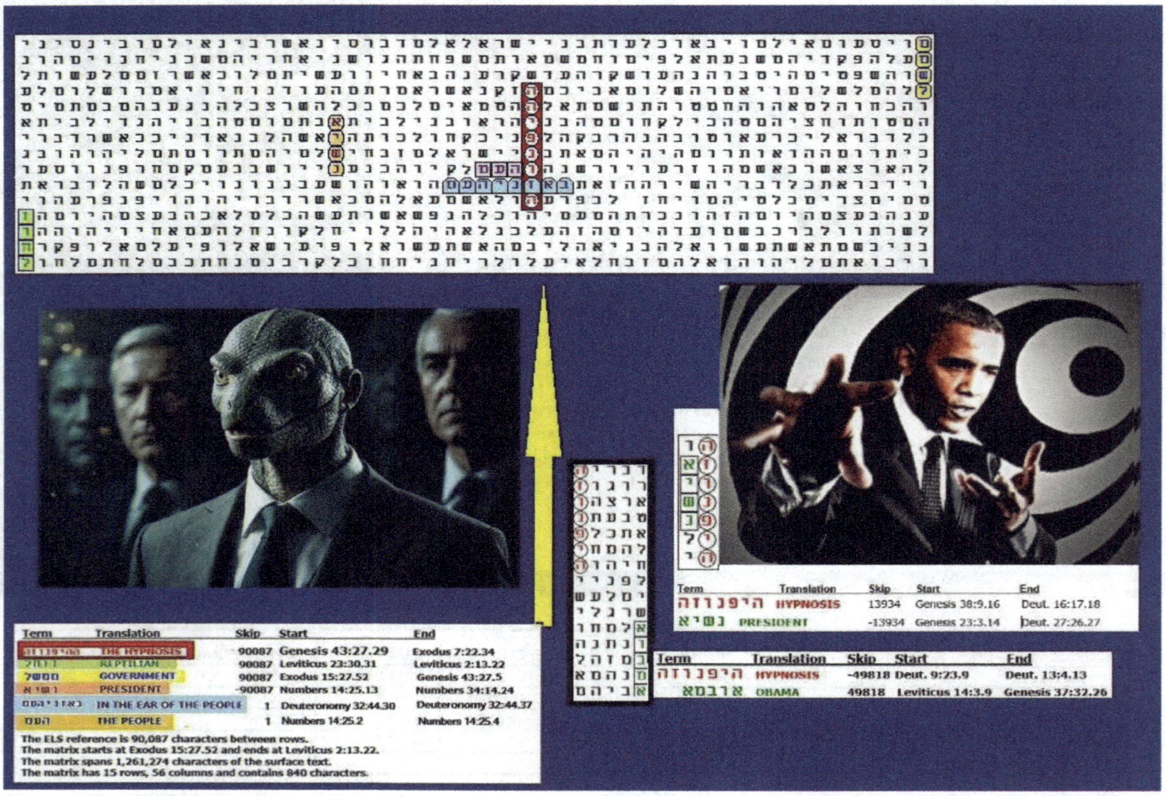

Figure 11-6 has an axis term of **THE HYPNOSIS** with **REPTILIANS, GOVERNMENT** and **PRESIDENT** all at the same absolute skip. It is a major find in that many liberal government officials are, in fact, shape-shifting reptilians who have successfully invaded our world. There are, in fact, many people, like Tampon Tim Walz, who may act like aliens because, although they were not born alien, they were abducted by aliens and hypnotized or brainwashed in an effort to keep them from having children. If ET can sterilize us, our population will shrink. That means we will be easier to conquer.

Online, I found this statement:

While there are no credible estimates of the number of people abducted by aliens, a 1991 Roper Poll found that 2% of those surveyed reported having experiences interpreted as indicating abduction, which extrapolated to approximately 3.7 million Americans. However, this interpretation and the validity of the poll results have been heavily criticized. Other researchers suggest that thousands or even millions of people across the world believe they have been abducted by aliens, according to Research Gate and ScienceDirect.com. For a more detailed understanding of these results, see https://www.researchgate.net/publication/33363035_A_Report_on_the_Demographics_and_Beliefs_of_Alien_Abduction_Experiencers.

At one point in my life, I was a registered Democrat. But by the time Obama was elected, I saw Democrats as suffering from mass insanity – so much so that President Trump had to remind us that there are only two genders – male and female. Why does Governor Tim Walz think that boys need to insert tampons?

A former Olympian running for the Senate has warned that Supreme Court nominee (now Supreme Court Justice) Judge Ketanji Brown Jackson's refusal to define a woman "could send the women's rights movement back decades."

Eli Bremer insisted that serious issues were raised by Jackson's insistence that she could not define a woman because she is "not a biologist."

"Judge" Ketanji Brown Jackson was confirmed. "I have no confidence in her ability to protect women," Bremer told Fox News Digital. "After all, how can you protect something that you cannot define?"

On **Figure 11-6**, the main axis term on the largest matrix is **THE HYPNOSIS**. At the same absolute skip are **REPTILIAN**, **GOVERNMENT**, and **PRESIDENT**. Both Obama and Biden are encoded in such a manner as to suggest that they are reptilian aliens, but **Figure 11-6** appears to clarify the issue by linking them to successful efforts to hypnotize or brainwash people. **Figure 11-6** also shows matches between **HYPNOTIZE** and **OBAMA** plus **PRESIDENT**. Of 33 encodings of **HYPNOTIZE** in Torah, three are encoded with **OBAMA** or **PRESIDENT**. **Figure 11-6** was found against odds of 2,114 to 1.

	A	B	C	D	E	F	G	H
1	Axis term: HYPNOSIS @ SKIP +90087	SKIP	NUMBER	DIVIDE BY	THE QUOTIENT	E QUOTIENT	POISSON EQUATION	EACH WORD OR
2	ELS RANK: 22	USED	IN	304805	EQUALS	X 840 LETTERS	PROBABILITY FOR	PHRASE HAS 1
3		ON	TORAH	LETTERS	FREQUENCY	EQUALS	THE WORD	CHANCE IN THE
4		FIGURE		IN THE	PER LETTER	WORD	APPEARING	IN THE FOLLOWING
5				TORAH		EXPECTANCY	AT LEAST ONCE	OF BEING HERE.
6	REPTILIAN @ Skips +/- 1 and +/- 90087	90087	2	304805	6.56157E-06	0.005511721	0.005496559	181.9320069
7	GOVERMENT @ Skips +/- 1 and +/- 90087	90087	15	304805	4.92118E-05	0.041337905	0.040495146	24.69431774
8	PRESIDENT @ Skips +/- 1 and +/- 90087	-90087	93	304805	0.000305113	0.256295008	0.226086376	4.423088284
9	THE PEOPLE @ Skip +1 only	1	343	304805	0.00112531	0.557028264	0.427090929	2.341421773
10	INITIAL COMBINED ODDS = (H6*H7*H8*H9)/22						0.000472837	2114.894569

Figure 11-7 OBAMA KILLS OUR MANNED SPACE FLIGHT PROGRAM.

"When I became the NASA administrator, Obama charged me with three things," NASA head Charles Bolden said in a recent interview with the Middle Eastern news network al-Jazeera. "Help me to help re-inspire children to want to get into science and math; to expand our international relationships; and perhaps foremost, he wanted me to find a way to reach out to the Muslim world and engage much more with dominantly Muslim nations to help them feel good about their historic contribution to science, math, and engineering." Why did he sabotage our space program. Figure 11-7 gives us some hints about his likely (Reptilian) motives. The axis term is **MARS DELAY** at skip +70,216. Two terms immediately jump out at us. The first is **ALIEN** at the absolute skip of the axis term. The second is the only ELS of **B. OBAMA** at skip +/-1. This matrix also includes NAZI at skip -1, **HITLER** at skip 70,215, **WAR** in the open text, **HIS ARMY** at skip +1, **DUST FROM THE SKY** (dust plays a major role in weather on Mars), and (separately) both **HAIM** and **ESHED**. **HAIM ESHED** is the retired Israeli Brigadier General who ran Israel's space program for 30 years. He claims that the U.S. military has a joint base underground on Mars with aliens. When we include the axis term, odds against this matrix are about 156 billion to 1.

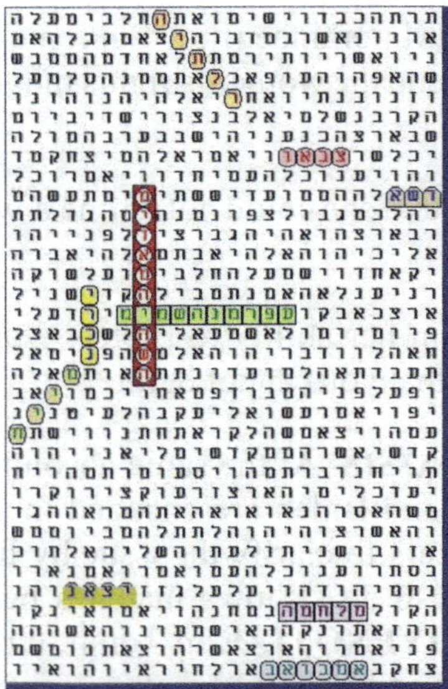

Term	Translation	Skip	Start	End
השהיהמאדים	MARTIAN DELAY	70216	Numbers 7:5.32	Leviticus 23:3.5
נכרי	ALIEN	-70216	Exodus 33:11.15	Numbers 20:12.3916
עפרמנהשמים	DUST FROM THE SKY	1	Deut. 28:24.21	Deuteronomy 28:24.30
היתלר	HITLER	70215	Leviticus 9:20.9	Exodus 34:16.28
נאצי	NAZI	-1	Genesis 38:12.50	Genesis 38:12.47
מלחמה	WAR	1	Exodus 32:17.36	Exodus 32:17.40
באובמא	B. OBAMA	-1	Genesis 24:62.12	Genesis 24:62.7
צבאו	HIS ARMY	1	Genesis 26:26.36	Genesis 26:26.39
חיים	CHAIM (life)	-70217	Exodus 18:7.21	Numbers 7:5.36
אשד	ESHED	-1	Leviticus 23:2.51	Leviticus 23:2.49

The ELS reference is 70216 characters between rows.
The matrix starts at Leviticus 9:19.42 and ends at Genesis 24:62.26.
It spans 2,317,152 characters of the surface text.
The matrix has 34 rows, 24 columns and contains 816 characters.

	A	B	C	D	E	F	G	H
1	AXIS TERM MARTIAN DELAY @ Skip +70,216	SKIP	NUMBER	DIVIDE BY	THE QUOTIENT	E QUOTIENT	Poisson Equati	EACH WORD
2	ELS RANK 1	USED	IN	304805	EQUALS	x 816 Letters	Probability for	OR PHRASE
3	AXIS TERM ELS RANK = 1	ON	TORAH	LETTERS	FREQUENCY	Equals	the Word	HAS 1 CHANCE
4	R FACTOR = 0.596 VALUE = 3.94573021	FIGURE		IN THE	PER LETTER	WORD	Appearing	IN THE
5				TORAH		EXPECTANCY	At Least once	FOLLOWING
6	DUST FROM THE SKY @ Skip +1 only	1	1	304805	3.28079E-06	0.002677121	0.002673541	374.0357623
7	B. OBAMA (2 spellings) +/- 1 and +/- 70216@ Skips	-1	2	304805	6.56157E-06	0.005354243	0.005339934	187.2682158
8	NAZI @ Skips +/- 1 and +/- 70216	-1	8	304805	2.62463E-05	0.021416972	0.021189257	47.19372714
9	ALIEN @ Skips +/-1 and +/- 70216	-70216	19	304805	6.23349E-05	0.050865307	0.049593325	20.16400381
10	WAR @ Skip +1 only	1	36	304805	0.000118108	0.096376372	0.09187784	10.88401732
11	ESHED @ Skips +/- 1 and /- 70216	-1	90	304805	0.000295271	0.240940929	0.214111952	4.670453891
12	ARMY @ Skip +1 only	1	108	304805	0.000354325	0.289129115	0.251084497	3.98272299
13	HITLER (2 spellings) @ Skips +70215 to +70400	70215	156	304805	0.000511803	0.417630944	0.341394754	2.929160413
14	HAIM @ Skips 70217 to 70560	70217	2197	304805	0.007207887	5.8816358	0.997209783	1.002798024
15	Combined odds against this matrix (not counting the axis term)						2.52276E-11	39639054687
16								
17	MARTIAN DELAY R FACTOR = 0.596 VALUE = 3.94573021						0.253438514	3.94573021
18	FINAL ODDS AGAINST MATRIX = H15*H17 (about 156 billion to 1)						6.39366E-12	1.56405E+11

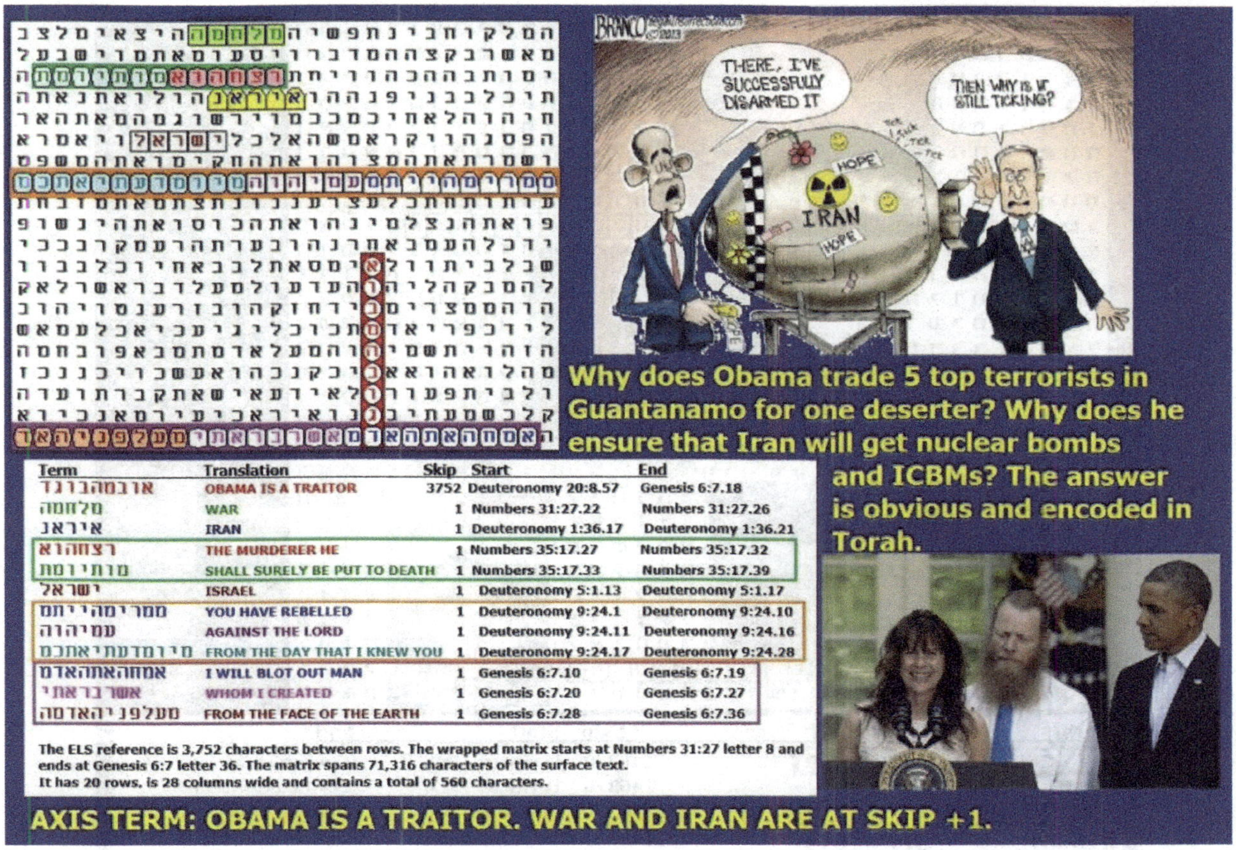

Term	Translation	Skip	Start	End
אובמהבוגד	OBAMA IS A TRAITOR	3752	Deuteronomy 20:8.57	Genesis 6:7.18
מלחמה	WAR	1	Numbers 31:27.22	Numbers 31:27.26
איראן	IRAN	1	Deuteronomy 1:36.17	Deuteronomy 1:36.21
רצחהוא	THE MURDERER HE	1	Numbers 35:17.27	Numbers 35:17.32
מותיומת	SHALL SURELY BE PUT TO DEATH	1	Numbers 35:17.33	Numbers 35:17.39
ישראל	ISRAEL	1	Deuteronomy 5:1.13	Deuteronomy 5:1.17
ממריםהייתם	YOU HAVE REBELLED	1	Deuteronomy 9:24.1	Deuteronomy 9:24.10
עםיהוה	AGAINST THE LORD	1	Deuteronomy 9:24.11	Deuteronomy 9:24.16
מיוםדעתיאתכם	FROM THE DAY THAT I KNEW YOU	1	Deuteronomy 9:24.17	Deuteronomy 9:24.28
אמחהאתהאדם	I WILL BLOT OUT MAN	1	Genesis 6:7.10	Genesis 6:7.19
אשרבראתי	WHOM I CREATED	1	Genesis 6:7.20	Genesis 6:7.27
מעלפניהאדמה	FROM THE FACE OF THE EARTH	1	Genesis 6:7.28	Genesis 6:7.36

The ELS reference is 3,752 characters between rows. The wrapped matrix starts at Numbers 31:27 letter 8 and ends at Genesis 6:7 letter 36. The matrix spans 71,316 characters of the surface text. It has 20 rows, is 28 columns wide and contains a total of 560 characters.

AXIS TERM: OBAMA IS A TRAITOR. WAR AND IRAN ARE AT SKIP +1.

Figure 11-8. **OBAMA IS A TRAIT**OR is the axis term. Obama used his power to betray the U.S. and Israel by reaching a bad nuclear deal with Iran. One of two Torah occurrences of **IRAN** at skip +1 is on the matrix. Obama tried to force Israel to go to war with Iran, and **WAR** is shown in the open text. On the line above, **OBAMA is a TRAITOR** runs into **I WILL BLOT OUT MAN**. In this phrase, BARACK is at skip -1. Sharing letters with **AND PUT AWAY** is **USA.** Obama wants Israel to return to its pre-1967 borders, and he wants Israel in Muslim-terrorist hands. On the matrix is **AND THEY ALSO POSSESS THE LAND**, right under **IRAN, FOR ALL ISRAEL** is under that. Obama spent millions of dollars trying to defeat Netanyahu.

Figure 11-9 below. JULY 6, 2015: Obama, Gay Marriage, Iran (4-letter transliteration), and An Atomic Holocaust.

Term	Translation	Skip	Start	End
מלחמתעלם	WORLD WAR	146320	Leviticus 13:52.71	Deuteronomy 12:2.17
אובמה	OBAMA	-146320	Leviticus 7:2.6	Leviticus 20:25.63
הלצאירן	THE FOOL IRAN	146320	Exodus 15:2.36	Genesis 42:11.25
תשעה	5775/2015	146320	Exodus 30.27.3	Deuteronomy 4:34.3
קרי	KERRY	1	Numbers 6:14.3	Numbers 6:14.5
תאבדוןאת	YOU SHALL SURELY DESTROY	1	Deuteronomy 12:2.4	Deuteronomy 12:2.11
כלהמקמות	ALL THE PLACES	1	Deuteronomy 12:2.12	Deuteronomy 12:2.19
אתאימתיאשלח	I WILL SEND MY TERROR	1	Exodus 23:27.1	Exodus 23:27.11
ישהומו	THERE IS A HOMOSEXUAL	-3	Genesis 35:10.16	Genesis 35:10.1
לאדםלאמצא	FOR MAN THERE WAS NOT FOUND	1	Genesis 2:20.44	Genesis 2:20.52
עזרכנגדו	A HELPER FIT FOR HIM	1	Genesis 2:20.53	Genesis 2:20.60

The ELS reference is 146,320 characters between rows.
The wrapped matrix starts at Leviticus 13:52 letter 58 and ends at Leviticus 7:2 letter 6
The matrix spans 3,950,676 characters of the surface text. It has 28 rows, 36 columns and 1,036 letters.

Figure 11-9 WORLD WAR, the Hebrew year (5775) and **OBAMA** are encoded at the same absolute skip. So is **THE FOOL IRAN**, although the transliteration of **IRAN** is not the most commonly used one. **YOU SHALL SURELY DESTROY ALL THE PLACES** in the open text crosses **WORLD WAR**. On the line below is **I WILL SEND MY TERROR**. Secretary of State **KERRY** (who worked hard to ensure that Iran would get the Bomb) has his name at skip +1. On the line above, **KERRY** is **THERE IS A HOMOSEXUAL**, but it isn't at a special case skip. However, in the open text on the matrix is **FOR ADAM (MAN) THERE WAS NOT FOUND A HELPER FIT FOR HIM**. This phrase in Genesis 2:2 refers to the idea that after Adam named all the animals, he didn't find one who could be his mate.

Running through the letter *vav* in **OBAMA** is Genesis 2:3, which describes God taking a rib from Adam so He could create a woman (Eve) as Adam's partner. But Obama, who Newsweek magazine described as America's first gay president, has pushed hard for gay marriage, and when a perverted court has backed him, he's defiled the White House by bathing it in gay movement rainbow colors. In Iran, gay behavior is a capital crime. American support for gay marriage provides an incentive for Iran and Islam to want us dead.

Note: 5775 did not necessarily imply that nuclear war would actually break out before the start of the next Hebrew year on the night of September 13, 2015. More likely, 5775 was the year of the causes of the eventual war—Obama agreeing to a nuclear deal allowing Iran 24 days to move to nuclear material before inspectors could get there, and the legalization of gay marriage. Overall, after factoring in ELS rank 2 of **OBAMA TRAITOR**, this matrix was found against odds of about 16,405 to 1.

Figure 11-9 JUNE 14, 2015: Obama again says he has no strategy for ISIS. On August 28, 2014, when the president was asked if he would seek Congressional approval for U.S. attacks on ISIS targets in Syria, he responded, "I don't want to put the cart before the horse. We don't have a strategy yet." On June 8, 2015, he again admitted that we don't yet have a complete strategy. While these statements dumbfounded most Americans, and, of course, his decision to release five top Taliban terrorists from Guantanamo in exchange for a known deserter (Bergdahl) was really weird, there really was no mystery here.

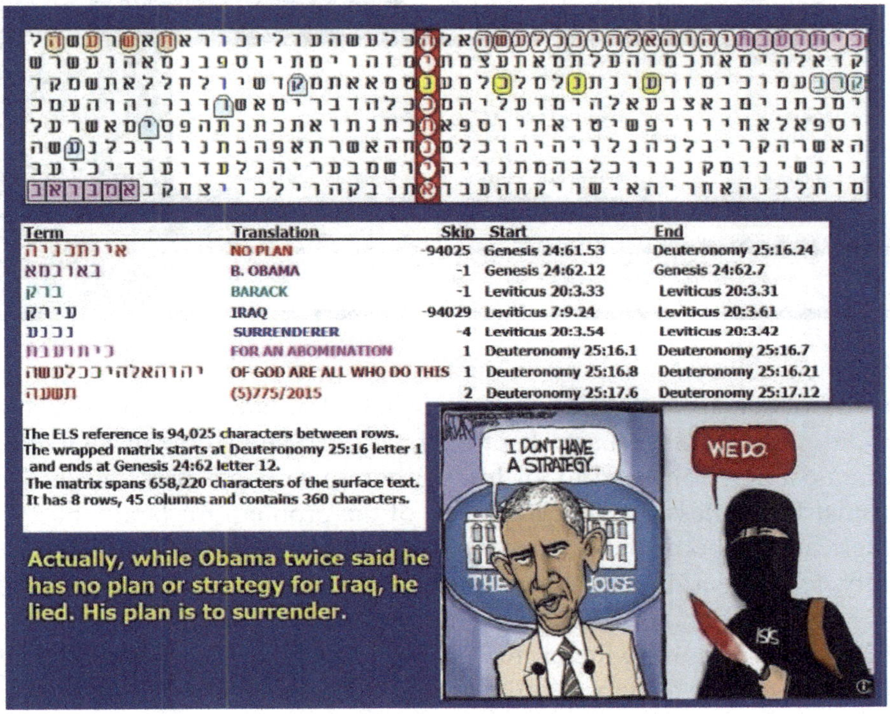

Term	Translation	Skip	Start	End
אינתכניה	NO PLAN	-94025	Genesis 24:61.53	Deuteronomy 25:16.24
באובמא	B. OBAMA	-1	Genesis 24:62.12	Genesis 24:62.7
ברק	BARACK	-1	Leviticus 20:3.33	Leviticus 20:3.31
עירק	IRAQ	-94029	Leviticus 7:9.24	Leviticus 20:3.61
נכנע	SURRENDERER	-4	Leviticus 20:3.54	Leviticus 20:3.42
כיתועבה	FOR AN ABOMINATION	1	Deuteronomy 25:16.1	Deuteronomy 25:16.7
יהוהאלהיככלעשה	OF GOD ARE ALL WHO DO THIS	1	Deuteronomy 25:16.8	Deuteronomy 25:16.21
תשעה	(5)775/2015	2	Deuteronomy 25:17.6	Deuteronomy 25:17.12

Above: Figure 11-10. The axis termF is NO PLAN. The best a priori term found with it was B. OBAMA.

Obama was a traitor who truly hated America, and his strategy was to use the race card to ensure the defeat of the United Stated on all fronts for as long as he was President. Overall, after factoring in ELS rank 21 of the axis term, the matrix was found against odds of about 52,946 to 1. The two most significant terms were **B. OBAMA** at skip -1 against odds of about 847 to 1, and **ABOMINATION** found in the open text against odds of about 106 to 1.

Figure 11-11: APRIL 6, 2015: OBAMA SURRENDER IN NUCLEAR NEGOTIATIONS WITH IRAN

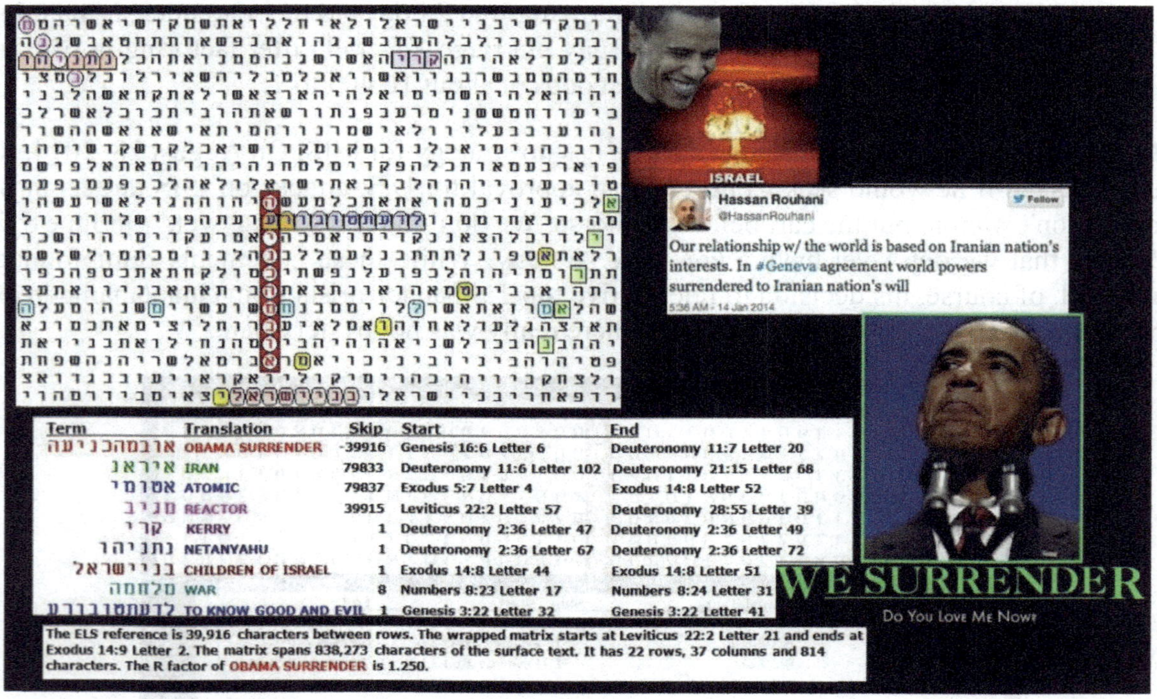

Figure 11-11 above. The deadline for an interim nuclear agreement with Iran was on March 31, 2015. That date came and went without an agreement, but Obama's hatred of Israel and, in particular, of Israeli Prime Minister Netanyahu did not pass. So, Obama kept Secretary of State Kerry in Switzerland until he got an agreement that will almost certainly lead to an Iranian bomb. This put pressure on Israel to use tactical nuclear weapons against Iran unless the U.S. took care of the problem for Israel. Fast forward to June 21, 2025, and seven American B-2 bombers dropped twelve 30,000-pound GBU-57A/B bombs at the Fordow Iranian nuclear facility deep underground, and two more at Netanz. At Fordow twelve bombs times 30,000 pounds each equals 360,000 pounds of explosive force. There are 2,000 pounds in a kiloton. Dividing 360,000 by 2,000 equals 180 kilotons, which is the size of a tactical nuclear weapon. The atomic bomb that destroyed Hiroshima had the force of only 13 kilotons. Nagasaki was destroyed with a 21-kiloton bomb.

On the matrix above, the axis term is **OBAMA SURRENDER**. It's 10 letters in Hebrew, which is long for an axis term. In the open text, the expression **TO KNOW GOOD AND EVIL** runs into it, with **EVIL** sharing a letter *ayin* with **SURRENDER**. **KERRY** and **NETANYAHU** are both at skip +1 and on the same line. **REACTOR** is at an ELS, crossing and sharing a letter with **NETANYAHU**. **ATOMIC** is at an ELS touching the last letter of **CHILDREN OF ISRAEL**. **WAR** is also at an ELS. Normally, I would just combine all the odds and state that the matrix was found against odds of about 27,915 to 1. But normally, I don't assign any value to the axis term. Where I make exceptions to this rule, the axis term must be at least 9 letters long. Again, here it's 10 letters. This term, **OBAMA SURRENDER**, only had about one chance in 17.782 to be found at all in the wrapped Torah. So, if we consider this fact, then we arrive at odds against the matrix really being about 497,415 to 1.

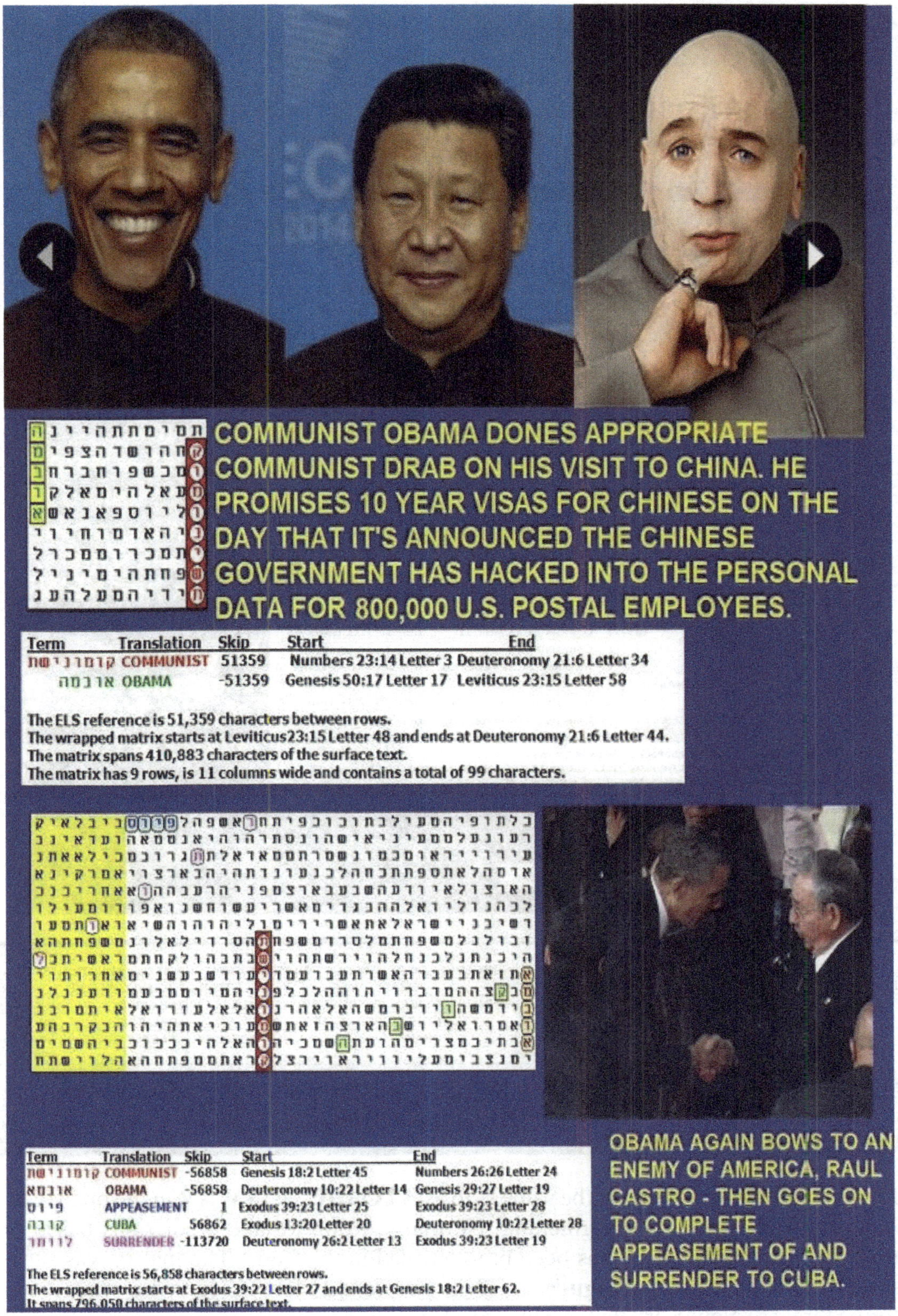

Figure 11-12 The Axis term **A COMMUNIST** is found twice in Torah Code. This one is at the same skip

as **OBAMA**. **APPEASEMENT** is at skip +1.

Figure 11-13 below SEPTEMBER 10, 2014: A nuclear nightmare is found in the Code. But is it just a dream? Once again, B. Obama is strongly encoded with it.

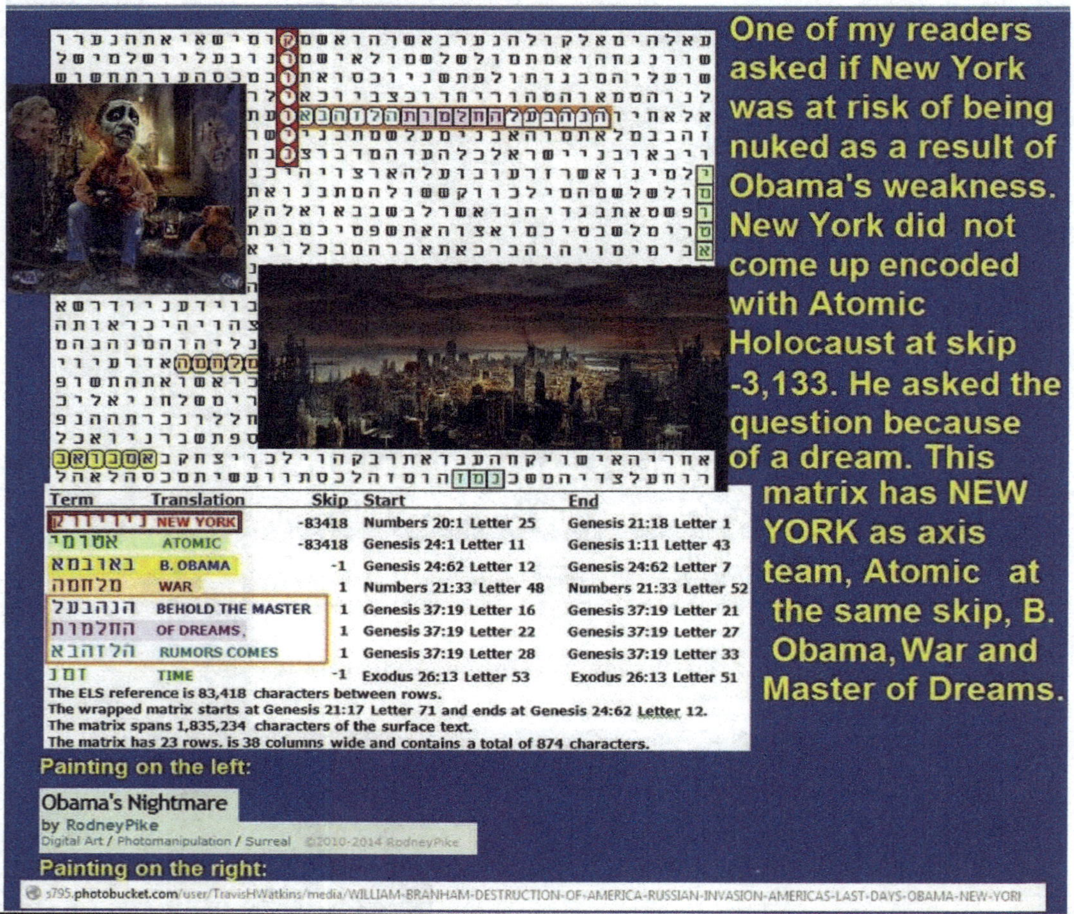

	A	B	C	D	E	F	G	H
1	AXIS TERM NEW YORK @ Skip -83,418	SKIP	NUMBER	DIVIDE BY	THE QUOTIENT	E QUOTIENT	POISSON EQUATION	EACH WORD
2	ELS RANK 233	USED	IN	304805	EQUALS	x 874 LETTERS	PROBABILITY FOR	OR PHRASE
3		ON	TORAH	LETTERS	FREQUENCY	EQUALS	THE WORD	HAS 1 CHANCE
4		FIGURE		IN THE	PER LETTER	WORD	APPEARING	IN THE
5				TORAH		EXPECTANCY	AT LEAST ONCE	FOLLOWING OF
6								BEING HERE
7	ATOMIC @ Skips +/- 1 and +/- 83418	-83418	1	304805	3.28079E-06	0.002867407	0.0028633	349.2473785
8	B. OBAMA @ Skips +/- 1 and +/- 83418	-1	1	304805	3.28079E-06	0.002867407	0.0028633	349.2473785
9	WAR @ Skips +1 only	1	36	304805	0.000118108	0.103226653	0.098077473	10.19602124
10	TIME @ Skips +/- 1 and +/- 83418	-1	27	304805	8.85812E-05	0.07741999	0.074498929	13.42301175
11	Combined odds against the matrix = (h7*h8*h9*h10)/233						1.39575E-05	71645.85843

Figure 11-13 above: **NEW YORK** is at the same skip as **ATOMIC**, with **WAR** in the open text. Sometimes a matrix looks beautiful, sometimes frightening. Normally, I don't make predictions based on the Torah Code; I just like to show how history has been pre-recorded in the Code. The matrix was shown earlier as **Figure 9-26** and is repeated here as **Figure 11-12.** It was found after I received a request on my blog to look at an atomic threat to New York that might result from Obama's weakness, or immigrants who had vanished from the public radar. Odds against this matrix were about 71,645 to 1.

Term	Translation	Skip	Start	End
קוראמחשבן	MIND READER	135088	Leviticus 23:30 Letter 51	Genesis 28:12 Letter 53
מחשב	COMPUTER	135088	Exodus 21:8 Letter 40	Numbers 12:5 Letter 9
אובמא	OBAMA	-135088	Leviticus 19:16 Letter 16	Numbers 31:16 Letter 16
ברקה	BARACK H.	-1	Leviticus 2:12 Letter 3	Leviticus 2:11 Letter 63
ובערת	AND PUT AWAY	1	Deuteronomy 17:7 Letter 40	Deuteronomy 17:7 Letter 44
הרעמקרבכ	THE EVIL IN YOUR MIDST	1	Deuteronomy 17:7 Letter 45	Deuteronomy 17:7 Letter 52
רכילבעמיכ	A TALEBEARER AMONG YOUR PEOPLE	1	Leviticus 19:16 Letter 6	Leviticus 19:16 Letter 14

The ELS reference is 135,088 characters between rows.
The wrapped matrix starts at Leviticus 23:30 Letter 40 and ends at Leviticus 19:16 Letter 34.
The matrix spans 2,431,613 characters of the surface text. It has 19 rows, 29 columns and a total of 551 characters.

Figure 11-14 above: AUGUST 28, 2014: Mind Reading Computer and Obama. MIND READER, **COMPUTER** and **OBAMA** are all at the same. Skip. The matrix also has **AND PUT AWAY THE EVIL IN YOUR MIDST** and warns about **A TALEBEARER AMONG YOUR PEOPLE.**

Figure 11-15 November 8, 2013: Obama apologizes for broken promise to keep your healthcare plan.

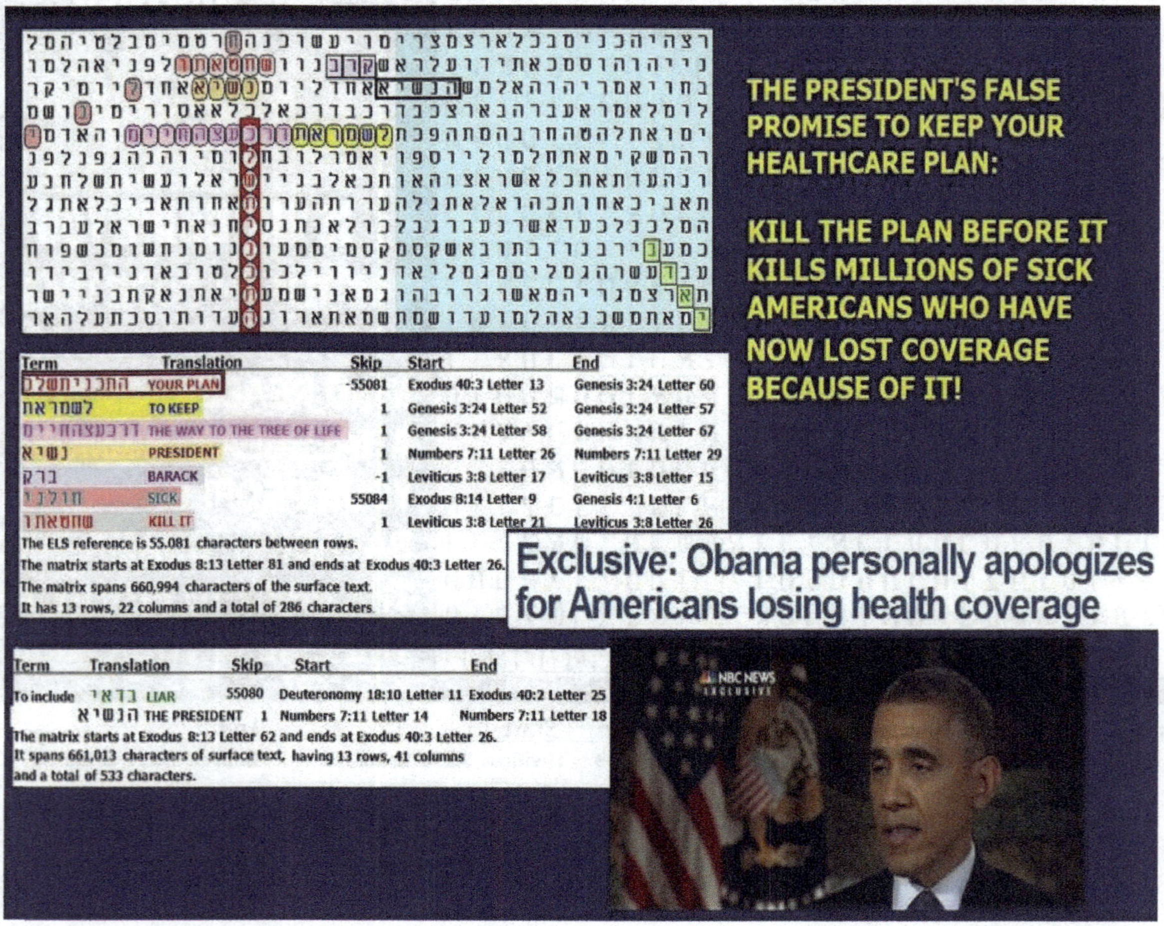

Figure 11-16 April 7, 2013: Atomic holocaust, Obama, and North Korea.

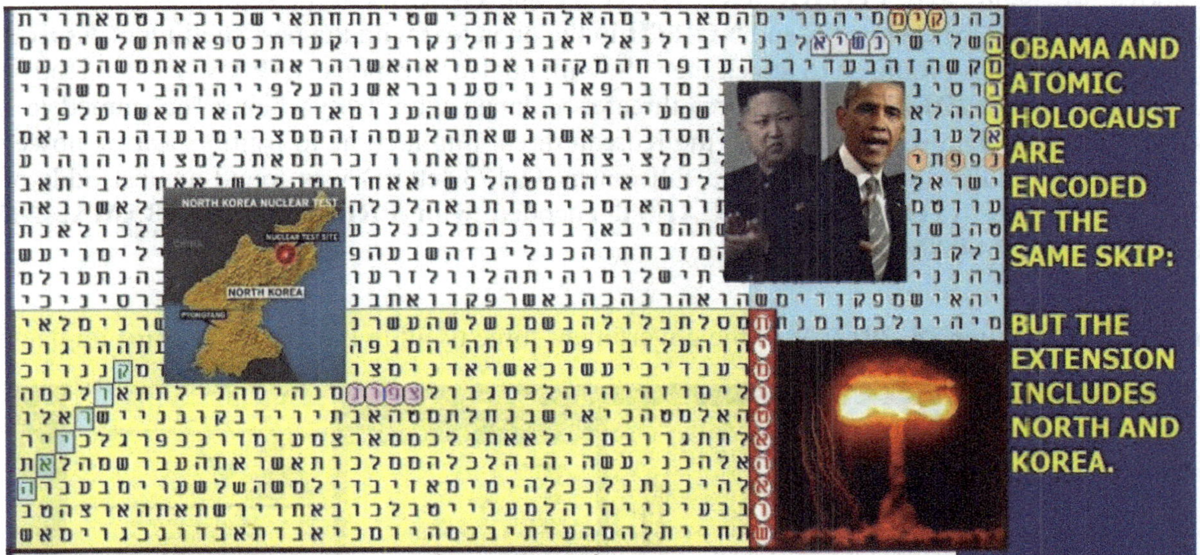

Term	Translation	Skip	Start	End
שואהאטומית	ATOMIC HOLOCAUST	-3133	Deuteronomy 8:19 Letter 52	Numbers 29:9 Letter 5
אובמה	OBAMA	-3133	Numbers 14:19 Letter 5	Numbers 7:24 Letter 5
צפונ	NORTH	1	Numbers 34:7 Letter 15	Numbers 34:7 Letter 18
קוריאה	KOREA	3134	Numbers 32:26 Letter 11	Deuteronomy 4:41 Letter 22
קים	KIM	1	Numbers 5:19 Letter 59	Numbers 5:19 Letter 61
נשיא	PRESIDENT	1	Numbers 7:24 Letter 11	Numbers 7:24 Letter 14
יפנ	JAPAN	-2	Numbers 15:38 Letter 68	Numbers 15:38 Letter 64

The ELS reference is 3,133 characters between rows.
The unwrapped matrix starts at Numbers 5:19 Letter 56 and ends at Deuteronomy 8:20 Letter 7.
The matrix spans 68,977 characters of the surface text, having 23 rows, 51 columns and a total of 1173 characters.

	A	B	C	D	E	F	G	H
1	*AXIS TERM ATOMIC HOLOCAUST @ Skip -3,133*	SKIP	NUMBER	DIVIDE BY	THE QUOTIENT	E QUOTIENT	POISSON EQUATION	EACH WORD
2	Axis Term ELS Rank 1 Full matrix	USED	IN	304805	EQUALS	X 1173 LETTERS	PROBABILITY FOR	OR PHRASE
3		ON	TORAH	LETTERS	FREQUENCY	EQUALS	THE WORD	HAS 1 CHANCE
4		FIGURE		IN THE	PER LETTER	WORD	APPEARING	IN THE FOLLOWING
5				TORAH		EXPECTANCY	AT LEAST ONCE	OF BEING HERE
6	KOREA @ Skips 3134 to 3407	3134	7	304805	2.29655E-05	0.026938534	0.026578928	37.62378912
7	NORTH @ Skip +1 only	1	11	304805	3.60886E-05	0.042331983	0.041448495	24.12632846
8	OBAMA @ Skips +/-1 and +/- 3133	-3133	3	304805	9.84236E-06	0.011545086	0.011478657	87.11789872
9	KIM @ Skip +1 only	1	80	304805	0.000262463	0.307868965	0.264988376	3.773750433
10	JAPAN @ skip +2 to +29	-2	612	304805	0.002007841	2.355197585	0.905125241	1.104819482
11	PRESIDENT@ Skip +1 only	1	68	304805	0.000223093	0.261688621	0.230249329	4.343117975
12	Final Combined Significance						6.98349E-07	1431948.126
13								
14	*AXIS TERM ATOMIC HOLOCAUST @ Skip -3,133*	SKIP	NUMBER	DIVIDE BY	THE QUOTIENT	E QUOTIENT	POISSON EQUATION	EACH WORD
15	Axis Term ELS Rank 1 Full matrix	USED	IN	304805	EQUALS	X 182 LETTERS	PROBABILITY FOR	OR PHRASE
16	This is window 1 (light blue background only)	ON	TORAH	LETTERS	FREQUENCY	EQUALS	THE WORD	HAS 1 CHANCE
17		FIGURE		IN THE	PER LETTER	WORD	APPEARING	IN THE FOLLOWING
18				TORAH		EXPECTANCY	AT LEAST ONCE	OF BEING HERE
19	OBAMA @ Skips +/-1 and +/- 3133	-3133	3	304805	9.84236E-06	0.001791309	0.001789706	558.751065
20	KIM @ Skip +1 only	1	80	304805	0.000262463	0.047768245	0.046645254	21.43838988
21	JAPAN @ skip +2 to +29	-2	612	304805	0.002007841	0.365427076	0.306099761	3.266908793
22	PRESIDENT@ Skip +1 only	1	68	304805	0.000223093	0.040603008	0.03978975	25.13210036
23	Final Combined Significance						1.01677E-06	983504.4376

Figure 11-16 above: Ever since Michael Drosnin published his first best-selling book (THE BIBLE CODE) in 1997, we have known that there is a 10-letter ELS for **ATOMIC HOLOCAUST** in the Code at skip -3,133. But what we did not know is that **OBAMA** is very close by, at the same skip. **PRESIDENT** is at skip +1. **NORTH** and **KOREA** are also on the matrix. The matrix has 23 rows, 51 columns, and 1,173 letters.

On January 1, 2012, I published a matrix with these terms and Iran. But I did not look for Korea. On the matrix above, we again have **ATOMIC HOLOCAUST** and **OBAMA** at the same skip, but we also have **KIM** (as in Kim Jong Un) at skip +1, and **NORTH** in the open text with the best spelling of **KOREA** very near it. Kim Jong Un has promised to attack the U.S. with nuclear weapons. **JAPAN** is also on the matrix at skip -2, touching **PRESIDENT**. North Korea threatened Japan with Tokyo being "consumed in nuclear flames." We don't yet know the year when the Korean threat might come to fruition. Dates, however, are rarely reliable on a matrix. As for odds, the full matrix (1,173 letters) was found against odds of about 1,431,948 to 1, with **ATOMIC HOLOCAUST** encoded with **PRESIDENT**, **OBAMA**, **NORTH**, **KOREA**, **KIM**, and **JAPAN**.

Figure 11-17 September 28, 2012: Will the government force us all to have identity chips implanted?

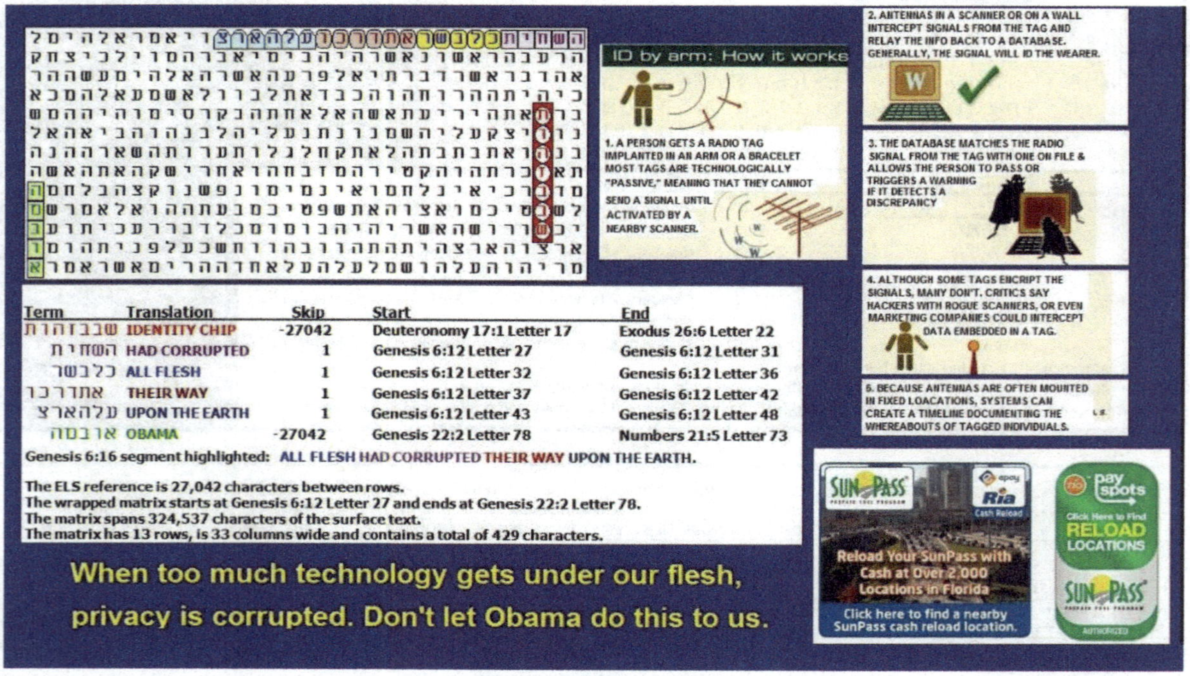

I once received a question from one of my readers about the possibility that President Obama would, if re-elected, ask U.S. military forces to have identity chips embedded in their skin. Before examining the matrix below, let me state that in attempting to look for any possible relationship in the above, because Obama is such a high-frequency term, I would only accept his name when it appears at skip +/-1 or the same skip as the axis term, here **IDENTITY CHIP**. On the matrix above, both **IDENTITY CHIP** and **OBAMA** are at skip -27,042. This match occurs at the 10th lowest ELS of the axis term. The reader's concern was that eventually all Americans (and in the end all of humanity) would be forced to have such a chip inserted into their flesh, thus making it easier for the leader of a world government to control all people. In the open text on this matrix is "**ALL FLESH HAD CORRUPTED THEIR WAY UPON THE EARTH**" (Genesis 6:12).

So, what is the status of identity chips in the U.S.? Elon Musk's company, **Neuralink**, has implanted brain chips in humans, and it's the first company to do so. The first human implant was performed in January, 2024 on a quadriplegic patient, and the patient was able to control a computer mouse using their thoughts. **Neuralink** is working to develop brain-computer interfaces that could potentially help people with paralysis or other neurological conditions regain lost function.

While no single country is officially implanting chips in humans, Sweden has seen a notable number of individuals voluntarily getting microchips implanted for various purposes, including gym access, e-tickets, and storing emergency contact information. Some Swedes have also used the chips to replace key cards and as a form of contactless payment. In 2024, it was estimated that between 50,000 and 100,000 people have been chipped.

Figure 11-18 September 13, 2012: Obama's Kenyan birth. There is an abundance of evidence that Obama was born in Mombasa, Kenya. Even his own first book backs this idea on page 307. There is reason to believe that even though Obama is no longer in office, his crimes were so great that he could still be impeached, but the Code is clear that he was born in Kenya and was thus never eligible to be our president. The possibility that his parents were aliens in discussed in Chapter 3. On Figure 11-18 **KENYA** is intersected and shares a letter hey with **HE WILL BE LIKE A FOREIGNER (alien).** Also, in sequence right before **OBAMA KENYA** is the same word for **FOREIGNER/ALIEN.**

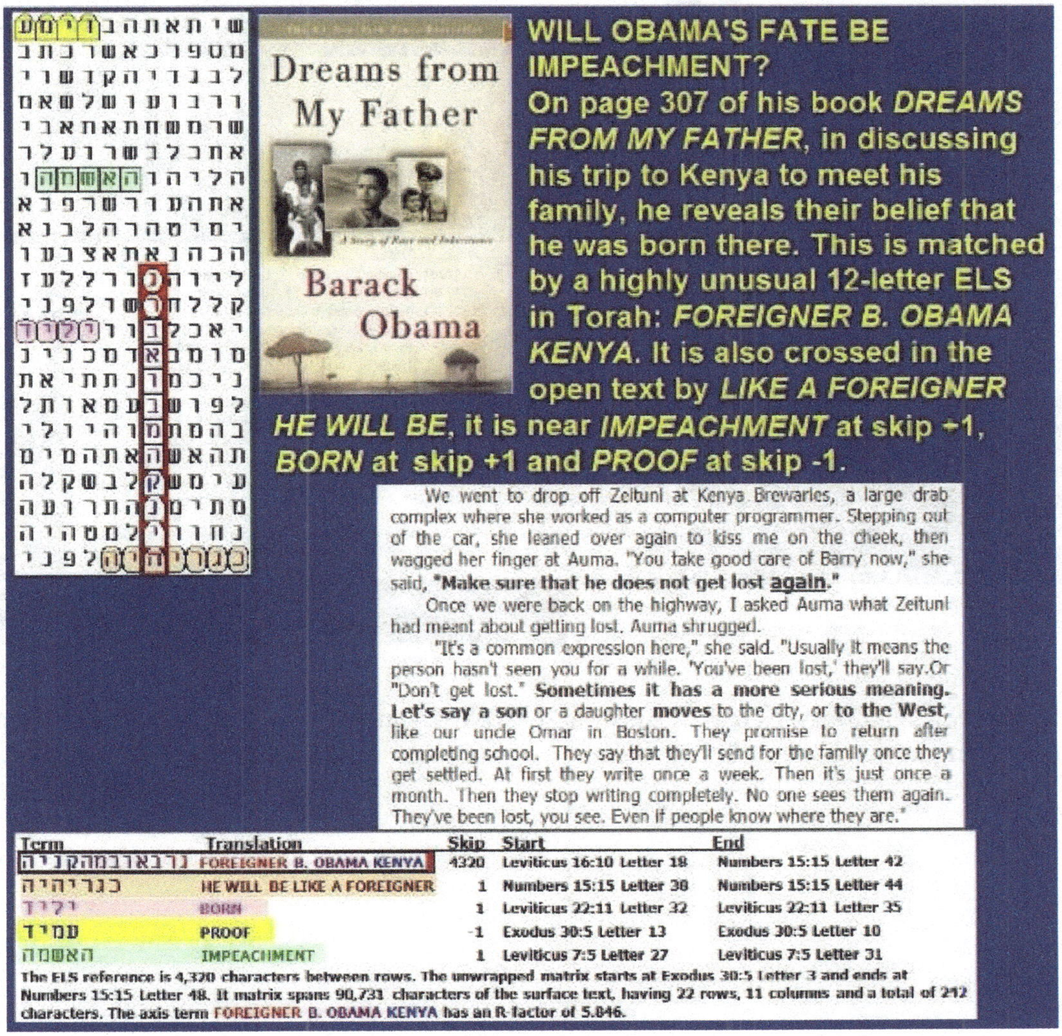

Term	Translation	Skip	Start	End
נכרבאבמהקניה	FOREIGNER B. OBAMA KENYA	4320	Leviticus 16:10 Letter 18	Numbers 15:15 Letter 42
כנכריהיה	HE WILL BE LIKE A FOREIGNER	1	Numbers 15:15 Letter 30	Numbers 15:15 Letter 44
יליד	BORN	1	Leviticus 22:11 Letter 32	Leviticus 22:11 Letter 35
עמיד	PROOF	-1	Exodus 30:5 Letter 13	Exodus 30:5 Letter 10
האשמה	IMPEACHMENT	1	Leviticus 7:5 Letter 27	Leviticus 7:5 Letter 31

The ELS reference is 4,320 characters between rows. The unwrapped matrix starts at Exodus 30:5 Letter 3 and ends at Numbers 15:15 Letter 48. It matrix spans 90,731 characters of the surface text, having 22 rows, 11 columns and a total of 242 characters. The axis term FOREIGNER B. OBAMA KENYA has an R-factor of 5.846.

Figure 11-18 above: Proof that Obama was born in Kenya.

On **Figure 11-19**, the axis term is **ILLUMINATI**. **OBAMA** and **HITLER** are at the same skip, **CLINTON** and **BUSH** are both at skip -1, and **KENNEDY** is there too, but not at a special case skip. An Illuminati goal of one world law/order is hinted at in the open text. Without including **CLINTON**, the matrix was found against odds of about 26 million to 1. The larger matrix with **CLINTON** was found against odds of about 5.6 million to 1.

Statistical Significance of Figure 11-19. I don't like to publish matrices that are over 1,000 letters in area. If we just look at everything from **BUSH** down through **KENNEDY**, that's just 37 columns by 12 rows (a 444-letter area). This turns out to be the most significant cut. Odds against it are about 37.7 million to 1. The bulk of this number is due to **HITLER** and **OBAMA**. **HITLER** and **OBAMA**, and **BARACK** are found with **ILLUMINATI** in 297 letters against odds of about 427,522 to 1.

The 740-letter section without CLINTON (is slightly more statistically significant than the full 1,184-letter matrix). Read ILLUMINATI and the ILLUMINATI DIAMOND inverted.

Figure 11-19 above. **ILLUMINATI, HITLER**, and **OBAMA** are all at the same absolute skip. Odds against this matrix are about 37,792,796 to 1.

33								
34	Axis Term: ILLUMINATI @ Skip -142,281	SKIP	NUMBER	DIVIDE BY	THE QUOTIENT	E QUOTIENT	POISSON EQUATION	EACH WORD OR
35	ELS Rank 3	USED	IN	304805	EQUALS	X 444 LETTERS	PROBABILITY FOR	PHRASE HAS 1
36	This calculation not not include CLINTON or	ON	TORAH	LETTERS	FREQUENCY	EQUALS	THE WORD	CHANCE IN THE
37	WORLD ORDINANCE/ORDER	FIGURE		IN THE	PER LETTER	WORD	APPEARING	IN THE FOLLOWING
38				TORAH		EXPECTANCY	AT LEAST ONCE	OF BEING HERE.
39	HITLER (2 spellings) @ Skips +/-1 and +/- 142281	142281	1	304805	3.28079E-06	0.001456669	0.001455609	686.9978691
40	OBAMA (Best spelling) @ Skips +/-1 and +/- 142281	142281	3	304805	9.84236E-06	0.004370007	0.004360472	229.3329467
41	EVIL @ Skip +1 only	1	71	304805	0.000232936	0.1034235	0.098254997	10.17759945
42	KENNEDY @ Skips 142283 to 142606	142284	76	304805	0.00024934	0.110706845	0.104798857	9.542088788
43	BUSH @ Skips +/-1 and +/- 142281	-1	302	304805	0.000990797	0.439914043	0.355908217	2.809713154
44	BARACK @ Skips +/-1 and +/- 142281	-1	329	304805	0.001079379	0.479244107	0.380748697	2.626404259
45	ENEMY (2 synonyms) @ Skips -2 to -175	-2	3769	304805	0.012365283	5.490185528	0.995872922	1.004144182
46	INITIAL SIGNIFICANCE = H39*H40*H41*H42*H43*H44*H45						8.82002E-09	113378389.7
47	FINAL SIGNIFICANCE = H46/3						2.64501E-08	37792796.56

Figure 11-20 below: The beating of Rodney King.

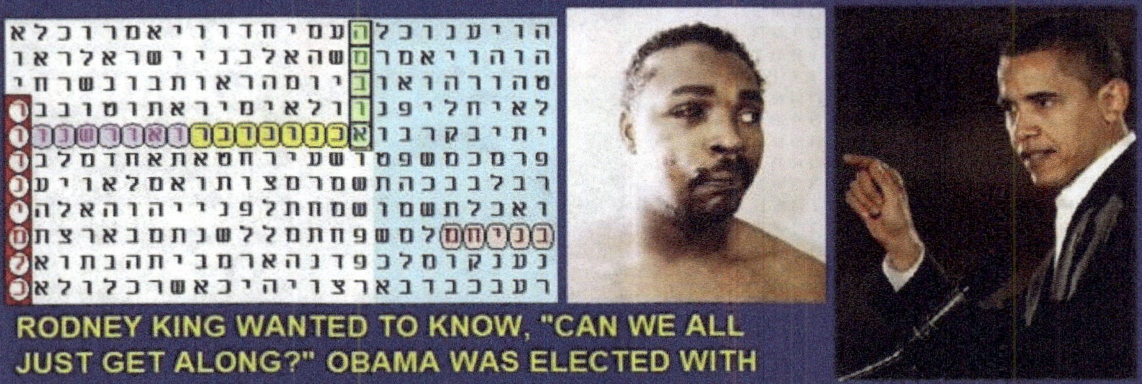

RODNEY KING WANTED TO KNOW, "CAN WE ALL JUST GET ALONG?" OBAMA WAS ELECTED WITH WHITE SUPPORT, BUT HE HAS GREATLY INJURED RACE RELATIONS.

Term	Translation	Skip	Start	End
רודניקינג	RODNEY KING	26385	Leviticus 27:10 Letter 24	Genesis 43:2 Letter 14
אובמה	OBAMA	-26385	Numbers 14:12 Letter 1	Exodus 19:8 Letter 8
אכנו בדבר	I WILL SMITE THEM WITH THE PESTILENCE	1	Numbers 14:12 Letter 1	Numbers 14:12 Letter 8
ואורשנו	AND DESTROY THEM	1	Numbers 14:12 Letter 9	Numbers 14:12 Letter 15
בני חם	SONS OF HAM	1	Genesis 10:20 Letter 4	Genesis 10:20 Letter 8

The ELS reference is 26,385 characters between rows.
The wrapped matrix starts at Exodus 19:7 Letter 55 and ends at Genesis 43:2 Letter 14.
The matrix spans 263,874 characters of the surface text.
The matrix has 11 rows, is 24 columns wide and contains a total of 264 characters.

The 4th lowest skip of RODNEY KING is encoded at the same absolute skip as OBAMA, but in the opposite direction. The two are connected by words of hatred. All three figures in this article refer to smiting and are reminiscent of the beating that King received in 1991.

	A	B	C	D	E	F	G	H	I
1	AXIS TERM RODNEY KING @ SKIP 26385	SKIP	NUMBER	DIVIDE BY	THE QUOTIENT	E QUOTIENT	POISSON EQUATION	EACH WORD	DIVIDE
2	(ELS RANK 4) FIGURE 1A	USED	IN	304805	EQUALS	X 264 LETTERS	PROBABILITY FOR	OR PHRASE	BY 4 DUE
3		ON	TORAH	LETTERS	FREQUENCY	EQUALS	THE WORD	HAS 1 CHANCE	TO AXIS
4		FIGURE		IN THE	PER LETTER	WORD	APPEARING	IN THE	TERM ELS
5				TORAH		EXPECTANCY	AT LEAST ONCE	FOLLOWING OF	RANK 4 =
6							(P VALUE)	BEING HERE	
7	OBAMA @ skips +/-1 and +/- 26385	26385	3	304805	9.84236E-06	0.002598383	0.00259501	385.3550145	96.33875

On March 2, 1991, police in Los Angeles arrested Rodney King for speeding while intoxicated. King resisted arrest, and then was shot with a taser, and repeatedly beaten and kicked while he was on the ground. The beating was filmed and broadcast on TV. On April 29, 1992, three police officers were acquitted of the beating, and nationwide rioting then ensued. My older son was walking home from Alameda High shortly after the riots began. A number of African Americans jumped him and started to beat him, but fortunately, some motorists stopped their cars and drove off the attackers. By the time it was all over, there were 53 dead, 2,383 injured, more than 7,000 fires, damages to 3,100 businesses, and nearly $1 billion in financial losses.

Mr. King then asked, "Can we all just get along?" In attempting to answer that question, many white Americans voted for Barack Obama in 2008, but after assuming office, he took racial

relations to lows not seen since the start of the Civil Rights Movement.

Figure 11-21 May 14, 2012: Newsweek magazine cover portrays Obama as a gay president who supports gay marriage.

On Figure 11-21 the axis term is **HOMO B OBAMA**. At the same skip are **HAVOC/DETRUCTION** and **IN ABOMINATION**. Leviticus 18:22 states:

*Thou shalt not lie with mankind, as with womankind: **it is abomination.***

Figure 11-21 also has **PRESIDENT** in the open text. **UNCLEAN** is at skip +1. **A WIFE UNDER HER HUSBAND AND IS DEFILED** is in the open text. This raises old questions about whether Michelle Obama was really a man.

At a Facts Check website (https://fullfact.org/online/michelle-obama-ex-boyfriend-man/) they state that:

Michelle Obama's ex-boyfriend David Upchurch has come forward to reveal that she is actually a man named Michael Robinson and only became 'Michelle' after meeting Barack Obama. However, they continue to say *Our verdict False. Mr Upchurch has always described Mrs Obama as female.*

Claims that the former US First Lady, Michelle Obama, is a man, are recirculating on social media, with one post on X, formerly Twitter, viewed more than 6.9 million times.

Posts have also been shared to Facebook, claiming her "ex-boyfriend David Upchurch has come forward to reveal that 'she' is actually a man named Michael Robinson and only became 'Michelle' after meeting Barack Obama and having prospects of becoming First Lady".

But there is absolutely no evidence for this claim, which appears to be part of a series of baseless theories that have been circulating doubting Mrs. Obama's gender.

Is this topic discussed in the Code? Possibly. Figure 11-22 explores the issue:

Term	Translation	Skip	Start	End
מישל אובאמה	MICHELLE OBAMA	-103950	Numbers Ch 31:15 Letter 10	Genesis Ch 15:17 Letter 47
האנשים	THE MEN	1	Genesis Ch 24:54 Letter 16	Genesis Ch 24:54 Letter 21
אשר עמו	WHO ARE WITH HIM	1	Genesis Ch 24:54 Letter 22	Genesis Ch 24:54 Letter 27
נשיא	PRESIDENT	1	Numbers Ch 34:25 Letter 14	Numbers Ch 34:25 Letter 17
ברק	BARACK	-1	Leviticus Ch 2:13 Letter 55	Leviticus Ch 2:13 Letter 53
הומו	HOMO	103950	Genesis Ch 15:17 Letter 56	Genesis Ch 20:14 Letter 53

The ELS reference is 103,950 characters between rows.
The matrix starts at Genesis Ch 15:17 Letter 44 and ends at Numbers Ch 34:25 Letter 22.
It spans 1,143,463 characters of the surface text.
The matrix has 12 rows, is 13 columns wide and contains a total of 156 characters.

	A	B	C	D	E	F	G	H
1	AXIS TERM MICHELL OBAMA @ Skip -103950	SKIP	NUMBER	DIVIDE BY	THE QUOTIENT	E QUOTIENT	POISSON EQUATION	EACH WORD
2	Axis Term ELS Rank 3	USED	IN	304805	EQUALS	X 156 LETTERS	PROBABILITY FOR	OR PHRASE
3		ON	TORAH	LETTERS	FREQUENCY	EQUALS	THE WORD	HAS 1 CHANCE
4		FIGURE		IN THE	PER LETTER	WORD	APPEARING	IN THE FOLLOWING
5				TORAH		EXPECTANCY	AT LEAST ONCE	OF BEING HERE.
6	MEN @ Skips +1 only	1	81	304805	0.000265744	0.041456013	0.040608465	24.62540772
7	HOMO @ Skips +/- 1 and +/- 103950	103905	81	304805	0.000265744	0.041456013	0.040608465	24.62540772
8	PRESIDENT @ Skip +1 only	1	68	304805	0.000223093	0.034802579	0.034203934	29.23640317
9	BARACK @ Skips +/- 1 and +/- 103950	-1	342	304805	0.001122029	0.175036499	0.160573618	6.227673099
10	Initial combined significance							110412.0846
11								DIVIDE BY 3 DUE TO
12								ELS RANK 3 =
13	Final Combined Significance						2.71709E-05	36804.02819

Figure 11-22 above: After conducting a homosexual marriage, comedian Joan Rivers made an accusation on TV that President Obama ia a homosexual, and Michelle Obama is transgender. Apparently the president (on at least two occasions) referred to his wife as Michael) rather than Michelle (see https://www.youtube.com/watch?v=sLiXFbht7s0. Films like this emphasize the length of the fingers, ratio

of height (5'11") to head size, presence of an Adam's Apple, and ratio of head to shoulder size, plus an apparent bulge where a woman should not have one. A few pictures looked like they might have been photo-shopped. While it will take far more than the matrix above to prove this sort of thing, on Figure 11-22 MICHELLE OBAMA has her name directly crossed in the open text by **THE MEN WHO WERE WITH HIM**. At the same absolute skip as the name, but opposite direction, is HOMO. BARACK is at skip -1, and directly below it is PRESIDENT at skip +1. Odds against this matrix occurring by chance alone are about 36,804 to 1. The person who sent the request asked me to check on other key words including proportions, shoulders, fingers, Sasha and Malia. Of these only Sasha came up at a special case skip near Michelle Obama, but the name of a daughter near that of a mother is not proof that the mother is, in fact, a man.

Figure 11-23 below February 3, 2011: Was Obama elected in Egypt, or does he see himself as a reincarnated pharaoh?

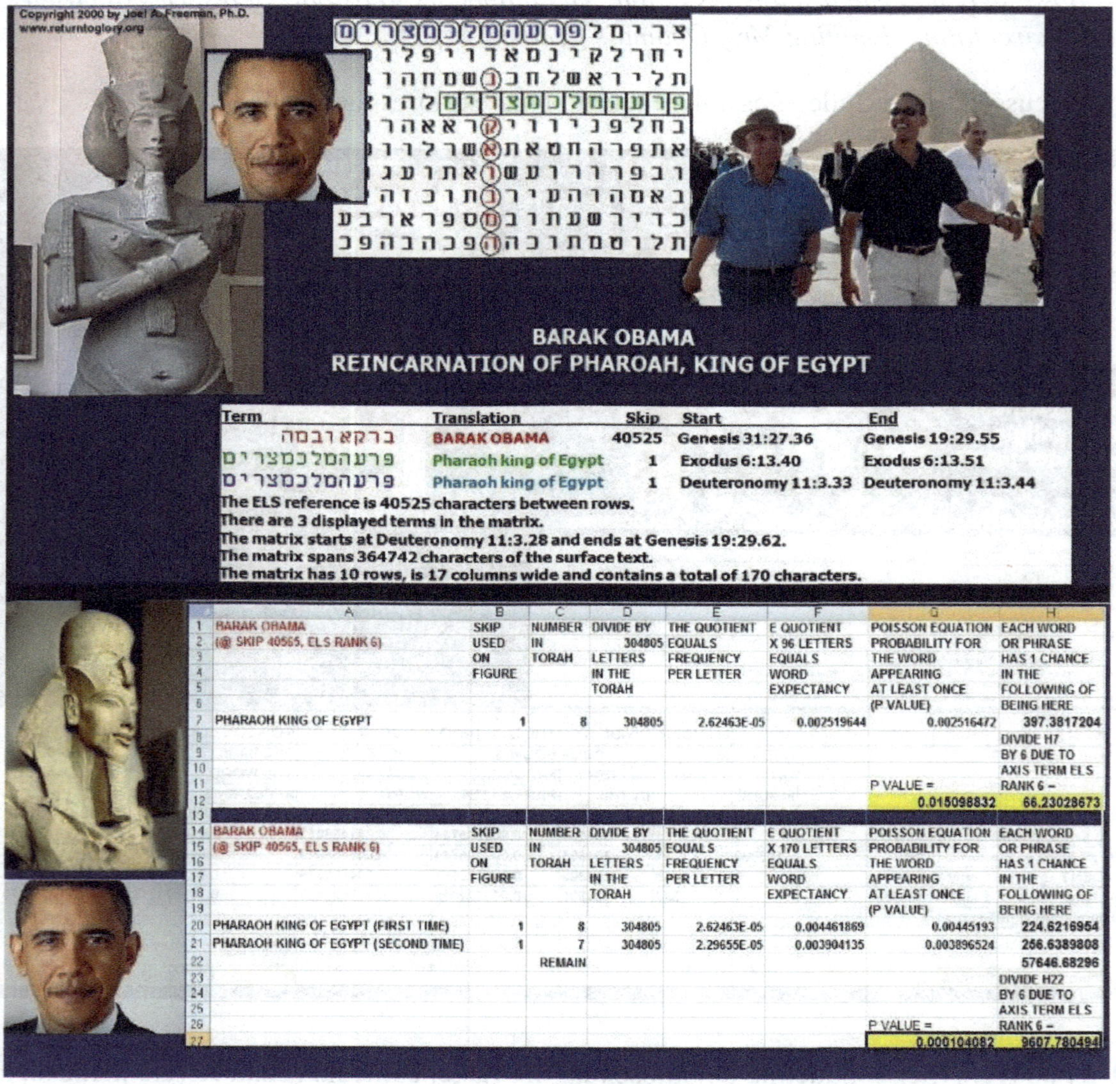

Figure 11-23 above was originally posted in conjunction with President Obama's overt hostility toward Israel. When the crisis arose in Egypt, where there was widespread opposition to President Mubarak, Obama

declared, "What is clear — and what I indicated tonight to President Mubarak — is my belief that an orderly transition must be meaningful, and it must begin now."

The demands made by the U.S. President had a tone that made it sound like he was also elected as President of Egypt. There seemed to be no concept of Egyptian sovereignty or the need for Egypt to solve its own problems internally. On the matrix, **BARACK OBAMA** is the axis term. His name is shown at its 6th lowest ELS in the wrapped Torah (which requires more than one computer pass through the 304,805 letters of Torah to find). It is directly crossed by one of the Torah's eight uses of the term **PHARAOH KING OF EGYPT**. Perhaps President Obama has a distant memory of being Pharaoh, King of Egypt, in a past life. Indeed, there is a statue that backs this idea in the Egyptian Museum in Cairo, which was attacked in the rebellion and which was at the heart of the rebellion in Cairo's Tahrir Square! Mubarak was toppled and arrested. Pharaoh Obama had laid out his vision for more change and included what Mitt Romney calls throwing Israel under the bus by driving it back to the pre-1967 borders.

Figure 11-24 October 9, 2009: Obama Nobel Prize

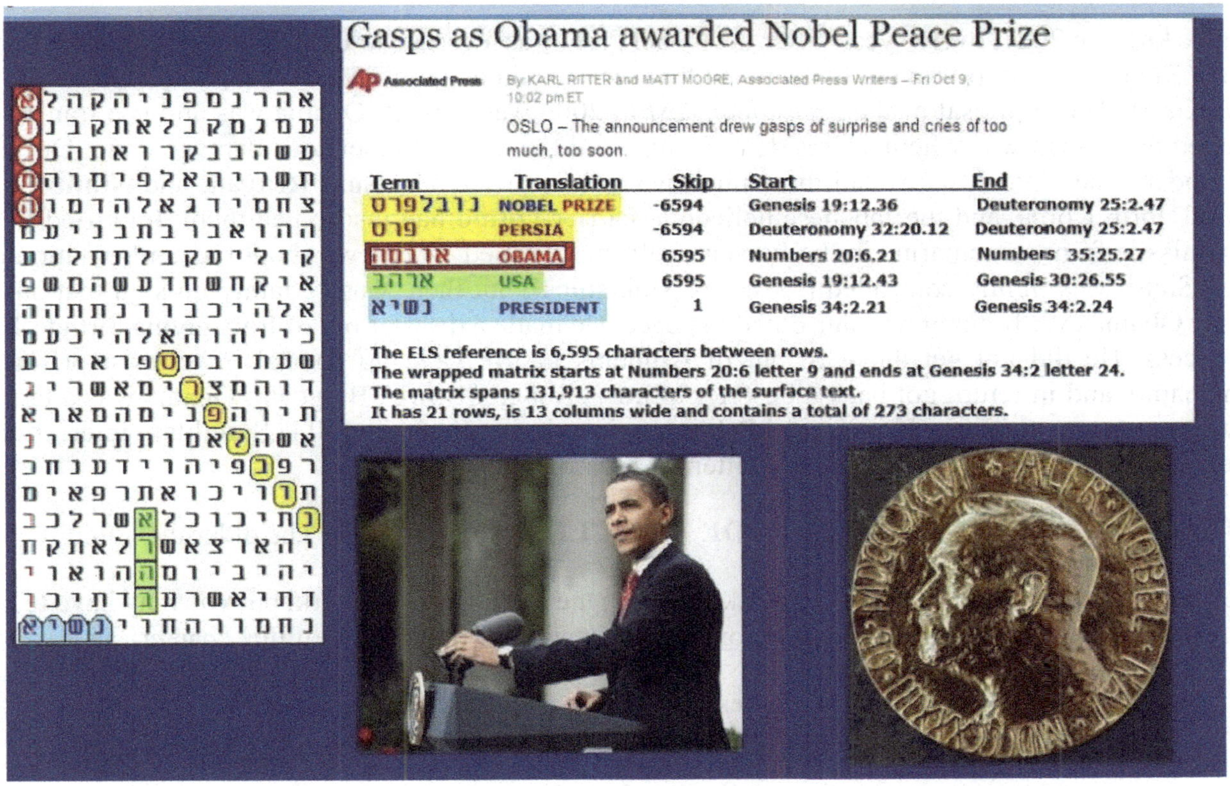

On this matrix, the axis term was **NOBEL PRIZE**. It is shown at its minimum (most significant) ELS. In the 273-letter matrix, in addition to the axis term at skip -6594, we find **OBAMA** at skip 6595, and also **USA** at skip 6,595. **PRESIDENT** was also on the matrix at skip +1. As the spreadsheet indicates, there was about one chance in 20 that **OBAMA** would be on the matrix at some skip, about one chance in 16 that **PRESIDENT** would be there at skip +1, and better odds than calling heads or tails (here about one chance in 1.8) that **USA** would be on the matrix. All combined, with no value at all assigned to the axis term, **NOBEL PRIZE**, the value of the matrix was computed at p = .00157 or about one chance in 634.

Using the 10 **Haralick** verse-permuted files, and just checking for **Nobel Prize** between skips - 35,000 and 35,000, I was not able to come up with a match as good as that seen in the Torah, but I did get close with one matrix having all the terms in a 300-letter (25 row, 12 column) matrix where the axis term, **Nobel Prize**, was at skip 4,631.

Political Aspects of My 2009 Article

Obama himself admitted that he was unworthy of receiving the prize. As for why he was nominated for it only 12 days into his presidency and then selected in October 2009, the *Jerusalem Post* speculated, "If Obama was unlikely to okay any military activity against Iran to keep it from getting nuclear weapons capabilities before the award, giving him the award now seems an attempt to handcuff him even further." I went looking for Iran on the Matrix and didn't see it at first. But then I realized that a Biblical synonym for Iran is Persia. The Hebrew for **Nobel Prize** is **Nun Bet Vav Lamed Peh Resh Samech**. Curiously, the last three letters are not only **Prize**, but also **PERSIA**!

Political and Codes Update of May 10, 2021

What we knew in 2021, twelve years after the Nobel Prize was awarded to Barack Hussein Obama, was that he gave $151.7 billion to Iran and guaranteed that they would acquire nuclear weapons or buy them from North Korea. The appraisal of the *Jerusalem Post* in 2009 was correct. Obama was and is a traitor. What remained to be seen was whether President Trump could undo the tremendous damage done. During my 2017 update, the United States had three nuclear aircraft carriers (**Vinson**, **Reagan**, and **Nimitz**) off the coast of North Korea, and the Japanese helicopter carrier **Izumo** had just joined them. Iran tried to fire a cruise missile from a submarine in the Persian Gulf, but it failed. The news about the confrontation in the United States was tightly controlled (loose lips sink ships), but there was certainly no sign that anything done by Obama ever brought or contributed to peace. He made a disaster out of Iraq, giving birth to ISIS in the process. He did not get the U.S. out of Afghanistan. He gave up control of terrorist prisoners at Guantanamo, and in return got back a deserter traitor—Robert "Bowe" Bergdahl. Obama was a terrorist's dream come true, and **Biden**, **Harris**, and **Pelosi** continued to work against all U.S. military interests. Much of the Torah Code appears to have been written as a warning about him.

LIST OF LINKS TO OLD OBAMA CODE ARTICLES (2009 to 2016). The older the article,

the less refined my techniques were. The links listed in the section should work, but you may have to access them via my https://roffmanmarsresearch.com website and the PDF version of this chapter.

- **MARCH 17, 2016:** Obama nominates Garland for Supreme Court. Oppose it.
- **NOVEMBER 3, 2015:** Is Obama only a Kenyan-born Muslim, or a Reptilian Alien?
- **JULY 6, 2015:** Obama, Gay Marriage, Iran, and An Atomic Holocaust.
- **MARCH 4, 2015: NETANYAHU GOES AFTER THE ENEMY OF ISRAEL – AND AMERICA – (OBAMA) IN HIS THIRD SPEECH BEFORE A JOINT SESSION OF CONGRESS.**
- **FEBRUARY 19, 2015: JOHN PODESTA, OBAMA, BUZZ ALDRIN, AND UFO DISCLOSURE.**

- **FEBRUARY 9, 2015: NANCY PELOSI PROPOSES DEMOCRATIC BOYCOTT OF NETANYAHU'S SPEECH BEFORE A JOINT SESSION OF CONGRESS.**

- **JANUARY 23, 2015: BOEHNER INVITES NETANYAHU TO JOINT SESSION OF CONGRESS.**

- **DECEMBER 7, 2014: OBAMA UFO DISCLOSURE UPDATE.**

- **NOVEMBER 5, 2014: REPUBLICANS MAY FILE IMPEACHMENT CHARGES AGAINST OBAMA IF HE USES EXECUTIVE ACTION TO KEEP MILLIONS OF ILLEGAL ALIENS.**

- **SEPTEMBER 23, 2014: EBOLA AND OBAMA.** It seems like this plague is well described in the Code.

- **SEPTEMBER 5, 2014 (Updated 9/11/2014): September 4 Bill O'Reilly show discusses if Obama is the enemy.** The Torah Code suggests an answer.

- **SEPTEMBER 1, 2014: Torah Code compares UK Prime Minister David Cameron and Obama.**

- **JULY 18, 2014: MALAYSIA FLIGHT 17 SHOT DOWN BY RUSSIAN REBELS IN UKRAINE. Obama's weakness is encoded as a cause.**

- **JULY 13, 2014: Is Obama Really Hamas?**

- **JUNE 2, 2014: White House Press Secretary Jay Carney Resigns.**

- **APRIL 25, 2014: Ranch War in Nevada.** Obama and the Federal Government would be wise to leave Bundy alone and not make him into a martyr.

- **JANUARY 30, 2014: Obama calls for an increase in the minimum wage.**

- **OCTOBER 31, 2013: Kathleen Sebelius & Obama struggle with the truth about the Obamacare malfunction.**

- **JULY 30, 2013: Fear of Obama and Kerry forces Netanyahu to release 140 terrorists from prison and to restart peace talks.**

- **MAY 1, 2013: Obama is up against a chemical, the red line he painted for Assad. Another war may lie ahead.**

- **APRIL 26, 2013: Obama, Mars Cover-up, and Life are encoded in 1 Column.**

- **MARCH 19, 2013: President Obama's visit to Israel.**

- **OCTOBER 7, 2012: Was Obama crushed in the first debate because he couldn't use his teleprompter, or was he distracted by something more important, like his real birth certificate?**

- **SEPTEMBER 25, 2012: Obama will lead us into a world war.**

- **SEPTEMBER 11, 2012: The Jerusalem vote from Obama at the Democratic convention leads to atomic Iran & destruction.**
- **SEPTEMBER 5, 2012: Obama as the king of Sodom at the Democratic convention.**
- **JUNE 8, 2012: U.S. military suicides hit one per day, the top rate of killed in action.**
- **MAY 30, 2012: Israel-made Flame computer virus hits Iran.**
- **MAY 22, 2012: Mormon concerns and election jitters.**
- **MAY 20, 2012: Congressman Akin introduces measure needed to protect U.S. military chaplains who oppose homosexual marriage.**
- **APRIL 18, 2012: The Colombian Obama security scandal.**
- **APRIL 6, 2012: First look at an Obama vs. Romney election in 2012.**
- **APRIL 4, 2012: Obama insults the Supreme Court by calling its members "unelected judicial activists."**
- **MARCH 28, 2012: President Obama promises Russia to kill our missile defense after the election.**
- **JANUARY 30, 2012: Arizona Governor Jan Brewer puts her finger in President Obama's face.**
- **JANUARY 29, 2012: Did a Saudi operative named Khalid Al-Monsour help Obama get into Harvard?**
- **JANUARY 24, 2012: A look at who the code says is Babylon.**
- **JANUARY 8, 2012: The link between Malcolm X and Barack Obama.**
- **DECEMBER 9, 2011: B. Obama & Netanyahu ordered attack on Iran's Isfahan nuclear site.**
- **SEPTEMBER 15, 2011: Obama, the Solyndra loan, and the corrupt.**
- **SEPTEMBER 15, 2011: Will Trump try again to impeach Obama?**
- **SEPTEMBER 8, 2011: Obama's economic failure: jobs.**
- **JULY 21, 2011: 42 years after Apollo 11, Obama presides over the death of the space shuttle and manned U.S. space flight.**
- **MAY 27, 2011: Accusation of a new alias for President Obama.**
- **MAY 20, 2011: Netanyahu says no to Obama!**

- **APRIL 14, 2011: Gaddafi.**
- **APRIL 14, 2011: Romney and Trump are both encoded with Obama's loss second time. Mike Huckabee does well, too.**
- **MARCH 10, 2011: Is Obama leading the U.S into another civil war?**
- **DECEMBER 19, 2010: Don't Ask, Don't Tell and New U.S. Military Gay Policy.**
- **NOVEMBER 3, 2010: Pharaoh Obama's Loss of Congress in 2010.**
- **SEPTEMBER 20, 2010: NBC TV series entitled "THE EVENT." Is a UFO Disclosure coming from the Obama Administration?**
- **APRIL 15, 2010: President Obama's plan for NASA and Buzz Aldrin's role in it.**
- **JANUARY 27, 2010: Will Obama end NASA?**
- **DECEMBER 14, 2009: Emanuel's Roman Hanukkah and Vatican plus Obama.**
- **JANUARY 20, 2009: Rahm Emanuel and Barack Obama.**

Chapter 12
Trump!

This chapter presents a few of the many matrices found during President Trump's first and early second terms. A more complete list of President Trump matrices is at https://roffmanmarsresearch.com/?page_id=2596.

Term	Translation	Skip	Start	End
דנלדתרמפ	DONALD TRUMP	-10109	Exodus 32:29.14	Genesis 38:14.29
למנות	NOMINATE	1	Genesis 38:14.10	Genesis 38:14.14

The ELS reference is 10109 characters between rows.
The unwrapped matrix starts at Genesis 38:14.10 and ends at Exodus 32:29.14.
The matrix spans 70,783 characters of the surface text.
The matrix has 8 rows, is 20 columns wide and contains a total of 160 characters.

Term	Translation	Skip	Start	End
הכלינתונ	H. CLINTON	-31439	Deuteronomy 28:56.71	Genesis 48:9.45
לקבוע	TO NOMINATE	-1	Exodus 19:5.18	Exodus 19:5.14

The ELS reference is 31,439 characters between rows.
The unwrapped matrix starts at Genesis 48:9.45 and ends at Deuteronomy 28:57.17.
The matrix spans 220,091 characters of the surface text.
It has 8 rows, 18 columns and 144 characters.

1	AXIS TERM DONALD TRUMP @ Skip -71760	SKIP	NUMBER	DIVIDE BY	THE QUOTIENT	E QUOTIENT	POISSON EQUATION	EACH WORD OR
2	Axis Term ELS Rank 1	USED	IN	304805	EQUALS	X 160 LETTERS	PROBABILITY FOR	PHRASE HAS 1
3		ON	TORAH	LETTERS	FREQUENCY	EQUALS	THE WORD	CHANCE IN THE
4		FIGURE		IN THE	PER LETTER	WORD	APPEARING	IN THE FOLLOWING
5				TORAH		EXPECTANCY	AT LEAST ONCE	OF BEING HERE.
6	TO NOMINATE AT @ Skips +1 only	1	4	304805	1.31231E-05	0.002099703	0.0020975	476.7579875
7	FINAL SIGNIFICANCE						0.0020975	476.7579875
8								
9	AXIS TERM H. CLINTON @ Skip -31,439	SKIP	NUMBER	DIVIDE BY	THE QUOTIENT	E QUOTIENT	POISSON EQUATION	EACH WORD OR
10	Axis Term ELS Rank 5	USED	IN	304805	EQUALS	X 144 LETTERS	PROBABILITY FOR	PHRASE HAS 1
11		ON	TORAH	LETTERS	FREQUENCY	EQUALS	THE WORD	CHANCE IN THE
12		FIGURE		IN THE	PER LETTER	WORD	APPEARING	IN THE FOLLOWING
13				TORAH		EXPECTANCY	AT LEAST ONCE	OF BEING HERE.
14	TO NOMINATE AT @ Skips +1 only	1	5	304805	1.64039E-05	0.002362166	0.002359378	423.8404746
15								DIVIDE ABOVE BY 5
16								DUE TO H. CLINTON
17								ELS RANK 5 =
18	FINAL SIGNIFICANCE						0.011796891	84.76809492

Figure 12-1. Trump and Clinton nominations in 2016. I found the above matrices in late 2015 or early 2016 when the two nominees were about to start their election campaigns. This experiment was simpler than those I typically run now. All I wanted to know was if the candidates were each found at an ELS and whether they could be paired with the word **NOMINATE**. If so, I also wanted to know who had the smaller, more significant matrix. The verb with **TRUMP** was found four times in Torah at skip +1. For **CLINTON**, I found **H. CLINTON.** The **H** initial for the first name was important because we already had a President Bill Clinton, and I needed a way to distinguish her from him. Both **DONALD TRUMP** and **H. CLINTON** have 8 letters in Hebrew. The H. CLINTON matrix (144 letters) was slightly smaller than the DONALD TRUMP matrix (160 letters in area), but it was not based on the lowest skip of **CLINTON.** Instead, it was a match that was only found at ELS rank 5 of H. CLINTON. That means that, by my standard protocol, I had to divide by 5 the odds of the H. CLINTON match with the verb found for TO NOMINATE (spelled lamed kuf bet vav ayin).

I found **DONALD TRUMP** and **H. CLINTON**. The **CodeFinder** dictionary offers two 5-letter words for **TO NOMINATE**. One of them, spelled *lamed mem nun vav tav*, is at skip +1 (the best- case scenario) and matches with **DONALD TRUMP** at **ELS rank 1** (also, best case scenario) in a 160-letter matrix. For a term at skip +1, I only search for word frequency at skip +1. As noted above, the verb with **TRUMP** was found four times at that skip. But for special case terms at skips

-1 or at the absolute skip of the axis term, I use the letter frequency at skips +/- 1 plus +/- the skip of the axis term. So, for **DONALD TRUMP** this works out to be that the match was found against odds of about 476 to 1, while for **H. CLINTON**, she was found with the verb against odds of about 423.81 divided by 5, which is about 84 to 1. This means that the **Trump** matrix was about 423.81/84.78 or 4.9989 (really, 5) times more significant than the **Clinton** matrix.

FIGURE 12-2 below. The effort to impeach Trump, in large part by an ignoramus known as Maxine Waters, begins. The axis term is **TRUMP IMPEACHED**. **CLINTON** is at skip -1.

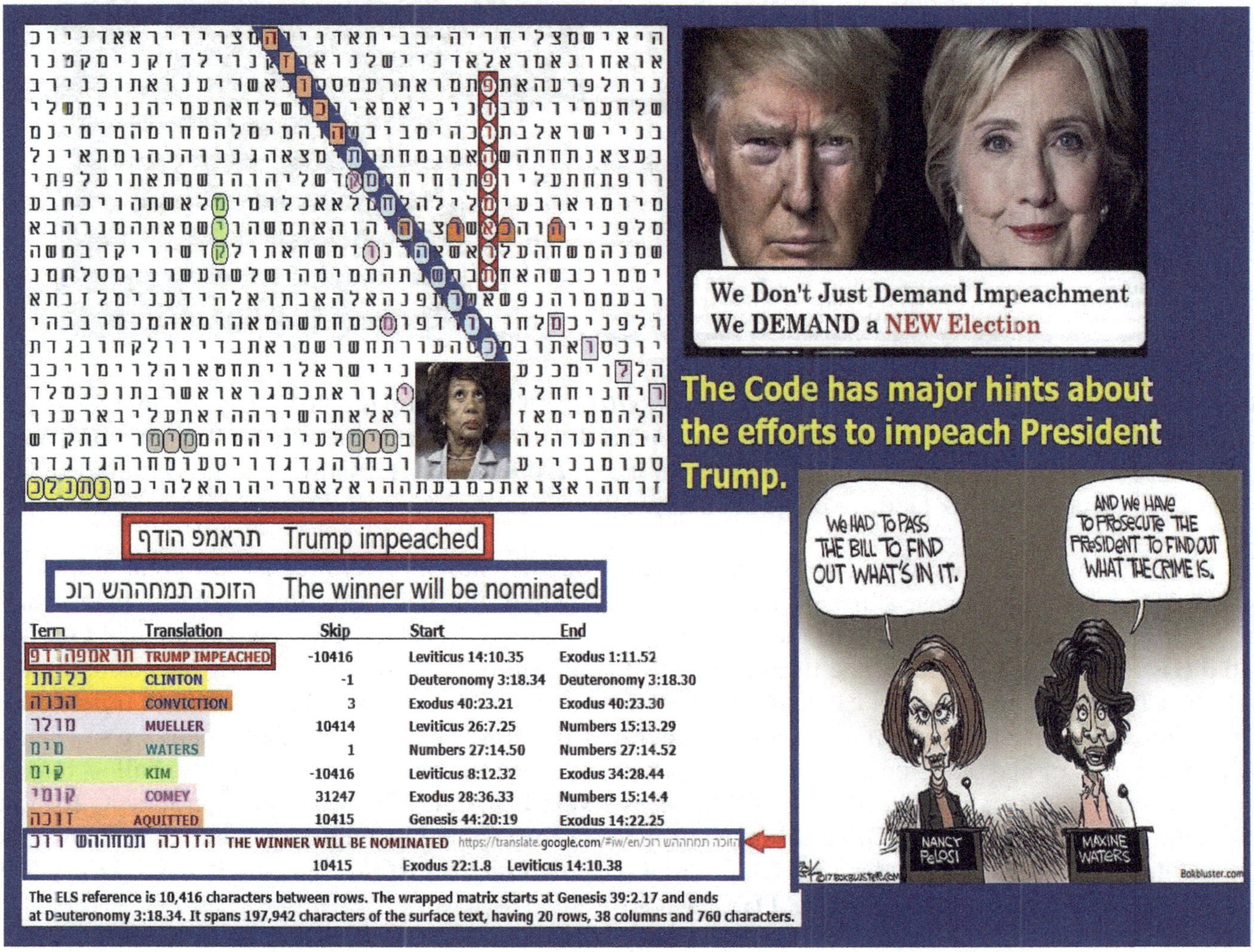

Figure 12-2 above: No sooner than Trump was sworn in for his first term, Congresswoman/gender-confused Maxine Waters called for his impeachment. More, she did it again when he won his second term. So how did the Code's Author know about this 3,300 years ago?

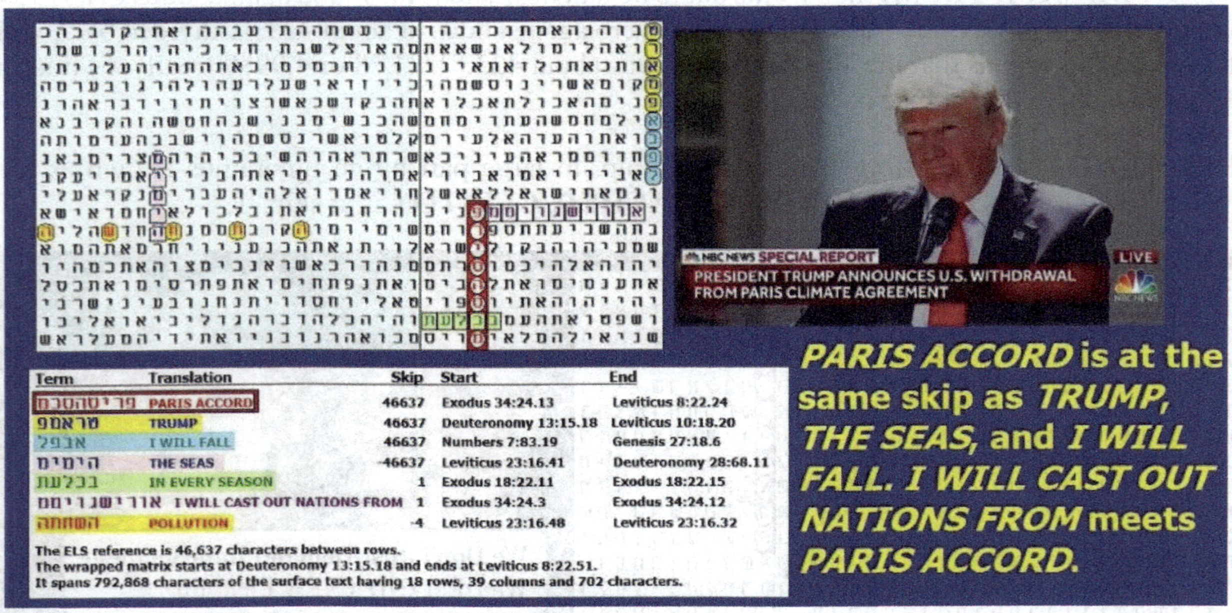

Figure 12-3. This page was posted on my site on 6/8/2017 and updated on 6/9/2017. My original focus was on the axis term: **PARIS ACCORD**. I saw **Trump** on the matrix at the same skip as Paris Accord, and I noted that following his name and in sequence with it was **I WILL FALL**. Sadly, that turned out to be true in the 2020 election. Thank God, he came back in the 2024 election.

For 28 years now I've worked on cracking the **Torah Code**. By now I pretty much know what to look for - *a priori* terms at special case skips +/- 1 or the absolute skip of the axis terms. On the matrix below there are three terms at the absolute skip of the axis term, which is **PARIS ACCORD**. Two of them – **TRUMP** and **THE SEAS** – are *a priori.* The third term was found *a posteriori,* which means it is not the kind of term that I normally include in a calculation for statistical significance. However, sometimes *a posteriori* finds offer an unexpected view of a possible future. I hoped that was not the case here, but as we saw in the 2020 election, it turned out to be true (even though new evidence shows that China sent false ballots to steal the election). The term, in sequence with **TRUMP**, is **I WILL FALL**. This was found on June 8, 2017, at about the time that former FBI Director James Comey was testifying in Congress in large part against the President. The matrix also includes **I WILL CAST OUT NATIONS FROM** touching **PARIS ACCORD**, and an **ELS** of **POLLUTION** touching **THE SEAS**.

As the spreadsheet below shows for **Figure 12-4**, the matrix was found against odds of 2,485,167 to 1, BUT. That number does not include the prediction that came true - **I WILL FALL**!

	A	B	C	D	E	F	G	H
1	Axis term PARIS ACCORD @ Skip 46637	SKIP	NUMBER	DIVIDE BY	THE QUOTIENT	E QUOTIENT	POISSON EQUATION	EACH WORD OR
2	Axis Term ELS Rank 1	USED	IN	304805	EQUALS	X 702 LETTERS	PROBABILITY FOR	PHRASE HAS 1
3	This calculation does NOT include I WILL FALL which was an	ON	TORAH	LETTERS	FREQUENCY	EQUALS	THE WORD	CHANCE IN THE
4	*a posteriori* event found via snooping.	FIGURE		IN THE	PER LETTER	WORD	APPEARING	IN THE FOLLOWING
5				TORAH		EXPECTANCY	AT LEAST ONCE	OF BEING HERE.
6								
7	TRUMP (best spelling) @ Skips +/- 1 and +/-46637	46637	1	304805	3.28079E-06	0.002303112	0.002300462	434.6953486
8	NATIONS @ Skip +1 only	1	55	304805	0.000180443	0.12667115	0.118976651	8.405010502
9	EVERY SEASON, SUMMER, AUTUMN, WINTER OR SPRING @ Skip + 1 only	1	23	304805	7.54581E-05	0.052971572	0.051593027	19.38246438
10	THE SEAS @ Skips +/- 1 and +/-46637	46496	48	304805	0.000157478	0.110549368	0.104657872	9.554943003
11	POLLUTION @ Skip -4 to -173	-4	138	304805	0.000452748	0.317829432	0.272273096	3.672783017
12	FINAL COMBINED SIGNIFICANCE = H7*H8*H9*H10*H11						4.02387E-07	2485167.242
13								
14	Axis term PARIS ACCORD @ Skip 46637	SKIP	NUMBER	DIVIDE BY	THE QUOTIENT	E QUOTIENT	POISSON EQUATION	EACH WORD OR
15	Axis Term ELS Rank 1	USED	IN	304805	EQUALS	X 270 LETTERS	PROBABILITY FOR	PHRASE HAS 1
16	This calculation does NOT include THE SEAS, or POLLUTION.	ON	TORAH	LETTERS	FREQUENCY	EQUALS	THE WORD	CHANCE IN THE
17	Nor does it include I WILL FALL.	FIGURE		IN THE	PER LETTER	WORD	APPEARING	IN THE FOLLOWING
18				TORAH		EXPECTANCY	AT LEAST ONCE	OF BEING HERE.
19								
20	TRUMP (best spelling) @ Skips +/- 1 and +/-46637	46637	1	304805	3.28079E-06	0.000885812	0.00088542	1129.407481
21	NATIONS @ Skip +1 only	1	55	304805	0.000180443	0.048719673	0.047551911	21.02964904
22	EVERY SEASON, SUMMER, AUTUMN, WINTER OR SPRING @ Skip + 1 only	1	23	304805	7.54581E-05	0.020373682	0.02016754	49.58462855
23	FINAL COMBINED SIGNIFICANCE = H20*H21*H22						8.49123E-07	1177686.642

Above: Spreadsheet for **Figure 12-3.**

Figure 12-4. Trump and the Media: One reason for Trump to temporarily fall in the 2020 election was that the Media was his enemy. On **Figure 12-4**, the axis term is **DONALD TRUMP** at skip -10109. At skip +10109, we find **MEDIA** and **ENEMY**.

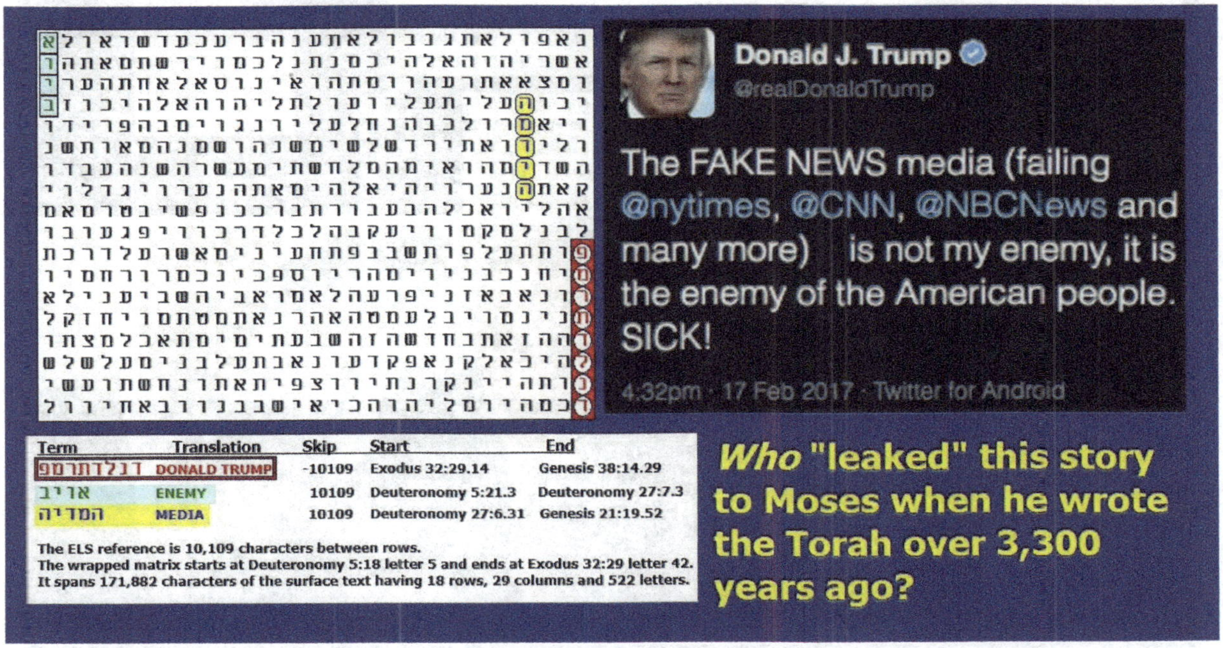

Odds against Figure 12-4 being found were about 5,786 to 1.

Figure 12-5 below. Shown earlier as Figure 11-1, the Code names Obama as an antisemite. When Trump was reelected, a major issue was how to reverse damage done to Jewish people in the Obama-Biden years.

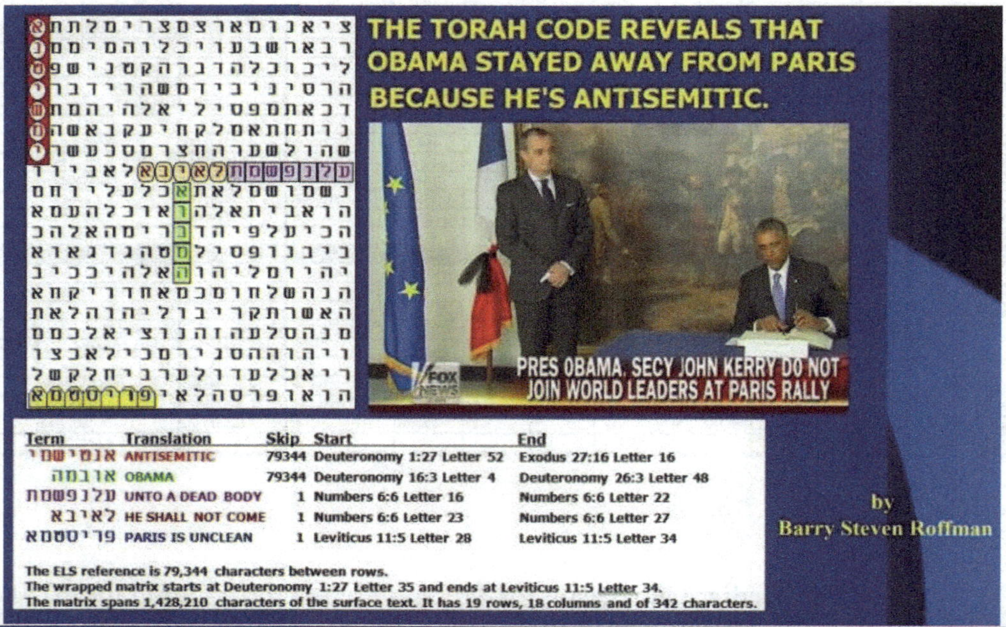

	A	B	C	D	E	F	G	H
1	Axis term: TRUMP best spelling @ Skip 77,392	SKIP	NUMBER	DIVIDE BY	THE QUOTIENT	E QUOTIENT	POISSON EQUATION	EACH WORD OR
2	Also required: The sons of God saw the daughters	USED	IN	304805	EQUALS	X 264 LETTERS	PROBABILITY FOR	PHRASE HAS 1
3	of Man	ON	TORAH	LETTERS	FREQUENCY	EQUALS	THE WORD	CHANCE IN THE
4		FIGURE		IN THE	PER LETTER	WORD	APPEARING	IN THE FOLLOWING
5				TORAH		EXPECTANCY	AT LEAST ONCE	OF BEING HERE.
6	PUTIN (best spelling) @ Skips +/- 1 and +/- 77392	77392	1	304805	3.28079E-06	0.000866128	0.000865753	1155.064466
7	ISRAEL @ Skips +1 only	77392	591	304805	0.001938945	0.511881367	0.400633112	2.496049306
8	COMBINED ODDS = H6*H7						0.000346849	2883.097859

Figure 12-6 above shows the relationship between **TRUMP, PUTIN**, and what sounds like people who were not human but were interbreeding with the **DAUGHTERS OF MAN** (Genesis 6:2).

The matrix was found against odds of about 2,883 to 1.

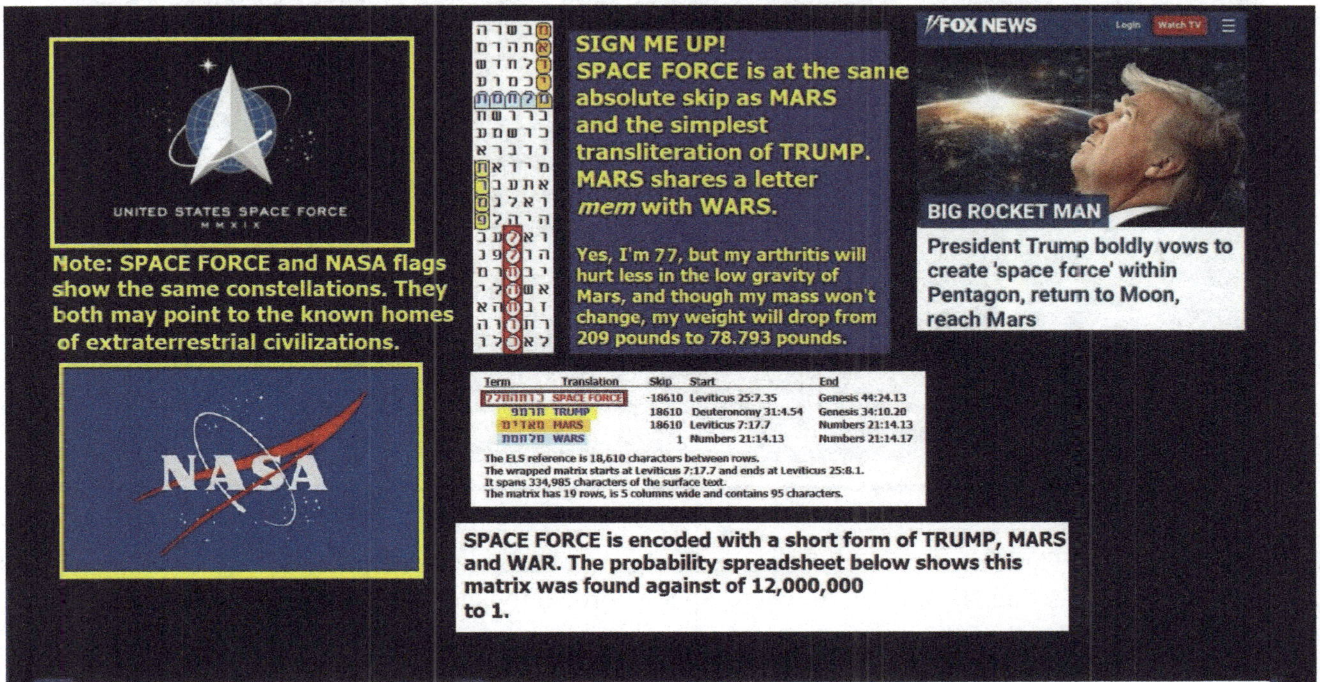

	A	B	C	D	E	F	G	H
1	Axis term SPACE FORCE @ Skip -18,610	SKIP	NUMBER	DIVIDE BY	THE QUOTIENT	E QUOTIENT	POISSON EQUATION	EACH WORD OR
2	ELS Rank 12	USED	IN	304805	EQUALS	X 95 LETTERS	PROBABILITY FOR	PHRASE HAS 1
3		ON	TORAH	LETTERS	FREQUENCY	EQUALS	THE WORD	CHANCE IN THE
4		FIGURE		IN THE	PER LETTER	WORD	APPEARING	IN THE FOLLOWING
5				TORAH		EXPECTANCY	AT LEAST ONCE	OF BEING HERE.
6								
7	MARS (5-letter Hebrew term) @ Skips +/- 1 and +/- 18610	18610	3	304805	9.84236E-06	0.000935024	0.000934587	1069.991306
8	TRUMP (4 & 5 letters, 4 spellings) @ Skips +/-1 and +/- 18610	18610	2	304805	6.56157E-06	0.000623349	0.000623155	1604.736894
9	WAR OR WARS @ Skip +1 only	1	38	304805	0.00012467	0.011843638	0.011773778	84.93450497
10	INITIAL SIGNIFICANCE = H7*H8*H9						6.85696E-09	145837176.1
11	FINAL SIGNIFICANCE = H10/12						8.22835E-08	12153098.01

Figure 12-7 above: The flags of NASA and U.S. Space Force share maps of the Pleiades and Alpha Draconis. But NASA's flag shows Orion, where Space Force only shows an exploded star in Orion (Betelgeuse). On the matrix, the axis term is **SPACE FORCE**. At the same absolute skip are **MARS** and a short transliteration of **TRUMP.** Sharing a letter mem with **MARS** is **WAR** in the open text. Both NASA and Space Force flags have the constellations Alpha Draconis (303 light years from Earth) and Pleiades (444.2 light years from Earth) on them, but NASA's flag also shows Orion while it is missing and replaced by a single bright star on the Space Force flag. Wikipedia points to Alpha Draconis as the possible home of Reptilian aliens. Betelgeuse may go supernova. The Space Force flag may indicate knowledge of what might happen.

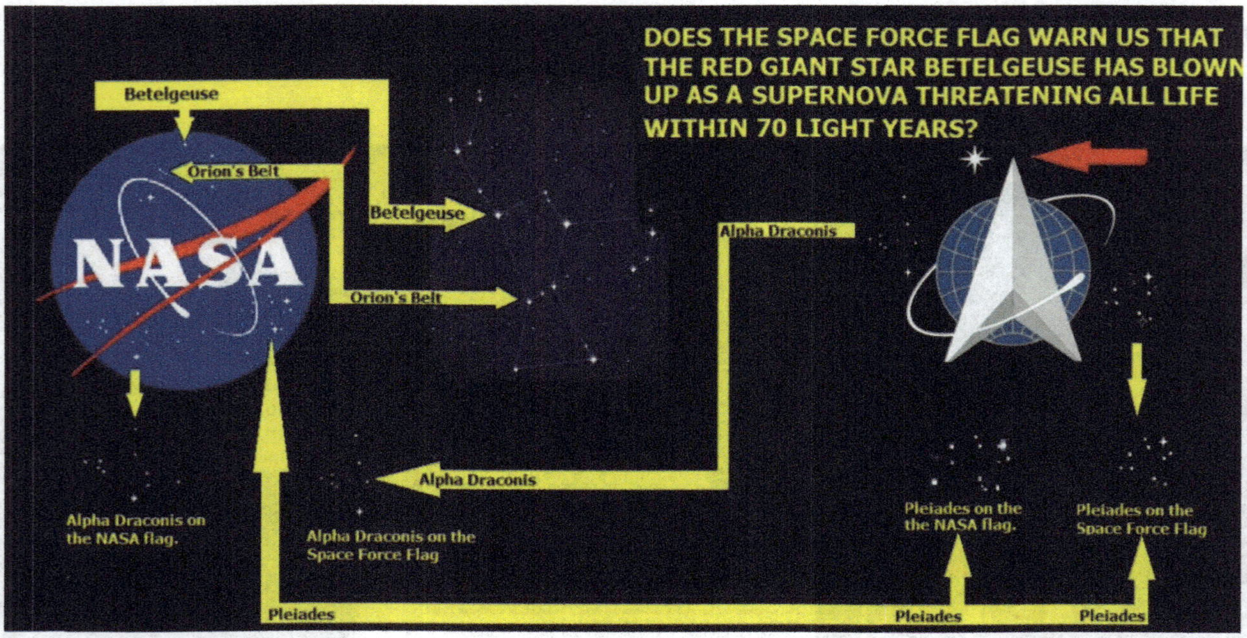

Figure 12-8 above: The flags of NASA and U.S. Space Force share maps of the Pleiades and Alpha Draconis. But NASA's flag shows Orion, where Space Force only shows an exploded star there.

Alpha Draconis is 303 light years from Earth, and the Pleiades are 444.2 light years from Earth. Wikipedia points to Alpha Draconis as the possible home of Reptilian aliens. This book also has speculation that Reptilian shapeshifting politicians (or leaders who were brainwashed and brain- damaged by them) control Earth. This includes Obama and befuddled Biden. Oddly enough, the phrase **OBAMA IS REPTILIAN** is encoded in the Torah with **PRESIDENT** and **UFO** at the same absolute skip of 75,716. Wikipedia also points to the Pleiades as the home for the Nordics. A woman (Rebecca) that I have known for 11 years claimed that she was an ET from the Pleiades, and she said that people from there are humans. I suppose she meant that at some point in our earlier existence, some people developed the ability to travel in space, and now they are finding their way back to Earth. However, after 11 years, we had a fight when the lady sounded more like a reptilian. Her stories were not consistent.

Term	Translation	Skip	Start	End
ישבנבוכב	THERE IS AN ASTEROID	-5189	Exodus 3:16.90	Genesis 31:30.37
הכלכלה	THE ECONOMIC	5189	Genesis 29:2.50	Genesis 43:15.19
ממטר	STANDARD	5189	Genesis 29:2.37	Genesis 37:26.9
הכסף	THE CURRENCY	1	Genesis 43:15.27	Genesis 43:15.30
ארהב	U.S.A.	5189	Genesis 24:4.26	Genesis 31:30.25
סין	CHINA	1	Genesis 37:26.37	Genesis 37:26.39
עולם	WORLD	1	Genesis 49:26.33	Genesis 49:26.36
כסף	SILVER	1	Genesis 31:30.19	Genesis 31:30.21
אוצר	MINE (n)	-1	Exodus 10:5.32	Exodus 10:5.29
כורה	MINER	-2	Exodus 10:5.33	Exodus 10:5.27
היתלר	HITLER	-5187	Exodus 20:10.39	Exodus 7:5.51

The ELS reference is 5,189 characters between rows.
The unwrapped matrix starts at Genesis 24:3.70 and ends at Exodus 20:10.57.
The matrix spans 77,884 characters of the surface text.
The matrix has 16 rows, is 49 columns wide and contains a total of 784 characters.

The ECONOMIC STANDARD for WORLD CURRENCY is not SILVER or gold. But there is reason to believe that THERE IS AN ASTEROID secretly being MINED by the USA, CHINA, and others in the WORLD. The metals from it may help set real exchange standards and HITLER may have played a role in this system.

	A	B	C	D	E	F	G	H
1	FIGURE 1 Axis term THERE IS AN ASTEROID @ Skip 5,189	SKIP	NUMBER	DIVIDE BY	THE QUOTIENT	E QUOTIENT	POISSON EQUATION	EACH WORD OR
2	ELS Rank 1	USED	IN	304805	EQUALS	X 784 LETTERS	PROBABILITY FOR	PHRASE HAS 1
3	The word for SILVER is the same as CURRENCY. This word is only enterred	ON	TORAH	LETTERS	FREQUENCY	EQUALS	THE WORD	CHANCE IN THE
4	into thte calculation once, but the frequency allows for the letter HEY for THE	FIGURE		IN THE	PER LETTER	WORD	APPEARING	IN THE FOLLOWING
5	up front.			TORAH		EXPECTANCY	AT LEAST ONCE	OF BEING HERE.
6								
7	ECONOMY @ Skips +/- 1 and +/- 5189	5189	4	304805	1.31231E-05	0.010288545	0.010235799	97.69632932
8	STANDARD Skips +/- 1 and +/- 5189	5189	11	304805	3.60886E-05	0.028293499	0.027896986	35.84616574
9	THE CURRRENCY @ Skip +1 only	1	20	304805	6.56157E-05	0.051442726	0.050141949	19.94338109
10	USA @ Skips +/- 1 and +/- 5189	5189	23	304805	7.54581E-05	0.059159135	0.057443236	17.40848998
11	CHINA @ Skip +1 only	1	39	304805	0.000127951	0.100313315	0.095446037	10.47712439
12	WORLD @ Skip +1 only	1	67	304805	0.000219813	0.172333131	0.158301269	6.317068733
13	MINE (noun) @ Skips +/- 1 and +/- 5189 (3 synonyms)	-1	53	304805	0.000173882	0.136323223	0.127439445	7.846864068
14	HITLER @ Skips -5187 to -5362 (2 spellings, tet and tav)	-5187	135	304805	0.000442906	0.347238398	0.293363153	3.408744385
15	MINER @ Skips -2 to -363	-2	3040	304805	0.00997359	7.819294303	0.999598095	1.000402067
16	FINAL COMBINED SIGNIFICANCE = H7*H8*H9*H10*H11*H11*H12*H13*H14*H15						4.43256E-11	22560347921

Figure 12-9 above: Asteroid-based economics – one reason President Trump established the Space Force. We may be mining asteroids now, but if the secret gets out, that would greatly reduce the value of what we extract.

Table 12-1 below has some estimates of asteroid mining values.

TABLE 12-1 - ASTEROID MINING VALUES				
Asteroid	Est. Value (US $billion)	Est. Profit (US $billion)	Delta V (km/sec)	Composition
Ryugu	83	30	4.663	Nickel, iron, cobalt, water, nitrogen, hydrogen, ammonia
1989 ML	14	4	4.889	Nickel, iron, cobalt
Nereus	5	1	4.987	Nickel, iron, cobalt
Bennu	0.7	0.2	5.096	Iron, hydrogen, ammonia, nitrogen
Didymos	62	16	5.162	Nickel, iron, cobalt
2011 UW158	7	2	5.189	Platinum, nickel, iron, cobalt
Anteros	5,570	1,250	5.440	Magnesium silicate, aluminum, iron silicate
2001 CC21	147	30	5.636	Magnesium silicate, aluminum, iron silicate
1992 TC	84	17	5.648	Nickel, iron, cobalt
2001 SG10	3	0.5	5.880	Nickel, iron, cobalt
Psyche	27.67	1.78	-	Nickel, iron, cobalt, gold [57]

Dr. Michael E. Salla asserts that not just the United States, but China, Russia, and the U.N. have secret space programs, and a major objective of these programs is to extract trillions of dollars of rare metals from asteroids. See page 191 of his book *The U.S. Navy's Secret Space Program & Nordic Extraterrestrial Alliance.* On **Figure 12-9**, I looked to see if any of this is encoded in Torah. It appears that it is. The axis term is **THERE IS AN ASTEROID**. At the same absolute skip are **THE ECONOMIC** or **THE ECONOMY**, **STANDARD**, and **USA**. Under **USA** are **CHINA** and

WORLD. The noun **MINE** at skip -1 is overlapped by **MINER** at skip -2. **SILVER/MONEY** is three rows directly above **THE CURRENCY**. Odds against this matrix appearing by chance are about 22,560,347,921 to 1.

Figure 12-10 ATTEMPTED ASSASSINATION: On **Figure 12-10** below, the axis term is **TRUMP ASSASSINATION** at skip +22,605. At the same absolute skip are **UFO, FALSE**, and **SPEAKER. OBAMA** is one skip off that of the axis term, and **CLINTON** is at skip -1. You say that the assassination attempts failed? The word **ATTEMPTER** is at skip +1. In 2021, there was another failed assassination attempt where the attempted assassin used a drone to shoot into Trump's bedroom window. This was discussed in a congressional interview of FBI Director Chris Wray at https://www.youtube.com/watch?v=Z46qSlU6K1A. On the matrix, **DRONE** is at skip - 22606 and **WINDOW** is at skip -22604. **WAR** is in the open text. The attempted killing, then, is part of an ongoing war against Trump and America. Who is trying to kill him? Besides **CLINTON** and **OBAMA**, the closest term at a special case skip is UFO at the same skip as **TRUMP ASSASSINATION**.

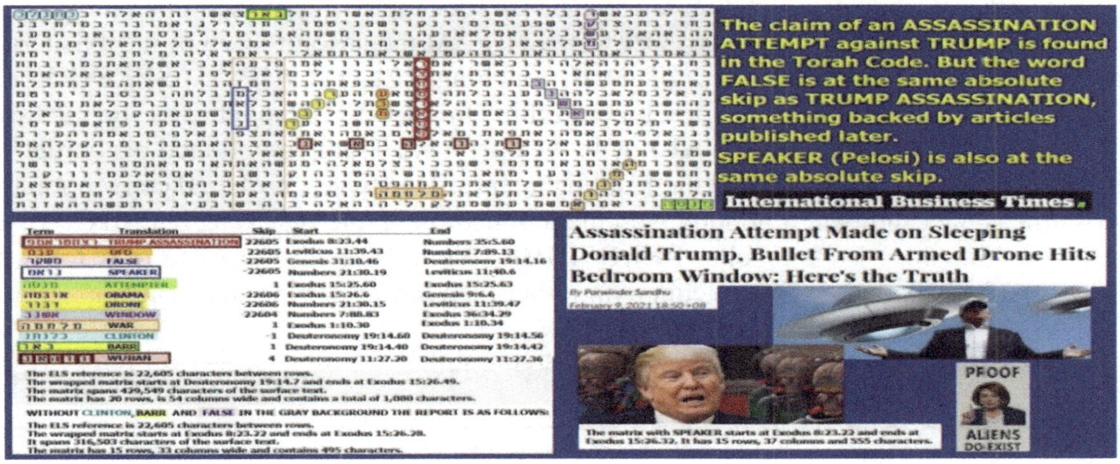

Figure 12-10 above. Odds against this matrix are 1,276,900,322 to 1; however, it is not yet proven that all the terms shown are going to be proven to be correct choices. It is of interest that the most significant *a priori* term was **ATTEMPTER**. All three assassination attempts were unsuccessful. There was only a 0.486 percent probability of finding this word. But for **WINDOW** (Trump's at Mar-a-Lago), there was a 92.15% chance that it would be on at a non-special case skip on the matrix. In a situation like this, it is best to look at individual probabilities rather than the combined joint probability of the matrix. As for suspects named in the attempt, here only two names came up. **CLINTON** (Hillary) was there at a special case skip against odds of about 11.3 to 1, and **OBAMA** was present against odds of about 5.56 to 1.

Above: Spreadsheet for Figure 12-10. Odds against this matrix were about 127,694,032 to 1. The most significant key word was **ATTEMPTER**. It was found against odds of about 295 tc 1 and it was likely used to indicate that the assassination ATTEMPT would fail.

FIGURE 12-11 below: PELOSI RIPS UP TRUMP SPEECH:

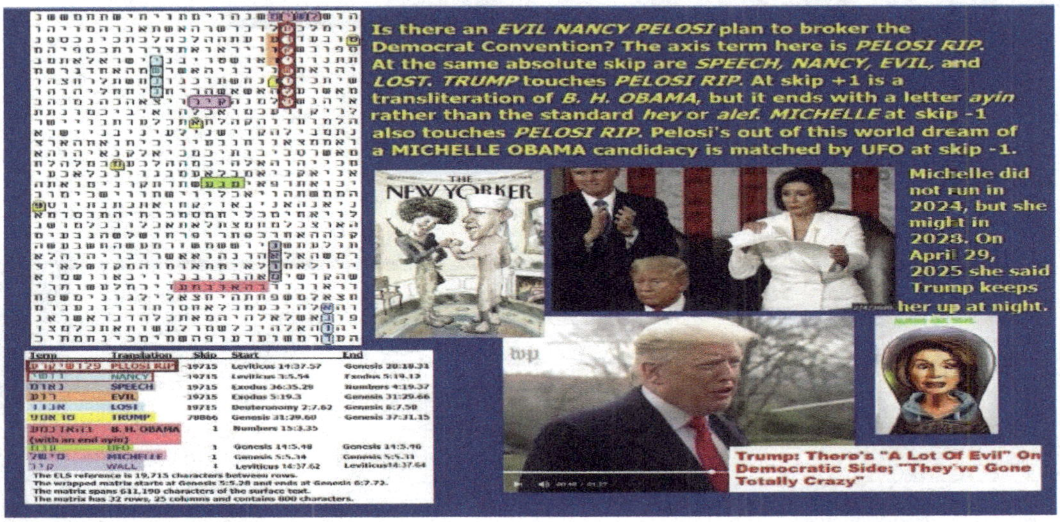

Perhaps the evilest manifestation of Pelosi's behavior was the Figure 12-11 image of her ripping up her official copy of President Trump's brilliant State of the Union speech while she was standing behind him on the night before her push to impeach him failed. The Codes capture this moment in almost photographic fashion. On the matrix above, the axis term is **PELOSI RIP**. At the absolute skip of the axis term are (1) a transliteration of **NANCY**, (2) every letter of **EVIL** touching **PELOSI RIP**, (3) **SPEECH**, and (4) the outcome of the next day's vote on her impeachment effort: She **LOST**. One letter of the best spelling for **TRUMP** touches **PELOSI RIP**. **THE WALL** is in the open text. Trump mentioned his border wall in his speech. This matrix may reveal the plan of the Establishment was to have a brokered convention that picked **Michelle Obama** as the candidate if nobody won the nomination on the first vote, but **MICHELLE** didn't want it even though at skip -1 she touches **PELOSI RIP**. While I usually forget to look for it, there is a transliteration of **OBAMA** that ends in a letter *ayin* rather than *hey* or *alef*. Using it, at skip

+1, see **B. H. OBAMA**. This matrix was found against combined odds of over 441 billion to 1.

	A	B	C	D	E	F	G	H
1	Axis term: PELOSI RIP @ Skip -19715	SKIP	NUMBER	DIVIDE BY	THE QUOTIENT	E QUOTIENT	POISSON EQUATION	EACH WORD OR
2	ELS Rank 1.	USED	IN	304805	EQUALS	X 800 LETTERS	PROBABILITY FOR	PHRASE HAS 1
3		ON	TORAH	LETTERS	FREQUENCY	EQUALS	THE WORD	CHANCE IN THE
4		FIGURE		IN THE	PER LETTER	WORD	APPEARING	IN THE FOLLOWING
5				TORAH		EXPECTANCY	AT LEAST ONCE	OF BEING HERE.
6	B. H. OBAMA (3 spellings) @ Skip +1 only	1	1	304805	3.28079E-06	0.002624629	0.002621188	381.5064687
7	TRUMP (best spelling) @ Skips 78886 to 79079	78886	2	304805	6.56157E-06	0.005249258	0.005235504	191.0035624
8	LOST @ Skips +/- 1 and +/- 19715	19715	9	304805	2.95271E-05	0.02362166	0.023344852	42.83599623
9	WALL @ Skip +1 only	1	11	304805	3.60886E-05	0.028870917	0.028458135	35.13933769
10	NANCY (2 spellings) @ Skips +/- 1 and +/- 19715	-19715	11	304805	3.60886E-05	0.028870917	0.028458135	35.13933769
11	SPEECH (4 synonyms) @ Skips +/- 1 and +/- 19715 only	19715	49	304805	0.000160759	0.128606814	0.120680366	8.286352036
12	MICHELLE @ Skips +/- 1 and +/- 19715	19715	97	304805	0.000318236	0.254589	0.224764946	4.449092351
13	UFO @ Skips +/- 1 and +/- 19715 only	-19715	189	304805	0.000620069	0.496054855	0.391071762	2.557075442
14	EVIL (two 3-letter words) @ Skips +/- 1 and +/- 19715 only	-19715	661	304805	0.0021686	1.734879677	0.823578573	1.21421323
15	COMBINED SIGNIFICANCE = H6*H7*H8*H9*H10*H11*H12*H13*H14						2.26666E-12	4.41177E+11

In May of 2025, we were shocked to learn that former Director of the FBI James Comey published a request for someone to assassinate President Trump. But it was no surprise to the Author of the Torah Code. Comey published a picture of seashells arranged to say 8647. The number 86 is a slang way to say eliminate or kill. The number 47 refers to President Trump, who is now the 47th President of the United States. See **Figure 12-12.**

Term	Translation	Skip	Start	End
חתמזכנומי	8647COMEY	-20424	Deuteronomy 28:69.69	Exodus 35:22.60
עבם	UFO	20424	Genesis 35:21/27	Exodus 10:17.19
שונא	ENEMY	-20424	Deuteronomy 14:26.22	Numbers 15:8.24
קשר	CONSPIRACY	-20424	Genesis 24:40.52	Deuteronomy 28:69.70
הנשיא	THE PRESIDENT	1	Numbers 27:2.34	Numbers 27:2.38
לשחתכלבשר	END OF ALL FLESH	1	Genesis 9:15.65	Genesis 9:15.73
זוחל	REPTILIAN	20429	Exodus 35:22.74	Numbers 4:12.37

The ELS reference is 20,424 characters between rows.
The wrapped matrix starts at Exodus 35:22.60 and ends at Exodus 10:17.24.
It spans 265,547 characters of the surface text.
The matrix has 14 rows, 35 columns and contains 490 characters

	A	B	C	D	E	F	G	H
1	*AXIS* TERM 8647 COMEY @ Skip -20,424	SKIP	NUMBER	DIVIDE BY	THE QUOTIENT	E QUOTIENT	POISSON EQUATION	EACH WORD
2	ELS RANK 1	USED	IN	304805	EQUALS	X 490	PROBABILITY FOR	OR PHRASE
3		ON	TORAH	LETTERS	FREQUENCY	EQUALS	THE WORD	HAS 1 CHANCE
4	R factor 1.69 for 8647 COMEY = 1 CHANCE IN 48.97788104	FIGURE		IN THE	PER LETTER	WORD	APPEARING	IN THE
5				TORAH		EXPECTANCY	AT LEAST ONCE	FOLLOWING OF
6								BEING HERE
7	END OF ALL FLESH	2	2	304805	6.56157E-06	0.00321517	0.003210007	311.5257781
8	THE PRESIDENT @ Skip +1 only	1	7	304805	2.29655E-05	0.011253096	0.011190017	89.36536924
9	R FACTOR 1.69 (value 48.97788104)						0.02041738	48.97788104
10	CONSPIRACY @Skips +/- 1 and +/- 20424	-20424	35	304805	0.000114828	0.056265481	0.054711854	18.27757484
11	REPTILIAN @ Skips 20429 to 20634	20429	77	304805	0.000252621	0.123784059	0.11642938	8.588897386
12	ENEMY (2 synonyms) @ Skips +/- 1 and +/- 20424	-20424	93	304805	0.000305113	0.149505421	0.138866231	7.201174796
13	UFO @ Skips +/- 1 and +/- 20424	20424	154	304805	0.000505241	0.247568117	0.21930296	4.559901973
14	ODDS AGAINST MATRIX = H7*H8*H9*H10*H11*H12						1.42273E-10	7028749731

FIGURE 12-12: Against odds of over 7 billion to 1, 8647 COMEY is near THE PRESIDENT, at the same absolute skip of CONSPIRACY, UFO, and ENEMY. REPTILIAN is here too.

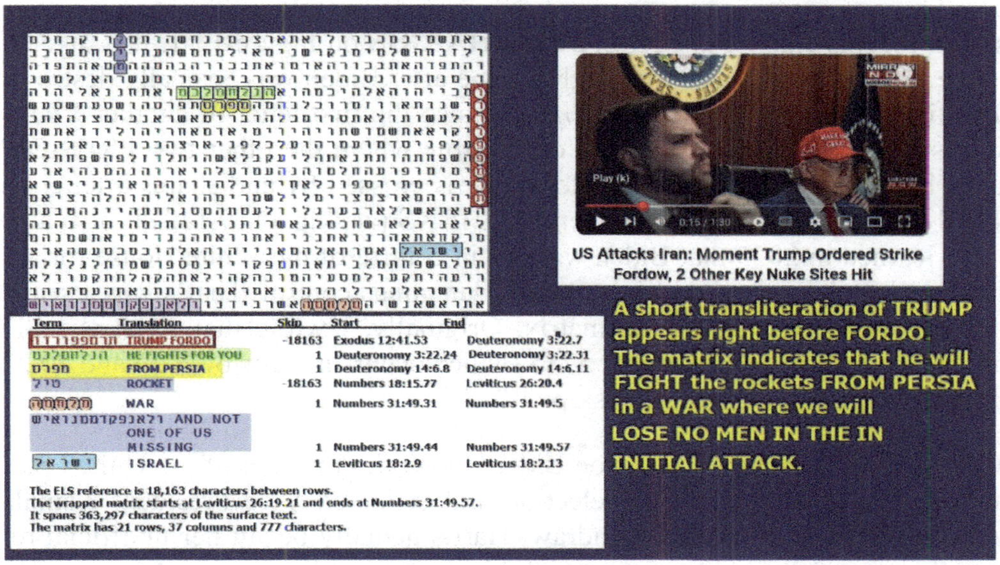

Figure 12-13 above. I will not offer a probability for this matrix because in the 9-letter axis term **TRUMP FORDO**, the transliteration of Trump is only 4 letters, which is shorter by a letter than the 5 letters that I usually use, but it is enough to recognize the name. At the same skip is **ROCKET.** America is helping **ISRAEL** shoot down rockets **FROM PERSIA** (Iran)**.** In the opening attack of this **WAR,** the **U.S. DID NOT LOSE A MAN**. In fact, Iran never even saw our planes at all. But the Author of the Torah saw it all, and recorded it 3,300 years ago.

Chapter 13
President Kamala Harris?

The main axis term for this chapter is **PRESIDENT HARRIS** on Figure 13-1. It is at skip +119,183. It's encoded only once in Torah and was only sought by me after Biden was basically destroyed in his debate with Trump. The first letter of **PRESIDENT** (*nun*) is also the first letter of another **PRESIDENT** in the open text. There is a third **PRESIDENT** also on the row where the other two PRESIDENTs appear or begin. The matrix hints at who controls her. Her name is crossed by and shares a letter alef with **NAZI.** Nazi at skip -119,184 is one skip off the axis term. **ALIENS** (at skip +119,182) is also at one skip off the axis term. We just learned in 2025 that Biden was not in control of the White House for a long time. Who was? Likely the Fourth Reich, which Nazis and Reptilian aliens control. That's why I searched for **NAZI, ALIENS,** and **UFO**, which are all on the matrix. Originally, I wrote that, "Harris will likely become President for at least part of the Hebrew Year 5785 (which occurs in parts of 2024 and 2025). Hebrew year of her partial 'rule' (5785) is on the matrix at skip -1."

On July 21, 2024 (15th of Tammuz 5784), Joe Biden, the 46th president of the United States, announced his withdrawal from the 2024 presidential election and endorsed Vice President Kamala Harris as his replacement. After Biden was forced to withdraw, Harris actually began her unofficial reign as "acting" President. During this period, the public was never told who was really running America. The Hebrew New Year began at sunset on Wednesday, October 2, 2024. On July 15, 2025 we still have not been told who was running the show, complete with thousands of pardons signed by autopen. So, she actually served in this capacity for about 73 days in 5784, and (per the Code) an additional 110 days in the Hebrew year of 5785.

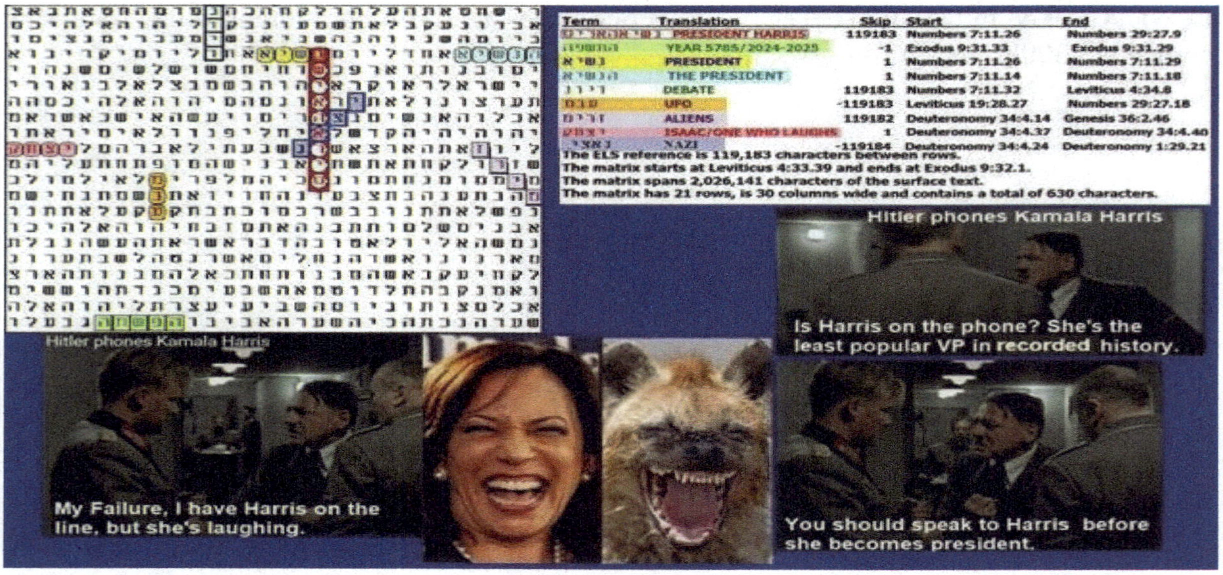

Figure 13-1 above: PRESIDENT HARRIS was predicted to be a **NAZI PRESIDENT.**

Odds against Figure 13-1 above are about 12,854,441 to 1.

	A	B	C	D	E	F	G	H
1	AXIS TERM: PRESIDENT HARRIS @ Skip 119,183	SKIP	NUMBER	DIVIDE BY	THE QUOTIENT	E QUOTIENT	POISSON EQUATION	EACH WORD
2	AXIS TERM ELS RANK: 1	USED	IN	304805	EQUALS	X 630 LETTERS	PROBABILITY FOR	OR PHRASE
3		ON	TORAH	LETTERS	FREQUENCY	EQUALS	THE WORD	HAS 1 CHANCE
4		FIGURE		IN THE	PER LETTER	WORD	APPEARING	IN THE
5				TORAH		EXPECTANCY	AT LEAST ONCE	FOLLOWING OF
6								BEING HERE
7	YEAR 5785 @ Skips +/- 1 and +/- 119183	-1	4	304805	1.31231E-05	0.008267581	0.008233498	121.455054
8	DEBATE @ Skips +/- 1 and +/- 119183	119183	22	304805	7.21773E-05	0.045471695	0.044453351	22.49549192
9	PRESIDENT @ Skip +1 only	1	68	304805	0.000223093	0.140548876	0.131118803	7.626671204
10	ADDITIONAL PRESIDENT (the president) @ Skip +1	1	67	304805	0.000219813	0.13848198	0.129321059	7.732692602
11	ISAAC (ONE WHO LAUGHS) @ Skip +1 only	1	101	304805	0.000331359	0.208756418	0.188407099	5.30765563
12	UFO @ Skips +/- 1 and +/- 119183	119183	163	304805	0.000534768	0.336903922	0.286022566	3.496227637
13	ALIENS @ Skips 119182 to	119182	265	304805	0.000869408	0.547727235	0.421737427	2.371143596
14	NAZI @ Skips 119184 to -119429	-119184	388	304805	0.001272945	0.801955349	0.551648772	1.813076287
15	COMBINED ODDS AGAINST THE MATRIX = H7*H8*H9*H10*H11*H12*H13*H14						7.77941E-08	12854441.97
16								
17	AXIS TERM: PRESIDENT HARRIS @ Skip 119,183	SKIP	NUMBER	DIVIDE BY	THE QUOTIENT	E QUOTIENT	POISSON EQUATION	EACH WORD
18	AXIS TERM ELS RANK: 1	USED	IN	304805	EQUALS	X 54 LETTERS	PROBABILITY FOR	OR PHRASE
19	This calculation shows the odds against an open text PRESIDENT intersecting PRESIDENT HARRIS	ON	TORAH	LETTERS	FREQUENCY	EQUALS	THE WORD	HAS 1 CHANCE
20	along with NAZI intersecting PRESIDENT HARRIS all in only 54 LETTERS	FIGURE		IN THE	PER LETTER	WORD	APPEARING	IN THE
21				TORAH		EXPECTANCY	AT LEAST ONCE	FOLLOWING OF
22								BEING HERE
23	PRESIDENT @ Skip +1 only	1	68	304805	0.000223093	0.012047046	0.011974771	83.50890152
24	NAZI @ Skips -119184 to -119197	119184	24	304805	7.87389E-05	0.004251899	0.004242872	235.6893975
25	COMBINED ODDS AGAINST A SECOND PRESIDENT PLUS NAZI IN 54 LETTER WITH PRESIDENT HARRIS = H23 *H24						5.08074E-05	19682.1627

STATISTICAL SIGNIFICANCE OF FIGURE 13-1. The full 630-letter matrix was found against odds of about 12.8 million to 1. The most significant term is the Hebrew year **5785**. It was found at a special case skip (-1) against odds of about 121 to 1. The most interesting combination of keywords has the 9-letter axis term (**PRESIDENT HARRIS**) beginning with a letter *nun* (N). What is this **nun** in the open text horizontal to the axis term? It is also the first letter of Nasey (*president*) in the open text! So, the Torah here translates itself from encoded to unencoded text. Further, while I often hear moronic Democrats compare Trump with Hitler (something never found in the Torah Code), **PRESIDENT HARRIS** is crossed by and shares a letter *alef* with **NAZI**. All of this is in 54 letters (6 columns by nine rows). Odds against such a quality match are about 19,682 to 1. To just show **PRESIDENT HARRIS** with **NAZI** only takes 36 letters (4 columns by nine rows) without considering contradicting letters. Actually, the matrix is even more significant because there was only about one chance in 2 that there would be any 9-letter ELS of PRESIDENT HARRIS in Torah.

ISAAC is also on Figure 13-1 in the open text. The name means **ONE WHO LAUGHS**. Harris is famous for being unable to stop laughing, even when there is nothing funny. People are probably more likely to laugh at her rather than with her. She is a pro-Palestinian, anti-Semitic idiot. I would have said, "God help America" when she moved up in office, but her boss (Biden) was also a Jew- hating Nazi who despises America and our Constitution.

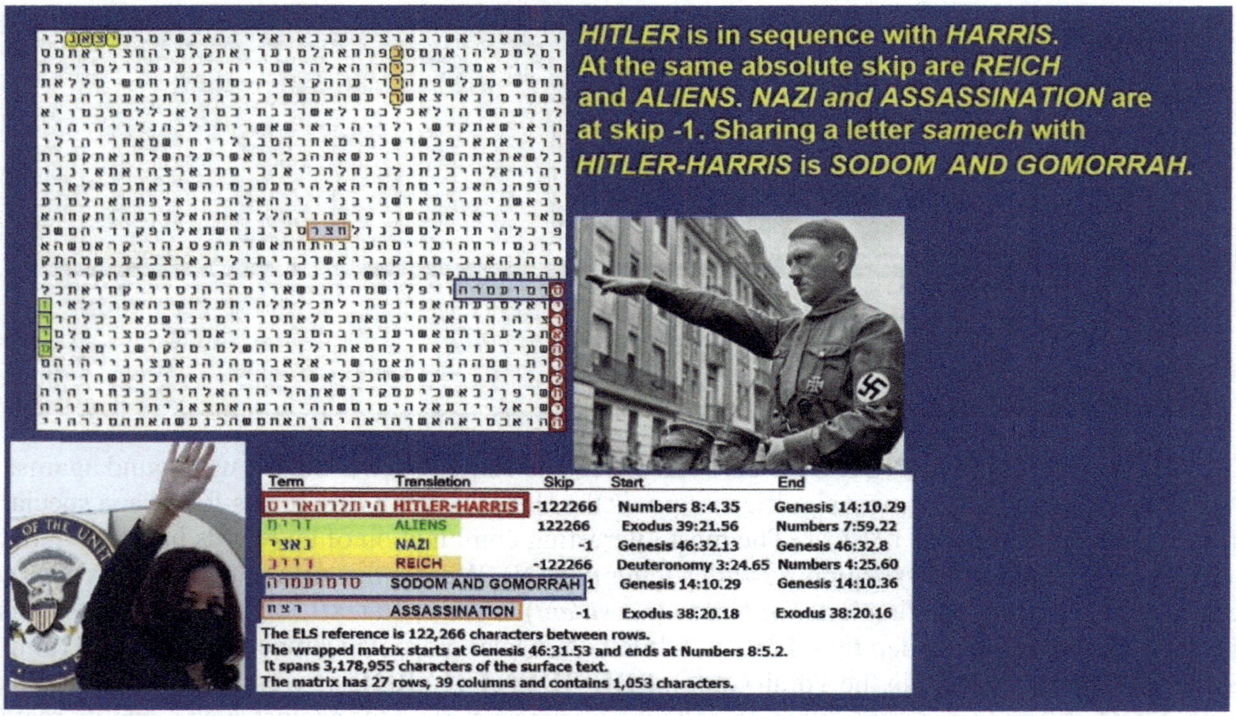

Figure 13-2 above. HITLER HARRIS is in sequence and with **ALIENS** at the same absolute skip. **HITLER HARRIS** shares the first letter with **SODOM AND GOMMORAH** in the open text. REICH is also at the same skip as HITLER HARRIS and ALIENS. NAZI is at skip -1.

Figure 13-3 below: Two congressmen (Doug Lamborn and Joe Wilson) call out Harris as antisemitic exactly two years and 1 day before the attack on Israel on October 7, 2023.

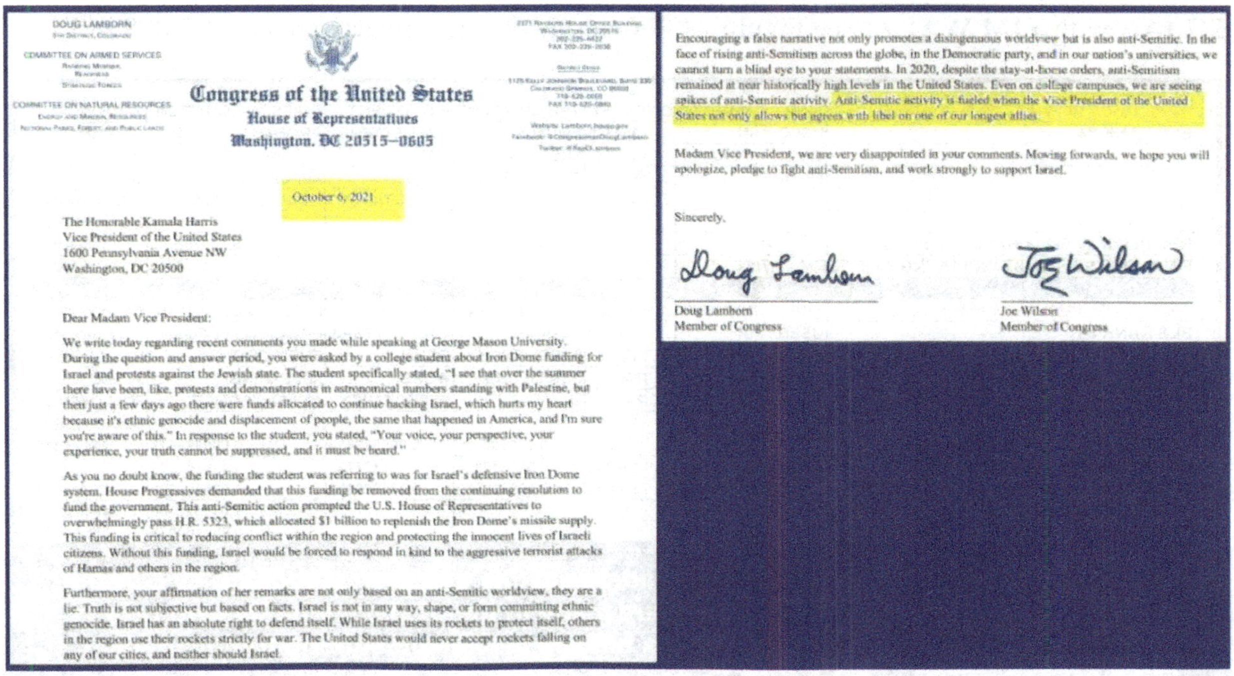

DOUG LAMBORN

COMMITTEE ON ARMED SERVICES

COMMITTEE ON NATURAL RESOURCES

Congress of the United States
House of Representatives
Washington, DC 20515–0605

October 6, 2021

The Honorable Kamala Harris
Vice President of the United States
1600 Pennsylvania Avenue NW
Washington, DC 20500

Dear Madam Vice President:

We write today regarding recent comments you made while speaking at George Mason University. During the question and answer period, you were asked by a college student about Iron Dome funding for Israel and protests against the Jewish state. The student specifically stated, "I see that over the summer there have been, like, protests and demonstrations in astronomical numbers standing with Palestine, but then just a few days ago there were funds allocated to continue backing Israel, which hurts my heart because it's ethnic genocide and displacement of people, the same that happened in America, and I'm sure you're aware of this." In response to the student, you stated, "Your voice, your perspective, your experience, your truth cannot be suppressed, and it must be heard."

As you no doubt know, the funding the student was referring to was for Israel's defensive Iron Dome system. House Progressives demanded that this funding be removed from the continuing resolution to fund the government. This anti-Semitic action prompted the U.S. House of Representatives to overwhelmingly pass H.R. 5323, which allocated $1 billion to replenish the Iron Dome's missile supply. This funding is critical to reducing conflict within the region and protecting the innocent lives of Israeli citizens. Without this funding, Israel would be forced to respond in kind to the aggressive terrorist attacks of Hamas and others in the region.

Furthermore, your affirmation of her remarks are not only based on an anti-Semitic worldview, they are a lie. Truth is not subjective but based on facts. Israel is not in any way, shape, or form committing ethnic genocide. Israel has an absolute right to defend itself. While Israel uses its rockets to protect itself, others in the region use their rockets strictly for war. The United States would never accept rockets falling on any of our cities, and neither should Israel.

Encouraging a false narrative not only promotes a disingenuous worldview but is also anti-Semitic. In the face of rising anti-Semitism across the globe, in the Democratic party, and in our nation's universities, we cannot turn a blind eye to your statements. In 2020, despite the stay-at-home orders, anti-Semitism remained at near historically high levels in the United States. Even on college campuses, we are seeing spikes of anti-Semitic activity. Anti-Semitic activity is fueled when the Vice President of the United States not only allows but agrees with libel on one of our longest allies.

Madam Vice President, we are very disappointed in your comments. Moving forwards, we hope you will apologize, pledge to fight anti-Semitism, and work strongly to support Israel.

Sincerely,

Doug Lamborn
Member of Congress

Joe Wilson
Member of Congress

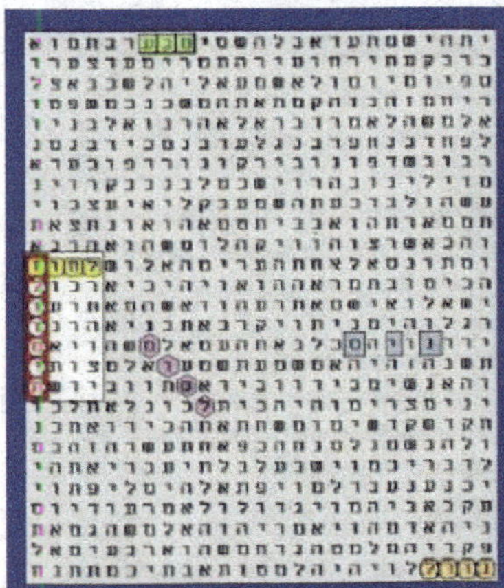

Harris's pick for Vice President TIM WALZ is encoded in 28 letters with REPTILIAN. They share a letter *zayin*. He is encoded with UFO at skip -1 and WEIRD at skip +1. Oddly, the Democrats have been trying to label Trump and Vance as being WEIRD.

Walz has been to CHINA over 30 times. He owes them a lot.

CHINA is at skip -2.

Term	Translation	Skip	Start	End
תימוולז	TIM WALZ	-58628	Genesis 19:16.38	Deuteronomy 19:12.6
עבם	UFO	-1	Numbers 33:49.31	Numbers 33:49.29
זוחל	REPTILIAN	-1	Deuteronomy 19:12.6	Deuteronomy 19:12.3
[illegible]	WEIRD	1	Numbers 33:54.74	Numbers 33:54.77
סין	CHINA	-2	Numbers 13:30.4	Numbers 13:29.67
מושל	GOVERNOR	58627	Numbers 13:30.15	Exodus 3:21.33

The ELS reference is 58,628 characters between rows.
The matrix starts at Numbers 33:49.15 and ends at Numbers 33:54.98.
It spans 1,524,353 characters of the surface text.
The matrix has 27 rows, 25 columns and contains 675 characters.
The part with a white background also has 58,628 letters between rows.
It starts at Deuteronomy 19:12.3 and ends at Genesis 19:16.38.
It spans 351,772 characters of the surface text.
The matrix has 7 rows, 4 columns and contains 28 characters.

Figure 13-4 above. **TIM WALZ** is encoded in 28 letters with REPTILIAN at skip -1. When Harris won the Democrat nomination for President, she chose TIM WALZ as her vice-presidential candidate**. WALZ** is strongly encoded with being a REPTILIAN. Amazingly, it only takes 28 letters to show **TIM WALZ** vertically at skip -58,628 and REPTILIAN horizontally at skip-1, sharing the letter zayin (Z) in

REPTILIAN with **TIM WALZ**. See Figure 13-4. Odds against this matrix are about 5,898 to 1.

	A	B	C	D	E	F	G	H
1	*AXIS* TERM TIM WALZ @ Skip *-58628*	SKIP	NUMBER	DIVIDE BY	THE QUOTIENT	E QUOTIENT	POISSON EQUATION	EACH WORD
2	ELS RANK 84	USED	IN	304805	EQUALS	x 28 LETTERS	PROBABILITY FOR	OR PHRASE
3		ON	TORAH	LETTERS	FREQUENCY	EQUALS	THE WORD	HAS 1 CHANCE
4		FIGURE		IN THE	PER LETTER	WORD	APPEARING	IN THE
5				TORAH		EXPECTANCY	AT LEAST ONCE	FOLLOWING OF
6								BEING HERE
7	REPTILIAN @ Skips +/- 1 and +/58628	-1	1	304805	3.28079E-06	9.1862E-05	9.18578E-05	10886.39286
8	FINAL ODDS AGAINST INCLUDING AXIS TERM = h7/84						0.007716054	129.5999151
9								
10	*AXIS* TERM TIM WALZ @ Skip *-58628*	SKIP	NUMBER	DIVIDE BY	THE QUOTIENT	E QUOTIENT	POISSON EQUATION	EACH WORD
11	ELS RANK 84	USED	IN	304805	EQUALS	x 120 LETTERS	PROBABILITY FOR	OR PHRASE
12		ON	TORAH	LETTERS	FREQUENCY	EQUALS	THE WORD	HAS 1 CHANCE
13		FIGURE		IN THE	PER LETTER	WORD	APPEARING	IN THE
14				TORAH		EXPECTANCY	AT LEAST ONCE	FOLLOWING OF
15								BEING HERE
16	REPTILIAN @ Skips +/- 1 and +/58628	-1	1	304805	3.28079E-06	0.000393694	0.000393617	2540.541699
17	GOVERNOR @ Skips 58627 to 58674	58674	400	304805	0.001312314	0.157477732	0.145704163	6.86322189
18	ENEMY (2 terms) @ Skips +/- 1 and /- 58628	'- 58628	91	304805	0.000298552	0.035826184	0.035192022	28.41553124
19	INITIAL COMBINED ODDS						2.01832E-06	495461.7673
20	FINAL ODDS FOR AXIS TERM RANK 84						0.000169539	5898.354372

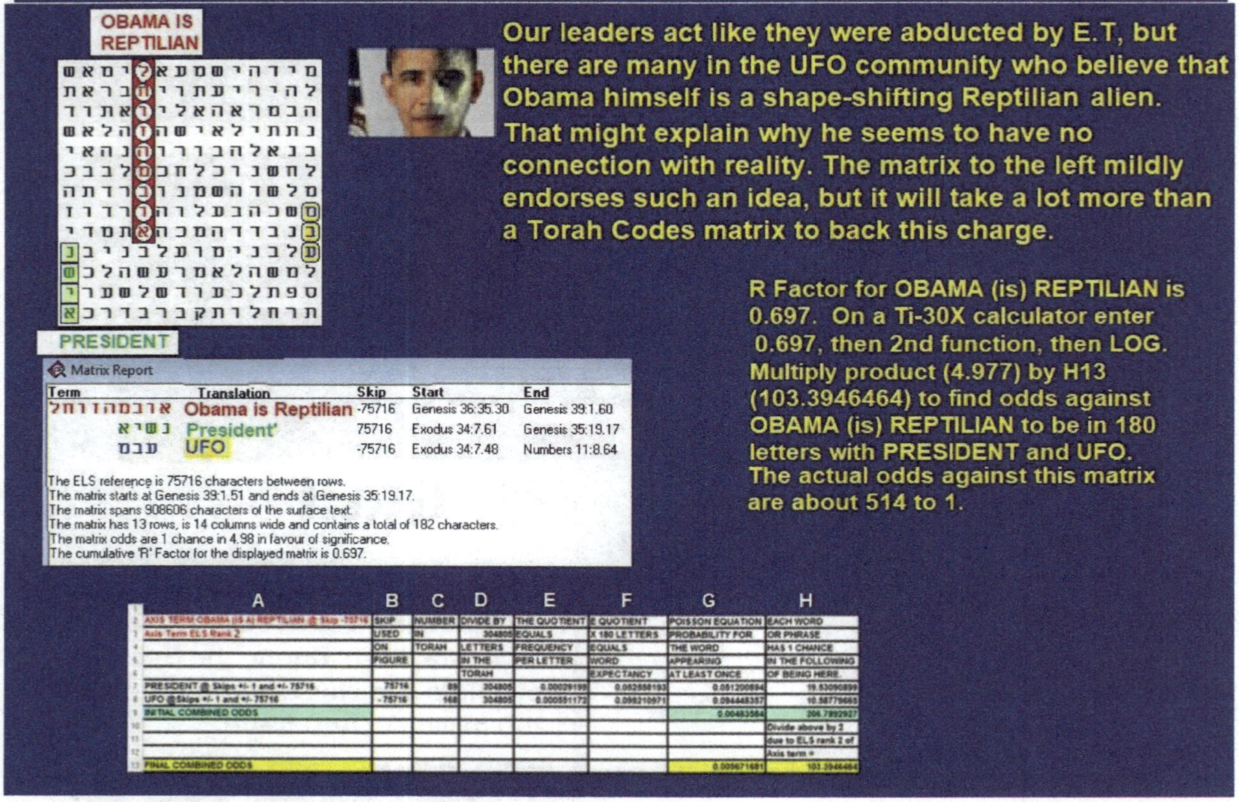

FIGURE 13-5 above: The axis term at skip -75,716 is **OBAMA IS REPTILIAN**. Without counting the axis term, odds against this matrix with **PRESIDENT** and **UFO** at the same skip as **OBAMA IS REPTILIAN** were about 103 to 1. The axis term has an R-factor of 0.697. This equates to one chance in 4.97737085 that the term would be found at least once in Torah, although this term ranks only second in ELS rank because it is found the first time at a lower skip, namely +67,569. So, if we do use the R factor for this 9-letter term, then 103.39 X 4.97737085 = 532.016. So, the matrix actually had only about one chance in 532 to be found.

If Obama is reptilian and Tim Walz is reptilian, it seems like what we are being warned about is that the leadership of the Democratic Party is firmly in alien hands. This message is again encoded with a warning that **BIDEN IS ALIEN** in Figure 13-6, which was found against odds of over 48 billion to 1.

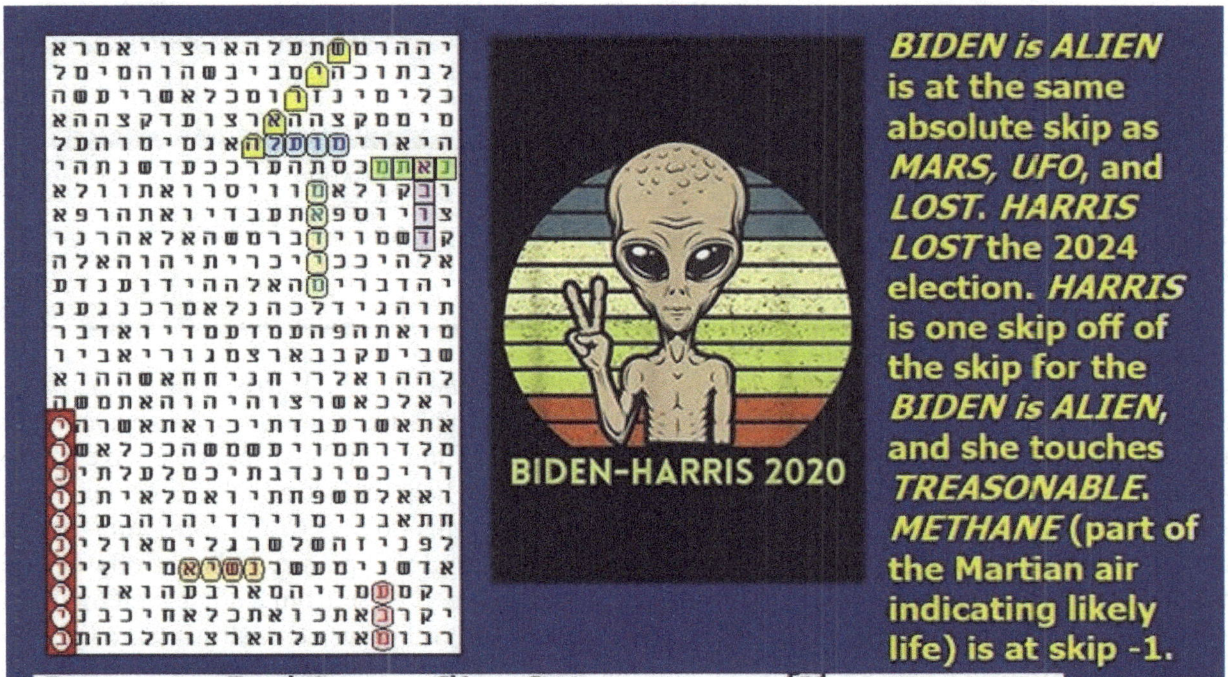

Term	Translation	Skip	Start	End
ביידננכרי	BIDEN (is) ALIEN	-97832	Genesis 7:18.31	Genesis 30:29.35
מאדים	MARS	97832	Deuteronomy 21:18.44	Genesis 43:7.68
האריס	HARRIS	-97833	Exodus 8:1.54	Genesis 1:28.81
נאתמ	METHANE	-1	Leviticus 27:23.13	Leviticus 27:23.10
מרעל	TREASONABLE	1	Exodus 8:1.5 0	Exodus 8:1.53
נשיא	PRESIDENT	1	Genesis 17:20.60	Exodus 27:16.547:20.63
עבמ	UFO	97832	Exodus 27:16.54	Genesis 7:18.16
אבוד	LOST	97832	Leviticus 27:23.11	Leviticus 21:23.60

The ELS reference is 97,832 characters between rows.
The matrix starts at Genesis 1:28.77 and ends at Genesis 7:18.31.
The matrix spans 2,445,818 characters of the surface text.
The matrix has 26 rows, 18 columns and 468 characters.

FIGURE 13-6 above: Here, the axis term is **BIDEN (is) ALIEN. MARS** is at the same absolute skip. HARRIS is at skip-97,833, while **BIDEN IS ALIEN** is at skip-97,832. The difference of 1 produces the diagonal display of **HARRIS** with all her letters touching each other, and the word **TREASONABLE** at skip +1 touching the first letter of her name. She is certainly treasonable, and she was not selected through the normal primary process. Instead, she was picked by the aliens.

Also, at the absolute skip of **BIDEN IS ALIEN**, we find **UFO**: There are allegedly several species that have made it to Mars. The life there gives off methane (CH_4). At skip +1 is **METHANE**. **PRESIDENT** is at skip +1. Finally, **LOST** is at the skip of **BIDEN IS ALIEN**. I looked for this word because in 2024, Biden got his confused backside thrown off the ticket, and **HARRIS** lost the election.

	A	B	C	D	E	F	G	H
1	AXIS TERM BIDEN IS ALIEN @ Skip -97,832	SKIP	NUMBER	DIVIDE BY	THE QUOTIENT	E QUOTIENT	POISSON EQUATION	EACH WORD
2	ELS RANK 1	USED	IN	304805	EQUALS	x 468 LETTERS	PROBABILITY FOR	OR PHRASE
3		ON	TORAH	LETTERS	FREQUENCY	EQUALS	THE WORD	HAS 1 CHANCE
4	Axis term R factor = 1.387, Value = 54.209. This is only used for	FIGURE		IN THE	PER LETTER	WORD	APPEARING	IN THIS OF
5	the second calculation			TORAH		EXPECTANCY	AT LEAST ONCE	BEING HERE
6	MARS @ Skips +/- 1 and +/- 97832	97832	3	304805	9.84236E-06	0.004606224	0.004595631	217.5979622
7	LOST @ Skips +/- 1 and +/- 97832	97832	10	304805	3.28079E-05	0.015354079	0.015236806	65.63055301
8	METHANE @ Skips +/- 1 and +/- 97832	-1	46	304805	0.000150916	0.070628763	0.06819225	14.66442296
9	TREASONABLE @ Skip +1 only	1	47	304805	0.000154197	0.072164171	0.069621857	14.36330539
10	PRESIDENT @ Skip +1 only	1	68	304805	0.000223093	0.104407736	0.09914209	10.0865334
11	HARRIS (2 spellings - samech & shin) @ Skips 97833 to 97947	-97833	114	304805	0.00037401	0.175036499	0.160573618	6.227673099
12	UFO @ Skips +/- 1 and +/- 97832	97832	155	304805	0.000508522	0.237988222	0.211788025	4.721702269
13	Combined odds against the matrix =H6*H7*H8*H9*H10*H11*H12						1.12087E-09	892167909.4
14		9E+08						
15	AXIS TERM BIDEN IS ALIEN @ Skip -97,832	SKIP	NUMBER	DIVIDE BY	THE QUOTIENT	E QUOTIENT	POISSON EQUATION	EACH WORD
16	ELS RANK 1. The full matrix was found against odds of over	USED	IN	304805	EQUALS	x 468 LETTERS	PROBABILITY FOR	OR PHRASE
17	48 bilion to 1.	ON	TORAH	LETTERS	FREQUENCY	EQUALS	THE WORD	HAS 1 CHANCE
18	Axis term R factor = 1.387, Value = 54.209. This is only used for	FIGURE		IN THE	PER LETTER	WORD	APPEARING	IN THE
19	this second calculation			TORAH		EXPECTANCY	AT LEAST ONCE	FOLLOWING OF
20								BEING HERE
21	VALUE FROM H14 ABOVE						1.12087E-09	892167909.4
22	ODDS AGAINST R FACTOR 54.209						0.018447121	54.209
23	Combined odds against the matrix = h21*H22						2.06767E-11	48363530201

The strong link of alien origin (or brainwashing) between Obama, Biden, Harris, and Walz, plus other Democratic leaders identified in Chapter 3 of this book, raises the question of just how far alien colonization of our planet has gone, as well as how do we fight what is happening to us? I suspect we have millions of ETs who are here leading more mundane lives, but are voting to help their kind take over our nation and planet. I have written a lot about one of them in two chapters of this book – Craig Ebrahimi – and shown that there is evidence pointing to him spreading COVID-19 around this planet. I also briefly described a woman (Rebecca) in Lakeland, Florida, who all but outright claimed to be from a solar system around the star Meia in the Pleiades. Her personality was much like that of a religious (terrestrial) Christian woman. Still, our 11-year friendship ended when she and her husband answered some questions and told some stories that made her (really – them) appear to be more reptilian than Pleiadian. If the three people that I speak about here are aliens (really, reptilians) in disguise, then our situation may be akin to what is true if you find one cockroach crawl out from under your refrigerator. The chances are that you have an infestation.

Before we can address such threats, we must know for sure that they exist. One of the many purposes of the Torah Code seems to be to warn us about the threat's reality, and even to tell us who to watch out for. The cover of this book is meant to imply that if asked to, I will lead any required legal battle to free our government of corruption, but such a battle would be as hard as trying to modify the Constitution in a way to reflect the battle at hand. I would prefer to take my role after President Trump agrees to (1) permit disclosure and (2) establish a new Department of UFO Disclosure Management with me heading it and serving on the President's Cabinet. Trump supposedly wanted to do a UFO disclosure during his first term, but was silenced. Just how much fight does he have in him? I hope it's enough, but he still faces threats against himself and his family. I have had my life threatened twice because of what I know. It is unconstitutional to oppose freedom of the press.

There may be an entire branch of our military that knows the truth about aliens. I speak about, of course, the United States Space Force, founded by a man who is trying hard to fulfill the role of Messiah as understood by Judaism: the peacemaker Donald Trump. While I want to serve in Trump's Cabinet, I am also willing to accept a commission in our Space Force, but I am 78 years old. I know ET well enough to know how to plan to defeat him. Father Time may be a tougher opponent, but I suspect that those who serve in positions that require interaction with friendlier types of visitors will have medical procedures that can help a lot of us live longer.

May we all live long and prosper!

Epilogue to God, Aliens, or Artificial Intelligence: Who Wrote the Torah Code?

This epilogue addresses the question raised in the title. Before studying Figure E-1, I recommend that you watch the last 38 seconds of the first MEN IN BLACK movie. It basically alludes to God or aliens with respect to who created everything. The film's conclusion is currently posted At

https://www.gA8t4Zkg3d6oogle.com/search?client=firefox-b-1d&q=VIDEO+MEN+IN+BLACK+END+OF+MOVIE#fpstate=ive&vld=cid:3d66ae28,vid:OKnpPCQyUec,st:0

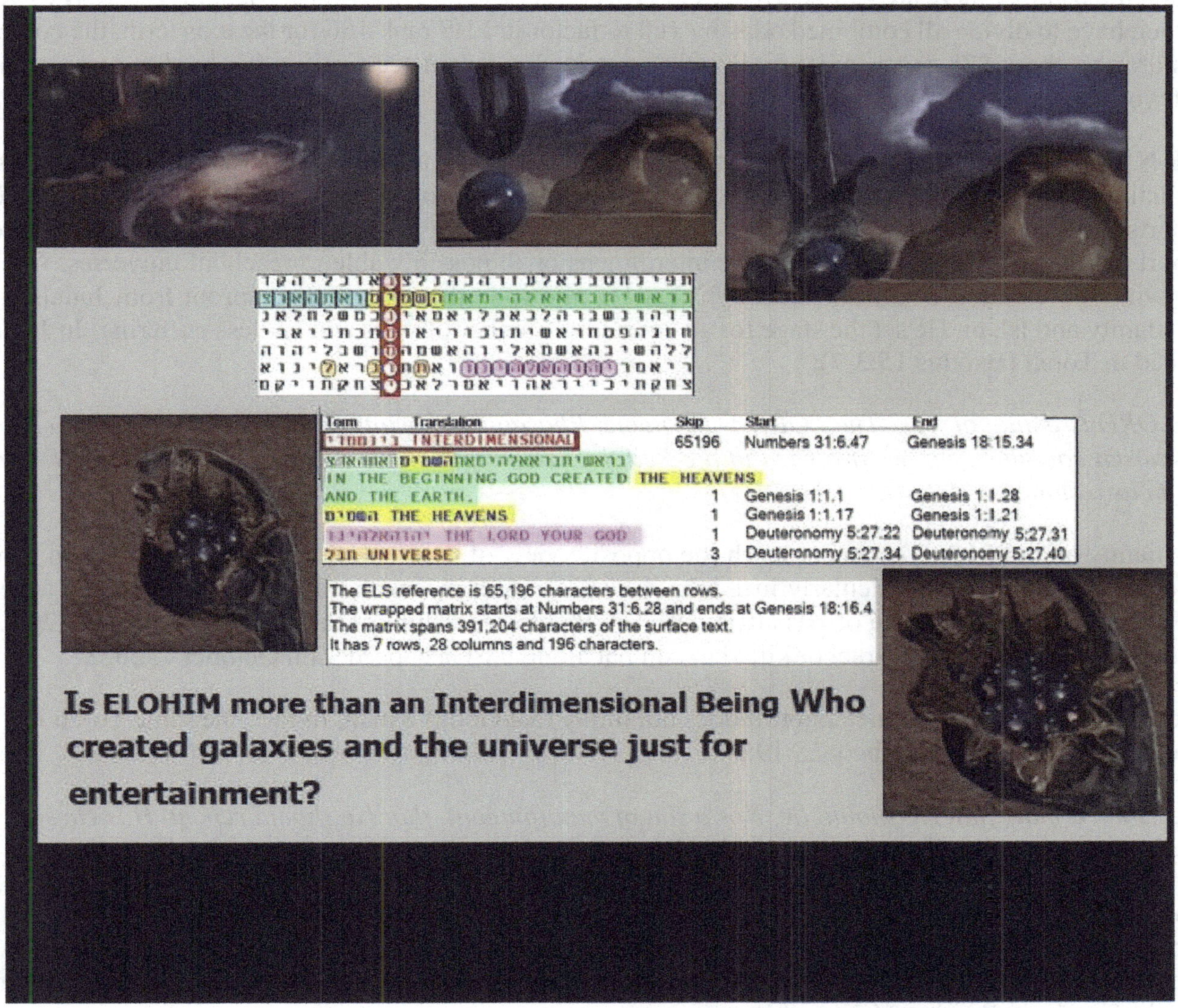

Figure E-1 above: The axis term is **INTERDIMENSIONAL**. The target word was THE HEAVENS only in Genesis 1:1 (the first verse in the Bible). The LORD YOUR GOD is only an a-posteriori find. **UNIVERSE** is at skip +3, touching **INTERDIMENSIONAL**.

	A	B	C	D	E	F	G	H
1	AXIS TERM INTERDIMENSIONAL @ Skip +65196	SKIP	NUMBER	DIVIDE BY	THE QUOTIENT	E QUOTIENT	POISSON EQUATION	EACH WORD
2	ELS RANK 160	USED	IN	304805	EQUALS	X 196 LETTERS	PROBABILITY FOR	OR PHRASE
3	No credit for THE LORD YOUR GOD because it was found a-posteriori	ON	TORAH	LETTERS	FREQUENCY	EQUALS	THE WORD	HAS 1 CHANCE
4		FIGURE		IN THE	PER LETTER	WORD	APPEARING	IN THE
5				TORAH		EXPECTANCY	AT LEAST ONCE	FOLLOWING OF
6								BEING HERE
7	GENESIS 1:1 (entire verse) @ Skip +1 only	1	1	304805	3.28079E-06	0.000643034	0.000642827	1555.627605
8	HEAVEN @ Skip +1 only	1	81	304805	0.000265744	0.05208576	0.050752544	19.70344585
9	UNIVERSE @ Skips +3 to + 29 only	1	1811	304805	0.005941504	1.164534703	0.687932167	1.453631691
10	COMBINED ODDS = (H7*H8*H9)/160						0.00359102	278.4724436

SPREADSHEET FOR FIGURE E-1 ABOVE. The odds against having **INTERDIMENSIONAL** intersect Genesis 1 and its Bible-opening statement about the creation of the UNIVERSE are about 1,555 to 1. However, the axis term **INTERDIMENSIONAL** seen here has a high ELS rank (160). When we include the fact that there are 81 occurrences of HEAVEN in Torah, throw in a 3-letter term **(UNIVERSE),** but then have to divide all combined odds by 160 to factor in ELS rank 160 for the axis term, the combined odds drop to about 278 to 1 against finding this match. We cannot see God, so it's hard to guess what the right words are to describe Him or His home.

THE NATURE OF GOD. Many writers speculate that angels are not extraterrestrial beings but are instead spiritual beings from another dimension that we cannot reach with a spaceship alone. The matrix above supports the contention that they are correct, for if angels are from an interdimensional home, they directly support God there. God is likely also to be interdimensional, possibly able to reach all universes. Does He play with His creation of all universes? In that He allowed other religions to sprout from Judaism, like Christianity and Islam, He set the stage for an enormous number of wars and endless suffering. In Judaism, we read in Torah (Exodus 15:3-7):

The LORD is a man of war: The LORD is his name. Pharaoh's chariots and his host he cast into the sea: His chosen captains also are drowned in the Red Sea. The depths have covered them: They sank into the bottom as a stone.

Christianity and its New Testament teach the opposite view of God: *"God is love"* is a Christian concept that appears in its Bible, particularly in 1 John 4:8 and 4:16. It means that love is a fundamental aspect of God's being, and not just one of His attributes. Tell that to the victim of the Holocaust, or the 1,200 Jews raped/and or slaughtered in Israel or killed in combat along the Gaza border on October 7, 2023.

Exodus 15:3 is, of course, a metaphor. It does not imply that God is a man. In fact, the same Torah includes this cautionary verse in Numbers 23:19:

"*God is not a man, that He should lie, Nor a son of man (human), that He should repent. Has He said, and will He not do? Or has He spoken, and will He not make it good?*

What is a "man o' war?" It is a metaphorical, nautical term. All ships of the line were men o' war, but not all men o' war were ships of the line. The latter were the largest category of sailing warships, with two or three decks fully armed with heavy guns. Obviously, God is not a ship. Rather than teaching that God is love, the Torah (in Exodus 34:14 to 16) states that:

"For you shall not prostrate yourself to an alien god, for the very name of HaShem is Jealous One. He is a jealous God. Lest you seal a covenant with the inhabitant of the land and stray after their gods, and he invite you and you eat of his slaughter. And you shall take their daughters to your sons, and their daughters stray after their gods and entice your sons to stray after their gods.

On many key concepts of God, both Christianity and Islam have found plenty of differences that serve as excuses for them to slaughter each other and Jews over the last 2,000 years. However, when we look at Exodus 34:14, it really is pertinent to the title of this book – God, Aliens, or Artificial Intelligence – Who Wrote the Torah? While the word alien in the verse refers to people from other lands outside of Israel, it is just as relevant when considering extraterrestrials. So why did God permit Muslims to slaughter so many Jews on October 7, 2023?

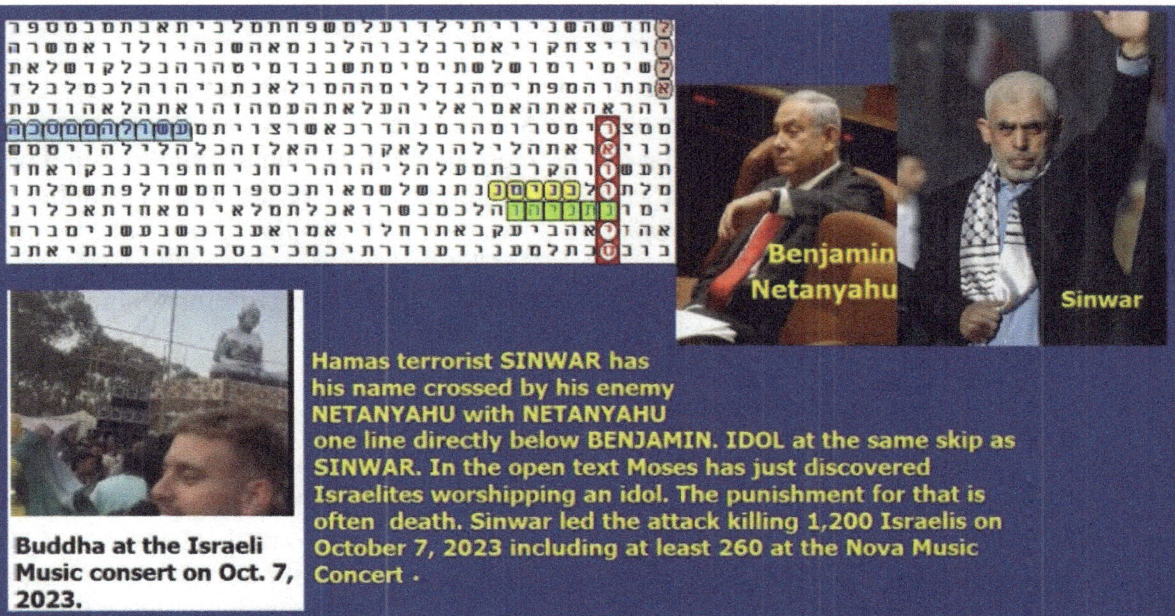

Term	Translation	Skip	Start	End
סינוואר	SINWAR	-137774	Leviticus 23:42.35	Deuteronomy 9:12.43
נתניהו	NETANYAHU	1	Numbers 11:18.76	Numbers 11:18.81
בנימנ	BENJAMIN	1	Genesis 45:22.23	Genesis 45:22.27
אליל	IDOL	-137774	Deuteronomy 29:2.23	Numbers 1:18.20
עשולהממסכה	THEY MADE A GRAVEN IMAGE	1	Deuteronomy 9:12.66	Deuteronomy 9:12.75

The ELS reference is 137,774 characters between rows.
The wrapped matrix starts at Numbers 1:18.20 and ends at Leviticus 23:43.30. It spans 1,515,550 characters of the surface text.
The matrix has 12 rows, 36 columns and contains 432 characters.

Figure E-2 above: With an axis term of **SINWAR, IDOL** is at the same skip. Seen in the photo on the left side of Figure E-2 is **BUDDHA** at the music festival where blood-thirsty Muslims slaughtered hundreds of Israelis. Strangely, there were no IDF troops present so close to the border to save the Jews. Instead, the Muslims were able to rape, kill, or take hostages all they wanted. The desecration of the law against allowing idols was done on the Jewish Sabbath. The Jews tempted a jealous God, and they paid a horrible price for it. Odds against this matrix are about 82,795 to 1, as is shown on the spreadsheet below. Notably, the most significant term (in open text) is **MADE A GRAVEN IMAGE** (the Buddha). It was found against odds of about 690 to 1.

	A	B	C	D	E	F	G	H
1	*AXIS TERM SINWAR @ Skip -133,774*	SKIP	NUMBER	DIVIDE BY	THE QUOTIENT	E QUOTIENT	POISSON EQUATION	EACH WORD
2	Axis Term ELS Rank 106	USED	IN	304805	EQUALS	X 442 LETTERS	PROBABILITY FOR	OR PHRASE
3		ON	TORAH	LETTERS	FREQUENCY	EQUALS	THE WORD	HAS 1 CHANCE
4		FIGURE		IN THE	PER LETTER	WORD	APPEARING	IN THE
5				TORAH		EXPECTANCY	AT LEAST ONCE	FOLLOWING OF
6								BEING HERE
7	MADE A GRAVEN IMAGE@ Skip +1 only	1	1	304805	3.28079E-06	0.001450107	0.001449057	690.1041932
8	NETANYAHU @ SKIP +/- 1 and +/- 133774	-1	16	304805	5.24926E-05	0.023201719	0.022934629	43.60218798
9	BENJAMIN@ Skips +1 only	1	31	304805	0.000101704	0.044953331	0.043957902	22.74903864
10	IDOL @ @ Skips +/- 1 and +/- 133774	-133774	56	304805	0.000183724	0.081206017	0.077996277	12.82112486
11	COMBINED SIGNIFICANCE (NO AXIS TERM) = (H7*H8*H9*H10)/106						1.2078E-05	82795.41018

Above: Spreadsheet for Figure E-2

With respect to labeling God as being jealous, an A.I. could be programmed to behave that way, but could it really *feel* that way? We are making tremendous progress in A.I. development, and time may prove me wrong, but unless we can construct a machine that can do more than mimic human emotions and what we call a soul, it really doesn't matter how fast it can do math. So, given what we are told in Numbers 23:19 ("*God is not a man, that He should lie, Nor a son of man [human], that He should repent")* that would seem to disprove the Christian concept that God *became* a man. If Christians are right, then the incarnated God should not need to worship anyone else. Yet the New Testament has him clearly point to someone else as God when he is quoted as saying in Matthew 27:46 and Mark 15:34:

My God, my God, why have ***you*** *forsaken* ***m****e?*

Christianity is content to excuse the above as a Trinitarian mystery, but clearly this quotation of Psalm 22:1 (Jewish Tenach/Christian Old Testament) seems out of place for a man who is supposed to be an all-knowing God. If he was trying to find support in Jewish Scriptures for later claims that he was God, or even the Messiah, this was the very worst verse that he could choose to quote as he suffered on the Cross. What was written about that later (in John 1:1 – 1:18) about his role in Creation, assuming it was not true, could only serve to enrage a Jealous God. Highlights of this portion of the New Testament are as follows:

1 In the beginning was the Word, and the Word was with God, and the Word was God. 2 He was with God in the beginning.

14 The Word became flesh and made his dwelling among us. We have seen his glory, the glory of the one and only Son, who came from the Father, full of grace and truth.

17 For the law was given through Moses; grace and truth came through Jesus Christ. 18 No one has ever seen God, but the one and only Son, who is himself God and is in closest relationship with the Father, has made him known.

I have a son, Dr. David Alexander Roffman. His B.S was in space physics from Embry-Riddle Aeronautical University, and his PhD is from the University of Florida in physics. He did his postdoc at Yale. I have written two major works before starting this book. The first was MARS CORRECT: CRITIQUE OF ALL NASA MARS WEATHER DATA. A lot of what is in this book (God, Aliens, A.I. – Who Wrote the Torah Code?) about Mars is taken directly from Mars Correct. The Basis Report is 249 pages long, but there are also 30 Annexes. They take the total length of the Report to about 1,200 pages. David was in charge of all the math and spreadsheets to explain the massive findings we had, but I wrote every single page. And yet, the cover page has his name before mine because people are more interested in what a PhD in physics has to say than a guy with a Bachelor's degree in Psychology. But this was my choice.

In the case of my other publication, Meteorological Implications: Evidence of Life, was published in the *Journal of Astrobiology and Space Science Reviews (1, 329-337, 2019*). Again, David did the math, and I included it in my paper, but I wrote every word. What made me really angry was when the Journal of Astrobiology and Space Science refused to list me as a writer on the cover page. It demanded that only David's name be allowed there. Further, they altered my words at the conclusion in a way that infuriated me. They would not publish anything that directly criticized NASA. I cannot be jealous of a son for getting the credit for the Implications paper, but I was angry as hell about the Journal denying me credit for what I wrote. You will find this paper in Chapter 8 of this book. There, at the Conclusion, I note:

Unfortunately, as stressed by Joseph et al. (2019), NASA has rammed down the throats of Joseph et al., plus my son and me, that, "the evidence is circumstantial, and similarities in morphology are not proof." They would not publish our work without this statement inserted against our will!

Why am I discussing my Mars writings? Because, when looking for who gets credit for creating the world (and Who wrote the Torah), I see that the New Testament removes credit from God and gives it to Jesus, a man who is not even accepted as a legitimate prophet by Judaism. This cannot be pleasing to a Jealous God. And, so we look at a Christian country like Ukraine (in October 2025), with two different types of Christianity, we find that what has often happened before in history – they are at war with each other. Where is the Christian love here?

Who is to blame for this chaos? In Deuteronomy 31:16 – 31:18 we read:

[16] And the Lord said to Moses: "You are going to rest with your ancestors, and these people will soon prostitute themselves to the foreign gods of the land they are entering. They will forsake me and break the covenant I made with them. [17] And in that day I will become angry with them and forsake them; I will hide my face from them, and they will be destroyed. Many disasters and calamities will come on them, and in that day they will ask, 'Have not these disasters come on us because our God is not with us?' [18] And I will certainly hide my face on that day because of all their wickedness in turning to other gods.

So, God promised (twice above) to hide his face from us, as he certainly did on October 7, 2023. I haven't seen any news commentators say a word about the Buddha, but that was likely one reason

why no help ever arrived. The Jews were guilty, but I contend that so was God, because after the Exodus, we never again saw the power of God up close again (until the Code was revealed in this book). Now we are in the age of technology, which leads many people to question the existence of God so long as He continues to hide His face. At best, in 2025, many of us think that our only Savior is President Trump. We need God to show up and be clear about His doctrines because billions of us have only conflicting (false) Scriptures (the New Testament, Book of Mormon, and Koran) to go by, often leading to wars. God (or aliens) allowed these scriptures to flourish, thus often causing these wars.

I have met one person (Craig Ebrahimi) who clearly claimed to be an alien. Craig did not identify his race or planet of origin. He is fully described in Chapters 5 and 6 of this book.

I have also met Rebecca living in Lakeland, Florida. She claimed to be from the Pleiades, but unlike Craig, she did not present any physical evidence, and her story changed over time. We were friends for eleven years before we had a fundamental disagreement. The things she said were more in line with her being a Reptilian than a Pleiadian. On February 1, 2025, my younger son first told me that she attempted to communicate with him by telepathy. He had a large mirror in his room that was surrounded by a floral-looking artwork frame.

He claims that he saw what looked like Grays in the mirror. David is very sick, supposedly from painful erythromelalgia, which is incurable.

Strange things have happened in this room before he had to move in with us. When I used it one night, something apparently threw me from the bed. I landed hard on my head far from bed – way too far to have just fallen asleep and just rolled off the bed. That night, my wife had a dream about two Gray aliens visiting her. Are some of us being abducted? I don't know, but on February 1, 2025, I decided to throw out David's mirror. I put it inside next to our front door while waiting for the Sabbath to end. Just before I actually took it to the garbage room, our dog, Einstein, suddenly got out from under the dining room table and ran toward the mirror. He was going nuts like it was a real threat (or a rocket launch – they can be deafening here in Cape Canaveral). He approached it from two directions, like he had discovered something really evil. That was when I actually threw it out. This is a very smart dog. When he is hungry, particularly at night, he can actually say, "I love you." He also brings his plate to us in his mouth when he wants to eat something. Dogs can sense things that people cannot. I can't say that he actually detected an alien or an interdimensional being, but it doesn't hurt to be careful. I didn't buy the mirror anyway. It came with the condo we bought.

Hint of the Torah Author in Chapter 6 of Genesis? This book explores three options for who created us and/or wrote the Torah with its ELS predictions of the future. The third option is an Artificial Intelligence or A.I. I did not see the word COMPUTER anywhere near the axis term below, nor have I ever seen a suggestion that a machine could fashion something as complex as a soul. So, the realistic choice of who to credit comes down to God or aliens, and the evidence backing both possibilities is vast. Given that Nephilim are mentioned so soon after Creation in the Bible and that we now have a species name to pin on them, Reptilian, we may be closer to an Answer to the Question posed in the title of this book. Here I refer to the woman (Rebecca) I know from Lakeland, Florida, who may be an alien.

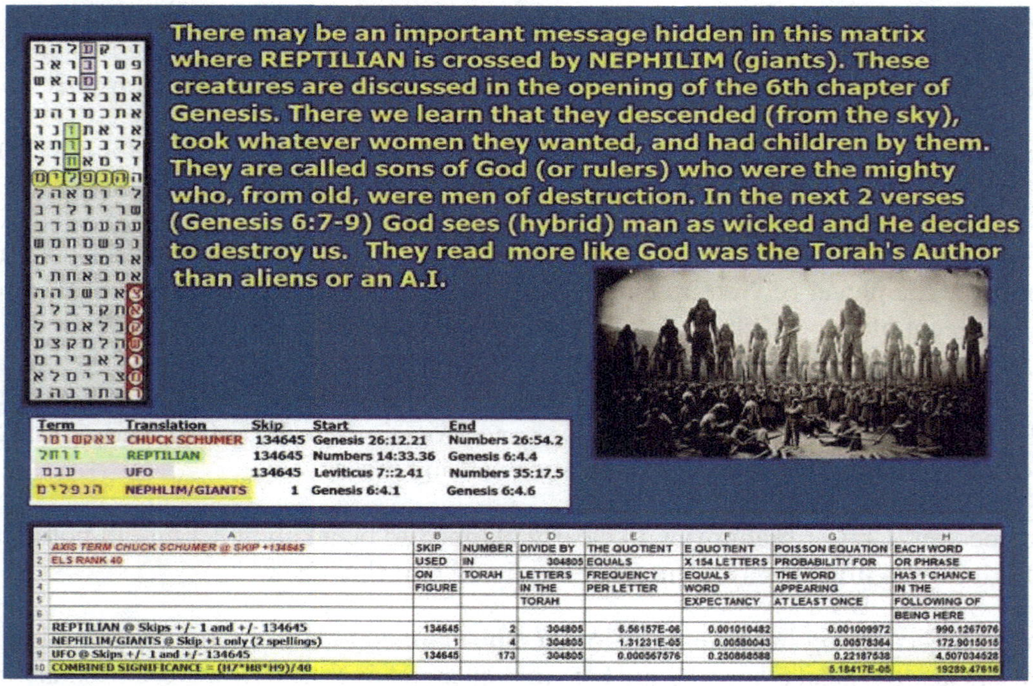

Term	Translation	Skip	Start	End
צאקשומר	CHUCK SCHUMER	134645	Genesis 26:12.21	Numbers 26:54.2
זוחל	REPTILIAN	134645	Numbers 14:33.36	Genesis 6:4.4
עבם	UFO	134645	Leviticus 7::2.41	Numbers 35:17.5
הנפלים	NEPHLIM/GIANTS	1	Genesis 6:4.1	Genesis 6:4.6

	A	B	C	D	E	F	G	H
1	AXIS TERM CHUCK SCHUMER @ SKIP +134645	SKIP	NUMBER	DIVIDE BY	THE QUOTIENT	E QUOTIENT	POISSON EQUATION	EACH WORD
2	ELS RANK 40	USED	IN	304805	EQUALS	X 154 LETTERS	PROBABILITY FOR	OR PHRASE
3		ON	TORAH	LETTERS	FREQUENCY	EQUALS	THE WORD	HAS 1 CHANCE
4		FIGURE		IN THE	PER LETTER	WORD	APPEARING	IN THE
5				TORAH		EXPECTANCY	AT LEAST ONCE	FOLLOWING OF
6								BEING HERE
7	REPTILIAN @ Skips +/- 1 and +/- 134645	134645	2	304805	6.56157E-06	0.001010482	0.001009972	990.1267076
8	NEPHILIM/GIANTS @ Skip +1 only (2 spellings)	1	4	304805	1.31231E-05	0.00580043	0.00578364	172.9015015
9	UFO @ Skips +/- 1 and +/- 134645	134645	173	304805	0.000567576	0.250868588	0.22187538	4.507034528
10	COMBINED SIGNIFICANCE = (H7*H8*H9)/40						5.18417E-05	19289.47616

Figure E-3 above. Nephilim were likely Reptilians. Note that **REPTILIAN** is at the same skip as Senator

CHUCK SCHUMER and **UFO.** This may hint at **SCHUMER** being a **REPTILIAN.** The **REPTLIAN-NEPHILIM** meeting and sharing a letter lamed may be telling us that the **NEPHILIM** were **REPTILIANs.**

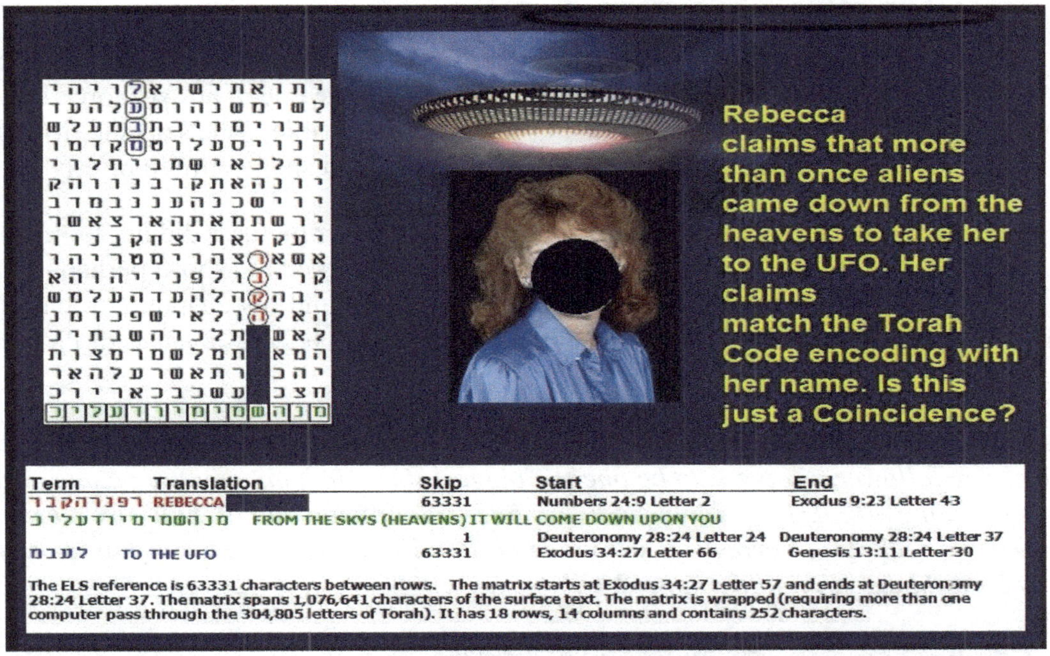

Figure E-4 above: The suspect has red hair. To preserve her privacy, I have blotted out most of her face next to the following matrix that I found about her.

Figure E-5 below: A fictional work that plays on the belief that Reptilians prefer redheads. Rebecca did claim to have some telepathic abilities and even some teleporting experience. Again, I saw no proof of either, although my son thinks he was affected by her alleged telepathic attempt to contact him.

Some Redheads are Aliens Paperback – 17 Jan. 2014
by Carl L Fike (Author)

Imagine driving down a wooded road in Maine and having an alien jump out in front of you. Now imagine that the alien is a pretty redheaded girl and is being chased by some very nasty creatures. She is nude. You take her to your home, feed her, bathe her, bind her wounds, clothe her and try to teach her some English. Various anomalies convince you that she is not of this planet. What do you do? The Nasty's have shot holes in your pickup while trying to shoot her. She's expecting her sister to land soon. You both arm yourselves and return to her landing place, to make it safe for sister, hoping to pick off a few Nasty's in the process. After all, they shot holes in your pickup. They should pay for that. Nasty's are gone sister is there, also nude. You introduce the girls to chocolate. Now you are a hero, but the Nasty's are still searching for them. Your house is fortified and has many explosive deterrents, so when they come, you blow them away. The girls shoot one who has clubbed you and given you a concussion. The girls take you to their home planet. It is solely populated by women and the planet is dying due to an invasion of bugs that ate everything green. They have also run out of frozen sperm and the population has fallen drastically. You are the first male they have ever seen. The Nasty's have followed you and bombard the girl's home which, luckily, has a great bomb shelter. You all return to Earth and are bombarded there too. You are starting to become seriously perturbed at them. Active measures are adopted to stop the Nasty's depredations, some of which involve teleporting and telepathy by children, your children. The Air Force and the Navy help too—and then there are always the Marines. That's when life gets exciting.....

Figure E-5 above. With respect to Rebecca, it bothered me a lot when she told me abcut Reptilians eating hybrid children. I felt that she should have issued some kind of protest, but it never happened. Initially, Rebecca came across as a Christian who was writing a book about her experiences with aliens, although she never wanted to call them abductions. She let me proofread the preface of her book. It read like nothing more than standard Christian theology. This led me to doubt how sincere her Christian beliefs were. However, she liked to sing Christian religious music. She had a good voice, and (unlike Craig) there was nothing about her physical appearance suggestive of an alien. Earlier photos did not indicate that there was any shape-shifting in her past, but she did profess to having many encounters when she was about seven years old, or younger, and she claimed to have previously seen a shape-shifted reptilian come out of a Publix supermarket when we visited her in Lakeland. She also claimed to see one at an Air Force Base in Florida. I think she indicated that the incident was at Tyndall (but Eglin is possible). She said she first had telepathic contact with him while he was indoors, and then he came outside and shape-shifted briefly for her.

With Craig Ebrahimi, the story was quite different. Some of his earlier pictures did not look much like him. His body shape was weird – too wide at his shoulders, but his legs looked too short, perhaps by a foot. More important, Craig put in writing to me via email that he is an alien, and his 27-pound meteorite had what looked like 16 alien (Reptilian) embryos in it. Again, on Friday, March 27, 2020 at 4:43 PM he sent me the following:

Craig Ebrahimi <craigebrahimi@hotmail.com> wrote: *"Guy your dumb ass shit keeps getting sent to me by one of your Hebrew followers, you got to be one the biggest dummie humans I ever came across you talk this shit in your writing ABOUT my biotechnology, once again your dumb!* ***You don't even know your dealing with fucking real extra- terrestrial self made multi millionaire since teens you idiot, if you looked into these eyes I would eat your soul puppet boy."***

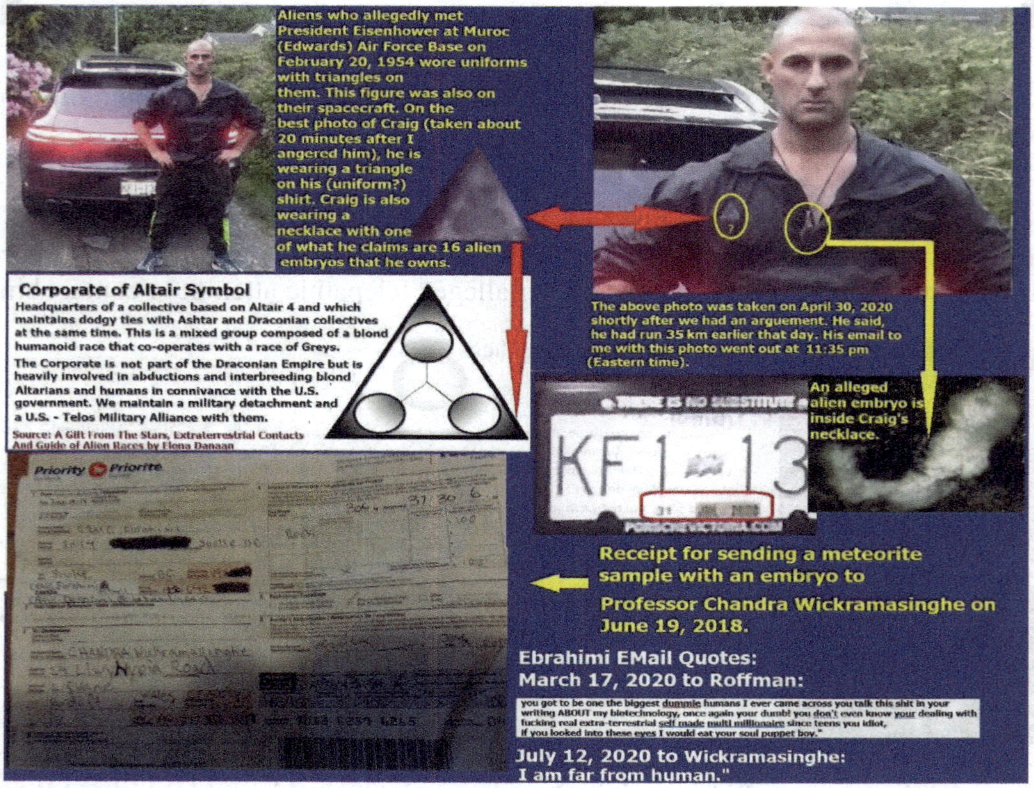

Figure E-6 above: Ebrahimi's proportions and former appearance are shown in the next set of photos. With regards to my desire to serve President Trump as the head of a New Department of Disclosure Management, the way I have gone after Ebrahimi is a sample of how I hope to pursue Reptilians who have used shape-shifting abilities to infiltrate our government with a goal of encouraging perverted legislation aimed at promoting homosexual behavior to keep our population from growing.

Figures E-7 and E-8 below – Craig's body proportions and appearance when younger (especially with hair).

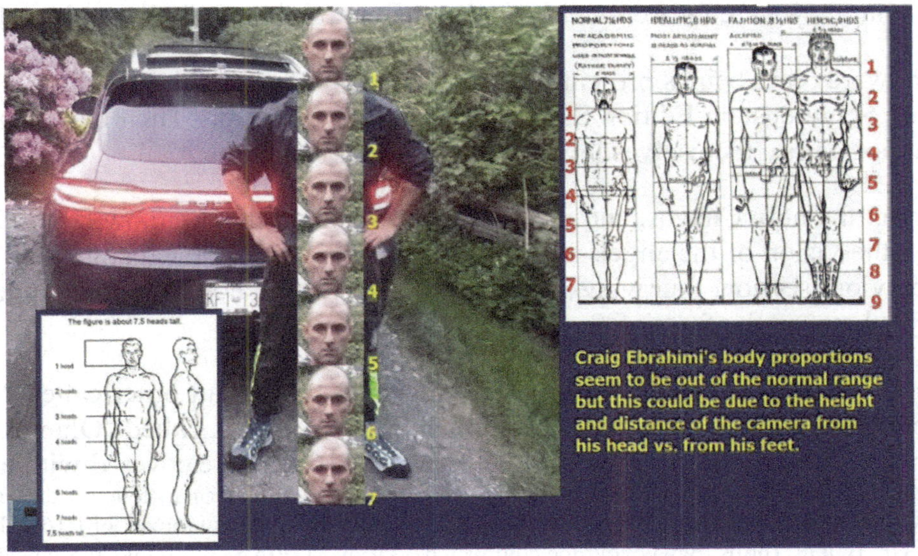

Figure E-8 above: Is Craig typical of what Reptilians are like when shifted in appearance? The insignia on his uniform above suggests that he is from Altair 4, so he's probably a member of another species, but if his email on March 27, 2020, is typical of his thought processes, there is no evidence that his kind were smart enough to write the Torah. Moreover, the behavior that is attributed to Nephilim/Reptilians in Genesis Chapter 6 was so wicked that God immediately wanted to destroy them and the whole world. However, He recanted a bit by saving Noah and his family, plus enough animals to prevent general extinction. Again, there was no evidence given to show that Reptilian thought processes went beyond rape and reproduction.

It appears that the Reptilians were, at best, (like God) interdimensional creatures, here before man, possibly with other species, but they were far from being God or a Creator, though they may have bred man as food.

The idea that there were other beings present soon after Creation may be hinted at in my (Jewish) Sabbath prayer book. In my Schottenstein Edition Siddur on page 327, we read, "Then the Ofanim, and the holy Chayos with great noise raise themselves toward the Seraphim facing them…"

Who are the Ofanim, the holy Chayos, and the Seraphim? Rabbi Jack Abramowitz writes, "The Rambam goes on to explain the vision of the angels in Ezekiel chapter 1. [III, 2] Each angel has the four faces we discussed in our previous section, four wings, and two hands. Overall, however, their appearance is human, as per Ezekiel 1:5, "they had the form of a man."

He goes on to state, "The angels we've been describing so far are of an order called "Chayos" ("living things"). From here, Ezekiel goes on to describe another order of angels, called "Ofanim" ("wheels"). The Ofanim are the color of beryl (a shade of green – see Ezekiel 1:16) and are covered with eyes (or, possibly, different colors – Ezekiel 1: 18). One part of them touches the Chayos, while another part touches the ground (verse Ezekiel 1:15). The Ofanim were somehow interconnected, described as "a wheel within a wheel" (Ezekiel 1:16)." In the past a 1978 TV show, PROJECT BLUE BOOK (https://www.youtube.com/watch?v=yVVADz0Afss), thought the wheels represented flying saucers.

Rabbi Abramowitz also says,

"The Talmud in Chagigah (16b) points out a contradiction between Isaiah's vision and Yechezkel's: our verse says that seraphim have six wings, but Ezekiel 1:6 says they have four! The Talmud reconciles this by saying that they have six wings when the Temple is standing and four when it is not. (Remember that angels are spiritual beings, so it is perfectly reasonable that their appearance would reflect the state of holiness in the world at a given time.)"

Is that clear? Not to me, so what do other scholars say on this topic? Wikipedia writes,

Its plural form, *seraphim*, occurs in both Numbers and Isaiah, but only in Isaiah is it used to denote an angelic being; likewise, these angels are referred to *only* as the plural *seraphim* – Isaiah later uses the singular *saraph* to describe a "fiery flying serpent", in line with the other uses of the term throughout the Tanakh. This sounds more like what we think Reptilians are like when not in a shape-shift mode.

For Judaism, Wikipedia writes, "Maimonides lists Ophanim as the closest of angels to God in his exposition of the Jewish angelic hierarchy.

The kedusha section in the morning prayer (in the blessings preceding the recitation of the Shema) includes the phrase,

"The ophanim and the holy living creatures (Chayos) with great uproar raise themselves; facing the seraphim they offer praise, saying, 'Blessed be God's glory from His place." The inspiration behind this particular passage is Ezekiel's vision (ch. 1). The theme of angels praising God was inserted into the passage by paytanim (Jewish liturgical poets)."

All we can clearly say here is that there was some kind of intelligent life before man.

So, who wrote the Code? My general impression is that it was God, and by God, I do not mean a Jew named Jesus. Rather, I mean what Christianity calls The Father. Did he use any alien technology to seed the universe with life? Likely, yes. Did he allow the destruction of other murderous or sinful worlds? Did he allow Satan to commit evil? Again, likely yes. But, as with Jewish slavery in Egypt and homosexual bestiality in Sodom and Gomorrah, we experienced the Holocaust after Germany gave rise to Reform Judaism. It rejects most of Jewish law. The October 7, 2023, Hamas slaughter of Jews came after they were dancing before a giant Buddha statue in Israel. God eventually comes around, but as with the Holocaust, millions of innocent people can suffer or die before. Then He gets around to doing right.

God, as I understand, is not perfect. In this book, you see many phenomenal matrices. But even when they exist against odds that are millions to one, there is generally space to take the odds up to billions or trillions to one. I recently posted a matrix that exists against odds of quadrillions to 1! However, we can forgive Him for such "shortcomings" when we realize that He must accommodate encoding for billions of people, PLUS preserve the history of the world and our people and all His laws, all in the same text of 304,805 letters.

The Torah is said to be the blueprint of the Universe. In it, the Author lays out what the future will be thousands of years in advance. Yet, when Moses failed to circumcise his son, read what God had to say about it:

At a lodging place on the way, the LORD met him and sought to put him to death. Then Zipporah took a flint and cut off her son's foreskin and touched Moses' feet with it and said, "Surely you are a bridegroom of blood to me!" So, he let him alone. It was then that she said, "A bridegroom of blood," because of the circumcision (4:24-26).

How could God, Who was in the process of transmitting the Torah to the world through Moses, even think about killing him? It may be because Moses was raising his son (Gershom) as a Midianite. The actual name Moses is not in the Hebrew text. The Hebrew only mentions him where the feet are being touched.

The Torah Code has many encodings of **MULTIVERSES**. It may be that fates vary from one world in our universe to others and/or to universes in interdimensional space. This may lead to occasional problems with translations or text interpretations in order to accommodate free will. A full set of universes may lie out there, but through our choices, we must decide which we choose to live in!

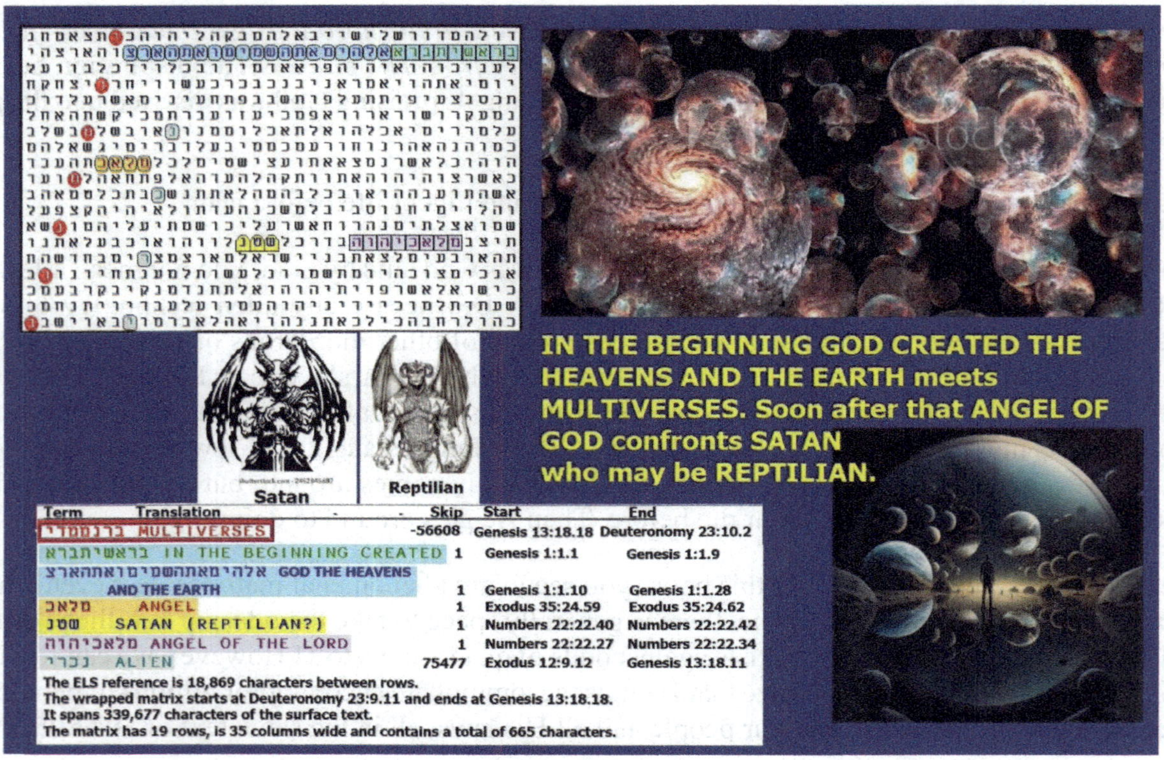

Figure E-9 above: Did God just create the heavens and the Earth, or did He create Multiverses (multiple universes), perhaps with different laws of physics where varying interpretations of Torah come true? And who is Satan (besides a fallen angel or alien)? Satan is often portrayed as both looking like and acting like Reptilians (also called Ciakahrr). Odds against this matrix were about 30,916 to 1, as is evident from the spreadsheet below.

	A	B	C	D	E	F	G	H
1	AXIS TERM MULTIVERSES @ SKIP +134645	SKIP	NUMBER	DIVIDE BY	THE QUOTIENT	E QUOTIENT	POISSON EQUATION	EACH WORD
2	ELS RANK 71	USED	IN	304805	EQUALS	X 665 LETTERS	PROBABILITY FOR	OR PHRASE
3		ON	TORAH	LETTERS	FREQUENCY	EQUALS	THE WORD	HAS 1 CHANCE
4		FIGURE		IN THE	PER LETTER	WORD	APPEARING	IN THE
5				TORAH		EXPECTANCY	AT LEAST ONCE	FOLLOWING OF
6								BEING HERE
7	GENESIS 1:1 @ Skip +1 ONLY	1	1	304805	3.28079E-06	0.002181723	0.002179345	458.8535653
8	SATAN @ Skip +1 ONLY	1	3	304805	9.84236E-06	0.006545168	0.006523795	153.2850066
9	ALIEN OF THE LORD @ Skips 1 ONLY	1	17	304805	5.57734E-05	0.037089287	0.036409904	27.46505444
10	ALIENS @ Skips +75477 to +75774	75477	972	304805	0.003188924	2.120634504	0.880044508	1.136306165
11	COMBINED SIGNIFICANCE = (H7*H8*H9*H10)/71						3.23451E-05	30916.56716

When I look back, not just at what I have learned in this study, but also at my experience in life, I feel that I have seen enough to rate what I have found in the Torah Code with respect to its likely Author. I see the probable Author as being God or like God in the sense that there is a logic to what He has produced that is clearly far advanced from what humans are capable of having. Yet he has not yet figured out how to guide us without risking the destruction of our world, as is hinted

at with ATOMIC HOLOCAUST on Figure E-11. So, on a scale from 1 to 100, I'll give God a score of 80 and show Figure E-11's spreadsheet. This matrix was found against odds of from about 2,782,626,903 to 1 up to about 1,319,638,507,000 (roughly 1.3 trillion to 1), with the higher number allowing for including the 10-letter axis term, **ATOMIC HOLOCAUST**.

I will admit that I might be underestimating God's ability to prevent war, as it may yet turn out to be that the Messiah (chosen by God) will be the one to prevent a nuclear war. The only man who it appears is trying to be such a peace-maker is President Donald Trump, but in October, 2025 it is still too early to know if he will succeed. It will take more than a nice matrix to name a man as The Messiah, but what does the Code have to say about Trump in this capacity? The preferred Hebrew spelling of Trump is *TET resh alef mem peh*. This spelling is not found in sequence with Messiah (*mem shin yud chet*) or **THE MESSIAH** (hey *mem shin yud chet*). However, with a TAV for the T letter in Trump, two 9-letter ELS chains mean **MESSIAH TRUMP** or **TRUMP MESSIAH.** The one at the lower skip does not require more than one computer pass through Torah to find. It is **TRUMP MESSIAH** spelled *at skip +2,844. But it is not found with any interesting keywords and its R factor is 2.143, meaning that it had about one chance in 139 to be found at least once in Torah. But without any meaningful keywords associated with the Messiah the find doesn't rate more than a yawn from me, even though he might have just prevented a potential nuclear war between India and Pakistan.*

The second find (**MESSIAH TRUMP**) does require more than one computer through Torah to find (meaning a *wrapped* search, and the right kind of software like CodeFinder is needed). It's at skip-77,952 and it is spelled *mem shin yud chet tav resh alef mem peh.* This find does come with an interesting 4-letter Hebrew *a priori* key term – USA – at the same skip as **MESSIAH TRUMP.** Although it only requires two columns and 15 rows to display (a 30-letter area), its cumulative R factor, according to CodeFinder software, is 0.706. This translates to it being found against odds of about 35.2 to 1, so it's actually less significant than **TRUMP MESSIAH** (because the skip is much higher and it requires more than one pass through Torah to find). However, if another *a priori* (named ahead of the search) key term can be found, then (combined with Trump's accomplishments), it becomes something that may be taken more seriously even if Trump is not (so far as we know) Jewish. Cyrus (a gentile Persian king) has the term for Messiah associated with his name in Isaiah 45:1. The matrix showing the match between **MESSIAH TRUMP** and **USA** is shown on Figure E-10.

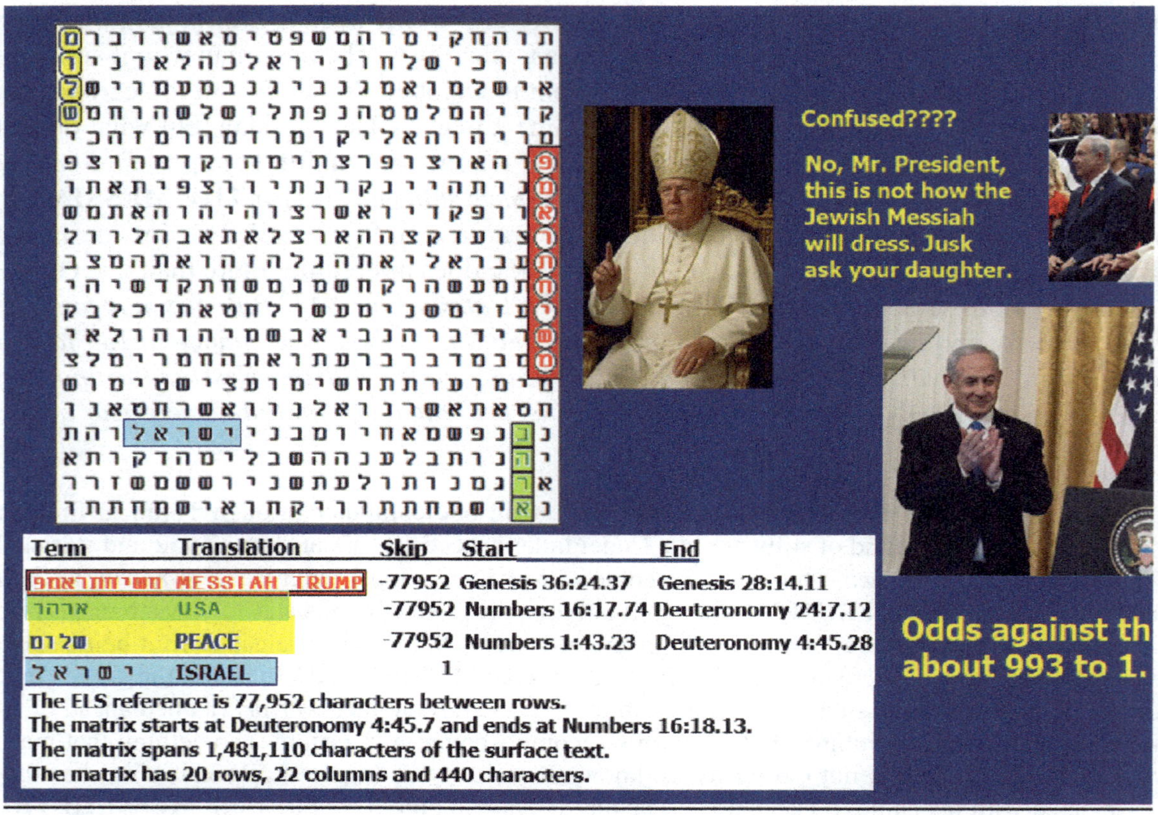

Figure E-10. Oy Guvult. The President's daughter, Ivanka Trump, converted to Judaism prior to marrying Jared Kushner in 2009**. MESSIAH TRUMP** is at the same absolute skip as **USA**. While Judaism does look for more from our Messiah, bringing peace in the Middle East does put him in the running as a messianic candidate. Orthodox Jews believe in resurrection. Does the Code address that? Yes. Trump loves surprises. He may decide to disclose the presence of aliens on Earth. If they are here, they likely possess time travel technology. Indeed, in my dealings with Fort Huachuca. I have seen strong indications that they have access to it, and that transmission of Torah Code details about our time is evidence for it. If it is used it will be evidence that the Torah Code was written not by aliens or a computer. It will be proof that the Torah Code was written by God. This is not just my idea. It forms a fundamental verse in the Jewish Bible in Ezekiel 37:13-14

[13] *And you shall know that I am the Lord, when I open your graves, and raise you from your graves, O my people.* [14] *And I will put my Spirit within you, and you shall live, and I will place you in your own land; then you shall know that I, the Lord, have spoken, and I have done it, says the Lord."*

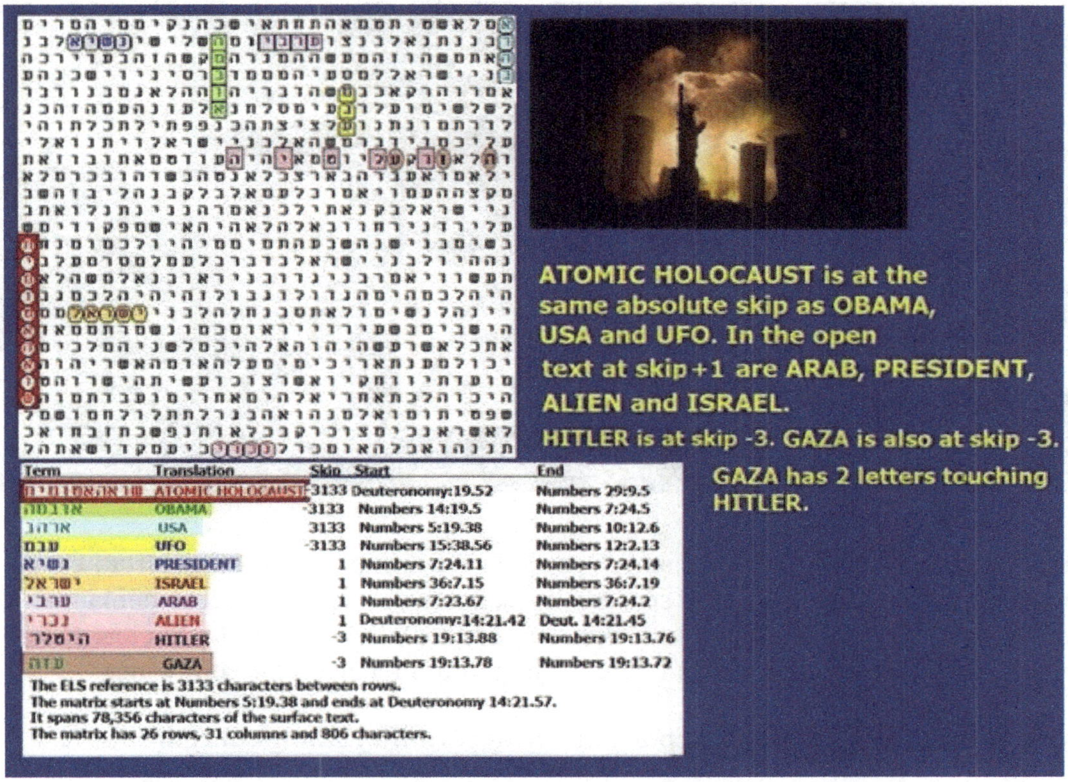

	A	B	C	D	E	F	G	H
1	Axis term: ATOMIC HOLOCAUST @ SKIP -3133	SKIP	NUMBER	DIVIDE BY	THE QUOTIENT	E QUOTIENT	POISSON EQLATION	EACH WORD OR
2	ELS RANK: 1	USED	IN	304805	EQUALS	X 806 LETTERS	PROBABILITY FOR	PHRASE HAS 1
3		ON	TORAH	LETTERS	FREQUENCY	EQUALS	THE WORD	CHANCE IN THE
4		FIGURE		IN THE	PER LETTER	WORD	APPEARING	IN THE FOLLOWING
5				TORAH		EXPECTANCY	AT LEAST ONCE	OF BEING HERE.
6	OBAMA @ Skips +/- 1 and +/- 3133	- 3133	3	304805	9.84236E-06	0.007932941	0.007901558	126.5573195
7	ALIEN @ Skip+1 only	1	10	304805	3.28079E-05	0.026443136	0.026096577	38.31920109
8	ARAB @ Skip +1 only	1	16	304805	5.24926E-05	0.042309017	0.041426481	24.1391491
9	HITLER (best spelling) @ Skips -3 to -196	-3	17	304805	5.57734E-05	0.044953331	0.043957902	22.74903864
10	USA @ Skips +/- 1 and +/- 3133	3133	20	304805	6.56157E-05	0.052886272	0.051512124	19.41290574
11	PRESIDENT @ Skip +1 only	1	68	304805	0.000223093	0.179813323	0.164573849	6.076299539
12	UFO @ Skips +/- 1 and +/- 3133	- 3133	105	304805	0.000344483	0.277652926	0.242440295	4.124726881
13	GAZA @ Skips -3 to -30	-3	237	304805	0.000777546	0.626702318	0.465548981	2.1475404
14	ISRAEL @ Skip +1 only	1	591	304805	0.001938945	1.562789324	0.79044925	1.265103358
15	INITIAL COMBINED ODDS = H6*H7*H8*H9*H10*H11*H12*H13						3.59373E-10	2782626903

Figure E-11 above: ATOMIC HOLOCAUST is at the same absolute skip as OBAMA, UFO and USA. PRESIDENT and ALIEN are at skip +1.

Figure E-12. below. Compare the symbol on Ebrahimi's uniform below with Danaan's portrayal of the Akhori uniform. The Akhori are, according to her, from planet Alkorin, which is the 4th planet out from Altair, a star 16 light years from Earth. She claims there is "a dodgy collaboration between the Orion Grey Collective and a Terran military presence (U.S.-Telos"). She also says, "They are heavily involved in abductions and in interbreeding with Terrans." Craig is married. As if to hint at possible interbreeding, Craig sent me the following email: On Fri, May 1st, 2020, at 12:06 PM, Craig <craigebrahimi@hotmail.com> wrote:

Barry, my wife she had two extremely healthy boys. Craig

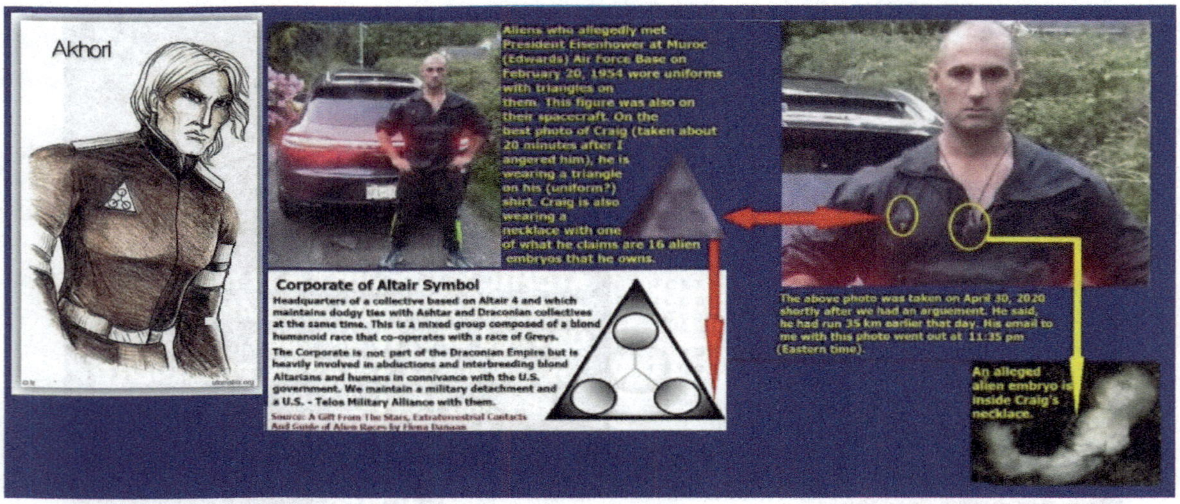

Above: **Figure E-13**. In Chapter 6, I presented over a year's worth of emails to and from Craig. During this period of intensive interrogation, I never sensed any concern for the welfare of humanity. Yes, he was willing to sell his meteorite, which likely held the source of COVID-19, but it was clear that this was only to get richer. When I angered him because I kept grilling him, I received the following:

On Fri, Mar 27, 2020, at 4:43 PM, Craig Ebrahimi <**craigebrahimi@hotmail.com**> wrote:

"Guy your dumb ass shit keeps getting sent to me by one of your Hebrew followers, you got to be one the biggest dummie humans I ever came across you talk this shit in your writing ABOUT my biotechnology, once again your dumb! You don't even know your dealing with fucking real extra- terrestrial self made multi millionaire since teens you idiot, if you looked into these eyes I would eat your soul puppet boy."

Notice that his insult to me included "dummie humans." He often distinguished himself from humanity, and he clearly thought his kind was superior to our kind. Often, I wrote him about my concern that his rock and a virus from it were killing millions of us. He never expressed any concern about it. As such, assuming that he is typical of his culture, I came to view them as inferior to us. Yes, he looked very strong. But he seemed to lack a soul, and he often seemed high (which he blamed on too much caffeine). Craig basically said that he received his meteorite's underwater coordinates by telepathy. If true, then we may conclude that his fellow aliens did not care about the millions of people who would die because of it. While I gave God a score of 80 out of 100 for the probability that He wrote the Torah Code, there was an enormous amount of detailed Code findings about aliens in general and specifically reptilians, and it was largely in the form of warnings.

At best, assuming that another race wanted to warn us, I could not give ET a score of more than 15 out of 100 points. As for a future, likely military artificial intelligence, the thought is due to the agent that I call Joseph *Sapphire,* who was once an Army intelligence officer at Fort Huachuca. His August 1, 2016, warnings to me and request for me to stop my writings were accompanied by a threat to have child pornography downloaded onto my computer so the government could prosecute me. This guy knew what would happen to my son about four months later. He indicated a U.S. interest in Torah Code findings, which is obvious to this day, but there was nothing he said that indicated that this was the source of the Code. However, even if I check my report statistics today, I see multiple hits from Fort Huachuca, which may be a joint base with ETs. If you how to know how to break the IP code, there is plenty there from U.S. and

British UKMOD intelligence. As an example, look at the print screen in Figure E-14. Any IP number with a six as the last number or immediately before the first decimal point can be traced to Fort Huachuca. The trick is that if the 6 is the last digit, then there is a need to write the reverse IP number. So, from today's report, there are a few IPs that lead to Fort Huachuca, but you cannot allow more than 6 to remain before the first decimal point. See Figure E-14 below.

Fort Huachuca is a heck of a top fan of my site. In addition to serving in this function since at least 2016, it may also be a joint base with ET. Elena Danaan writes:

"ARIZONA: Fort Huachuca: Here is one of the main underground hangars for spaceships. This facility is a landing point for incoming Ciakahrr (Reptilian) contingents and a central hub deserving of a vast amount of other facilities. It serves as well as a training facility." The most interesting of the people who read my site every day is Fort Huachuca, an Army Intelligence base in Arizona. There are plenty of stories online about them actually having captured or visited alien spacecraft on the base. There are also stories about them delivering children to reptilians for slavery and even for food purposes. Elena Danaan writes on pages

163 to 164 in her sensational (2020) book – **A GIFT FROM THE STARS, Extraterrestrial Contacts and Guide of Alien Races**:

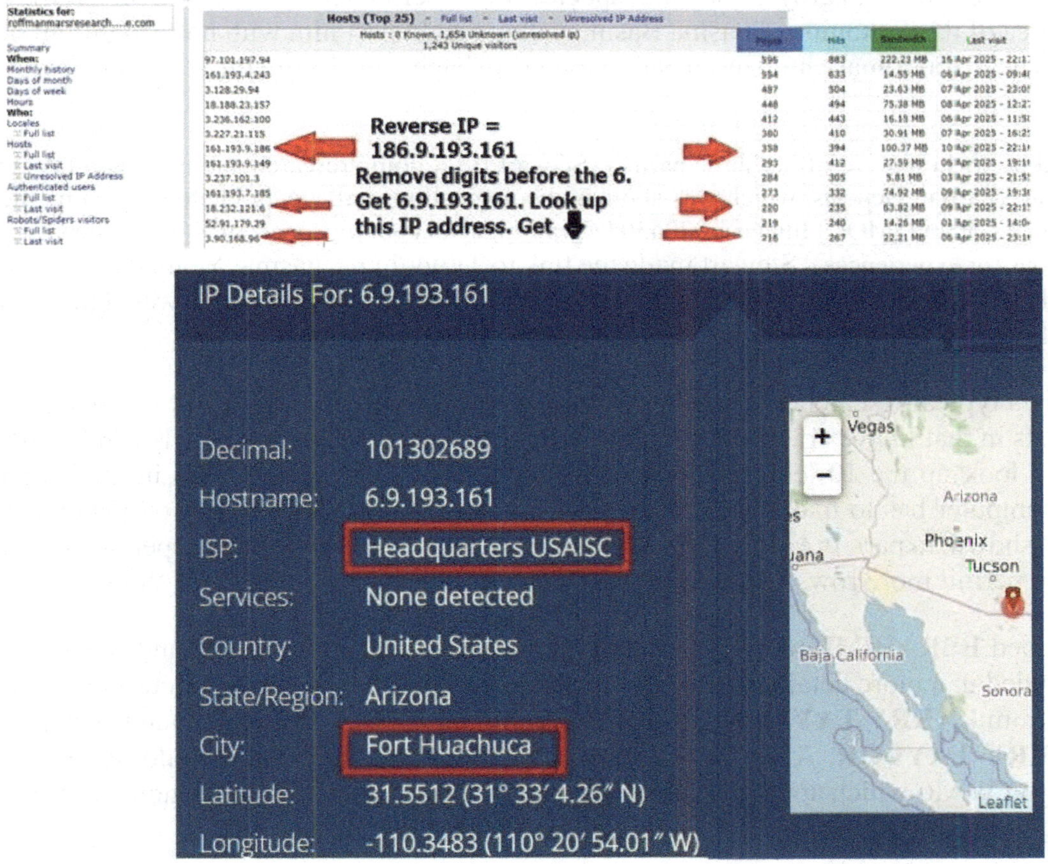

Figure E-14 above.

Larry Taylor and Mars

Many times, I have been a guest speaker on a blog talk radio show called NIGHT SHADOWS with Larry Taylor and Stewart Best. On several of these shows (which generally lasted about an hour to two hours), I have discussed Mars, its weather, possible alien Craig Ebrahimi, and how he brought COVID-19 to Earth. Chapters 5 and 6 in this book largely discuss what I have learned from Craig and about him. I also had all of our interviews posted on my website until their program was shut down. Probably the best show I had was with Rebecca, a possible alien, as a guest with me. I mentioned her before and provided a partly censored photo and matrix (Figure E-4). On the show, I questioned her about her claim that reptilians took her to a ship where she was shown terrified, screaming human-hybrid children just before they were eaten (alive) by reptilians. She said that it was important that they be terrorized before being eaten because the adrenaline that they produced enhanced their flavor.

I was on Larry's show a lot over a decade. But in all that time, he never mentioned being taken by ET. So, I was shocked when he sent me a flyer with a story of his visions and possible experiences being transported to Mars. At first, it sounded like Larry was experiencing out-of-body sensations that dreams might explain. But on questioning him, he said that at least once he remembered being taken aboard an alien craft. In all, he said he was possibly taken to Mars seven times, where he

saw round spacecraft and two species of aliens. One of them looked like a typical big- eyed grey, but they were taller and browner than greys. The other species was described as being a "HUMANOID CREATURE (tall) in the area of the Tithonium Chasma Basin, walking right past him with no recognition of him." Larry wrote that, "The extraordinary humanoid had a strangely shaped head, with the rest of an obvious humanoid-shaped frame."

Larry claimed he was unaware of his show partner, Stewart Best, doing research and making a DVD video entitled Project Redstar, Martian Genesis, which was about archeological ruins allegedly photographed by the Mars Global Surveyor Orbital Camera (MOC) Image m0204304 (http://gigapan.com/gigapans/192466)/. When Larry told Stewart about his visions (or experiences), Stewart made the link to Tithonium Chasma. Years later, Larry read a book, MIND TREK by Joseph McMoneagle, who claims to be a CIA-hired remote viewer. The book showed the same type of creature with the strangely shaped head that Larry saw.

If I were only a typical researcher, I would not put Larry's story in this book. But I have an advantage over other scientists in that I have cracked the Torah Code (likely better than anyone else on Earth). So, what do I find when I look up the 10-letter Hebrew name for **LARRY TAYLOR**? It's in the Torah Code once, though the computer has to make about four passes through the 304,805 letters of Torah to find it at skip 138,549. It is shown, in part, in Figure E-14. In reality, the matrix has an infinite repeating pattern that grows in size as you permit more rows and columns of Torah to be shown. Figure E-15 shows more.

I have described EBRAHIMI in detail on many of the shows done with Larry and Stewart. This name is actually included in a more detailed view of the matrix, where one of two transliterations is found less than three letters from **LARRY TAYLOR**. Further, the minimum skip of **MARS** is found at skip -2 three letters right of **LARRY TAYLOR.** Also, on the matrix is LIKE A STAR (which could describe Mars as seen from Earth) and WAR, which apparently destroyed or drove inhabitants underground, according to Dr. John Brandenberg.

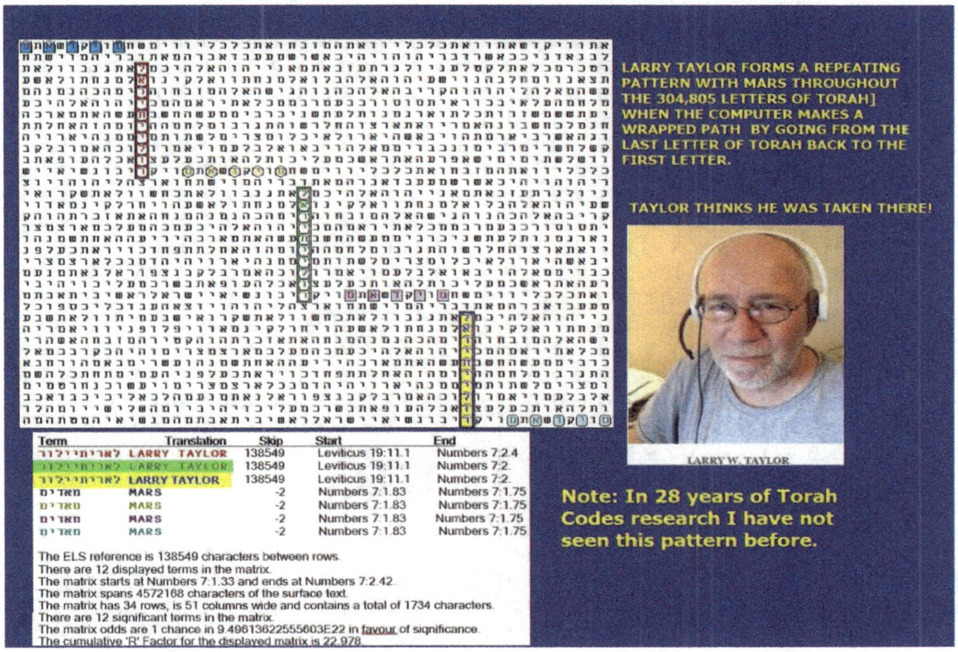

Figure E-15 above. Here, the repeating pattern of **LARRY TAYLOR** is shown with **MARS**. Figure E-16 below shows a more detailed picture of how **LARRY** is encoded. In 28 years of Torah Code research, I have never before experienced a matrix like this that repeats the axis term and its key words. The matrix provides a tangible reason to take Larry at his word. Larry is not the only person that I ever met who claimed to have been transported to Mars. On my website I discuss similar claims by Andrew Basiago. I also remember claims of teleportation by Rebecca, the woman in Lakeland, Florida who claimed to be an ET, but I can't recall if Mars was one of her destinations. Here is the bigger picture of the **LARRY TAYLOR** matrix:

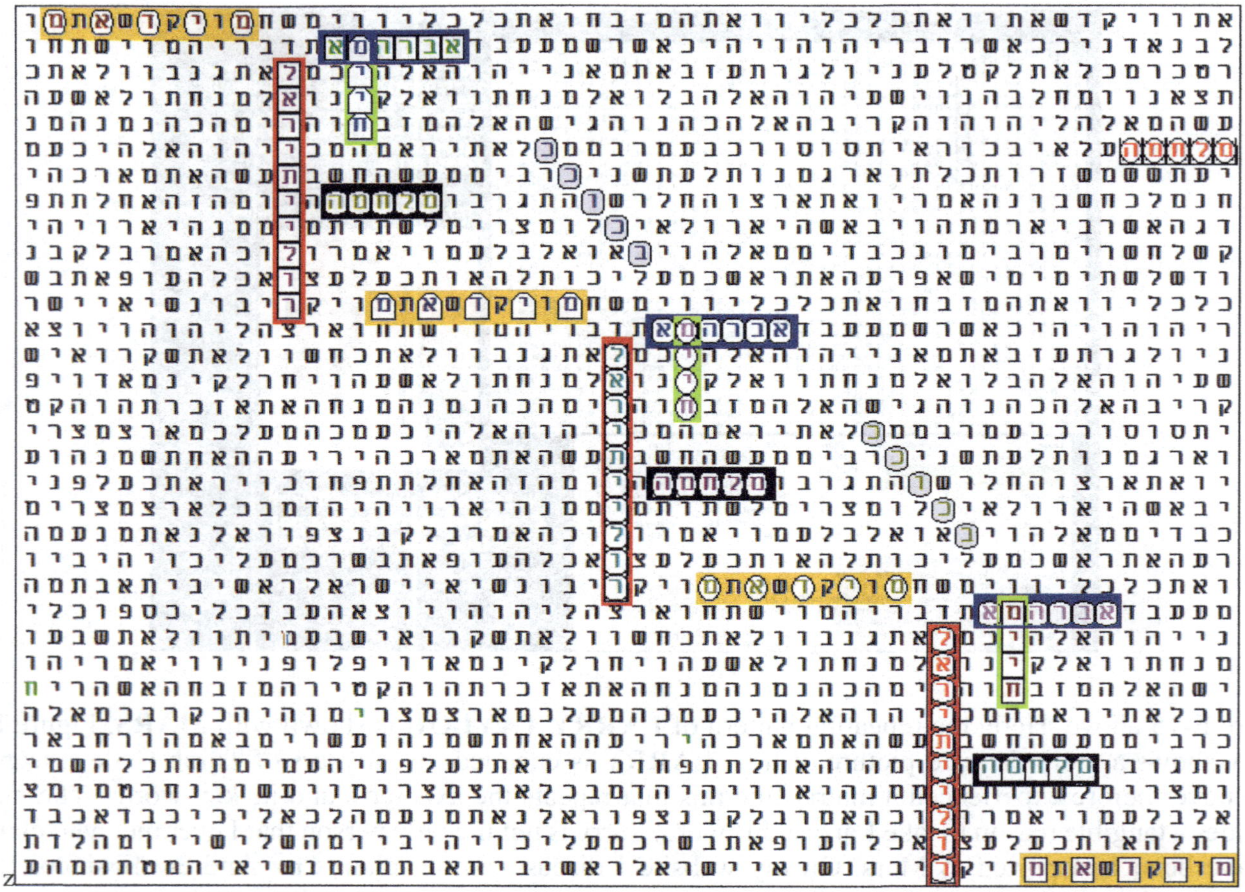

Figure E-16 above. Here the matrix includes the terms **MARS, LIKE A STAR, EBRAHIMI, LIFE** and **WAR. LIFE** is at the absolute skip of **LARRY TAYLOR** and it intersects and shares a letter mem with **EBRAHIMI.**

Ah, you say, if I can find my friend Larry Taylor, what about me? I have found matrices about my future several times. The newest one is shown on Figure E-17 below. It was found on July 6, 2025, shortly before formal publication of this book. **THIS MATRIX ACTUALLY DESCRIBES THE COVER OF THIS BOOK!**

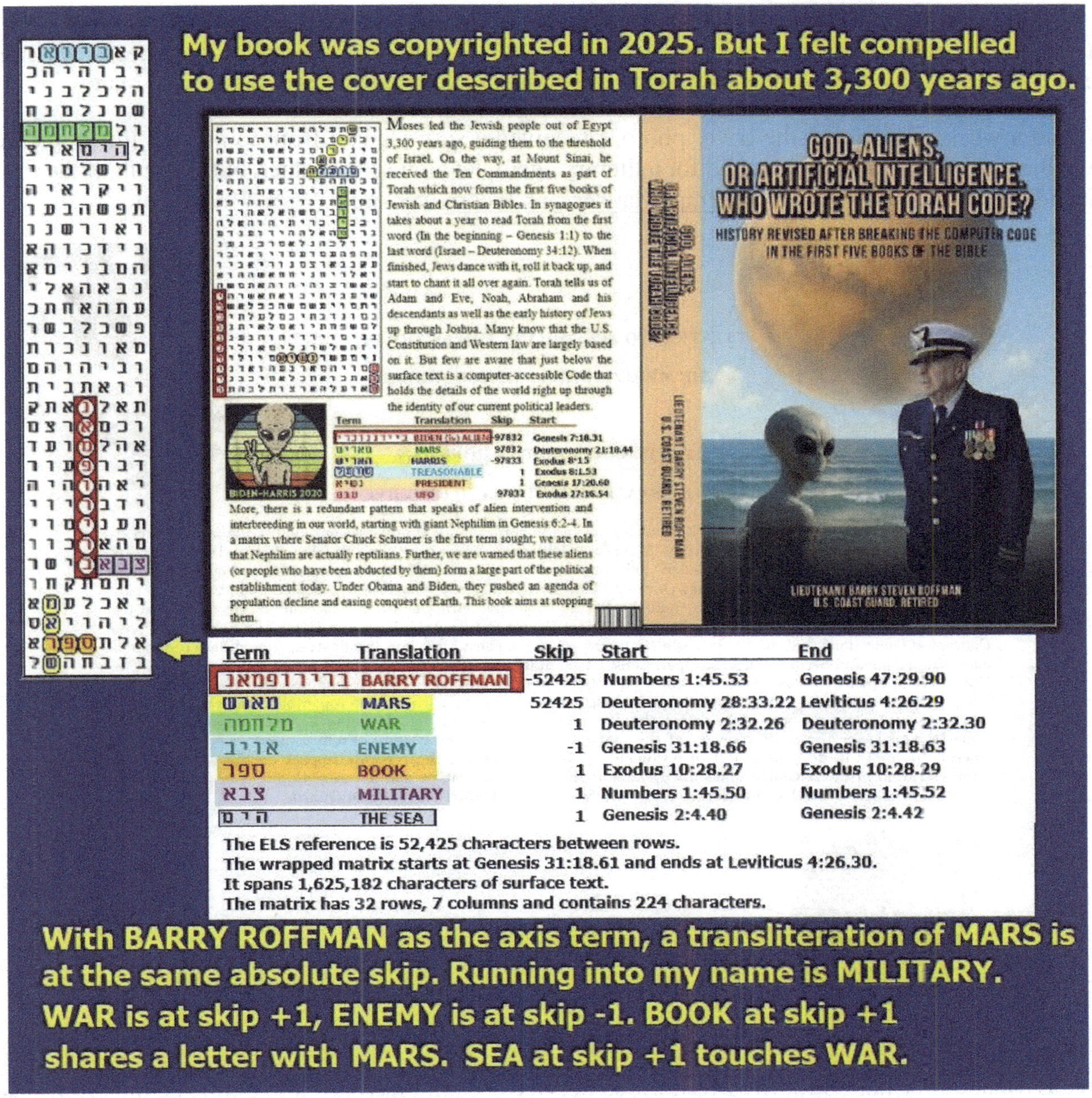

Figure E-17 is shown above. Its spreadsheet is shown below. I found the cover of my new book described in the Code. See Figure E-17 which adds a matrix naming me. The first term (axis term) is the best transliteration of **BARRY ROFFMAN**. Touching the first letter of my name is **MILITARY**. I am shown in my military uniform. At the same absolute skip as my name is a transliteration of **MARS**. Chapters 5 and 6 of this book are about Mars. **WAR** is in the open text. This book might set off a war with ET, and if so, I want to write plans about how to defeat them. The medals on my chest show my ability to do so. See the citation on Figure E-18 below for the Meritorious Service Medal given to me on my 60th birthday Continuing, on Figure E-17 above, **ENEMY** is at skip -1. I show the cover of my book. The word **BOOK** at skip +1 runs into and shares a letter with **MARS.** The threat of such a war is why President Trump established the Space Force. Finally, on May 16, 2026 I took a cruise specifically to have an excuse to wear my old uniform on formal night.

Why? I wanted to be seen gazing out to **THE SEA** (at skip +1 touching **WAR**) where I would photoshop an alien that I envision being prepared to fight one day. I wanted **MARS** shown rising over the sea to suggest, as Israeli Brigadier General Haim Eshed says, that there is now a joint U.S. military-alien base underground base on Mars.

Less than a week after taking the cover photo, I received a surprise call from to publish this book. At the time of the photo I had not even looked for a literary agent, let alone a publisher, but the bulk of the material in this book was up on my https://RoffmanMarsResearch.com website.

CITATION TO ACCOMPANY THE AWARD OF

THE MERITORIOUS SERVICE MEDAL

TO

LIEUTENANT BARRY S. ROFFMAN

UNITED STATES COAST GUARD RESERVE

Lieutenant ROFFMAN is cited for meritorious service in the performance of duty as Contingency Planner, Exercises Staff Branch, Coast Guard Pacific Area, Alameda, California, from July 2003 to May 2007. Demonstrating exceptional leadership, expert knowledge, and remarkable skills, he effectively planned and coordinated the development of the Pacific Area Military Outload Plan, the Contingency Response Operations Plan, two supporting District Contingency Operations Plans, and the Pacific Area Standard Operating Procedures. He facilitated the establishment of the U.S. Navy Third Fleet/U.S. Coast Guard Pacific Area Maritime Homeland Security/Defense Concept of Operations that now serves as a single reference for Third Fleet and Pacific Area subordinate commands executing Homeland Security/Defense Operations. He adapted this plan to serve as the template for a parallel U.S. Navy Second Fleet/U.S. Coast Guard Atlantic Area Concept of Operations. This method provided the Coast Guard, Navy, federal, state, and local agencies a credible means to affect a rapid, interoperable, and comprehensive maritime response to reduce risk; detect threats; deter, disrupt and defeat enemy actions; minimize incident impacts; and foster the swift recovery of the maritime transportation system. From concept to completion, he worked tirelessly to develop a host of documents that had a direct, theater-wide impact on the operation and administration of Coast Guard units. Lieutenant ROFFMAN's dedication and devotion to duty are most heartily commended and are in keeping with the highest traditions of the United States Coast Guard.

Figure E-18 above: Citation for my Meritorious Service Medal. I have tried to duplicate the effort made in official war plans that I wrote here in this book. Although I am no longer on active duty, I think and act like I am when researching sensitive information like that presented in this book. I strongly endorse adding Torah Code Studies as a class at the Defense Language Institute Foreign Language Center (DLIFLC) in Monterey, California. I would be willing to teach the class too.

	A	B	C	D	E	F	G	H
1	AXIS TERM *BARRY ROFFMAN** @ Skip- 52,425	SKIP	NUMBER	DIVIDE BY	THE QUOTIENT	E QUOTIENT	POISSON EQUATION	EACH WORD
2	ELS RANK 1	USED	IN	304805	EQUALS	x 224 LETTERS	PROBABILITY FOR	OR PHRASE
3	AXIS TERM ELS RANK = 1	ON	TORAH	LETTERS	FREQUENCY	EQUALS	THE WORD	HAS 1 CHANCE
4	R FACTOR = 0.475. VALUE 2.985	FIGURE		IN THE	PER LETTER	WORD	APPEARING	IN THE
5				TORAH		EXPECTANCY	AT LEAST ONCE	FOLLOWING OF
6								BEING HERE
7	WAR @ skips +1 only	1	36	304805	0.000118108	0.026456259	0.026109358	38.30044375
8	ENEMY (2 synonyms) @ Skips _/- 1 and +/- 52425	-1	38	304805	0.00012467	0.027926051	0.027539723	36.31118522
9	MARS (3 spellings) @ Skips +/- 1 and +/- 52425	52425	49	304805	0.000160759	0.036009908	0.035369264	28.2731356
10	BOOK @ Skip +1 only	1	104	304805	0.000341202	0.076429192	0.073581491	13.59037432
11	MILITARY @ Skip +1 only	1	108	304805	0.000354325	0.079368777	0.076300777	13.1060264
12	THE SEA @ Skip +1 only	1	447	304805	0.001466511	0.328498548	0.279996027	3.571479254
13	COMBINED ODDS AGAINST EXCLUDING AXIS TERM = h7*h8*h9*h10*h11*H12						3.99789E-08	25013173.5
14								
15	BARRY ROFMAN R FACTOR 0.475 = 2.9853826						0.334965441	2.9853826
16	FINAL ODDS AGAINST MATRIX = H13*H15						1.33916E-08	74673892.93
17								
18	*Note: Hebrew names rarely uses double consanants. A rare exception is							
19	for Isaachar. The transliteration of BARRY ROFFMAN does not use a							
20	second letter *resh* in BARRY or a second letter *feh* in ROFFMAN,							

Above: Spreadsheet for the Roffman Book Cover Figure E-17. This matrix was found against odds of about 74,673,892 to 1 As such, it is extremely significant, not only in a statistical way. This matrix is evidence supporting time travel and pre-destination. The Author of the Code was well aware of the future existence of this book over 3,000 years before it would be published. The book presents a reality much different than that found in schools, history and text books, churches, non- Orthodox synagogues and mosques around the world. To be truly free and sovereign, it's time to have complete disclosure. For any extraterrestrials who read this book, the only card that I am leaving you is with complete disclosure of your medical technology. I am talking about the cure for cancer, and the means to extend our life expectancies. If you cannot do that, get the hell off our world.

As the Charlie Kirk Figure E-19 below (found against odds of 39,743 to 1) indicates, it's likely that Reptilian aliens are disastrous everywhere. **CHARLIE KIRK** is at the same absolute skip as **REPTILIAN**, but that doesn't equate to him being reptilian, The first name of his killer, **TYLER**, is also at the same absolute skip as his name. **ASSASSINATION** is at skip -1.

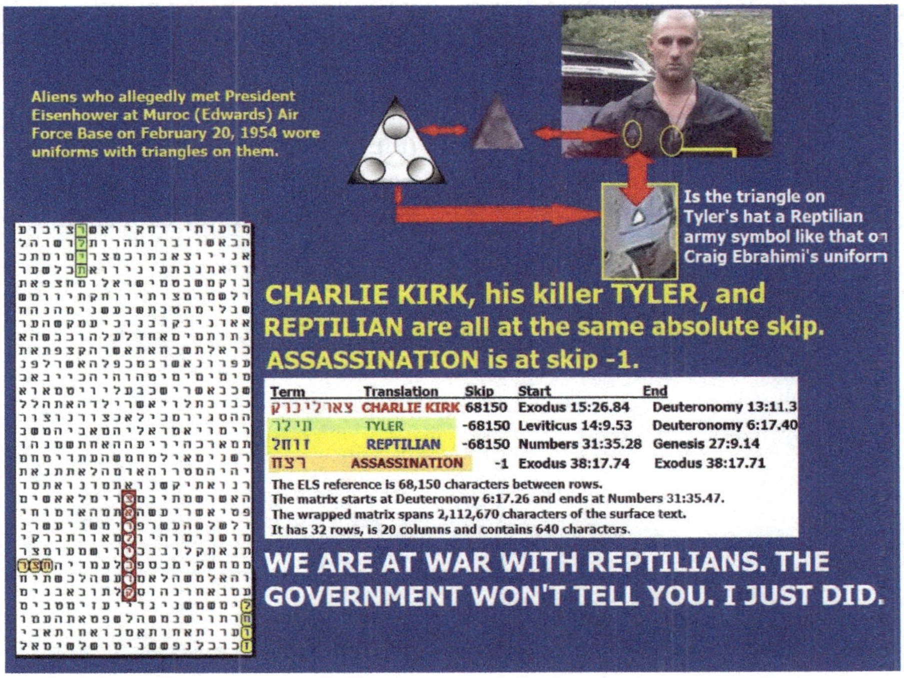

Figure E-19 - The Code names Tyler as the assassin of Charlie Kirk. It is not known if Captain Kirk of Star Trek was named after the Kirk on the matrix, but much of the Star Trek series and the U.S. Space Force are reportedly modeled on each other.

Annex A to God, Aliens or Artificial Intelligence – Who wrote the Torah Code? Torah Code Basics

The Torah Code hypothesis suggests that historically related words can be found encoded together at Equidistant Letter Sequences (ELS) within Torah matrices, these being significantly more compact than similar patterns found in non-religious texts. I have refined this hypothesis to account for the fact that most names requiring more than eight Hebrew letters are unlikely to appear in the Torah Code. However, shorter combinations—such as a brief first and last name, just a last name, or a last name paired with a first initial—often do appear within the Code.

While there is rabbinical tradition supporting the notion of such encoding in the Torah (the first five books of the Bible), no similar tradition exists for the remainder of the Bible. Although many hope to use the Torah Codes as a kind of prophetic tool or crystal ball, making predictions is difficult. The main challenge is not knowing in advance which words or names should be used as key search terms. As a result, many researchers limit their searches to events that have already occurred.

However, once someone accepts the non-predictive nature of such searches, their findings related to past events must be viewed with caution. Positive results may be misleading if we only see what was found, and not what was searched for but yielded nothing. When only successes are presented, the results may seem impressive. But when failures are also considered, we can more accurately evaluate the true value of those successes.

As will be shown later in relation to Ark Code research, this study addresses statistical significance. But ultimately, the test of success lies in predictive clarity: Do the encoded maps eventually lead to the recovery of the Ark of the Covenant in the area around 31°9′ North, 33°4′ East—specifically on the Bardawil or Zuqba Peninsula in Northern Egypt?

ELS Map 1 (shown below) provides insight. While the general course of 252.35° True runs from **JERUSALEM** (starting with the letter *yud)* to the Zuqba Peninsula (beginning with the letter *tsadeh*) in Egypt, the more refined course angle of 251.565° runs from the last letter of **TEMPLE** (shin) to the first letter of **ARK OF THE COVENANT** (*alef*).

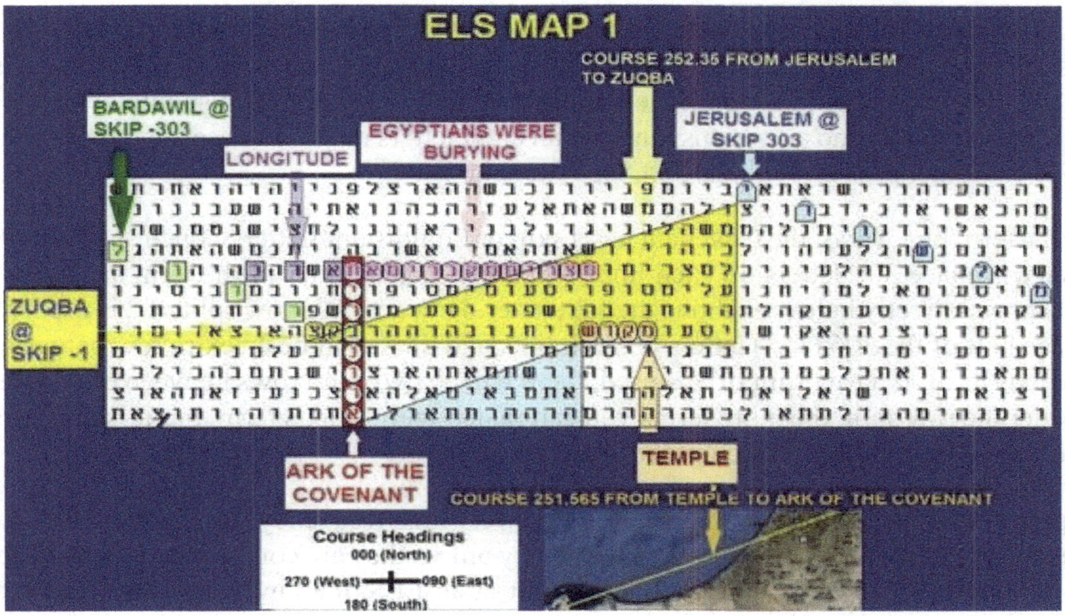

Figure A-1 above: ELS Map 1. ELS maps involving the Ark of the Covenant that housed the 10 Commandments generally have a course of 251.565°, true from the first encoded letter to the last. This results from a ratio of one row down and three columns left, portrayed with the interval, or ELS, being equal to the number of letters shown on each line, unless a "row skip" (explained below) is used. Note that the spaces between words are removed. So, the example of an ELS above would appear in a "matrix" as follows in Figure A-2:

M	A	N	W	A	L	K
E	D	O	U	T	H	E
E	D	O	O	R	I	N
T	O	T	H	E	S	T
A	N	D	S	T	H	A
T	S	E	A	T	E	D
8	0	0	0	F	A	N
S						

Figure A-2 above - example of how to read an ELS text. Further, it should also be pointed out that the really proper way to display the above would be to wrap it into a cylinder with a circumference of 7 letters.

Axis Term: The first term sought is the "axis term." The axis term appears vertically. Hebrew is normally read right to left. The code is really based on a spiraling cylinder. The ELS or "skip" of the axis term = the number of letters on each line. In general, the minimum length of an axis term should be at least six letters. Most often, it is difficult to find axis terms longer than eight letters unless one uses a wrapped search, making more than one computer pass through the 304,805 letters of Torah.

Row Split: If a term like Ark of the Covenant has a skip or Equidistant Letter Sequence (ELS) of 306 letters, the computer will place 306 letters on each line. When the second letter of the term arises (without a row split), it will appear directly below or above the first letter. However, if a row split of 2 is used, the computer will only place half of the 306 (153) letters on each line, and there will be an extra row between each of the letters of the axis term. If a row split of 3 is used, there will be two extra rows between each letter of the axis term. The larger the row split, the more terms you can match with an axis term, but it is also true that as the row split increases, matrix significance generally decreases.

Word Frequency: Just as in English, some words have very high frequency (like THE or AND). In Hebrew, two or three letter words are extremely high in frequency unless they are composed of somewhat rare letters. The higher the frequency of a word in the open text or at an ELS, the lower the significance of its match with an axis term.

Matrix Size: The bigger the matrix, the more likely it is that short or moderate length ELSs (up to 6 letters) will be found somewhere on the matrix. This is like looking in a phone book. If you rip a page out at random and look for a single last name, it probably will not be there. But if you use the entire phone book, the chances of finding it are greatly improved.

Spelling: Even within Torah, there are multiple spellings of the same word. Code adversary Dr. Brendan McKay points out that on the Rabbi Abulafia synagogue in Tiberias, Israel, the rabbi's name is spelled four different ways on the same building. The more spellings one can use to find a match, the less significant the match will be. Such multiple spellings must be factored into any probability calculation describing the significance of a matrix.

Wrapped vs. Unwrapped Matrices: The best Torah Codes programs allow the computer to make multiple passes through the 304,805 letters of the Torah to locate a term at an Equidistant Letter Sequence (ELS). This multi-pass approach is especially important when searching for axis terms that are eight or more letters long, as many such terms can only be found with more than one Torah pass. A wrapped matrix simulates how Jews read the Torah—finishing the last word of Deuteronomy each year on Simchat Torah in October, and then immediately beginning again with the first word of Genesis.

Some Codes researchers, particularly in Israel, do not use software like CodeFinder, which can

locate wrapped matrices. These researchers are limited to unwrapped matrices where the software only makes a single pass through the Torah. Although I believe CodeFinder is the best Torah Codes tool available, I have not been able to persuade many of my Orthodox Jewish colleagues to use it. Their hesitation stems from the fact that CodeFinder includes files for the New Testament. Many Orthodox Jews strictly avoid any mention of the Nazarene's name and will not allow any version of the New Testament into their homes.

While the New Testament files can be deleted, my colleagues are unwilling to take even that step. I have repeatedly asked Kevin Acres—the gentile creator of CodeFinder software—to offer a version without the

New Testament files. Unfortunately, more than two decades of such requests have gone unanswered. This has significantly limited access for Israeli researchers, who arguably have the best understanding of the Hebrew used in Torah Codes matrices, and thus the best chance of recognizing the true nature of the Torah Code, assuming it is not merely a technological masterpiece engineered by aliens or a future human military civilization.

Torah Code Restrictions and Modification to Probability Calculations.

My basic protocol for calculating the significance of matrices is found in my web article on Skip Tables. With time, the value of short ELSs that were not at skips +1, -1, N, or -N, where N is the skip of the axis term, has come into question by me and a number of other Torah Codes experts. Therefore, the following modifications have been built into my work for most of the last 28 years:

(1) Emphasize keywords found at skip +1 by just using their frequency at skip +1 alone. This usually (but not always) equates to their frequency in the open text, the exception being when two sequential words make up one larger word with a different meaning.

(2) Emphasize key words found at skip -1, N, and -N where N is the axis term skip by just using their frequencies at -1, N, -N, and also +1.

(3) Reject any matrix with an axis term less than six letters in length.

(4) Reject any matrix with no axis term that is just a mix of short 3 to 5-letter ELSs.

(5) Reject any 3-letter ELS that does not have its letters within three letters of each other.

(6) While I may show them and while I often discuss them, I reject all *a posteriori* findings for calculation purposes.

(7) I would never do a matrix based on a year as an axis term because it is too short, and because I have seen many thousands of times that years are not statistically important. Or to phrase it another way, there is no evidence seen that dating events was a purpose of the Code. This fits in with the concept that God, in His mercy, hides the date of death for most people.

(8) Axis terms that can be found at a single ELS like **Ark of the Covenant** (in Hebrew *alef resh vav nun bet resh yud tav*) are never split into two spatially separated words like Ark and Covenant. The term must appear as eight letters in sequence as it appears

in Torah, or theoretically as nine letters in sequence as it appears in the 3rd to 6th chapters of Joshua as Ark of **THE** Covenant (*alef resh vav nun **HEY** bet resh yud tav*). There are no such ELS maps found with text outside of Torah.

(9) Because many names require a wrapped search (more than one computer pass through the Torah's 304,805 letters) to find, the wrapped search is the method used to find the name rather than splitting it.

(10) Where a full first and last name cannot be found at an ELS even in a wrapped search, the first name initial and last name are sought. This generally occurs where a name has any of the following letters:

multiple *samechs, tets, gimels,* and *zayins.* In such cases, if the name is just a transliteration, a *shin/sin* may be substituted for a *samech,* and a *tav* may be substituted for a *tet.* I do not usually assign any significance to the axis term unless very long (although it is extremely rare that I ever find one over 10 letters in length).

A 10-LETTER AXIS TERM: I was asked to match COVID-19 with the name of a man who found a meteorite that might have been the source of the virus – EBRAHIMI. Here is what I wrote:

The trick is to find what is actually there. For COVID-19, it takes 7 Hebrew letters, which would more likely be an axis term than an *a priori* key word. There are four transliterations that would work. They are (with frequency in Torah in unwrapped and wrapped finds) as follows:

kuf vav vet yud dalet yud tet (0/8) *caf vav vet yud dalet yud tet* (7/35) *kuf vav vav yud dalet yud tet* (1/15) *caf vav vav yud dalet yud tet* (14/58)

So, without more than one computer pass through Torah (an unwrapped search), there are, in 304,805 letters of Torah, 22 possible matches for an axis term. With a wrapped search, there are 116 possible matches. Usually, I look for an axis term to be eight letters long. Eight is OK, especially when there is a relatively low frequency letter like *tet,* but to find a 7-letter key term that has a letter like *tet, samekh,* or *zayin* is extremely rare. Likewise, to put COVID-19 first and expect to find a 10-letter Hebrew word like one spelling of EBRAHIMI is basically unheard of. Results for keywords that are not axis terms are generally as follows:

1 letters – too short and easy to find. Rarely of great value unless at skip +1 and with rare letters.

2 letters – still short. Rarely of great value unless they are in a special case skip (+/- 1 or the absolute skip of the axis term).

3 letters – good but only excellent if they are in a special case skip or have rare letters.

4 letters – good to great.

5 letters – very good.

6 letters – rare (except with ARK OF THE COVENANT)

7 or more letters – exceedingly rare. Basically, not seen.

R VALUES: I generally don't rely on Rotenberg R-values unless I'm working with a 10-letter axis term—or in rare cases, a 9-letter term containing infrequent Hebrew letters like *tet, samech, gimel,* or *zayin.* Although R-values are included in CodeFinder reports, I routinely remove them because I disagree with how the program applies them. As a rule, I assign no statistical weight to axis terms that are eight letters or fewer. Additionally, the higher the ELS rank of an axis term, the more I penalize the combined odds of a matrix.

In technical terms, p is the probability that the characters of an ELS will randomly match the characters of the text at a given position. N is the number of possible placements for an ELS at the specified skip. The R-value is defined as 1/E, where E = pN. E represents the expected number of times an ELS with an absolute skip smaller than the found skip would randomly appear in a permuted version of the Torah. For further explanation, see CodeFinder: Scores or Probabilities by Dr. Robert Haralick.

Here's an example of an R-value calculation: suppose an axis term was found twice in wrapped Torah, but the matrix displays the instance at the higher skip. This term has an R-value of 0.484, which indicates there is roughly a 32.8% chance of finding it based on letter frequency. (To convert an R-value from the CodeFinder report—such as 0.484—into a percentage chance, enter .484 into a scientific calculator like the TI-30XA, press "2nd function," then "LOG," followed by "1/x," and multiply the result by 100.)

Not Everything That Looks Impressive Is Impressive

The recurring question that arises whenever a new matrix is discovered is (or should be): *"Is this finding statistically significant, or simply the result of high probability?"* Professional researchers often turn to the Monte Carlo technique to answer this. Dr. Robert Haralick, who wrote the foreword to my Ark Code book, discusses the concept of probability on his website at http://www.torah-code.org/probability/probability.shtml. He also explains the idea of scrambling Torah texts to create what are known as "Monkey Texts" at http://www.torah-code.org/monkey_texts.shtml. Both discussions are central to understanding the Monte Carlo method.

Applied to Torah Codes, this technique involves scrambling the Torah text 10,000 or even 100,000 times, then comparing the occurrence of key term meetings in the real Torah with those in the scrambled texts. The level at which the Torah is scrambled—letter, word, or verse—matters greatly. Since much attention is given to keywords appearing in specific phrases, scrambling at the verse level is likely the most appropriate.

For instance, if you scramble the Torah 10,000 times and find 1,000 instances where the Monkey

Texts produce tighter or smaller term meetings than the original, that suggests a one-in-ten chance the original matrix occurred by chance, statistically insignificant. But if only 100 such occurrences are found, then the original matrix might have just a one-in-a-hundred probability of happening by chance, which becomes more compelling.

When I began my research, the Monte Carlo method posed two major challenges: (a) it was often very slow due to the sheer volume of data—304,805 letters per scrambled Torah—and (b) it required specialized software that was not available to the general public. Because of these limitations, I developed an alternative method.

My approach focuses on matrix size (number of rows and columns), the number of ways a key term can appear within that matrix (vertically, horizontally, diagonally, forwards, and backwards), and how frequently that term appears at an ELS given the constraints of the matrix. It also accounts for the percentage of the Torah represented within the matrix, the chi-square value, and the combined total probability for everything found. Importantly, these results must be adjusted to reflect the impact of terms that were searched for but not found.

For more details, see Ark Code, Appendices A and B (pages 183–221). Note: In the case of ELS maps, the probabilities must also be adjusted to account for the requirement that the discovered terms appear at precise course angles corresponding to real-world map coordinates.

Roffman Skip Formula

The number of ways that a term (either forwards, backwards, or diagonal) can fit into a matrix is determined as follows:

(1) Let the number of skips possible in a forward direction on a row of length (r), where r = the number of columns in the matrix, be equal to "Sr."

(2) Likewise, let the number of skips possible in a vertical direction on a column of length (c), where c = the number of rows in the matrix, be equal to "Sc."

(3) The Roffman Skip Formula for total skips is as follows: Skips = 2(Sr + Sc + 2[Sr][Sc]) = 2Sr + 2S + 4SrSc.

Where do we find values for Sc or Sr? Use the following ROFFMAN SKIP TABLES found on my website:

As an example, for 8-letter axis terms use the table below

TABLE 1F - 8 LETTER WORDS

COLUMNS (C) OR ROWS (R)	SKIPS POSSIBLE FOR 8-LETTER WORDS	COLUMNS (C) OR ROWS (R)	SKIPS POSSIBLE FOR 8-LETTER WORDS
1	0	50	7
2	0	51	7
3	0	52	7
4	0	53	7
5	0	54	7
6	0	55	7
7	0	56	7
8	1	57	8
9	1	58	8
10	1	59	8
11	1	60	8
12	1	61	8
13	1	62	8
14	1	63	8
15	2	64	9
16	2	65	9
17	2	66	9
18	2	67	9
19	2	68	9
20	2	69	9
21	2	70	9
22	3	71	10
23	3	72	10
24	3	73	10
25	3	74	10
26	3	75	10
27	3	76	10
28	3	77	10
29	4	78	11
30	4	79	11
31	4	80	11
32	4	81	11
33	4	82	11
34	4	83	11
35	4	84	11
36	5	85	12
37	5	86	12
38	5	87	12
39	5	88	12
40	5	89	12
41	5	90	12
42	5	91	12
43	6	92	13
44	6	93	13
45	6	94	13
46	6	95	13
47	6	96	13
48	6	97	13
49	7	98	13

We apply the Poisson Equation to see the probability that they are present at least once. This is necessary to determine a true probability for each word. Just because a word is likely to appear once per plot does not imply it will always be there. Words may average out to many times per plot area without actually being in a given plot of that area. Of course, if the expected frequency is sufficiently high, we eventually reach a probability like 0.9999999, which we simply round off as 1.0.

How to Find the Chance of a Term Appearing At Least Once

1. FIND THE PROBABILITY IT DOES NOT OCCUR BY THE POISSON EQUATION.

   x(-lambda)

   f(x) = Lambda e x = 0; lambda = expected frequency per matrix x!

2. 1 - f(0) = THE PROBABILITY OF OCCURRING AT LEAST ONCE. (where f(0) = the probability it will not occur)

On an Excel or Works spreadsheet, place head columns as follows:

A: Whatever identifies the calculation; B: Skips Used on the Matrix;

C: Number of hits (on CodeFinder or similar software) in Skip Range; D: Divide by 304,805 Letters in Torah or Control;

E: The Quotient Equals Frequency Per letter;

F: E Quotient Multiplied by Letters on Matrix = Word Expectancy;

G: Poisson Equation = 1-EXP(-F#) where # equals the row number of the item in Column F in question on the spreadsheet. If you want to know the chance for the item to be on the matrix, head to Column H accordingly. The value of Column H will be the reciprocal of the value found in Column G by the Poisson Equation.

Note: While this author (Barry S. Roffman) discovered the Roffman Skip Formula, my son (an MIT geophysics graduate), Rabbi/Air Force Major Robert Roffman, is the author of the spreadsheets and the man who first introduced the use of the Poisson Equation into my research.

Row Split and Wrapped Matrices

There is some indication that when using a row skip—or row split—function for the axis term, the true value of an open-text match should be calculated by dividing the standard probability by the row split value. For example, the lowest ELS for the Ark of the Covenant was found at skip -306 (with a cylinder circumference of 306 letters). This term had approximately a one-in-2,931 chance of appearing in a 104-letter matrix alongside the phrase "Egyptians were burying." At skip -306, the row split function was not enabled in CodeFinder.

Had the row split been enabled and set to 2 (thus halving the cylinder circumference to 153 letters), and assuming the matrix area remained the same, I would have divided the 2,931 probability by 2,

yielding a revised chance of about one in 1,465. However, in this specific case, the matrix with a 153-letter circumference would have been larger, since the match found using the 306-letter circumference was already nearly as tight as possible with the row split function turned off.

There is also discussion about adjusting a matrix's statistical value based on the number of Torah passes made by CodeFinder during a wrapped (rounded torus) search before locating an axis term. For more on this, see the permutation experiment available at: https://www.arkcode.com/custom3_53.html.

Special Case Skips

Finally, when computing the value of *a priori* open text terms on a matrix, it is my practice to only employ the frequency of this term at Skip +1 (in unwrapped Torah) on my spreadsheet in column

C. However, if the *a priori* term appears at skip -1, N (parallel to, in the same direction, and at the skip of the axis term), or -N (parallel to, in the opposite direction, and at the skip of the term) in column C, I list the frequency (number of hits) as the total hits at skips +1, -1, N, and -N (with wrapped Torah allowed if that was required to find the axis term). These skips are considered special because they seem to leap out at the eye of the researcher and make the case for deliberate encoding seem more plausible.

The Talmud May Indicate Something About Why We Find Names Encoded.

In a discussion of names and prophecy, the Talmud (Gemara for Berachos 7b) indicates that God inspires parents to select a name that has significance unbeknownst to the parent. Many years later, the aptness of the name becomes apparent to all.

The Longer the Name, The Less Likely That It Will Be at an ELS.

Many Codes advocates have been proceeding under the rabbinical (Vilna Gaon) claim that everyone's life is predicted in the Torah – birth, death, all details, etc. But the full 10-letter Hebrew name for the United States (Alef Resh Tsadeh Vav Tav Hey Bet Resh Yud Tav, pronounced Artzot HaBrit) is not present at an ELS in unwrapped or wrapped Torah (or even if we search the entire Tenach/Old Testament). We can, however, find transliterations of the United States as are shown on Figure A-3 below.

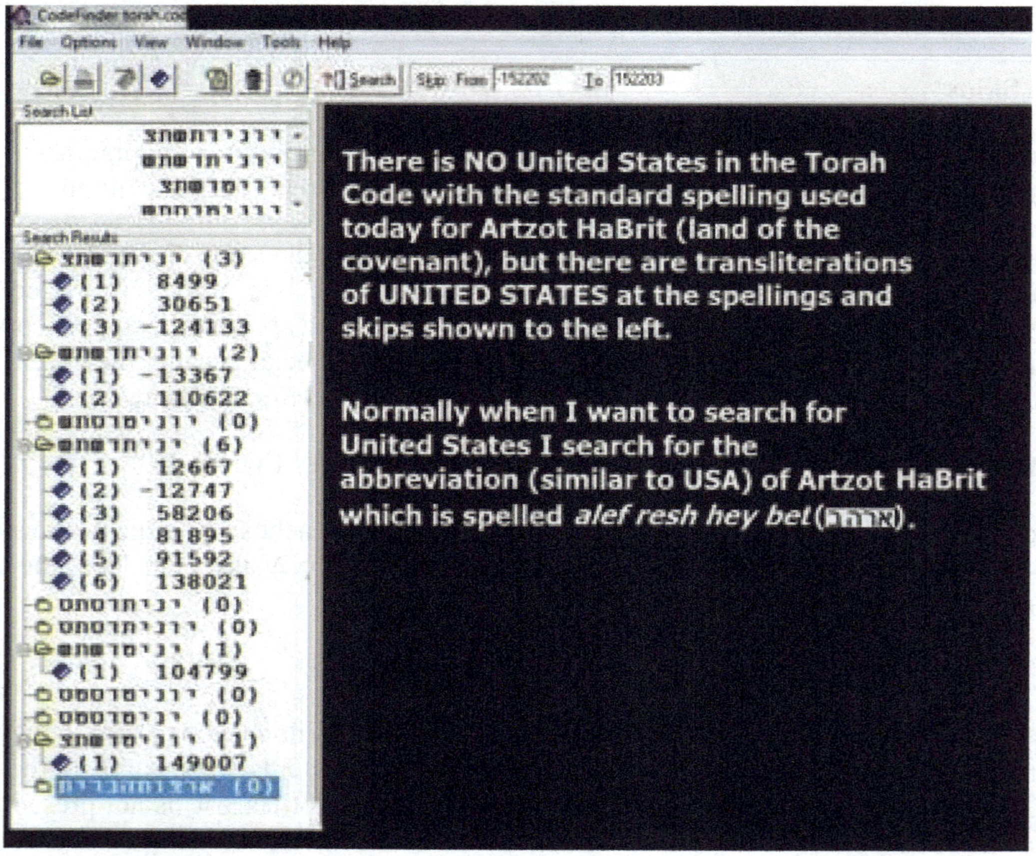

These are several possible spellings for UNITED STATES or its abbreviation. Because USA is only four letters in Hebrew, I generally don't use it unless it is at a special case skip (+/- 1 or +/- the skip of the first term (axis term) used to start the matrix). Alternately, we could look for America, but due to time constraints, I seldom do unless the national location of an event is critical.

The most common ceiling of letters found at ELS without snooping for additional words before or after an axis term (first term sought) is eight letters. Nine-letter ELSs arise, perhaps 5% of the time, while 10-letter ELSs are probably found much less than 1% of the time. For example, suppose we just look for an ELS that has eight letters – A*lef Bet Alef Bet Alef Bet Alef Bet*. In wrapped Torah (where the computer makes several passes through Torah), it's there 48 times, but in unwrapped Torah (one computer search of the Torah), it's there only 5 times. What happens if we add a letter *Alef* at the end to make a 9-letter ELS to hunt for? We do not find this 9-letter ELS even once in the wrapped Torah. We hit the ceiling at eight letters. However, if instead of adding an *Alef* at the end of the first 8-letter ELS sought, we add a *Bet* at the beginning to get the nine letters of *Bet Alef Bet Alef Bet Alef Bet Alef Bet*, we do find it at ELS once. But the nine letters here are the limit, and it is only found in the wrapped Torah. It does not exist in one pass through the Torah. There, the limit remains at eight letters.

My son, Dr. David Roffman, and I have performed many experiments looking for names of people, whether his school colleagues or U.S. presidents, and the 8-letter limit was almost always there. So what about people with names that require more than eight letters? We generally kick back to initials, or forms that eliminate vowel letters, but even then, we don't always find everybody. Further, even if we did find everybody, it would generate serious questions about free will if their fates were prerecorded. This is a problem for Judaism, as our faith strongly backs the concept of free will.

So what is really going on? To understand, we need to consider another Code written by God (though possibly modified by aliens) – the Genetic Code. I have written in my book and to Dr. Rips before about DNA-structured Code matrices, but this is not the point I wish to make here.

Consider, if you will, the cases of murderers who have murderers up their family trees. There are many of them. But having fathers, grandfathers, and great-grandfathers who were criminals does not mean that their descendants must be criminals too. There is free will, and people can overcome bad genes (as with modern Australians who often trace their ancestry to prisoners brought there from England). People can learn from mistakes made by their ancestors; however, we are not shocked when they do not. So, it may be with the Codes. Not everyone's name is at an ELS. But when a name is at an ELS, there may be a tendency for events to unfold in that life that match what is seen in the text that is close to the ELS name. A bad ELS location may imply a *tendency* to have something negative happen, while a positive environment in the Torah text may be a blessing. Therefore, parents should check the ELS location before settling on a name. However, it is more than possible that some names they might choose have no ELS – especially when the names are(a) very long, or (b) contain hard-to-find letters like *zayin (Z), tet* instead of *tav* for *T,* or *samech* instead of *shin* for *S.*

This also makes sense when we consider the issue of how a text as beautiful and coherent as Torah could possibly be written while the Author must consciously alter the proposed text to accommodate the fates of billions of people. No adjustment in text design was required. Rather, people adjust to their ELS location. Earlier, I had written that, "*There might be cases, however, where the proximity of names and events is meant to serve as a warning. While there have not yet been enough cases to prove such links, if they exist, they probably are similar in nature to what is seen on a matrix like that under Vatican Rule seen on Figure A-4.*" However, I now believe that an exception is with respect to a warning that top Democrats under and including Obama and Biden are, in fact, reptilian invaders or brainwashed people hostile to our species.

During the 2008 Presidential campaign, a record $745,000,000 was raised for the election of Barack Obama (Kamala Harris spent 1.5 billion dollars in 2024). Where did it come from? Before Trump, all major candidates (including Obama and McCain) except Ron Paul were members of the Council on Foreign Relations (CFR). Who on Biden's team was a member of the CFR? Harris is. Of the CFR Group on the Biden Team. Jewish members of the CFR include Antony Blinken, Secretary of State, Janet Yellen, Secretary of the Treasury, Ron Klain, Chief of Staff, and Jeffery Zients, Counselor to the President. Non-Jewish CFR members of Biden's cabinet include Vice President Kamala Harris, Lloyd Austin, Secretary of Defense, Linda Thomas-Greenfield, UN Ambassador, Cecilia Rouse, Council of Economic Advisors, Alejandro Mayorkas, Secretary of Homeland Security, Jake Sullivan, National Security Advisor, John Kerry, Special Envoy for Climate, Susan Rice, Chief of Domestic Council, William J. Burns, Director of the CIA, Kurt M. Campbell, Indo-Pacific Tsar, Thomas Vilsack, Secretary of Agriculture, Gina Raimondo, Secretary of Commerce, and Eric S. Lander, Director of Office of Science and Technology.

Sojourners Magazine wrote in January 2021 that among Biden's nominations for his Cabinet, there were 15 heads of executive departments, as well as 10 other Cabinet-level positions. In this group, there are at least eight Catholics and seven Jews, who control the CFR. In fact, who is the real power behind many politicians? Many think it's (self-hating) Jews on Biden's staff that would be the Secretary of State, the Secretary of the Treasury, the Chief of Staff, and a Counselor to the President. Yet, Biden was a true enemy of Israel. But the CFR is also suspected by many to be none other than the Vatican itself. As such, I searched for two words in sequence: VATICAN RULE. See Figure A-4. I found it at its minimum ELS of -24686, and in a two-column by 15 rows matrix with OBAMA. If I expanded the matrix to four columns, BARAK was there at skip-1 (from Numbers 7:13, letters four back to 2) touching OBAMA.

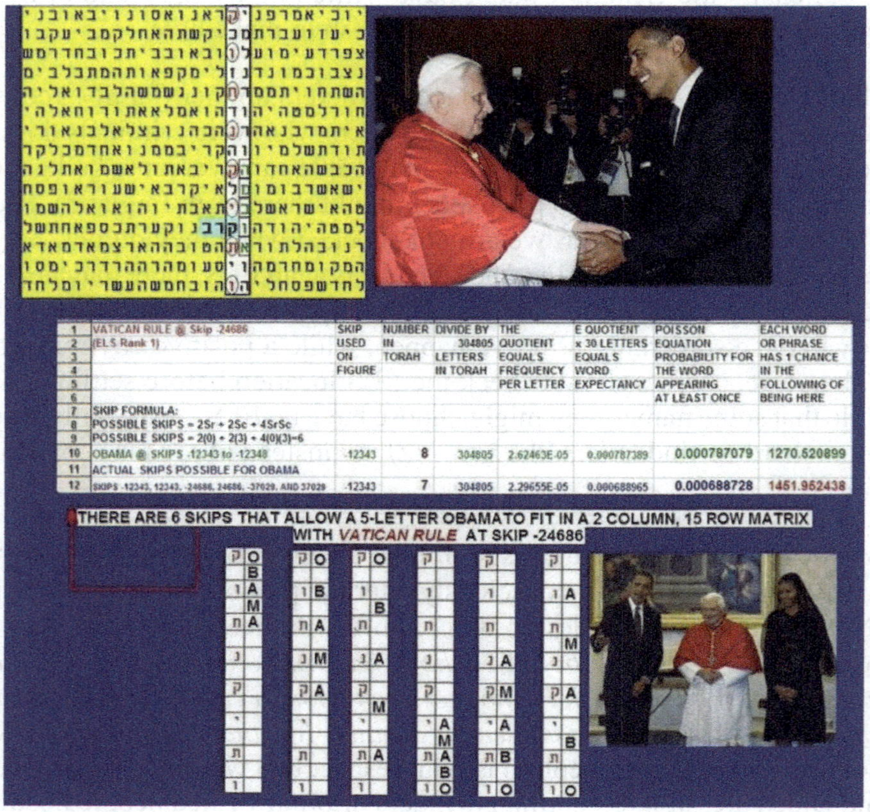

Note how the letters of **VATICAN** fit in with those of **OBAMA**. This kind of match may suggest that such a link exists in the real world.

Sometimes we find descriptions of places encoded. In Figures A-5, A-6, and A-7, **PHILADELPHIA** shares a letter *hey* with **THE CITY** in the open text below. Figures A-5, A-6, and A-7 are below. I grew up in Philadelphia, and though it is known as the City of Brotherly Love, I hated it because it was also a city of mass murders. When I saw that with no row split employed, **PHILADELPHIA** touched and was, in the open text, just one letter below **CITY** (see Figures A-5 to A-7), I was surprised, but I wanted to know if the murderous link was also included on the matrix. It is, indeed:

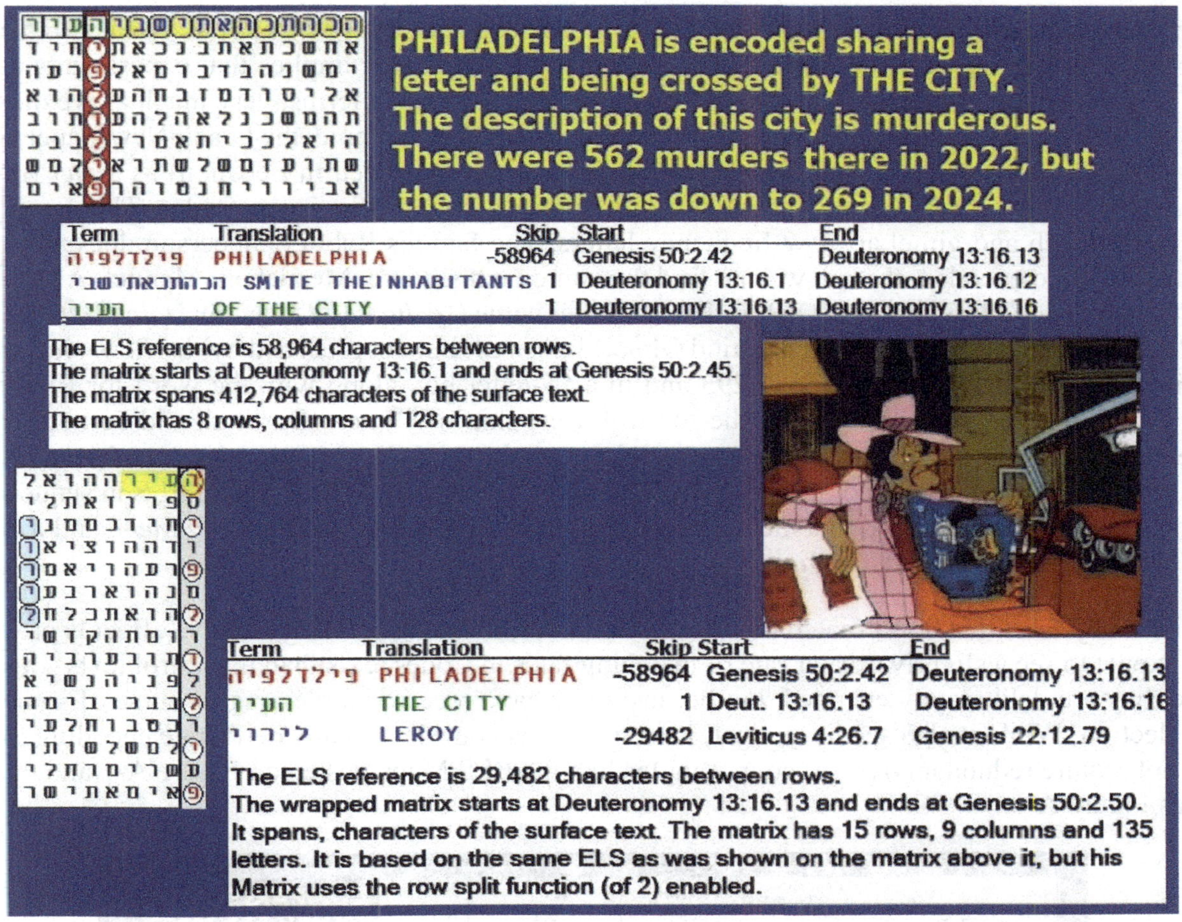

Figures A-5, A-6 and A-7 above. The word **CITY** is at the end of the words **SMITE THE INHABITANTS** OF THAT CITY (Figure A-6). This set me off singing an old, popular song about Bad, bad**, LEROY** Brown. So, I looked for **LEROY** and found him with the above material. Although his letters are vertical with no line separations, as is normally seen in an axis term that has no row split, here there is a row split of 2 used for the Philadelphia axis term. While it still stands out, this is not a special case skip.

The Leroy I Knew. Back in 1968, I worked during weekends at the Temple University emergency room on weekends. For many nights, we had a repeat, black, drunk patient. I don't remember his last name. For at least the first two weeks, we sewed up knife wounds. Then we took a bullet out of him. The next weekend, he walked in holding his abdomen in his hands. In this way, he kept his intestines from spilling out of a bad cut. When two doctors rushed in to treat him minutes after his arrival, I will never forget that one of them proclaimed, "You sure have a lot of guts, Leroy." The second doctor immediately followed up with, "Yea, Leroy, all over the table." A good story like this often has a matrix behind it.

Suppose we do find a statistical tendency to support the Codes with respect to the fates of people. In that case, my hypothesis is that it will usually not be because they have been encoded to definitely do something. Rather, we are seeing the result of the analogy of good or bad genes with good or bad location of the names at ELS. The location of the ELS does NOT require the person to experience that fate, but it makes it more likely that they will experience the fate suggested than someone who has no such ELS location in Torah. I went looking for Philadelphia. In unwrapped text, it was only there once. But in wrapped text, Philadelphia is there 4 times. The first thing to look for? **CITY.** It matches perfectly with **PHILADELPHIA** at skip-58,964 (ELS rank 3) in wrapped Torah (see image above). All of this implies that experiments designed should perhaps aim at matching names to a pre-specified set of related terms, very much as we did before. But we should not expect to find all names or all fates encoded. For example, the Israeli Newspaper HaAretz spells the name Goldwasser (a prisoner of Hezbollah who was murdered in Lebanon) as *gimel vav lamed dalet vav vav samech resh,* and this name does not appear at an ELS in Torah Codes. This is only eight letters, but samech and gimel are low-frequency letters. The famous Rabbi Schneerson does not have his 10-letter last name encoded, though we can find the name on his grave, Menachem Mendel. Also missing: former Israeli Prime Minister Ehud Olmert. His 10- letter name (*spelled Alef Hey Vav Dalet Alef Vav Lamed Mem Resh Tet*) is not at an ELS. We can find Olmert followed (but not preceded by) by *Resh Hey Mem,* the abbreviation for Prime Minister, at skip-293 and in a 55-letter box along with the word for **President** or Prince (*Nun Sin Yud Alef*). Skip -293 is the fourth lowest skip of Olmert. But it should be noted that the three letters to form the abbreviation for Prime Minister are the third (Hey), fifth (Mem), and seventh (Resh) most common letters in Torah, comprising over 23% of it. We also find Olmert at an ELS in sequence with the same three letters scrambled in a meaningless fashion after his name as *Hey Resh Mem,* and *Resh Mem Hey*, if we search the wrapped Torah.

There are many fates that don't match the names. However, if such a blessing or curse effect is present, we should expect to see a slightly greater rate of matching fates in Codes texts than in control texts – again, not because they are deliberately encoded, but because they have received a blessing or curse of some sort at name selection. While it also seems to exist, I am not yet as confident of any deliberate encoding for terms that do not require redundant *a priori* course angles like the ELS Maps posted on Figure A-1 and on the list below as Figure A-6.

MAP		TERM RANK (#1 = SHORTEST)	FROM	TO	ANGLE
2	ירושלם (JERUSALEM)	Skip –30, #3	ירושלם (JERUSALEM)	צקב (ZUQBA)	251.565° T
3	מקומארונ POSITION (of) ARK	Skip 2,628, #2	ירושלם (JERUSALEM)	מקומארונ POSITION (of) ARK	251.565° T
4	מקומהארנ POSITION (of) THE ARK	Skip 3,373, #1	מירושלם (FROM JERUSALEM)	מקומהארנ POSITION (of) THE ARK	251.565° T
4	*ALTERNATE AXIS TERM* מירושלם (FROM JERUSALEM)	843, #15	מירושלם (FROM JERUSALEM)	זהב (GOLD)	251.565° T
5	עגרתשלימה (UGRET SELIMA)	Skip 37,187, #1	ירושלם (JERUSALEM)	עגרתשלימה (UGRET SELIMA)	251.565° T
7	קתבאלכלש KATIB EL QALS	Skip –10,560, #1	ירושלם (JERUSALEM)	קתבאלכלש KATIB EL QALS	251.565° T
11	ARK MAP ארונמפה	-523, #1	ארונמפה ARK MAP	שובמצרימ RETURN TO EGYPT	251.565° T
11	ARK MAP ארונמפה	-523, #1	ארונמפה ARK MAP	ארונברית (ARK OF COVENANT)	252.12° T
12	ארונרופמנ ROFFMAN ARK	=2041, #1	ירושלים (JERUSALEM)	צקב (ZUQBA)	251.565° T
13	ARK 251 565 ארונבהאהוה	88,418, @ #1	רופמנ ROFFMAN -7, #1	ARK 251 565 ארונבהאהוה	251.565° T
13	ARK 251 565 ארונבהאהוה	88,418, @ #1	ירושלם (JERUSALEM)	לגד (31-9 – SUSPECT LATITUDE)	251.565° T

COMPARISON OF OTHER ELS MAPS. NOTE THE LOW ELS RANKS OF AXIS TERMS

Figure A-8 above. ELS maps exist in many places in Torah. The course angle is redundantly encoded as 251.565 degrees from the site where the Ark was kept in Jerusalem to a spot on the El Zuqba peninsula in the northern Sinai. There is a salt marsh near it called Ugret Selima. The coordinates of the site are 31 degrees 9 minutes North, 33 degrees 4 minutes East. To arrive at this course, the ratio from the first position encoded to the possible current location is one row down and three columns left on the matrix.

Some examples follow

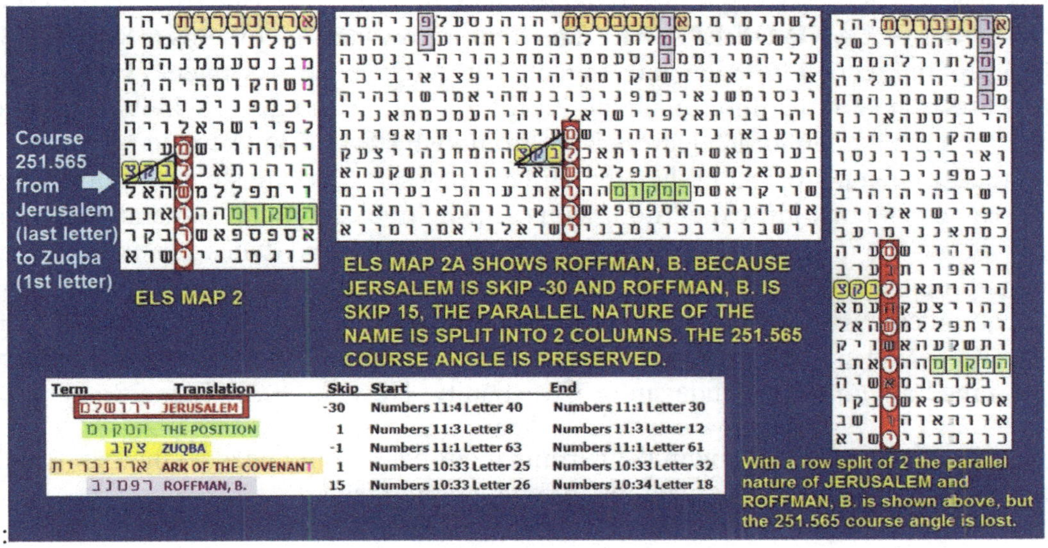

:

Figure A-10: ELS Map 3. On this matrix, the axis term is POSITION *(of)* ARK. The angle from JERUSALEM to POSITION (of) ARK again equals a course heading of 251.565 degrees True. Running into POSITION (of) ARK is the LAND OF EGYPT. POSITION (of) ARK is at its second lowest skip in Torah.

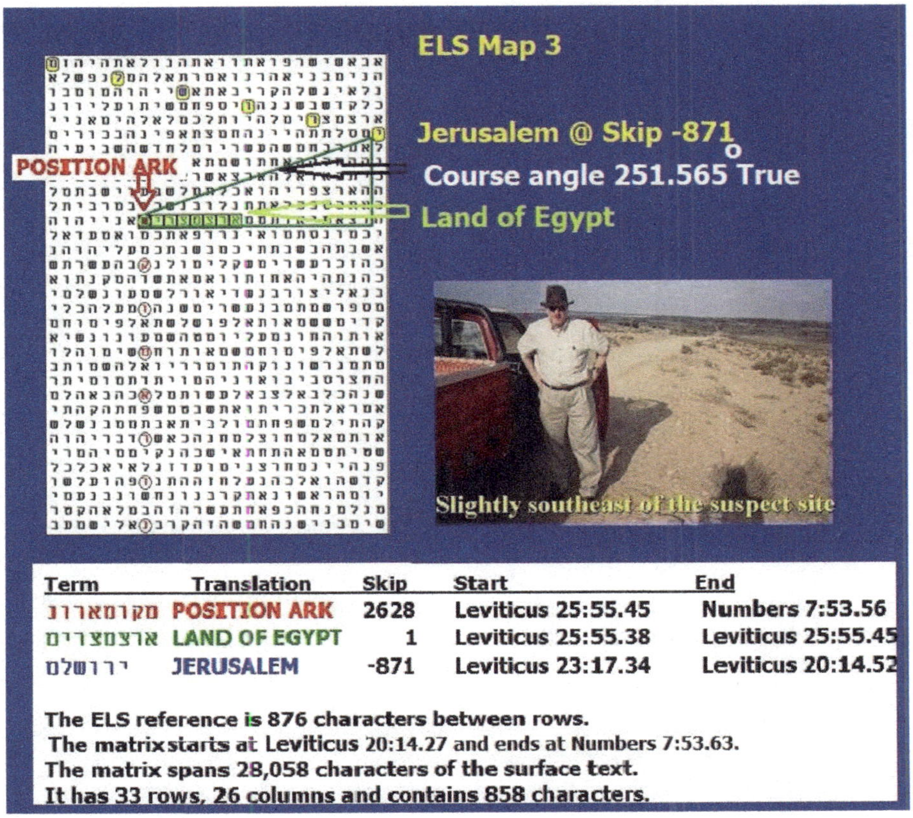

How is the course angle calculated? The ordinate is six rows down from the first letter (*yud*) of **JERUSALEM.** The abscissa is 18 columns to the left with the first letter (mem) of **POSITION** (of) **ARK** (it's also the last letter of **LAND OF EGYPT**). The ratio of rows (6) to columns (18) is 0.33333. On a TI-30XA calculator, hit the second function and tangent. This angle is the arctangent (18.43494882 degrees). Next, since the direction is in the third quadrant, hit the +/- button and get -18.43494882 degrees. To this add 270 degrees and get 251.5650512 degrees True from **JERUSALEM** (Temple Mount) to **POSITION** (of) **ARK** on the El Zuqba peninsula in the Northern Sinai desert. The site sits in a military restricted zone forbidden to foreigners. On my last expedition, with some help via Egyptian General Benil Shaddah, I got within a mile of it, but the men that I paid to drive me there (over sand dunes in a truck with low-inflated tires) did not want me to get out of the truck, and I was not permitted to use my metal detector. Nor did I have any GPS equipment with me. The driver said I might be shot if I were spotted in my Indiana Jones hat. At that time, Dr. Zawi Hawass was in charge of Egyptian antiquities. He was adamantly opposed to helping any expedition aimed at supporting Judaism or Christianity. Now, in 2025, I am 78.

Even if I could raise the money and obtain the permits required, I would need a younger, healthy partner. If you want to be that person, please contact me through my publisher.

Made in the USA
Monee, IL
08 November 2026

67979981R00188